Collins

# PHYSICS

## AQA A-level Year 1 and AS

Student Book

Dave Kelly

William Collins' dream of knowledge for all began with the publication of his first book in 1819.

A self-educated mill worker, he not only enriched millions of lives, but also founded a flourishing publishing house. Today, staying true to this spirit, Collins books are packed with inspiration, innovation and practical expertise. They place you at the centre of a world of possibility and give you exactly what you need to explore it.

Collins. Freedom to teach

---

HarperCollins*Publishers*

The News Building, 1 London Bridge Street

London SE1 9GF

Browse the complete Collins catalogue at www.collins.co.uk



First edition 2015

10 9 8 7 6 5 4 3

ISBN 978-0-00-759022-3



A catalogue record for this book is available from the British Library

Authored by Dave Kelly
Commissioned by Emily Pither
Project managed and developed by Jane Roth
Edited by Geoff Amor
Proofread by Sophia Ktori and Tony Clappison
Artwork and typesetting by Jouve
Cover design by We are Laura
Printed by Grafica Veneta S.p.A.

The publisher would like to thank Frank Ciccotti, Gurinder Chadha, Sue Fletcher, Sue Glover and Lynn Pharoah.



## Approval Message from AQA

This textbook has been approved by AQA for use with our qualification. This means that we have checked that it broadly covers the specification and we are satisfied with the overall quality. Full details for our approval process can be found on our website.

We approve textbooks because we know how important it is for teachers and students to have the right resources to support their teaching and learning. However, the publisher is ultimately responsible for the editorial control and quality of this book.

Please note that when teaching the AS and A-level Physics course, you must refer to AQA's specification as your definitive source of information. While this book has been written to match the specification, it cannot provide complete coverage of every aspect of the course.

A wide range of other useful resources can be found on the relevant subject pages of our website: www.aqa.org.uk

# CONTENTS

# TO THE STUDENT

The aim of this book is to help make your study of advanced physics interesting and successful. It includes examples of modern applications, of new developments, and of how our scientific understanding has evolved.

Physics is our attempt to understand how the Universe works. Fortunately, there are some deep, underlying laws that simplify this ambitious task, but the concepts involved are often abstract and will be unfamiliar at first. Getting to grips with these ideas and applying them to solving problems can be daunting. There is no need to worry if you do not 'get it' straight away. Discuss ideas with other students, and of course check with your teacher or tutor. Most important of all, keep asking questions.

There are a number of features in the book to help you learn:

- Each chapter starts with a short outline of what you should have learned previously and what you will learn through the chapter. This is followed by a brief example of how the physics you will learn has been applied somewhere in the world.
- Important words and phrases are given in bold when used for the first time, with their meaning explained. There is also a glossary at the back of the book. If you are still uncertain, ask your teacher or tutor because it is important that you understand these words before proceeding.
- Throughout each chapter there are many questions, which enable you to quickly check your understanding. The answers are at the back of the book. If you get really stuck with a question, check the answer before you carry on.
- Similarly, throughout each chapter there are checklists of Key Ideas that summarise the main points you need to learn from what you have just read.
- Where appropriate, Worked examples are included to show how important calculations are done.
- There are many Assignments throughout the book. These tasks are designed to consolidate or extend your understanding of a topic. They give you a chance to apply the physics you have learned to new situations and to solve problems that require a mathematical approach. Some refer to practical work and will encourage you to think about scientific methods. The relevant Maths Skills (MS) and Practical Skills (PS) from the AQA AS Physics specification are indicated.
- Some chapters have information about the 'Required practical' activities that you need to carry out during your course. These sections (printed on a beige background) provide the necessary information about the apparatus, equipment and techniques that you need to carry out the required practical work. There are questions about the use of equipment, techniques, improving accuracy in practical work, and data analysis.

This book covers the requirements of AQA AS Physics and the first year of AQA A-level Physics. There are a number of sections, questions, Assignments and Practice Questions that have been labelled 'Stretch and challenge', which you should try to tackle if you are studying for A-level. In places these go beyond what is required for your exams but they will help you to expand your knowledge and understanding of physics.

Good luck and enjoy your studies. We hope this book will encourage you to study physics further after you have completed your course.

# PRACTICAL WORK IN PHYSICS

Practical work is a vital part of physics. Physicists apply their practical skills in a wide variety of contexts: from nuclear medicine in hospitals to satellite design; from testing new materials to making astronomical observations. In your AS or A-level physics course you need to learn, practise and demonstrate that you have acquired these skills.

*Physicists need to solve problems, such as design problems. This machine weaves superconducting wire into cable to produce powerful superconducting electromagnets for accelerators.*

## WRITTEN EXAMINATIONS

Your practical skills will be assessed in the written examinations at the end of the course. Questions on practical skills will account for about 15% of your marks at AS and 15% of your marks at A-level. The practical skills that will be assessed in the written examinations are listed below. Throughout this book there are questions and longer assignments that will give you the opportunity to develop and practise these skills. The contexts of some of the exam questions will be based on the 'required practical activities' (see the final section of this chapter).

### Practical skills assessed in written examinations:

**Independent thinking**

- Solve problems set in practical contexts
- Apply scientific knowledge to practical contexts

**Use and application of scientific methods and practices**

- Comment on experimental design and evaluate scientific methods

*Physicists need to apply their knowledge when using practical equipment. This is a laser deposition chamber, in which a laser beam evaporates material in order to coat another surface.*

- Present data in appropriate ways
- Evaluate results and draw conclusions with reference to measurement uncertainties and errors
- Identify variables, including those that must be controlled

**Numeracy and the application of mathematical concepts in a practical context**

- Plot and interpret graphs
- Process and analyse data using appropriate mathematical skills
- Consider margins of error, accuracy and precision of data

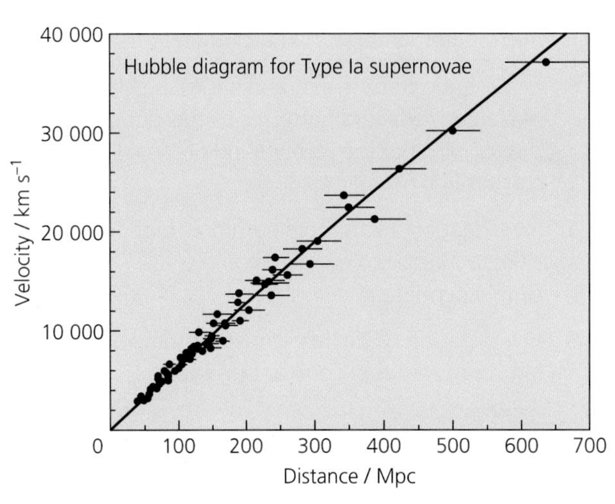


*This graph of velocity against distance for supernova events, similar to that originally produced by Edwin Hubble, plots the distances with error bars because of the uncertainty in the values.*

**Instruments and equipment**

- Know and understand how to use a wide range of experimental and practical instruments, equipment and techniques appropriate to the knowledge and understanding included in the specification

*You will need to use a variety of equipment correctly and safely.*

## ASSESSMENT OF PRACTICAL SKILLS

Some practical skills, such as handling materials and equipment and making measurements, can only be practised when you are doing experiments. For A-level, the following *practical competencies* will be assessed by your teacher when you carry out practical activities:

- Follow written procedures
- Apply investigative approaches and methods when using instruments and equipment
- Safely use a range of practical equipment and materials
- Make and record observations
- Research, reference and report findings

You must show your teacher that you consistently and routinely demonstrate the competencies listed above during your course. The assessment will not contribute to your A-level grade, but will appear as a 'pass' alongside your grade on the A-level certificate.

These practical competencies must be demonstrated by using a specific range of *apparatus and techniques*. These are as follows:

- Use appropriate analogue apparatus to record a range of measurements (to include length/distance, temperature, pressure, force, angles and volume) and to interpolate between scale markings
- Use appropriate digital instruments, including electrical multimeters, to obtain a range of measurements (to include time, current, voltage, resistance and mass)
- Use methods to increase accuracy of measurements, such as timing over multiple oscillations, or use of a fiduciary marker, set square or plumb-line
- Use a stopwatch or light gates for timing
- Use calipers and micrometers for small distances, using digital or vernier scales
- Correctly construct circuits from circuit diagrams using dc power supplies, cells and a range of circuit components, including those where polarity is important
- Design, construct and check circuits using dc power supplies, cells and a range of circuit components
- Use signal generator and oscilloscope, including volts/division and time-base

- Generate and measure waves, using microphone and loudspeaker, or ripple tank, or vibration transducer, or microwave/radio wave source
- Use laser or light source to investigate characteristics of light, including interference and diffraction
- Use ICT such as computer modelling or data logger with a variety of sensors to collect data, or use software to process data
- Use ionising radiation, including detectors

For AS, the practical competencies will not be assessed, but you will be expected to use these skills and these types of apparatus to develop your manipulative skills and your understanding of the processes of scientific investigation.

## REQUIRED PRACTICAL ACTIVITIES

During the A-level course you will need to carry out 12 *required practical activities*. These are the main sources of evidence that your teacher will use to award you a 'pass' for your competency skills. If you are studying the AS course, you will need to carry out the first six in this list.

1 Investigation into the variation of the frequency of stationary waves on a string with length, tension and mass per unit length of the string

2 Investigation of interference effects to include the Young's slit experiment and interference by a diffraction grating

3 Determination of $g$ by a free-fall method

4 Determination of the Young modulus by a simple method

5 Determination of resistivity of a wire using a micrometer, ammeter and voltmeter

6 Investigation of the emf and internal resistance of electric cells and batteries by measuring the variation of the terminal pd of the cell with current in it

7 Investigation into simple harmonic motion using a mass–spring system and a simple pendulum

8 Investigation of Boyle's (constant-temperature) law and Charles's (constant-pressure) law for a gas

9 Investigation of the charge and discharge of capacitors; analysis techniques should include log-linear plotting, leading to a determination of the time constant $RC$

10 Investigate the relationship between magnetic flux density, current and length of wire using a top-pan balance

11 Investigate the effect on magnetic flux density of varying the angle using a search coil and oscilloscope

12 Investigation of the inverse-square law for gamma radiation

Information about the apparatus, techniques and analysis of required practicals 1 to 6 are found in the relevant chapters of this book, and information on required practicals 7 to 12 will be given in Book 2.

You will be asked some questions in your written examinations about these required practicals.

**Practical skills are really important. Take time and care to learn, practise and use them.**

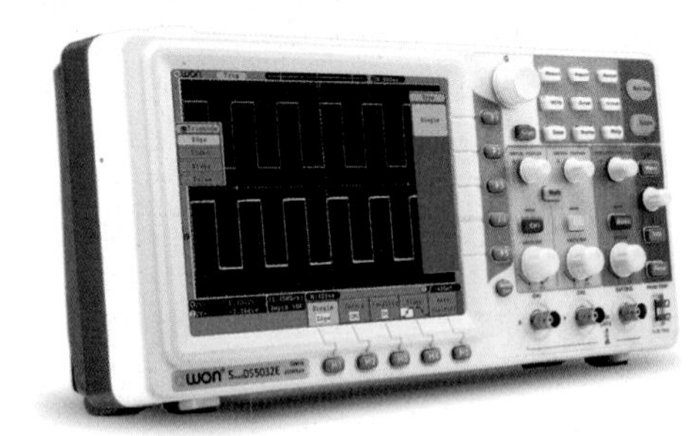


*An oscilloscope*

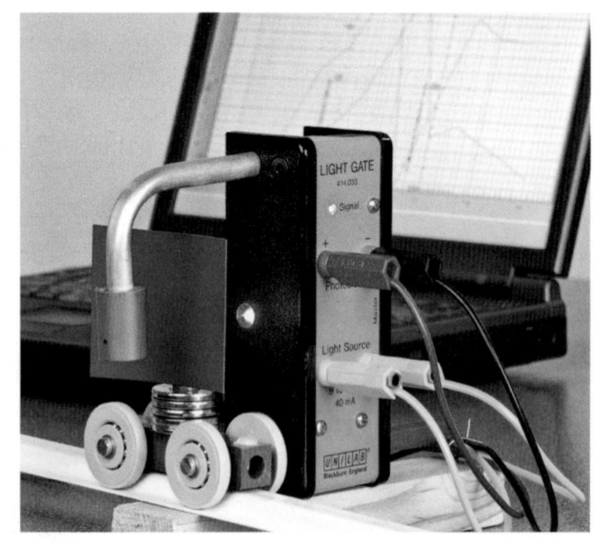


*A motion experiment using a light gate*

# 1 MEASURING THE UNIVERSE

## PRIOR KNOWLEDGE

You will have carried out experiments and made measurements in your previous studies of science, so you will know something about the scientific method.

## LEARNING OBJECTIVES

In this chapter you will find out how to get a rough idea of atomic size by a simple experiment. You will learn about physics experiments and measurements in general: what units to use and how they are defined; how errors can occur; and how to estimate the uncertainty in your experimental results.

**(Specification 3.1.1, 3.1.2, 3.1.3, 3.2.1.1 part)**

One of the big questions in physics is: "What is the Universe made of?" Until 1998, most physicists would have said "matter and energy" and been reasonably confident what that meant. "How much matter and energy?" seemed the more pertinent question (Figure 1). Albert Einstein had shown that mass and energy are interchangeable. Their combined amount determines the 'energy density' of our Universe, a quantity that will decide its ultimate fate.

In 1929 Edwin Hubble published measurements that showed that the Universe was expanding. Physicists knew that gravity would act to slow the rate of expansion. If the energy density was low, then the Universe would keep expanding, but at a slower and slower rate. A high value of energy density would eventually stop the expansion, and the Universe would begin to contract, eventually ending in a 'Big Crunch'.

In 1998 observations of distant supernovae showed that the expansion was not slowing at all, but speeding up. The measurements were reproduced by independent research teams, some using different methods. These results suggested that something unknown must be pushing the Universe apart. This is now called 'dark energy' and it seems to make up almost 70% of the Universe. Much of the remaining 30% is also mysterious. Work by Vera Rubin in the 1970s on the rotation of galaxies had shown that there must be a significant amount of mass in the Universe that we cannot observe – now known as 'dark matter'. Current thinking is that the Universe contains a mere 4.9% 'ordinary' matter and energy, and researchers are aiming to discover what the mysterious 'dark' quantities might be.

Physicists are often labelled either theoretical or practical. Einstein was firmly in the theoretical camp. Saul Perlmutter shared the 2011 Nobel Prize in Physics for practical measurements of the expansion rate of the Universe. Both aspects of physics are equally important. As Robert Millikan (Nobel Prize winner in 1923 for his work on the elementary charge) put it:

"Physics walks forward on two feet, namely theory and experiment. ... Sometimes it is one foot first; sometimes the other, but continuous progress is only made by the use of both."

***Figure 1*** *(background) How much matter and energy is there in the Universe? This Hubble Ultra Deep Field view taken by the Hubble Space Telescope shows a vast number of distant galaxies.*

## 1.1 MEASUREMENT IN PHYSICS

Towards the end of the 20th century, just before the discoveries of dark matter and dark energy, it was suggested that the 'big questions' in physics had been answered. In a remarkably similar way 100 years earlier, some eminent physicists felt that physics was almost complete. Newton's laws described forces and motion, Faraday had linked electricity and magnetism, and Maxwell's equations described electromagnetic waves. Michelson, famous for measurements of the speed of light, went as far as to say:

"The more important fundamental laws and facts of physical science have all been discovered, the possibility of their ever being supplanted (by) new discoveries is exceedingly remote."

But then, as now, physics was turned upside down by experimental discoveries. Radioactive decay, for example, proved hard to explain for 19th-century scientists, who were still arguing about whether atoms really existed. Observations, measurements and the analysis of recorded data provide the basis for discoveries and advancement in physics.

## 1.2 THE SCALE OF THINGS

### Scientific notation

Physicists investigate matter and energy in the Universe on every scale, from infinitesimally small measurements of subatomic particles to inordinately large ones, like galaxies. These measurements generate very large and very small numbers. We need a concise way of writing the numbers, to avoid strings of zeros across the page. Large numbers are written as a number from 1 to 10, multiplied by a power of 10. For example, the speed of light, $c = 300\,000\,000\,\mathrm{m\,s^{-1}}$, can be written as $3.0 \times 10^8\,\mathrm{m\,s^{-1}}$. In a similar way, small numbers are written as a number between 1 and 10, multiplied by a negative power of 10. In this way the wavelength of red light, $\lambda = 0.000\,000\,650\,\mathrm{m}$, would be written as $\lambda = 6.50 \times 10^{-7}\,\mathrm{m}$. This method of writing large or small numbers is known as **scientific notation**, often referred to as **standard form** in the UK.

It is usual to use powers of 10 that go up in steps of 1000, or $10^3$, so the wavelength of the red light would probably be written as $650 \times 10^{-9}\,\mathrm{m}$. When using SI units (Système International d'Unités) (see the next subsection), the powers of $10^3$, $10^6$, $10^9$ and so on are given names, such as kilo or mega. These have abbreviations used as prefixes, so a distance of 1000 m ($10^3$ m) is known as a kilometre and is written 1 km. The names and prefixes that you may come across at AS and A-level are shown in Table 1.

### QUESTIONS

1. Satellite TV signals are transmitted on a frequency of 27 000 000 Hz. Rewrite this number using scientific notation.
2. The mean distance from the Earth to the Sun is about 149 600 000 km. Rewrite this in scientific notation. (*Careful!* The distance is given in km in the question.)
3. How long does it take light to travel across the room you are in? (Distance = speed × time, speed of light = $3.0 \times 10^8\,\mathrm{m\,s^{-1}}$.)

| Multiplication factor | | Prefix | Symbol | Example length |
|---|---|---|---|---|
| 1000 000 000 000 | $10^{12}$ | tera | T | Radius of Pluto's orbit (5.9 Tm) |
| 1000 000 000 | $10^{9}$ | giga | G | Mean Earth–Moon distance (0.4 Gm) |
| 1000 000 | $10^{6}$ | mega | M | Mean radius of Earth (6.37 Mm) |
| 1000 | $10^{3}$ | kilo | k | Distance from Manchester to London (320 km) |
| 0.001 | $10^{-3}$ | milli | m | Microwave wavelength (~mm) |
| 0.000 001 | $10^{-6}$ | micro | μ | Wavelength of visible light (~μm) |
| 0.000 000 001 | $10^{-9}$ | nano | n | Approximate atomic diameter (~nm) |
| 0.000 000 000 001 | $10^{-12}$ | pico | p | Wavelength of a gamma ray (~pm) |
| 0.000 000 000 000 001 | $10^{-15}$ | femto | f | Approximate diameter of an atomic nucleus (~fm) |
| 0.000 000 000 000 000 001 | $10^{-18}$ | atto | a | Range of weak nuclear force (~am) |

***Table 1*** *SI prefixes and symbols*

## Choosing the units

Physics describes the world in terms of the values of physical quantities. A physical quantity is something that can be measured, such as speed, energy or mass. Each measurement needs a unit, a standard value that is well defined. Giving your height as '1.65' means nothing, but giving it as 1.65 metres makes it clear. The numerical value of a measurement depends on the unit that is used (Figure 2).

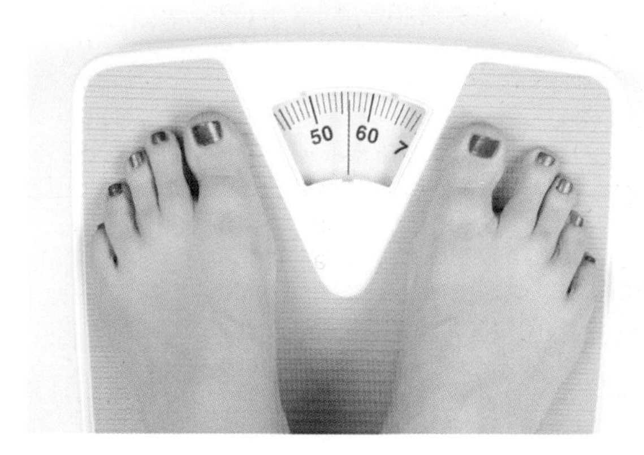


***Figure 2*** *Your mass might be 56, 8.8, 123 or 1.1, depending on whether you measured it in kilograms, stones, pounds or hundredweight. British school science lessons have not used imperial units (feet, stones, pounds and so on) since the early 1970s, but most students still give their height in feet and inches.*

Units were often chosen in the past to be a convenient size for the measurement, like the 'grain' (64.8 mg) traditionally used for the mass of medicines and gunpowder, or the carat (200 mg) used for precious stones. But it is awkward to have many different units for each physical quantity. Every unit has to be defined in terms of a standard, so that the entire world can agree on its magnitude. It is not feasible to maintain several standard definitions for each physical quantity like length or mass, and there is the problem of converting between units, which has been known to have disastrous results (Figure 3).

The Systême International d'Unités (**SI**) for defining units of measurement was established in 1960 and is now almost universally accepted – by the scientific world at least. SI units are defined in terms of seven **base units**. These are shown in Table 2. All other units, known as **derived units**, can be defined in terms of these base units.

***Figure 3*** *In 1999, the Mars Climate Orbiter probe was destroyed because one of the control systems used imperial units (feet and inches), but the navigation software used metric (SI) units.*

| Base quantity | Name | Symbol |
|---|---|---|
| length | metre | m |
| mass | kilogram | kg |
| time | second | s |
| electric current | ampere | A |
| thermodynamic temperature | kelvin | K |
| amount of substance | mole | mol |
| luminous intensity | candela | cd |

***Table 2*** *SI base units*

The base units are now almost all defined in terms of physical constants. For example, a length of one metre is defined in terms of the speed of light:

One metre is the length of the path travelled by light in vacuum during a time interval of $\frac{1}{299\,792\,458}$ of a second.

This rather arbitrary time is chosen to match the older definition of the metre. This modern definition of length depends on specialised equipment, but in principle every country can have the same standard metre. However, we also need an independent definition of the second.

> One second is the time taken for 9 192 631 770 complete oscillations of the radiation corresponding to the transition between the two hyperfine levels of the ground state of the caesium-133 atom.

The atomic clocks on the satellites that make up the Global Positioning System (GPS) (Figure 4) are stable to 1 part in $10^{12}$; in other words, they are accurate to within 1 second in 32 000 years.

The only base unit not yet defined in terms of a universal constant is the kilogram. This is still defined as the mass of a particular cylinder of platinum–iridium alloy, the International Prototype Kilogram (IPK), which is kept in a vault in Paris.

***Figure 4*** *To use the GPS system, the receiver must be able to see a minimum of four satellites.*

The kilogram is also the only base unit with kilo in its name. Logically, the base unit of mass should be the gram, but a historical quirk meant that the kilogram was chosen. This is important for calculations. If you need to put a value of mass into an equation, you must use kilograms, so a mass of one kilogram = 1 kg, whereas a mass of one gram = $1 \times 10^{-3}$ kg.

## QUESTIONS

4. The IPK is kept in a controlled atmosphere, and is only rarely taken from its vault. Why?
5. The IPK is a right-circular cylinder (height = diameter) of 39.17 mm. Why is it this shape? Why is the choice of metal important?
6. Use the formula density $= \frac{\text{mass}}{\text{volume}}$ to find the density of the standard (IPK) kilogram. The SI derived unit for density is kg $m^{-3}$. Remember that 1 mm = $1 \times 10^{-3}$ m, so 1 $mm^3$ = $1 \times 10^{-9}$ $m^3$.
7. The earliest units for length were based on the human body, for example the cubit in ancient Egypt was defined as the distance from the tip of the forefinger to the elbow. Give an advantage, and a disadvantage, of this system.
8. The speed of light is now exactly 299 792 458 m $s^{-1}$. Why 'now' and why 'exactly'?

Units for all the other physical quantities, such as velocity, acceleration, force and energy, are derived from these base units. For example, the unit for velocity is metre per second (m $s^{-1}$). Derived units are sometimes given names, like newton (N) for force, and joule (J) for energy. As a rule the named unit is written in full with a lowercase initial letter, but the abbreviation begins with an uppercase letter. For example, the SI unit for frequency is the hertz, abbreviated to Hz.

All named derived units can be expressed in terms of the base units. For example, one newton is defined as the force that will accelerate a mass of one kilogram by one metre per second, every second. This definition can be represented by the equation $F = ma$.

As the units have to be the same on both sides of this, or any other, equation (you could not have metres = kilograms, for example), then, in terms of units, $F = ma$ becomes

$$1\,\text{N} = 1\,\text{kg} \times \frac{1\,\text{m}}{(1\,\text{s} \times 1\,\text{s})} = 1\,\text{kg}\,\text{m}\,\text{s}^{-2}$$

## QUESTIONS

9. One pascal (1 Pa) is the SI derived unit of pressure. Since pressure $= \frac{\text{force}}{\text{area}}$, write 1 Pa in terms of SI base units.
10. Express the joule in terms of base units. *Hint:* What equation links energy or work to other quantities?

Physicists do not always quite stick to the rules for using SI units. Sometimes it is just too clumsy to use the SI base unit. For example, the kilogram is rather large when it comes to the mass of an atom, so the atomic mass unit is used. The metre is too small for interstellar or intergalactic distances, so astronomers use light years or megaparsecs. You might have used kilowatt hours, rather than joules, to measure the electrical energy used in a house. It is of course possible to convert all these to the relevant SI unit.

In the assignment you can find a value for the size of an atom in metres.

## ASSIGNMENT 1: FINDING A MAXIMUM SIZE FOR AN ATOM

**(MS 0.1, MS 0.2, MS 2.3, MS 4.3, PS 1.1, PS 3.2)**

There is a way to find a rough value for the size of an atom using ordinary school laboratory equipment. Olive oil, a very clean tray, a magnifying glass and scale, a ruler and some fine powder, such as lycopodium (that is, pollen – a potential allergen) are needed.

The general idea is to let a small drop of olive oil fall onto the surface of some water (Figure A1). The drop will spread out into a very thin film. If the surface of the water is coated with powder first, it will allow the oil film to be seen more clearly.

The volume of the film will be the same as the volume of oil in the drop. In theory the film of oil should be a circle. (In practice it is not!) Imagine that the film is roughly cylindrical in shape (a very thin cylinder). Then if the volume and the area of the film are known, its thickness can be calculated. The oil molecules cannot be bigger than this thickness, and an atom must be smaller still, so we can arrive at the maximum size for an atom.

Figure A2 is a scale drawing of typical results. If you are able to do this experiment yourself, you can use your own results. Otherwise, use measurements from Figure A2.

### Questions

**A1** Find the volume of the drop (volume of sphere $= \frac{4}{3}\pi r^3$).

**A2** Find the area of the film (area of circle $= \pi R^2$).

**A3** Find the thickness of the film, $h$.

**A4** An olive oil molecule is 10 atoms long. What is the maximum diameter for an atom in metres?

**A5** What have you assumed in this calculation?

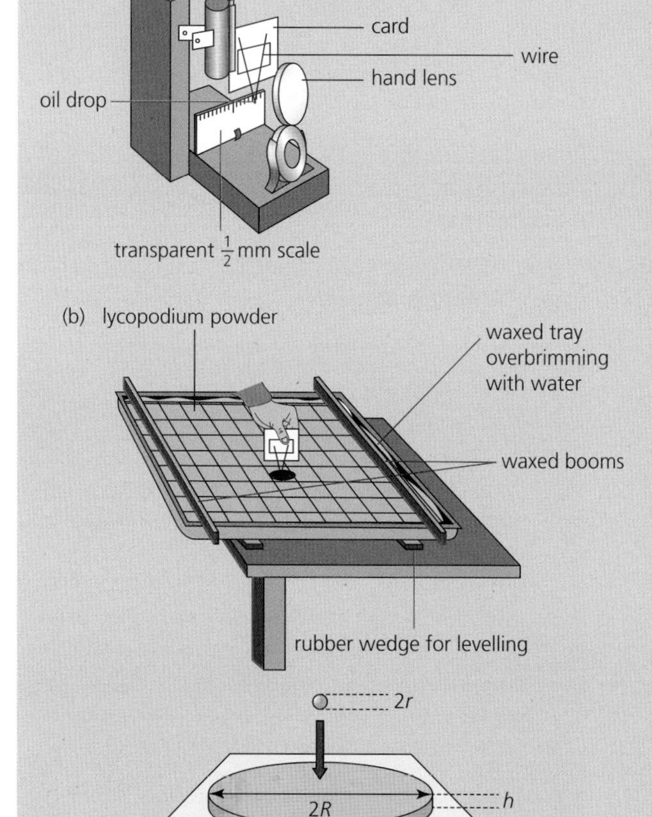


***Figure A1*** *The oil drop experiment*

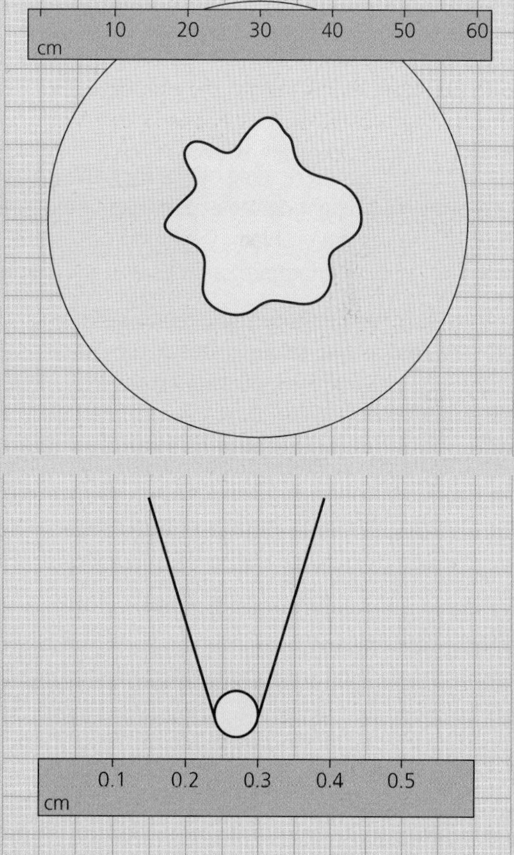


***Figure A2*** *Volume of the drop = volume of the oil film = area of the oil film × thickness of the film*

**KEY IDEAS**

- We use SI units. There are seven base units. All other units are derived from these.
- Large and small numbers are expressed in standard form, for example $3.0 \times 10^8$.
- Multiples of units in powers of 10 increasing in threes, for example, $10^3$, $10^6$, and so on, are given standard prefixes, for example, kilo (k), mega (M).

# 1.3 EXPERIMENTS IN PHYSICS

## Experimental error

It is surprising that we can get an estimate for the size of an atom using such simple apparatus as that used in Assignment 1. After all, we have managed to measure something that is far too small to see. How do we know that our answer is right? What errors might we have made? Can we correct them, reduce them or at least account for them?

The term 'experimental errors' does not generally refer to the sort of blunders we all make from time to time, such as forgetting to connect the battery, misreading a scale or failing to take a reading at the right time. These are annoying, but repeating the experiment with more care usually solves the problem. Experimental errors fall into two types: random errors and systematic errors.

**Random errors** can cause readings to be too high or too low. Just as the name suggests, the readings fluctuate about the mean. Random errors may arise due to a number of different causes.

- Observation or reading errors: perhaps when timing the oscillations of a pendulum, or when trying to read the flickering needle on the dial of an analogue meter.
- Environmental: perhaps the temperature of the room is fluctuating, or the supply voltage keeps changing.

The crucial thing about a random error is that it is equally likely to give you a result that is too high as one that is too low. Repeating the readings and calculating a mean value is useful because the more readings you have, the more the random fluctuations will be averaged out.

**Systematic errors** on the other hand lead to results that are consistently wrong. Repeating these readings is pointless, since the error occurs in the same way each time. Systematic errors may also occur due to a number of reasons.

- Instrument error: a poorly calibrated thermometer, for example, or a top-pan balance that has not been zeroed correctly.
- Reading error: perhaps due to parallax error (Figure 5) when reading the scale.
- Poor experimental design: for example, ignoring the effect of an external factor like magnetic field, temperature or pressure.

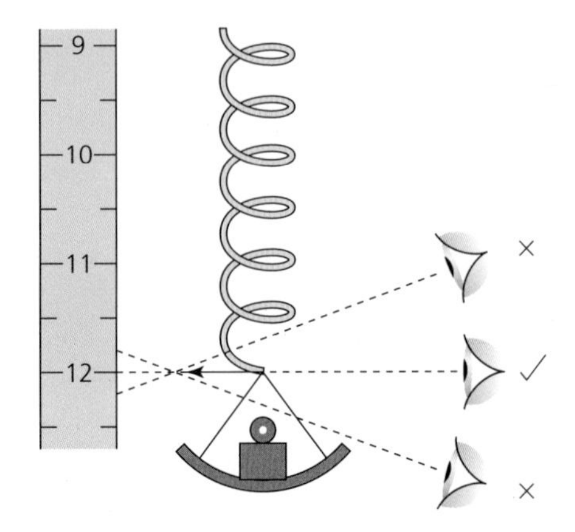


***Figure 5*** *Try to position your eye close to the scale and look in the correct direction to avoid parallax errors.*

Ernest Rutherford (Figure 6), who discovered the atomic nucleus, believed that experiments should have a clear outcome:

"If your experiment needs statistics, you ought to have done a better experiment."

***Figure 6*** *Ernest Rutherford*

## QUESTIONS

11. a. You intend to use a top-pan balance to find the mass of a particular ball bearing. Is it worth repeating the measurement several times and taking an average?
    b. Suppose you need to find the mass of a typical ball bearing. How would you make your result as accurate as possible?
12. When you are measuring the diameter of a wire, it is good practice to take readings at several points along the length of the wire. The readings should also be taken in different orientations. Explain why.

***Figure 7*** *In 2011 physicists at the Gran Sasso laboratory in Italy measured the speed of neutrinos emitted by the accelerator at CERN and found they were travelling faster than the speed of light. Special relativity says it is impossible for a particle to reach the speed of light, as this would give it an infinite mass. Faster-than-light travel also raises the possibility of time travel. Physics Professor Jim Al-Khalili promised to eat his boxer shorts on live TV if the measurements were shown to be right. In fact, the measurements were wrong ... they were caused by a loose fibre-optic cable!*

### Accuracy, precision and uncertainty

How sure of our measurements can we be? This can be a difficult question to answer, especially if you happen to be one of the first to make the measurement. In practice, the experiment is repeated by the experimenter to check that it gives consistent results. If so, then the measurement is said to be **repeatable**. If other experimenters get similar results, preferably in different laboratories using different techniques, then the measurement is said to be **reproducible**.

A result is said to be **accurate** if it is close to the true value, that is, the standard or accepted value. In exceptional cases, of course, the new measurement may not agree with the accepted value, because the accepted value is wrong. However, you need to be very sure before making a claim like that. The standard 'textbook' answer will have been repeated many times, probably in different laboratories and using different methods. If the new results prove to be repeatable and reproducible, it may mean that an established theory could be wrong (Figure 7).

'Precision' does not mean that the measurements are right; it merely tells you whether the results are numerically close together. For example, suppose that a measurement was made five times and the results were 3.223, 3.222, 3.223, 3.221 and 3.223. These results vary through a range of 0.002, from the lowest to the highest value recorded. The mean of the five readings is 3.222 (correct to four significant figures), so we can say that the **uncertainty** in this mean value due to the scatter of results is ±0.001. This seems a small uncertainty but it depends on the size of the measurement. We need to compare this uncertainty in the readings with the overall result by expressing it as a percentage. The percentage uncertainty is $(0.001/3.222) \times 100\% = 0.031\%$. This is a very small percentage uncertainty, so the results could be said to be very **precise**. That does not mean that the results are correct or even accurate. They could all be wrong in the same way. Figure 8 gives a visual summary of the meanings of precision and accuracy.

When recording your results, you should be careful not to overstate the precision by writing an excessive number of digits in your answer. Suppose that three teachers have timed the school 100 m sprint race and have recorded times of 12.3, 12.5 and 12.6 s for the winner. The average time was 12.466 66 s. But each teacher's reading had an uncertainty of 0.1 s at least, and the range of their readings was 0.3 s, or ±0.15 s, so the result

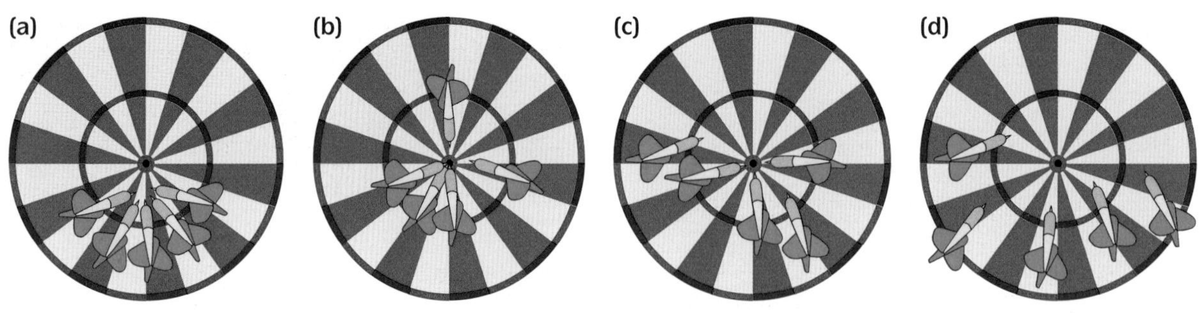


***Figure 8*** *Precision and accuracy: (a) precise but not accurate; (b) precise and accurate; (c) accurate but not precise; (d) neither accurate nor precise.*

should be quoted to a similar level of precision: (12.5 ± 0.15) s is more reasonable but, since we usually err on the side of caution, (12.5 ± 0.2) s is probably appropriate.

On a similar theme, it would be a mistake to record this set of readings, all taken with the same equipment:

| Time *T* / s | 1.214 | 1.20 | 0.800 | 0.5 |
|---|---|---|---|---|

If the readings are all made with the same precision, they should all be quoted to the same number of **significant figures**. The trailing zeros are important!

## QUESTIONS

**13. a.** Describe a real situation where measurements could be precise, but not accurate.

**b.** Describe a real situation where measurements could be accurate, but not precise.

Some readings do not vary widely but are still not precise, simply because they are measured with a device with low **resolution**. The resolution of a measuring device is the smallest increment in the measured quantity that can be shown on the device. For example, you could find the mass of an object using a digital balance with a resolution of (that is, gives readings in) grams, tenths of grams, or hundredths of grams. Suppose you were using the balance that measured to the nearest gram to find the mass of a mango; your readings could be out by 0.5 g. In fact, it is worse than this because when you zeroed the balance, it could also have been out by 0.5 g. So your reading is said to have an uncertainty of ±1 g. As a rule of thumb you can estimate the uncertainty associated with taking a reading to be ± the smallest scale division.

Measurements should always be written with a value, the associated uncertainty and an appropriate unit, for example, mass of a mango = (142.3 ± 0.1) g. This is not too much of a problem, with less than 0.1% uncertainty. But if you were finding the mass of a blackcurrant, you would probably want a balance with a higher resolution.

You can improve the precision of measurement of repeatable events by just doing a lot of them. Galileo is said to have timed the oscillations of a pendulum (candelabras hung from the ceiling of a church) using his pulse as a time keeper. This is not a very high-resolution instrument. But by timing 10 oscillations he could share the uncertainty among all 10 oscillations, and arrive at a more precise answer.

Whether you are measuring the speed of neutrinos, or the area of an oil film, it is important to know how precise your result is. Every experimental result should be accompanied by an estimate of its uncertainty. For example, the currently accepted value for the mass of an alpha particle is given as

$$(6.644\,656\,75 \pm 0.000\,000\,29) \times 10^{-27}\ \text{kg}$$

sometimes written as

$$6.644\,656\,75(29) \times 10^{-27}\ \text{kg}$$

The numbers in brackets indicate the uncertainty in the last two digits. That is a high-precision measurement, an uncertainty of 29/664 465 675, or less than $5 \times 10^{-6}$%. That is equivalent to knowing the distance from London to New York to within 25 cm.

## QUESTIONS

**14.** Estimate the resolution of Galileo's time keeper. Suppose the candelabra took 2 s to complete one oscillation. What would Galileo's result be if he timed one oscillation? What would the uncertainty be? What would the percentage uncertainty be? How would these answers be changed if he now timed 10 oscillations and used that to calculate the time for one oscillation?

**15.** If you did not have a high-resolution balance, how could you find the mass of a blackcurrant more precisely? (Assume that you have a large number of blackcurrants to hand!)

**16.** Suppose you took three oranges and found their mean mass, using a balance with a resolution of 0.1 g. Why would it be wrong (and certainly misleading) to write your answer as 121.333 333 g? How would you write it?

**17.** **a.** What is the uncertainty associated with measuring the width of this book? (Suppose that you used a 30 cm ruler with mm divisions.) Write your answer as result ± uncertainty, followed by the correct unit.

**b.** Why is the uncertainty more of a problem if I asked you to use the same ruler to measure the thickness of the book?

**c.** How would you find the thickness of one page of this book? How precise do you think you could be?

## KEY IDEAS

- Experimental errors can be systematic. These tend to affect all readings in the same way. Repeating the readings does not help. Try to improve the method.
- Experimental errors can be random. These fluctuate above and below the mean value. Repeated readings will improve the precision of these readings.
- No measurement of a physical quantity is ever exact. There is always an uncertainty associated with it.
- An accurate measurement is one that is close to the accepted value.
- A precise set of measurements are closely grouped together, with little spread or uncertainty.

# 1.4 COMBINING UNCERTAINTIES

The final result of an experiment is often a combination of several measurements. That means that the overall uncertainty will depend on a combination of the precision of each measurement.

What if you are *adding or subtracting* two quantities? The general rule is:

If you are adding or subtracting quantities, you need to add their absolute uncertainties.

An absolute uncertainty is the possible deviation from the mean value in the unit of measurement. Finding the difference of two measurements can lead to large *percentage* uncertainties.

### Worked example 1

In a situation like that in Figure 9, we might have these readings:

Reading A : Mass of bowl = (200 ± 1) g, which is a percentage uncertainty of 0.5%.

Reading B: Mass of bowl *and* flour = (220 ± 1) g, a percentage uncertainty of ≈0.5%.

Mass of flour = reading B − reading A = (20 ± 2) g

The uncertainty in this difference between two measured values is 2 g since reading B might be higher by 1 g and reading A might be lower by 1 g, and vice versa. The percentage uncertainty is now 10% (compared with 0.5% in the measured values).

***Figure 9*** *Finding a difference in two quantities, for example finding the mass of sugar in a bowl, can lead to large percentage uncertainties.*

What if you are *dividing or multiplying* two quantities? Suppose you have been given a small metal cube, which you suspect is made of lead. You might decide to measure the density of the cube to see if it could indeed be lead. The density of a material is defined as the mass of a given volume. In SI units this should be measured in kilograms per cubic metre, but grams per cubic centimetre is also commonly used. Lead has a density of $11.34\,\text{g}\,\text{cm}^{-3}$.

As density equals mass divided by volume,

$$\text{density} = \frac{\text{mass}}{\text{volume}}$$

you might begin by using a top-pan balance to measure the mass, and a ruler marked in millimetres to measure the dimensions of the cube.

Mass of metal cube = $(89 \pm 1)\,\text{g}$

Length of metal cube = $(2.1 \pm 0.1)\,\text{cm}$

Width of metal cube = $(1.9 \pm 0.1)\,\text{cm}$

Depth of metal cube = $(2.1 \pm 0.1)\,\text{cm}$

The volume of the cube = $2.1 \times 1.9 \times 2.1 = 8.379\ \text{cm}^3$

But how precise is this measurement? It is possible that all the dimensions have been underestimated by 0.1 cm, so the volume could be as large as $2.2 \times 2.0 \times 2.2 = 9.68\ \text{cm}^3$. Similarly, the volume could be as small as $2.0 \times 1.8 \times 2.0 = 7.20\ \text{cm}^3$. Possible values for the volume of the cube are from 7.20 to $9.68\ \text{cm}^3$, a range of $2.48\ \text{cm}^3$, so the uncertainty is approximately $\pm 1.24\ \text{cm}^3$. The volume of the cube is therefore $(8.38 \pm 1.24)\ \text{cm}^3$, an uncertainty of almost 5%.

It is possible to find the uncertainties in calculated values by inserting the largest and smallest values of your data into the relevant formulae. But this can be time-consuming and there is a better way. You saw in Worked example 1 that to find the uncertainty in the difference of two masses you simply add the individual uncertainties together. But if you are *multiplying* or *dividing* two quantities, the general rule is:

If you are multiplying or dividing quantities, then you add the *percentage uncertainties* together.

## Worked example 2

Using the data given above for the metal cube, the percentage uncertainties in each measurement of length are $0.1/2.0 = 5\%$. So the uncertainty in volume is $5 + 5 + 5 = 15\%$.

To find the density, we need to divide the mass by the volume. The percentage uncertainty in the measurement of mass = $1/89 = 1.1\%$. The overall uncertainty in the density value is therefore $15 + 1.1 = 16\%$.

Density of the cube = $89/8.37 = 10.6 \pm 16\%$ or $(10.6 \pm 1.7)\ \text{g cm}^{-3}$.

Since the accepted value for the density of lead, $11.34\ \text{g cm}^{-3}$, falls within the range of uncertainty, the metal could be lead. The measured value for density is not precise enough, however, to rule out other metals. We could improve the precision by using instruments with better resolution to measure length and mass. Two instruments commonly used in the laboratory to measure length precisely are shown in Figures 10 and 11. They both use a **vernier scale** – a movable scale that allows a fractional part on the main scale to be determined.

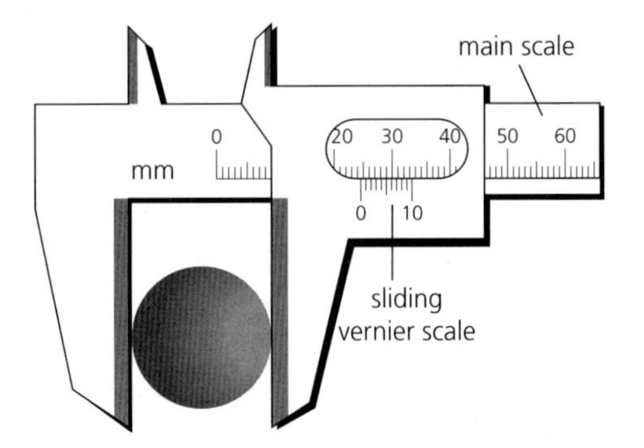


The reading in mm is taken from the position of the zero on the sliding scale. Here this is between 24 and 25. The next significant figure (to 0.1 mm) is found by judging which scale mark on the sliding scale is perfectly aligned with a mark on the main scale. Here this is 5.
The reading is 24 + 0.5 = 24.5 mm.

***Figure 10*** *Vernier callipers can measure length to one-tenth of a millimetre.*

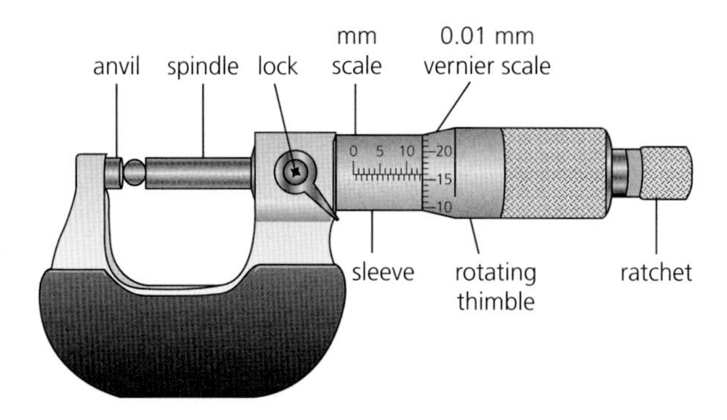


Turning the ratchet moves the spindle until it just touches the object. The ratchet then slips to avoid deforming the object.

The reading to the nearest 0.5 mm is taken where the thimble meets the sleeve. Here this is 12.5 mm. The final significant figures are given by judging which mark on the rotating scale coincides with the horizontal line on the sleeve. Here this is 16.

The reading is 12.5 + 0.16 = 12.66 mm.

***Figure 11*** *A micrometer screw gauge can measure length to one-hundredth of a millimetre.*

## QUESTIONS

**18.** Suppose you could improve the precision of either the measurement of length or the measurement of mass, for the small metal cube considered in the text. Which would most improve your final answer?

**19.** The cube may not be perfect, so that the dimensions may differ at different points. How would you allow for this?

**20.** You have been asked to find the density of a liquid, which you suspect is ethanol, which has a density of around 80% that of water. Suppose that you measure the volume using a measuring cylinder and its mass on a top-pan balance. By estimating the values of the mass and volume of ethanol you would use, and the resolution of the measuring instrument, calculate an approximate value for the uncertainty in your answer.

## KEY IDEAS

- If two physical quantities are to be added or subtracted, then their uncertainties must be added.
- If two physical quantities are to be multiplied or divided, then their percentage uncertainties must be added.

## ASSIGNMENT 2: FINDING THE UNCERTAINTY IN THE ATOMIC DIAMETER MEASUREMENT

**(MS 0.4, MS 1.1, MS 1.5, PS 1.1, PS 2.1, PS 2.3, PS 3.2, PS 3.3)**

It would be useful to estimate the uncertainty in our measurement of atomic size in Assignment 1.

- First you need to estimate the uncertainty in all your measurements.
- Then calculate the percentage uncertainty for your readings.
- Estimating the uncertainty in the area of the oil film is difficult. One way would be to estimate the largest and smallest area that the film could be. This will give you a spread of results. Halve this to find the uncertainty.
- Combine the uncertainties.

For example, the uncertainty in the diameter of the oil drop could be ±0.1 mm. This is a significant uncertainty, since the drop only has a diameter of 0.5 mm, so that is a ±20% uncertainty. The radius = (0.25 ± 0.05) mm since we divide the uncertainty by 2 as well. But the uncertainty in the volume will be larger than that because volume depends on the radius cubed, $V \propto r^3$, so that is three times the percentage uncertainty. In this case volume has a percentage uncertainty of 60%.

### Questions

**A1** What is the final uncertainty in your value for atomic diameter?

**A2** How would you improve the experimental method to try to reduce the uncertainty in this answer?

## 1.5 USING GRAPHS

A common way of reducing the uncertainty in a measured quantity is to repeat the reading, using a set of different values of the independent variable, and then plot a graph. Suppose you were asked to find the mass of a raindrop. You have an electronic top-pan balance with resolution of 0.01 g. Assume that you can count the raindrops! You could use any one of the following methods:

A Catch one raindrop and find its mass.

B Repeat the above method lots of times and find the mean mass.

C Collect 100 drops, find the mass and divide by 100.

D Collect 100 drops, recording the mass after every 10 drops. Then plot a graph of your answers.

Which method will give the most precise, and the most useful, results?

Method A will give a very large percentage uncertainty. The average mass of a raindrop depends on the type of rain (it varies from mist to downpour!) but is unlikely to be much more than 100 mg. The reading would have a percentage uncertainty of

$$\left(\frac{0.01}{0.1}\right) \times 100 = 10\%$$

Method B is better, but tedious! Theoretically, the precision is equivalent to that of method C, which would give a percentage uncertainty of

$$\left(\frac{0.01}{10}\right) \times 100 = 0.1\%$$

In practice, drying the container between each drop would be ridiculous. Method D gives the same uncertainty as method C, but allows you to spot any results that do not fit the pattern and ignore them if they are genuinely anomalous results. The results of such an experiment are shown in Table 3 and the graph obtained is shown in Figure 12.

| Number of raindrops | Accumulated mass / g |
|---|---|
| 10 | 3.60 |
| 20 | 3.70 |
| 30 | 3.80 |
| 40 | 3.90 |
| 50 | 4.01 |
| 60 | 4.11 |
| 70 | 4.20 |
| 80 | 4.30 |
| 90 | 4.40 |
| 100 | 4.49 |

*Table 3* Mass of every 10 raindrops

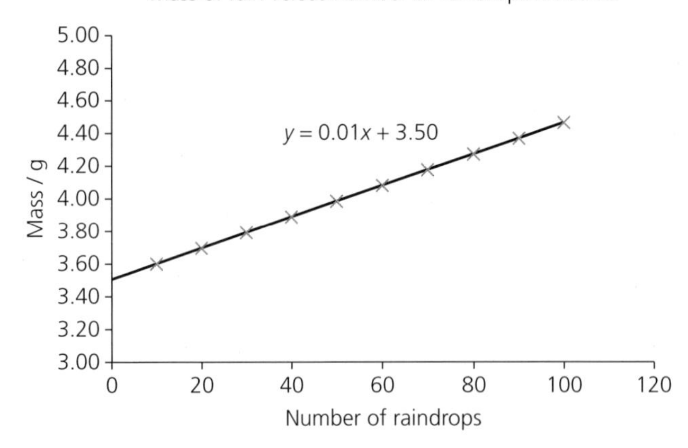


***Figure 12*** *The equation of a straight line is always of the form $y = mx + c$, where $y$ is the variable plotted vertically, $x$ is the variable plotted horizontally, $m$ is the gradient and $c$ is the intercept on the $y$-axis. If the equation is to be a straight line, $m$ must be a constant (fixed number). In this case the equation says mass collected ($y$) = average mass of a raindrop ($m$) × number of raindrops ($x$) + any other mass (perhaps the container or a zero error on the balance).*

The gradient of the line is found by calculating $\frac{\text{difference in } y}{\text{difference in } x}$, which in this case is

$$\frac{\text{mass}}{\text{number of drops}} = \text{mass of one raindrop}$$

The intercept of the line on the $y$-axis gives a mass reading before any raindrops are collected. This value, 3.50 g in this case, is a zero error, which might not be noticed without the graph.

### QUESTIONS

**21.** Look at the graph in Figure 13. It shows another set of results from the raindrop experiment. What do you think happened? Could you still use the results?

Mass of rain versus number of raindrops collected

***Figure 13*** *Plot of another set of results*

## Plotting graphs of experimental results

Graph-plotting in physics is not quite the same as in mathematics, where you are often plotting a function with perfectly accurate points that lie on an ideal curve. Physics data is often taken from real-life experiments and has some uncertainty associated with it. The data points will be scattered rather than being a perfect fit to a function. We often do not know whether the results are following a mathematical law or not. Indeed, that is often what we are trying to find out. In practice, it is difficult to draw quantitative conclusions from a curve, so we try to draw straight-line graphs to test relationships between quantities. This may mean that we plot a function of the variables, for example, $x^2$ or $1/x$, instead of the raw data.

Suppose that you were studying a falling object (Figure 14) and took a series of measurements showing how far the object had fallen after certain periods of time, say after 1, 2, 3, ... seconds. You would get a graph like the one shown in Figure 15(a). It may *look* as if distance depends on time squared but we cannot be sure from a curved graph. The distance fallen, $s$, and the time taken, $t$, could be related by a quadratic (squared) equation like $s = At^2 + B$, where $A$ and $B$ are constants (just a fixed number). This needs to be compared with the equation of a straight line, $y = mx + c$:

$$\begin{array}{ccccc} s & = & At^2 & + & B \\ \downarrow & & \downarrow\downarrow & & \downarrow \\ y & = & mx & + & c \end{array}$$

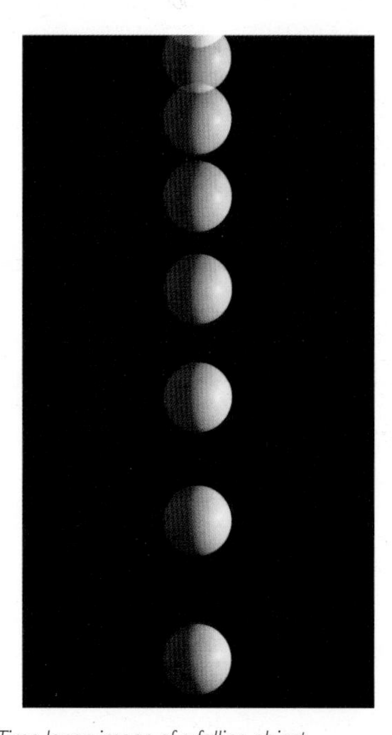

***Figure 14*** *Time-lapse image of a falling object*

Plotting $s$ on the $y$-axis and $t$ on the $x$-axis gives a curve (Figure 15a), but if we plot $s$ on the $y$-axis and $t^2$ on the $x$-axis, we should get a straight line (Figure 15b). If the points are a good fit to the straight line, we can deduce that the experimental data follows the relationship. The gradient will equal the constant $A$ and the $y$-intercept will equal the constant $B$.

Table 4 shows a few examples of what to plot in order to confirm a relationship.

## Good practice in graph drawing

Accuracy in graph work is important, not least because it often accounts for a significant number of exam marks. So here are a few tips on good practice:

- Choose your scales on each axis so that your data spreads over at least half of the axis. Use a false origin if necessary. You do not need to start the graph at (0, 0), unless you have a measured data point to plot there.
- Use a sharp pencil and a ruler to draw the axes.
- Label each axis with the quantity and unit separated by a solidus (slash) /, for example $T^2$ / $s^2$, $F$ / N, $\lambda$ / m, and so on.

| Variables | Constant(s) | Possible relationship | Rearrange to | $y$ | $x$ | Gradient (constant for a straight line) | $y$-intercept |
|---|---|---|---|---|---|---|---|
| $m, T$ | $k$ | $T = 2\pi\sqrt{\frac{m}{k}}$ | $T^2 = 4\pi^2\left(\frac{m}{k}\right)$ | $T^2$ | $m$ | $\frac{4\pi^2}{k}$ | 0 |
| $f, \lambda$ | $c$ | $c = f\lambda$ | $f = \frac{c}{\lambda}$ | $f$ | $\frac{1}{\lambda}$ | $c$ | 0 |
| $V, i$ | $E, r$ | $V = E - ir$ | $V = E - ir$ | $V$ | $i$ | $-r$ | $E$ |
| $F, r$ | $G, M, m$ | $F = \frac{GMm}{r^2}$ | $F = \frac{GMm}{r^2}$ | $F$ | $\frac{1}{r^2}$ | $GMm$ | 0 |

***Table 4*** *Examples of what to plot to confirm a relationship*

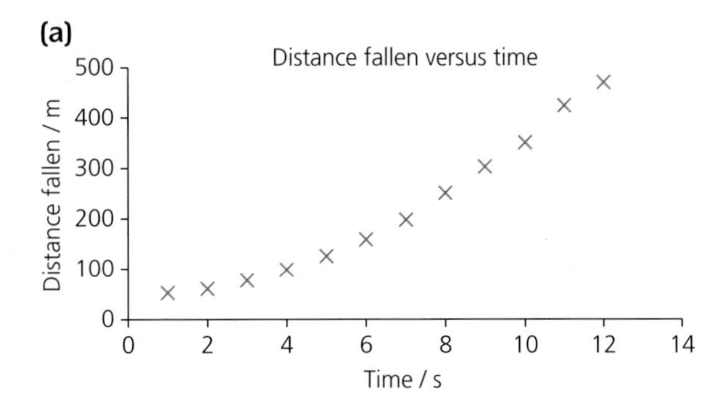


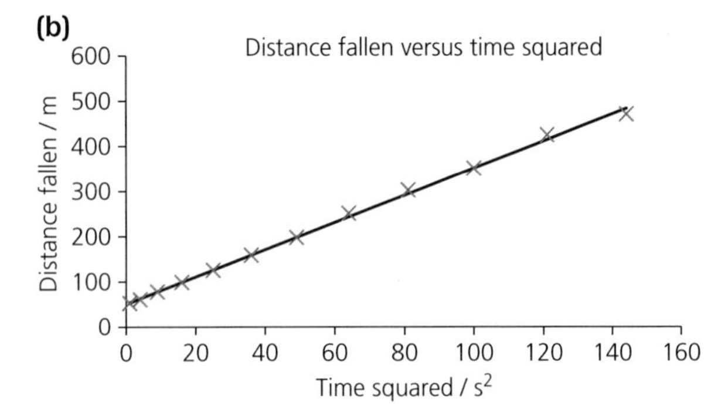


***Figure 15*** *Finding the relationship between distance fallen and time taken*

- Plot points (using a sharp pencil) with a small cross.
- Give the graph a meaningful title.

## Drawing a best-fit straight line and calculating the gradient

If the points look as if they may fall close to a straight line, you may opt to draw a 'best-fit' straight line. When a computer does this mathematically, it chooses the straight line that minimises the total distance of the points from the line. You should aim to do the same. You have two advantages over the computer:

- You can use your discretion and ignore any **outliers**, especially if you have practical reasons to suspect their accuracy. An outlier may pull a computer's best-fit line way off course. Try to identify these anomalous results, repeat them or at least try to explain why they are going to be ignored.
- You may know that the line must go through a given point, (0, 0) for example, and you can pivot your ruler about that point. (Make sure you have a 30 cm clear plastic ruler so that you can see the points through it.)

The gradient of a graph in physics often represents an important physical quantity. For example, if you plot velocity ($y$-axis) against time ($x$-axis), the gradient at a particular point gives the value of the acceleration at that time. You will often need to find the gradient of a best-fit line. Choose a large section of the graph, covering at least two-thirds of each axis (Figure 16). This will reduce the effect of any uncertainties in reading the points. Choose your two points, say $(x_1, y_1)$ and $(x_2, y_2)$; then

$$\text{gradient} = \frac{(y_2 - y_1)}{(x_2 - x_1)}$$

It is useful to include the unit when quoting the value of a gradient. The unit will be that of the quantity on the $y$-axis divided by that of the quantity on the $x$-axis. So for a velocity against time graph, the gradient will have a unit of $(\text{m s}^{-1})/\text{s} = \text{m s}^{-2}$.

### Uncertainties in graph plotting

Uncertainties on graphs may arise in two ways:

- There may be a large uncertainty in each measurement.
- It might be difficult to choose the best-fit line to find the gradient.

The first problem can be dealt with using 'error bars'. Plot the reading with a small cross as before. Then use bars through the point to show the horizontal and vertical extent of the uncertainty (see Figure 17).

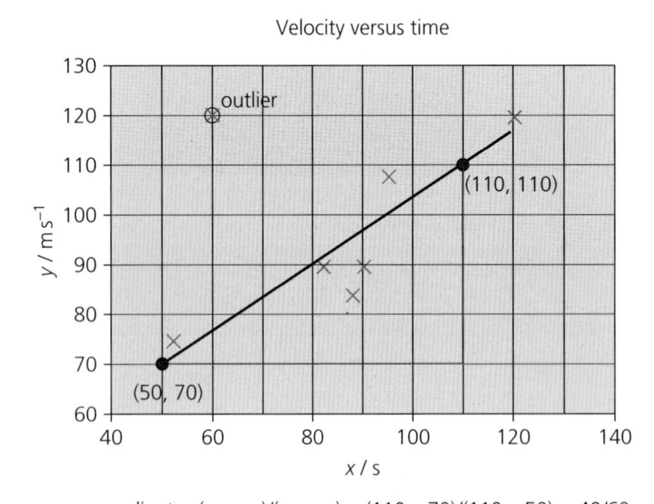


***Figure 16*** *Calculating the gradient of the best-fit line*

The second problem can be dealt with by drawing a best-fit line and then a 'worst-case' best-fit line (see Figure 17). Find the gradient and intercept of both lines. Your answer can be quoted as best-fit gradient ± (difference between the values).

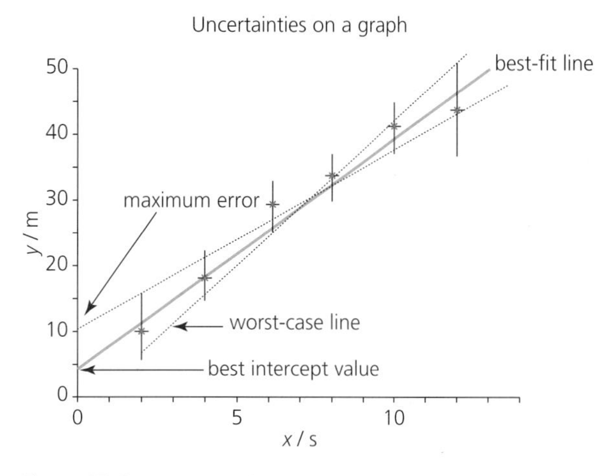


*Figure 17* *Best and worst fits through points with error bars*

## KEY IDEAS

- The equation of a straight line is of the form $y = mx + c$, where $m$ is the gradient and $c$ is a constant equal to the $y$-intercept.
- Uncertainty in a data point can be shown on a graph by drawing error bars.
- Best-fit and worst-fit lines can be drawn through the error bars to estimate the uncertainty in the gradient and intercept values.

## ASSIGNMENT 3: PLOTTING A GRAPH

**(MS 0.1, MS 1.1, MS 1.5, MS 3.1, MS 3.2, MS 3.3, MS 3.4, PS 1.1, PS 2.3, PS 3.1, PS 3.2, PS 3.3)**

An experiment has been carried out to measure the time it takes for a pendulum to complete 10 oscillations. Theory suggests that the time for one oscillation, $T$, depends on the length of the pendulum, $l$, according to the following equation:

$$T = 2\pi\sqrt{\frac{l}{g}}$$

The results are shown in Table A1.

| Time for 10 oscillations / s | Length of pendulum / cm |
|---|---|
| 8.7 | 20.0 |
| 12.0 | 30.2 |
| 13.7 | 40.4 |
| 14.2 | 49.8 |
| 14.3 | 60.2 |
| 16.7 | 70.0 |
| 17.0 | 80.2 |
| 19.2 | 89.9 |
| 19.0 | 100.2 |
| 22.0 | 110.0 |
| 25.0 | 120.0 |

*Table A1*

The uncertainty in the time measurement was estimated to be ±0.1 s. The uncertainty in the length was estimated at ±0.1 cm.

### Questions

**A1** Plot a suitable graph to test the relationship.

**A2** Find the gradient of the best and 'worst' case lines.

**A3** Find the value of $g$ that this gives you.

**A4** Estimate the uncertainty in your answer. Comment on the precision of this result.

**A5** The accepted value for $g$ is 9.81 m s$^{-2}$. Is your result accurate?

## 1.6 MAKING AN ESTIMATE

Finding a rough value for the diameter of an atom, and the uncertainty in this value, is an example of estimation, helped by a practical measurement or two. Estimation is a very useful skill in physics, and indeed in life!

It is possible to estimate the answers to questions such as: How much water is there in the reservoir? How many people could it supply? How many cars are there on the M25 at a given time? How many wind turbines would be needed to provide all the electricity for a small town? You do not always need an exact answer – sometimes an order of magnitude will do. It is good enough to know whether the answer is tens, hundreds, thousands or millions. An estimate is also useful for checking your answer in an exam question, for example, "Could the radius of Earth really be 6000 m or have I made an error?"

Enrico Fermi, a physicist who built the world's first nuclear reactor in a squash court, used to ask his students questions like these, which could be solved, very approximately, on the back of an envelope.

- How many grains of sand are there on Earth's beaches?
- How many piano tuners are there in Chicago?
- How many atoms are in your body?

(This sort of question has also become popular in some university interviews – popular with the interviewers rather than with candidates!)

If you are faced with one of these, such as, "How many cows could you fit in a barn?":

- Do not panic!
- Think of something you know, or can reasonably guess or find out about the situation.
- Break the problem down into smaller, hopefully easier, questions.
- Make simplifying assumptions, for example treat the cow as a cube!

Fermi gave this example. When asked, "What is the diameter of the Earth?", he reasoned like this:

1. I pass through three time zones when I fly from New York to Los Angeles.
2. I know that it is about 3000 miles from New York to Los Angeles.
3. That is 1000 miles per time zone, on average.
4. There are 24 hours in a day, so there must be 24 time zones around the world.
5. 24 time zones × 1000 miles per time zone = 24 000 miles.

So that is a circumference of about 24 000 miles. Circumference = $\pi$ × diameter. Take $\pi$ as approximately equal to 3, which gives the diameter of the Earth as about 8000 miles. An accurate value is 7926 miles, so 8000 is not a bad estimate!

### ASSIGNMENT 4: MAKING AN ESTIMATE

**(MS 0.4, MS 1.4)**

Here are a few 'Fermi-type' questions for you to try. At this stage, the way you approach this is just as important as the result, so record your method as well as your answer. It would be useful to work with a partner, or in a small team, at first so that you can discuss different approaches to the problem. You can look up some of the basic facts if necessary, but try to make as much progress as you can by reasoning from what you know.

#### Questions

**A1** How many Jelly Babies could you fit into a supermarket carrier bag?

**A2** "If all the mobile phone chargers in the UK were unplugged when not in use, we could save enough energy to boil 1 000 000 kettles every year." Could this be true? If it is true, is it important?

**A3** In his book *Sustainable Energy*, David MacKay says that trying to save energy by unplugging mobile phone chargers is like "trying to bail out the Titanic with a tea-spoon". How long would that take?

**A4** What mass of plastic is used every year in the UK to hold bottled water?

**A5** Make up your own estimation question and swap with another group. (You should have an answer, or at least a way of getting there.)

### Moving on: the start of the atomic age

In the first few years of the 20th century, the arguments over atomic reality were quickly forgotten. In 1897 the scientific debate was shifted by two discoveries: that of the electron (by J. J. Thomson in Cambridge) and radioactivity (by Henri Becquerel in Paris). These showed not only that the atom was real, but also that it had a structure and could be taken apart. The search to understand the composition of the atom had begun. But it was not until the 1950s that we could actually see images of atoms – search for 'field ion microscope'.

## PRACTICE QUESTIONS

1. You have been asked to measure the thickness of a sheet of printed paper.
   a. Describe how you would do this as precisely as possible.
   b. Estimate the uncertainty of your reading.
   c. The average density of the paper is quoted as 120 g $m^{-3}$. How would you verify this?
2. The timing of races for a school sports day is done manually. Time keepers for the 100 m race stand at the finish line with stopwatches. They start their stopwatches when they hear the starting pistol and stop them as the runners cross the finish line.
   a. The physics teacher points out that the time keepers start their stopwatches some time after the runners have started because of the time taken for the sound of the pistol to reach them. Given that the speed of sound in air is around 340 $m\,s^{-1}$, calculate the size of this delay.
   b. Is this a systematic error or a random error? Explain your answer.
   c. The time for the winner is given by the time keeper as 12.72 s. The physics teacher is critical of this. Explain why and rewrite the time in a way that can be justified scientifically.
3. An electric kettle is used to bring water to the boil. The temperature of the water is measured with an electronic thermometer every 30 s. The results are shown in Table Q1.
   a. A student has made a number of mistakes in recording the results. Suggest **two** corrections.
   b. Plot a graph of temperature ($y$-axis) against time ($x$-axis).

| Time | Temperature |
|---|---|
| 0 | 10 |
| 30 | 35.3 |
| 60 | 54.7 |
| 90 | 72.4 |
| 120 | 87 |
| 150 | 95 |
| 180 | 100.2 |
| 210 | 100.2 |

***Table Q1***

   c. Use the graph to calculate the greatest rate of increase of temperature.
   d. Explain the shape of the graph.
4. A metal cube of side length 4.0 cm is manufactured to a tolerance of ±0.1 cm. Its volume will be:
   **A** (64.0 ± 0.1) $cm^3$
   **B** (64.0 ± 0.2) $cm^3$
   **C** (64 ± 5) $cm^3$
   **D** (64 ± 7.5) $cm^3$
5. The speed limit on British motorways is 70 mph. In SI units this would be written as:
   **A** 31.1 $m\,s^{-1}$
   **B** 1.87 $km\,min^{-1}$
   **C** 43.8 $km\,h^{-1}$
   **D** 43.8 $m\,s^{-1}$
6. Density is measured in kilograms per cubic metre. Water has a density of 1000 $kg\,m^{-3}$. What is the mass of 1 litre of water?
   **A** 100 kg
   **B** 10 kg
   **C** 1 kg
   **D** 100 g

7. Estimate the mass of a five-door family hatchback car. Which of these values is closest to the actual value?
   A 100 kg
   B 500 kg
   C 1000 kg
   D 5000 kg

8. Pressure is defined as the force on a certain area. Which of these would be the correct unit to measure pressure?
   A pound per square inch
   B kilogram per square metre
   C newton per cubic metre
   D newton per square metre

9. Estimate how many footballs you could fit into your (empty) classroom. Choose from:
   A 300 000
   B 30 000
   C 3000
   D 300

10. An experiment using polarised light requires a sugar solution of strength 100 g of sugar per litre of water. You are provided with a measuring cylinder of capacity 50 $cm^3$, marked in $cm^3$, and an electronic balance sensitive to 1 g. The maximum strength of your solution could be:
    A 100.3 g $cm^{-3}$
    B 100.2 g $cm^{-3}$
    C 102 g $cm^{-3}$
    D 103 g $cm^{-3}$

11. In an experiment a student measures the wavelength, $\lambda$, of different frequencies, $f$, of sound. The velocity of sound, $v$, is given by velocity = frequency × wavelength, $v = f \times \lambda$. To find a value for the velocity from the gradient of a graph, what should the student plot? Choose the correct row from Table Q2.

| | $y$-axis | $x$-axis | Gradient |
|---|---|---|---|
| A | $f$ | $\lambda$ | $v$ |
| B | $\lambda$ | $f$ | $v$ |
| C | $f$ | $\frac{1}{\lambda}$ | $v$ |
| D | $\frac{1}{\lambda}$ | $\frac{1}{f}$ | $v$ |

*Table Q2*

12. A student needs to measure the dimensions of a mobile phone as precisely as possible. Which of the rows in Table Q3 would be the most appropriate measuring devices?

| | Length | Width | Thickness |
|---|---|---|---|
| A | Ruler | Vernier callipers | Micrometer |
| B | Ruler | Micrometer | Vernier callipers |
| C | Vernier callipers | Vernier callipers | Micrometer |
| D | Micrometer | Micrometer | Vernier callipers |

*Table Q3*

13. Two students are measuring the current through a circuit. Student A has a digital meter, which reads 0.1 A when the circuit is off. Student B has an analogue meter, which he views from an angle, leading to a parallax error. Which row in Table Q4 correctly describes the nature of these errors?

| | Student A | Student B |
|---|---|---|
| A | Random | Random |
| B | Systematic | Random |
| C | Random | Systematic |
| D | Systematic | Systematic |

*Table Q4*

**14.** Which row in Table Q5 correctly names the part of the micrometer screw gauge in Figure Q1 and correctly identifies its function?

| | Part | Name | Function |
|---|---|---|---|
| **A** | 6 | Spindle | To clamp the specimen tightly |
| **B** | 7 | Ratchet | To slip, rather than over-tighten and deform the specimen |
| **C** | 4 | Ratchet | To lock the jaws |
| **D** | 1 | Sleeve | To measure the specimen |

*Table Q5*

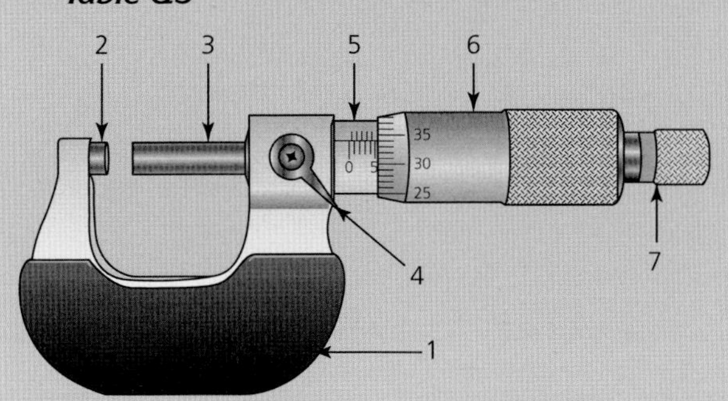


*Figure Q1*

**15.** The micrometer in Figure Q1 is reading:

**A** 5.34 mm

**B** 5.534 mm

**C** 5.84 cm

**D** 5.34 cm

**16.** A micrometer like that in Figure Q1 has:

**A** A range of 25 cm and a resolution of 0.1 mm

**B** A range of 25 mm and a resolution of 0.01 mm

**C** A range of 25 mm and a resolution of 0.1 mm

**D** A range of 2.5 mm and a resolution of 0.01 mm

**17.** Vernier callipers are to be used to measure a short pipe.

Which of the following statements is **false**?

**A** Vernier callipers can be used to measure the internal and external diameter of the pipe.

**B** Vernier callipers have better resolution than a micrometer.

**C** Vernier callipers have a larger range than a micrometer.

**D** Vernier callipers can measure to the nearest 0.1 mm.

**18.** Look at the calliper scales in Figure Q2.

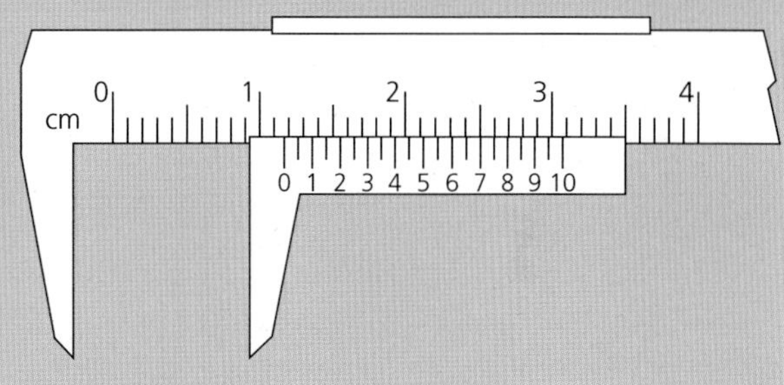


*Figure Q2*

What is the reading on the callipers?

**A** 1.07 cm

**B** 1.15 cm

**C** 7.25 cm

**D** 1.17 cm

# 2 INSIDE THE ATOM

## PRIOR KNOWLEDGE

You will be familiar with the nuclear atom and you probably know something about radioactivity. You should be aware of the forces between electric charges and know what the term 'potential difference' means.

## LEARNING OBJECTIVES

In this chapter you will learn more detail about the model of the atom and how this model evolved. You will learn more about nuclear decay and the four fundamental interactions will be introduced.

**(Specification 3.2.1.1, 3.2.1.2 part, 3.2.1.4 part)**

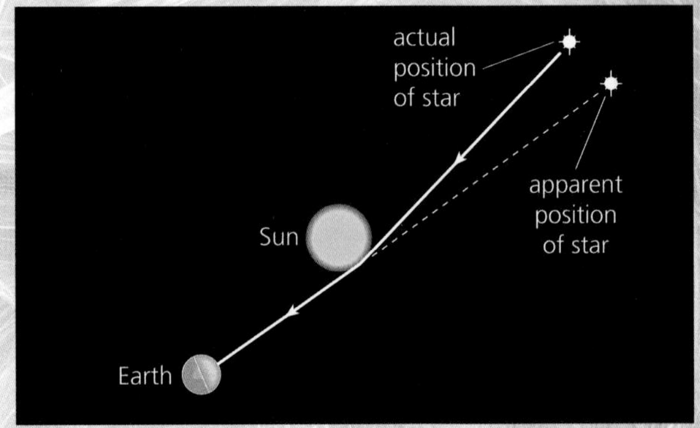


***Figure 1*** *General relativity predicts that light from a star will bend as it passes close to the Sun. The Sun's own light makes this impossible to detect, except during a solar eclipse. Arthur Eddington photographed the 1919 eclipse from the west coast of Africa. He sent this message to the mathematician Bertrand Russell: "Einstein's theory is completely confirmed. The predicted angular deflection was 1.72° and the observed angle 1.75° (±0.06")."*

Modern physics is built on two outstandingly successful theories: quantum theory and general relativity. Quantum theory describes phenomena on the atomic and subatomic scale. It has been tested to extraordinary levels of precision. For example, electrons have a magnetic property due to their spin, called the magnetic moment. Quantum theory predicts its value to be 1.001 159 652 181 13 (±8) (in certain units). Recent experimental measurements put this figure at 1.001 159 652 180 73 (±28). The uncertainty in brackets after these figures is in the least significant digit(s). The precision of agreement is better than 1 part in a trillion ($10^{12}$), like knowing the distance to the Moon to within the thickness of a human hair.

General relativity is used to explain behaviour on an astronomical scale. It also has been tested thoroughly and has come through unscathed (Figure 1). The theory, which predicts that clocks will run slower in a stronger gravitational field, is essential to the accuracy of navigation using the GPS system.

Attempts to unite the two theories mathematically have so far failed. The most likely candidate for a unified 'Theory of Everything' is string theory (Figure 2). This has succeeded in explaining many of the phenomena observed by particle physicists. However, it does not make any unique new predictions that can be tested by experiment. This has led some to suggest that string theory is not science at all. Richard Feynman, the Nobel physicist whose work on quantum theory described the interaction between electrons and light, said,

"It doesn't matter how beautiful your theory is, it doesn't matter how smart you are. If it doesn't agree with experiment, it's wrong."

***Figure 2 (background)*** *String theory treats fundamental particles, such as electrons, as vibrating loops of energy.*

# 2.1 ATOMIC STRUCTURE

Before 1900, scientists considered the atom a fundamental particle – a particle without structure or constituent parts. A few suggested that there could be something inside the atom, which would help to explain atomic spectra (see Chapter 5), but nobody suspected that the atom itself could be taken apart. J. J. Thomson's discovery of the electron changed that – he realised that the electron was a part of the atom that had somehow broken free.

## The electron

Physicists in the 19th century had two wonderful new pieces of technology: the vacuum pump and the induction coil. The vacuum pump could create a

***Figure 3*** *Thomson at work in the Cavendish Laboratory, Cambridge*

very low pressure in a glass tube. The induction coil could generate a large potential difference (voltage) across electrodes placed in the tube. This technology led to the demonstration of radio waves by Hertz, the transmission of wireless telegraph signals by Marconi, the discovery of X-rays by Röntgen and the discovery of the first subatomic particle, the electron, by Thomson at the Cavendish Laboratory in Cambridge (Figure 3).

In Thomson's 'discharge tube' (Figure 4), a large potential difference was applied across two metal plates inside a partially evacuated glass tube. This caused a fluorescent glow from the tube.

Thomson was working on the hypothesis that **cathode rays** (Figure 5) were in fact a stream of electrically charged particles. He thought that these were emitted from the negatively charged plate (the cathode). A large potential difference accelerated particles; they travelled at high speed across the tube until they struck the coated glass, causing the glow. Thomson showed that he could deflect the beam with a magnet (Figure 6). He could also deflect the beam towards a positively charged plate. This proved that the cathode rays carried a negative electric charge.

## Thomson's balancing act

Thomson knew that the cathode rays were associated with negative charge, but to prove that they were a stream of identical particles he needed to show that they had a unique mass and charge. Initially, Thomson could not find either of these quantities separately but he was able to measure the charge-to-mass ratio for the cathode rays. This ratio is called the **specific charge**. For the electron written $e/m$, where $e$ is the charge carried by the electron and $m$ is its mass. Specific charge is measured in coulombs per kilogram, C kg$^{-1}$. A constant value of $e/m$ would prove that cathode rays were in fact a stream of identical particles.

***Figure 4*** *Thomson discovered the electron using this discharge tube at the Cavendish Laboratory, in 1897. In the following year, there was a formal toast at the laboratory's annual dinner:* "The electron: may it never be of use to anybody."

***Figure 5*** *Discharge tubes, or Crookes tubes, were the particle accelerators of their day. The fluorescence, first seen in the glass, became brighter with fluorescent coatings. This modern tube shows a metal object, a Maltese cross, casting a shadow by blocking the cathode rays.*

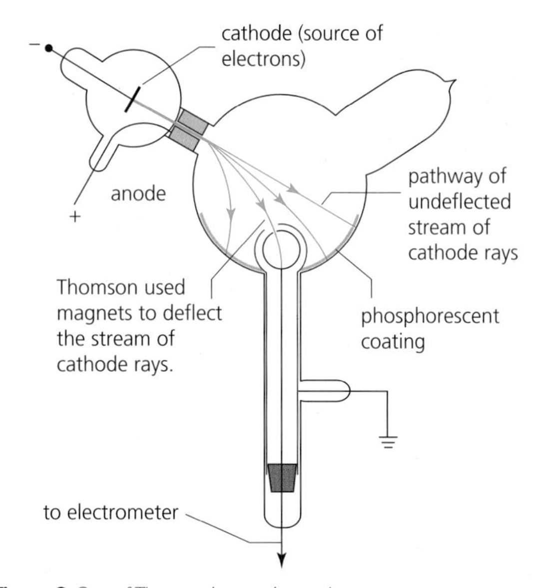


***Figure 6*** *One of Thomson's experimental set-ups*

Thomson was able to find the specific charge of the electron using electric and magnetic fields at right angles to one another. His apparatus (Figure 6) had electric and magnetic fields arranged so that the forces from each field exactly cancelled out and so the cathode rays carried on in a straight line (Figure 7). By equating the forces from the two fields, Thomson was able to calculate a value for $e/m$ (see Assignment 2).

Thomson then applied his method to the particles emitted by the photoelectric effect (see Chapter 8) and found that they had the same value for $e/m$ as cathode rays. In 1899, Thomson concluded that cathode rays were "a splitting up of the atom, a part of the mass of the atom getting free and becoming detached from the original atom". Thomson had discovered the electron.

Later, Thomson measured the electron's charge, by finding the velocity of falling charged water droplets, to be of the order of $10^{-19}$ C. He then arrived at a value for the mass of the electron of $3 \times 10^{-29}$ kg. This is respectably close to the modern measurement of $9.109 \times 10^{-31}$ kg.

Further experiments by Robert Millikan found a more accurate result for the charge on an electron. The currently accepted value is

$$e = -1.602\,176\,57 \times 10^{-19}\ \text{C}$$

The value of the specific charge of the electron is

$$e/m = -1.758\,820\,088(\pm 39) \times 10^{11}\ \text{C kg}^{-1}$$

There is no evidence to suggest that the electron has any internal structure, or has any other constituents. It is still considered to be a fundamental particle.

## QUESTIONS

1. **a.** Thomson repeated the experiment using different metals for the cathode. Why?

   **b.** Changing the metal made no difference to the path of the cathode rays. What does this suggest?

(a) Electrons are attracted to the upper plate. They follow a parabolic path while they are in the electric field.

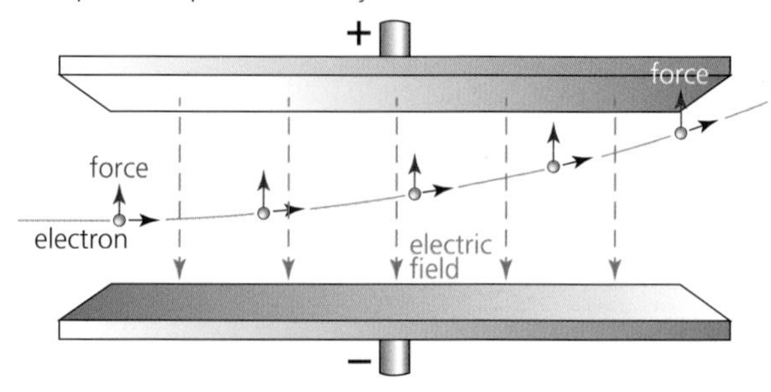


(b) In a magnetic field electrons are made to travel in a circular path.

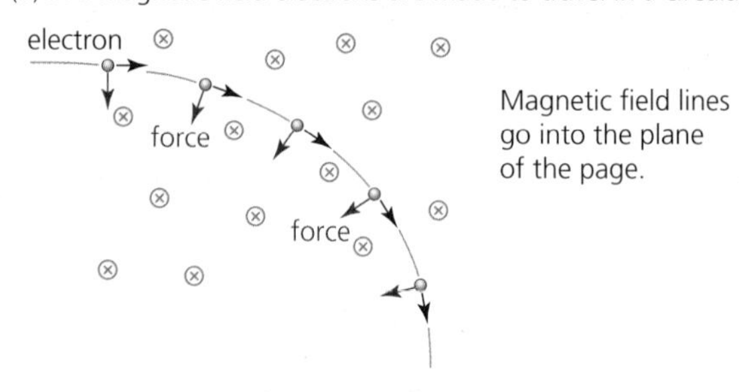


(c) Thomson balanced the effects of the magnetic and electric fields so that the electrons travelled in a straight line.

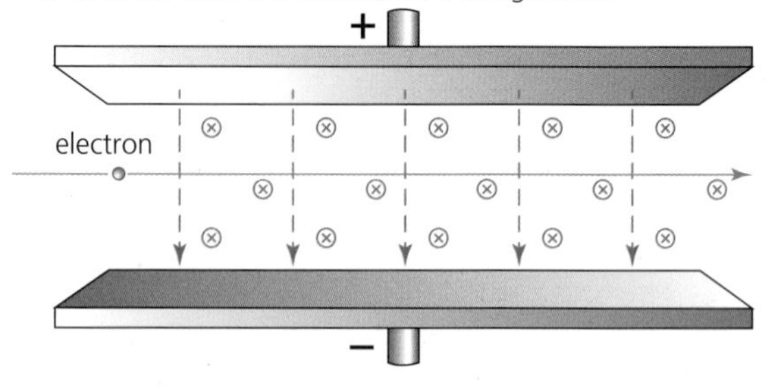


***Figure 7*** *Electric deflection (a) and magnetic deflection (b) of cathode rays could be balanced (c).*

## CATHODE RAYS

*BY J. J. THOMSON, M.A., F.R.S.,*
*Cavendish Professor of Experimental Physics, Cambridge.*

THE EXPERIMENTS discussed in this paper were undertaken in the hope of gaining some information as to the nature of cathode rays . . . According to the almost unanimous opinion of German physicists they are due to some process in the aether[1] . . . another view of these rays is that . . . they are in fact wholly material, and that they mark the paths of particles of matter charged with negative electricity . . . The following experiments were carried out to test some of the consequences of the electrified-particle theory.

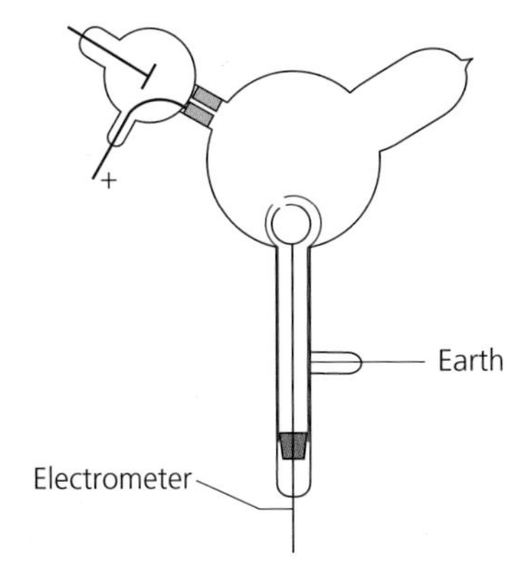


Two coaxial cylinders with slits in them are placed in a bulb connected with a discharge tube; the cathode rays . . . do not fall upon the cylinders unless they are deflected by a magnet. The outer cylinder is connected with the Earth, the inner with the electrometer. When the cathode rays (whose path was traced by the phosphorescence on the glass) did not fall on the slit, the electrical charge sent to the electrometer . . . was small and irregular; when, however, the rays were bent by a magnet so as to fall on the slit there was a large charge of negative electricity sent to the electrometer. I was surprised at the magnitude of the charge . . . Thus this experiment shows that however we twist and deflect the cathode rays by magnetic forces, the negative electrification follows the same path as the rays, and that this negative electrification is indissolubly connected with the cathode rays.

[Thomson goes on to discuss the effect of an electric field on cathode rays . . .]

*Deflexion of the Cathode Rays by an Electrostatic Field*

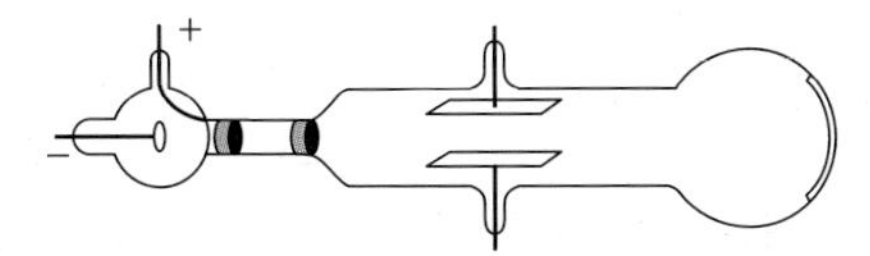


At high exhaustions (a strong vacuum or low pressure) the rays were deflected when the two aluminium plates were connected with the terminals of a battery of small storage-cells; the rays were depressed when the upper plate was connected with the negative pole of the battery, the lower with the positive, and raised when the upper plate was connected with the positive, the lower with the negative pole. The deflexion was proportional to the difference of potential between the plates, and I could detect the deflexion when the potential difference was as small as two volts. It was only when the vacuum was a good one that the deflexion took place.

***Figure 8*** *Article in* The London, Edinburgh and Dublin Philosophical Magazine and Journal of Science, *October 1897*

[1] The aether was thought to be an invisible substance that filled the vacuum.

## ASSIGNMENT 1: UNDERSTANDING THOMSON'S EXPERIMENTS

**(PS 1.2, PS 2.1)**

J. J. Thomson was awarded the 1906 Nobel Prize for Physics "for his theoretical and experimental investigations on the conduction of electricity by gases". Read the extract from J. J. Thomson's paper *Cathode Rays* in Figure 8. You may need to do some brief Internet searches to help you to answer the questions below.

### Questions

**A1** In Thomson's first experiment (detail shown in Figure 6), how were the cathode rays deflected? How could Thomson see what path they took? What did he use to detect the cathode rays?

**A2** How did Thomson's view of cathode rays differ from that of the 'German physicists' that he mentions?

**A3** In the second experiment described by Thomson he says that a good vacuum (low pressure) is needed to see the cathode rays. What does this suggest about cathode rays?

**A4** Thomson says that the deflection is proportional to the difference in the potential of the (horizontal) plates. What does this suggest about cathode rays?

**A5** The Crookes tube apparatus had been around for at least 40 years before J. J. Thomson used a version of it to identify the electron. Why did it take so long?

## ASSIGNMENT 2: MEASURING THE SPECIFIC CHARGE (*e/m*) OF THE ELECTRON

**(MS 2.3, PS 1.2, PS 2.4, PS 3.2, PS 3.3, PS 4.1)**

It is possible to carry out or observe a modern version of Thomson's experiment using the apparatus in Figure A1. If you are not able to see this first-hand, there are simulations on the Internet. Your aim in this assignment is to understand how the apparatus works and to get a value for *e/m*, known as the specific charge.

*Figure A1* A school deflection tube

You should be able to vary three independent factors:

- The accelerating voltage, $V_a$. Increasing this makes the electrons travel faster (see Figure A2).
- The deflecting voltage, $V_d$. Increasing this increases the force on the electron as it passes between the plates.
- The current in the magnetic field coils, $I$. Increasing this makes the magnetic field stronger and so increases the force on the electron.

Start with $I = 0$ and $V_d = 0$. The beam should be straight.

Keeping $V_a$ constant:

1 Vary $V_d$ while $I = 0$

2 Vary $I$, while $V_d = 0$

3 Find a pair of non-zero values for $V_d$ and $I$ so that the beam is straight.

When the beam is straight, the electric force due to the deflection plates balances the force due to the magnetic field. By equating these forces, you can find the ratio *e/m*. It is given by:

$$e/m = \frac{V_d^2}{2V_a B^2 d^2}$$

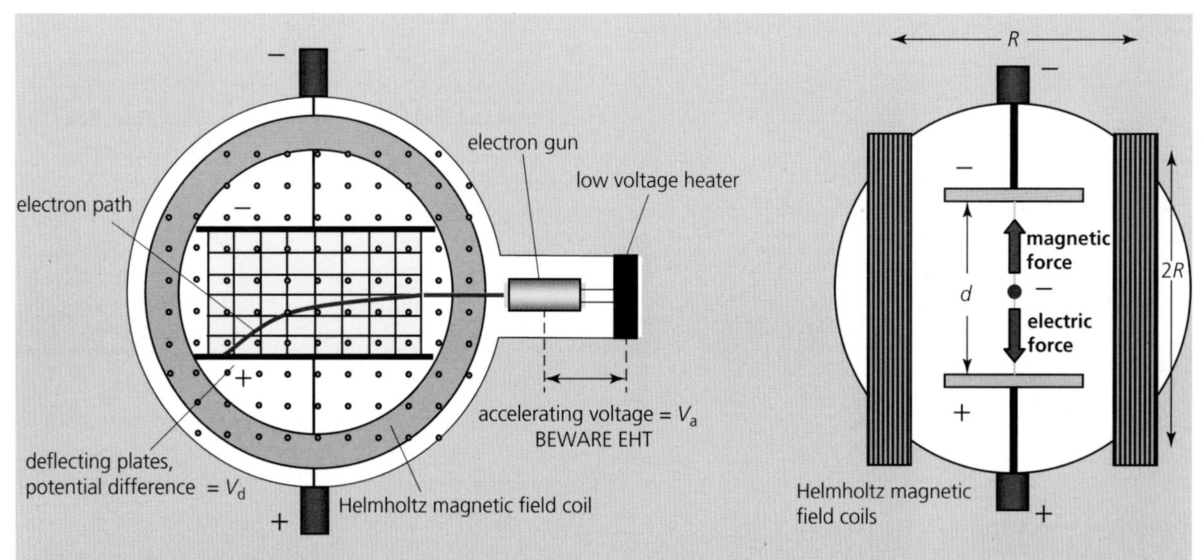


**Figure A2** *Details of the deflection tube*

You should have values for $V_a$ and $V_d$.

$d$ is the distance (m) between the deflection plates.

$B$ is the magnetic flux density between the coils.

For Helmholtz coils the magnetic flux density is:

$$B = \frac{8\mu_0 NI}{5R\sqrt{5}}$$

where

$R$ is the average radius of the coils = separation between the coils

$N$ is the number of turns in each coil

$I$ is the current through the coils

$\mu_0 = 4\pi \times 10^{-7}$ N A$^{-2}$ = permeability of free space

### Questions

**A1** What shape is the path of the cathode rays when there is a potential difference across the horizontal plates? What shape is the path when there is only a magnetic field?

**A2** What factors did Thomson need to control in his experiment?

**A3** Use the formula for *e/m* on the previous page to calculate a value for *e/m*. Compare this with the accepted value.

**A4** How might the precision and accuracy of the school experiment be improved?

## The plum pudding atom

Physicists construct models that describe some aspect of the real world, such as the kinetic theory of gases, or the Big Bang model for the beginning of the Universe. Naturally, physicists began to suggest models for the atom.

A first thought (Thomson's) was that atoms were made entirely of electrons. He suggested that a hydrogen atom might consist of around 1000 electrons. However, there had to be some positive charge to cancel out the negative charge carried by all those electrons. This was necessary or else electrostatic repulsion would push apart the whole atom. How much mass did this positive charge have and where was it? Thomson's plum pudding model (Figure 9) pictured the atom as a uniform, positively charged cloud with tiny electrons embedded in it, like plums in a pudding.

The plum pudding model did not stand the test of time, though. Its fate was finally decided by experiments carried out at Manchester University in 1909.

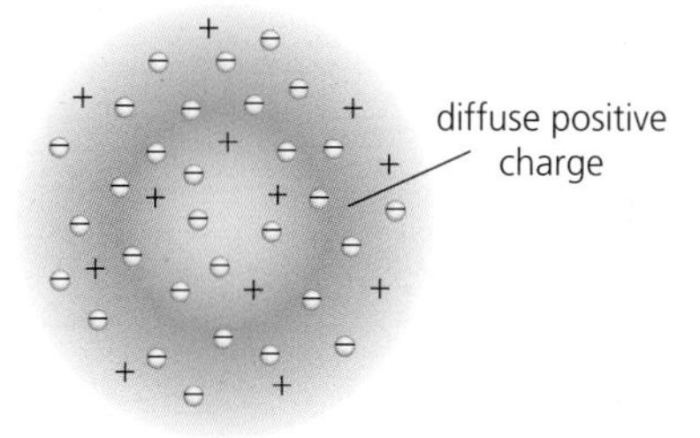


**Figure 9** *The plum pudding model. The term 'pudding' is rather misleading; the low density of the positive part of the atom makes it more like a cloud than a pudding.*

**KEY IDEAS**

- The electron is a fundamental particle.
- The electron has a mass of $9.1 \times 10^{-31}$ kg and carries a negative charge of $1.6 \times 10^{-19}$ C.
- The charge-to-mass ratio, or specific charge, is very high for the electron:
  $e/m = -1.8 \times 10^{11}\ \mathrm{C\,kg^{-1}}$
- The plum pudding model represented atoms as tiny electrons embedded in a positively charged background of uniform density.

## 2.2 THE DISCOVERY OF THE NUCLEUS

The first steps in nuclear physics were taken early in 1909. Hans Geiger and his research student Ernest Marsden were conducting an experiment to investigate the scattering of alpha particles as they collided with gold atoms (Figure 10). The experiment was overseen by Ernest Rutherford. He had shown (see section 2.4) that an alpha particle is a tightly bound group of two neutrons and two protons – a helium nucleus – that is emitted at high velocity from some radioactive isotopes.

Geiger and Marsden used radium as the source of the alpha particles, which were beamed towards a thin gold foil. The alpha particles, after passing through the foil, hit a zinc sulfide screen. When each particle hit the screen there was a flash of light (scintillation). In hours of painstaking observations, they recorded the number of scintillations, and hence alpha particles deflected at each angle. Rutherford suggested that they should look for alpha particles reflected from the metal surface. To their great surprise, a small fraction of the alpha particles, about 1 in 8000, bounced back from the gold foil.

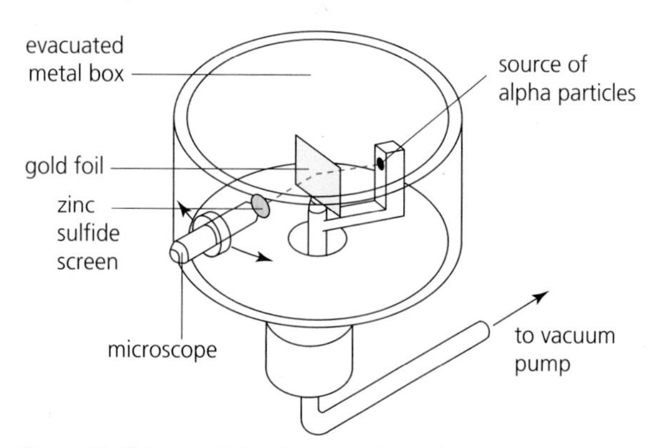


*Figure 10* Geiger and Marsden's experimental set-up

Alpha particles have a positive charge, known to be twice the magnitude of the charge on the electron. They are also relatively heavy, about 8000 times as massive as an electron. An alpha particle travelling at $10\,000\ \mathrm{km\,s^{-1}}$ should not have been bounced back by a gold atom that consisted of a few tiny electrons stuck into a positive 'pudding'. Rutherford was amazed. He said later, "It was quite the most incredible event that has ever happened to me. It was almost as incredible as if you fired a 15 inch shell at a piece of tissue paper and it came back and hit you."

The plum pudding model of the atom had to go. Rutherford deduced from the scattering results that all the positive charge, and almost all the mass, of an atom must be concentrated in the centre of the atom. He called this the **nucleus**.

Rutherford suggested that the electrons carried all the negative charge and that they orbited the nucleus through empty space, a relatively long way from the nucleus. Most of the alpha particles passed through the gold foil with small or zero deflections; the particles were simply too far away from the nucleus of a gold atom to be affected by it. Very occasionally, an alpha particle passed so close to a gold nucleus that it would be repelled by the positive charge and suffer a large deflection (Figure 11).

Rutherford used the results of these scattering experiments to calculate the size of the **nuclear atom**. The nucleus has a radius of the order of $10^{-15}$ m, compared with the radius of the atom, which is of the order of $10^{-10}$ m. Rutherford's model of the atom is often pictured as a miniature solar system: the electrons orbit the nucleus rather like planets orbiting the Sun. This image does not really reflect the true scale of the atom. Picture the atom scaled up, so that the nucleus is the size of the Sun; the electrons would orbit ten times further from the nucleus than Pluto is from the Sun. The atom is almost all empty space with an extremely small, dense nucleus.

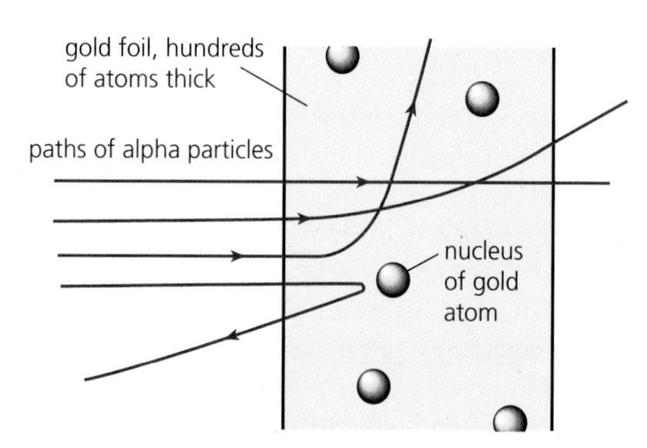


*Figure 11* Rutherford scattering

QUESTIONS

2. In the film *The Men Who Stare at Goats*, General Hopwood tries to run through a wall. Scientists had told him that the atoms in his body and those in the wall are almost all empty space. This is true. Explain to the General why he still cannot run through a wall.
3. If the atomic nucleus were the size of a football placed on the centre circle at Wembley stadium, where might you find the electron's orbit?

## 2.3 INSIDE THE NUCLEUS

Rutherford's model of the nuclear atom has a positively charged nucleus orbited by negatively charged electrons. The particles that carry the positive charge in the nucleus are protons. The number of protons is the **proton number**, or the atomic number, and its symbol is $Z$. In a neutral atom – an atom that carries no overall charge – the number of protons in the nucleus is equal to the number of electrons orbiting the nucleus, since these have equal but opposite charge. The atomic number of an atom is therefore also the number of electrons in the neutral atom. The atomic number for each element indicates its place in the Periodic Table. Hydrogen, with $Z = 1$, has one proton in its nucleus and one electron in orbit. Helium, with $Z = 2$, has two protons and two electrons, and so on through the Periodic Table, to the heaviest naturally occurring atom, uranium, which has 92 protons and 92 electrons.

### The strong nuclear force

A nucleus composed entirely of positive charges would not hold together. Positive charges repel one another. At the very small separations inside the nucleus, the electrostatic forces pushing the protons apart are very large. There must be another force, acting inside the nucleus, that holds the nucleus together. This force is the **strong nuclear force** or strong interaction.

The strong nuclear force has a very short range; it has little effect at separations greater than about 3 fm ($3 \times 10^{-15}$ m). When two protons are 3 fm apart or closer, the strong interaction acts as an attractive force, pulling the protons closer together. This happens until the separation is about 0.5 fm. Closer than this, the strong interaction becomes highly repulsive (Figure 12). The overall effect of the force is to pull the nucleus together, but the repulsive action prevents it from collapsing to a point.

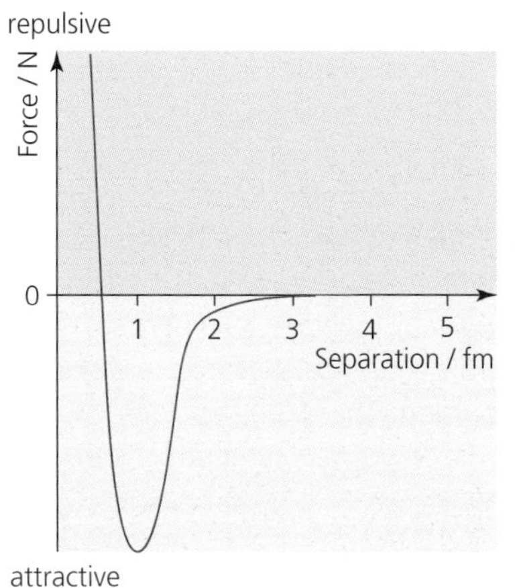


***Figure 12*** *The strong nuclear force versus distance between two protons*

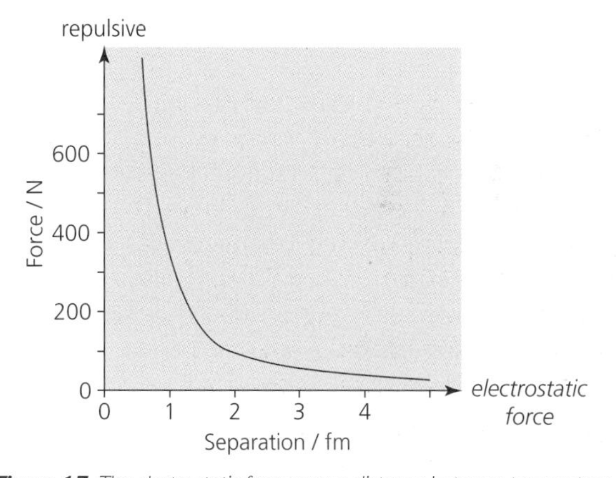


***Figure 13*** *The electrostatic force versus distance between two protons*

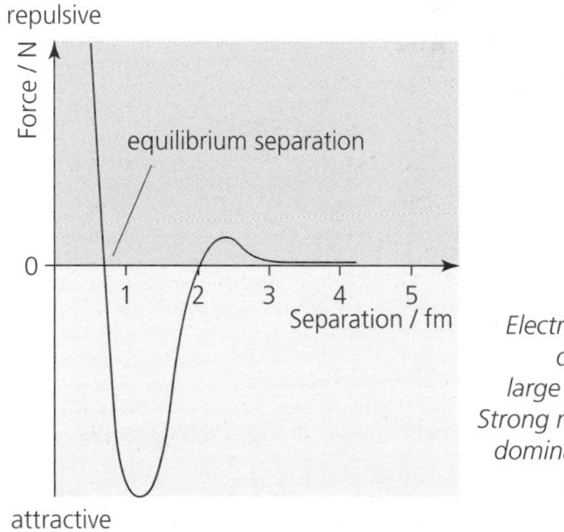


***Figure 14*** *The combined effect of the electrostatic and the strong nuclear force versus distance between two protons*

At distances of less than about 2 fm, the strong nuclear attraction between two protons is larger than the electrostatic repulsion (Figure 13) so the nucleus is held together (Figure 14).

## The neutron

For larger nuclei, there is still a problem. The strong nuclear force acts over a much shorter range than electrostatic repulsion. It is just not possible to get all of the protons close enough together for the strong nuclear force to overcome the electrostatic repulsion. There must be some other particle, or particles, in the nucleus that help to glue it together. The particle is the neutron; it adds extra strong nuclear force, without any electrostatic repulsion. The strong nuclear force between two neutrons, or between a neutron and a proton, acts in the same way as it does between two protons (see Figure 12).

The neutron is a particle with a mass of similar value to that of the proton (Figure 15), but the neutron carries no electric charge (Table 1). It does exert a strong nuclear attraction on protons and on other neutrons. The number of neutrons in a nucleus is known as the **neutron number**, $N$.

Protons and neutrons are the only particles in the nucleus. They are referred to as **nucleons**. The strong nuclear force acts between *any* pair of nucleons, whether that is two protons, two neutrons or a proton and a neutron. Electrostatic repulsion acts only between two protons, because of their positive charge.

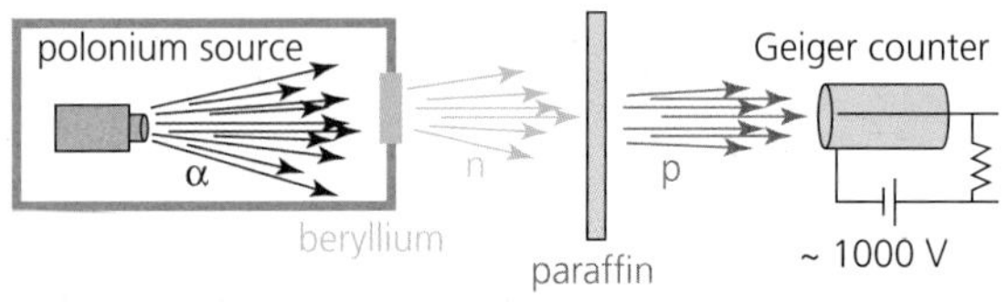


*Figure 15* *Chadwick discovered the neutron in 1932. Neutrons are hard to detect, so his ingenious experiment allowed neutrons to collide with protons, which were detected. He was also able to show that neutrons and protons have a very similar mass.*

### QUESTIONS

**4.** Work out the specific charge (charge-to-mass ratio) of the proton. Use the data in Table 1.

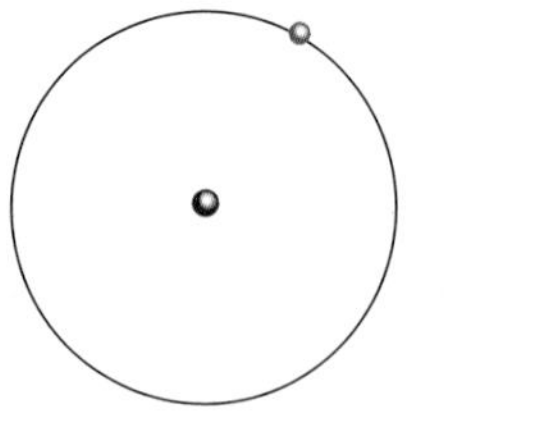

An atom of hydrogen has one proton in its nucleus and one electron in orbit around it.

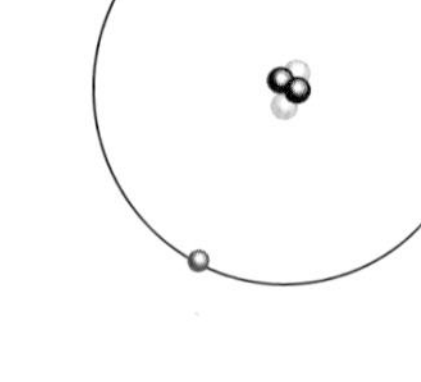

An atom of helium has two protons and two neutrons in its nucleus, and two electrons in orbit around it.

***Figure 16*** *Hydrogen and helium atoms*

## Building nuclei

The simplest atom of all is hydrogen. It has one proton in its nucleus and no neutrons. It has only one electron orbiting the nucleus. The most common form of helium has two protons and two neutrons in its nucleus and, therefore, there are two electrons in orbit around it (Figure 16).

The total number of nucleons in the nucleus of an atom is referred to as its **nucleon number**, or mass number, $A$. This is the number of protons plus the number of neutrons, so $A = Z + N$.

The nuclear composition of an atom can be described using symbols. The most common form of carbon has six protons and six neutrons in its nucleus. It can be written as $^{12}_{6}C$. The upper number is the nucleon (mass) number, $A$; the lower number is the proton (atomic) number, $Z$.

In general, an element X with an atomic number $Z$ and nucleon number $A$ is written as $^{A}_{Z}X$. Using this system, hydrogen is represented as $^{1}_{1}H$, and helium is represented as $^{4}_{2}He$. The nucleon number is also the atomic mass number. The atomic mass number is always an exact integer, since it counts the number of nucleons in the nucleus. The **relative atomic mass** of an element is slightly different. It is found by comparing the mass of the atom to that of the most common form of carbon atom, $^{12}_{6}C$, whose mass is $19.93 \times 10^{-27}$ kg. This atom is defined as having a mass of exactly 12 'units'. This 'unit' is known as the unified **atomic mass unit**, u, and has the value $1.661 \times 10^{-27}$ kg.

| | Proton | Neutron | Electron |
|---|---|---|---|
| Symbol | $^{1}_{1}p$ | $^{1}_{0}n$ | $^{0}_{-1}e$ |
| Charge / C | $+1.602 \times 10^{-19}$ | 0 | $-1.602 \times 10^{-19}$ |
| Mass / kg | $1.6726 \times 10^{-27}$ | $1.6749 \times 10^{-27}$ | $9.1094 \times 10^{-31}$ |

***Table 1*** *Proton, neutron and electron data*

| Element | Symbol | Proton number, $Z$ | Neutron number, $N$ | Nucleon number, $A$ | Atomic mass / u |
|---|---|---|---|---|---|
| Hydrogen | H | 1 | 0 | 1 | 1.0078 |
| Helium | He | 2 | 2 | 4 | 4.0026 |
| Lithium | Li | 3 | 4 | 7 | 7.016 |
| Beryllium | Be | 4 | 5 | 9 | 9.012 |
| Boron | B | 5 | 6 | 11 | 11.009 |

***Table 2*** *The first five elements in the Periodic Table*

The mass of an atom, in atomic mass units, is not exactly equal to the nucleon number. This difference, and its implications, is for discussion in Book 2. However, the relative atomic mass of hydrogen is 1.0078 u, which is almost the same as its nucleon number, 1. Similarly, helium, with nucleon number 4, has a relative atomic mass of 4.0026 u. Indeed *all* nuclei have relative atomic masses that are very close to their nucleon numbers (see Table 2). To an accuracy of three significant figures the mass of the proton, the mass of a hydrogen atom and the mass of a neutron are all equivalent to 1 atomic mass unit.

All nuclei carry a net positive charge equal to their proton number multiplied by the proton charge. We can calculate the specific charge of a nucleus, in C $kg^{-1}$, as follows:

$$\text{specific charge} = \text{proton number} \times \frac{1.602 \times 10^{-19}}{\text{nuclear mass}}$$

where the nuclear mass is *approximately* equal to nucleon number $\times$ $1.66 \times 10^{-27}$ kg.

For example, to a first approximation the specific charge of an uranium-238 nucleus, $^{238}_{92}U$, is:

$$\frac{92 \times 1.602 \times 10^{-19}}{238 \times 1.61 \times 10^{-27}} = 3.728 \times 10^{7}\ \text{C kg}^{-1}$$

(Using a more precise value for the nuclear mass of uranium-238, 238.05076 u, would make only a very small difference to this result.)

## Isotopes

Elements can exist in more than one atomic form. Though the simplest atom, hydrogen usually has only one proton as its nucleus, some hydrogen atoms have one or two neutrons in their nucleus. These different forms of hydrogen are **isotopes** (Figure 17).

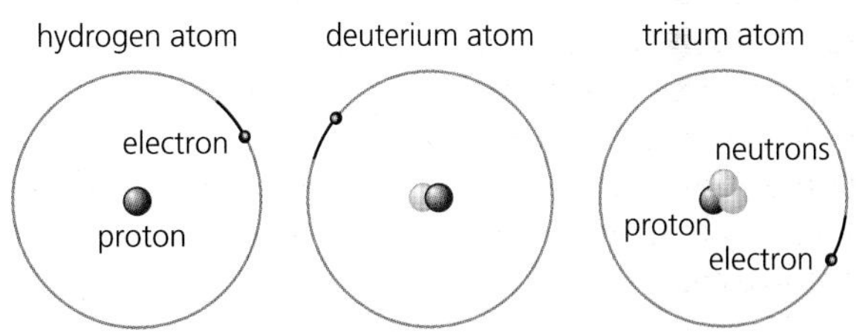


***Figure 17*** *Hydrogen isotopes*

The different isotopes of an element are chemically identical. This is because their atoms have the same number of electrons. Isotopes also have the same number of protons in their nucleus. The difference between the isotopes of an element is simply the number of neutrons that they have. This makes some isotopes heavier than others. The most common form of carbon has six protons and six neutrons in its nucleus; this isotope is carbon-12. Carbon-13 has six protons and seven neutrons; carbon-14 has six protons and eight neutrons (see Table 3).

| Isotope | Proton number, $Z$ | Number of electrons | Neutron number, $N$ | Nucleon number, $A$ | % abundance |
|---|---|---|---|---|---|
| Carbon-12 | 6 | 6 | 6 | 12 | 98.89 |
| Carbon-13 | 6 | 6 | 7 | 13 | 1.11 |
| Carbon-14 | 6 | 6 | 8 | 14 | $< 0.001$ |

***Table 3*** *Isotopes of carbon*

## QUESTIONS

5. An isotope of uranium, atomic number 92, has 235 nucleons in its nucleus. How many protons, neutrons and electrons are there in a neutral atom of uranium-235?
6. Uranium-238, $^{238}_{92}U$, is another isotope of uranium. How does it differ from uranium-235?
7. Calculate the specific charge of each of the nuclei in Table 2.
8. The lighter elements, like carbon and oxygen, tend to have equal numbers of protons and neutrons in their nuclei. In heavier elements, there are always more neutrons. Explain why this is so.
9. List the properties of the strong nuclear force.

## KEY IDEAS

- The diameter of a nucleus is of the order of $10^{-15}$ m. The diameter of an atom is 100 000 times larger.
- The nucleus contains two different particles of approximately equal mass, the neutron and the proton.
- Protons are positively charged. Neutrons are uncharged.
- Protons and neutrons are affected equally by the short-range strong nuclear force. This is a short-range force of attraction, up to a distance of about 3 fm, but is highly repulsive at distances closer than about 0.5 fm. This holds the nucleus together.
- The proton number, $Z$, is the number of protons in a nucleus. The nucleon number, $A$, gives the total number of nucleons in a nucleus. The nucleus of element X can be represented by $^{A}_{Z}X$.
- Isotopes of an element have the same number of protons and electrons. They are chemically identical. Different isotopes have different numbers of neutrons in their nuclei.

## 2.4 RADIOACTIVITY

The discovery of radioactivity owed something to chance. In 1896, Henri Becquerel was researching the action of light on fluorescent materials. He was investigating the possibility that light might cause these fluorescent materials to emit X-rays and so darken photographic plates. Becquerel's technique was to wrap the unexposed film in black paper and place a thin copper cross between the film and the fluorescent material that he was investigating. He would then expose the fluorescent material, one of which was uranium salts, to sunlight. For some weeks his experiments had produced only negative results and, when the skies turned overcast, he put all the apparatus, uranium salts, screen and film, into a dark cupboard to await sunnier weather. After four days of continuous cloud cover, Becquerel grew tired of waiting and decided to develop the film anyway. To his astonishment, the photographic plate had darkened strongly, with the image of the copper cross standing out white against a dark background.

Becquerel reported that he had discovered a natural radiation that could penetrate paper. He also found that these rays could cause a gas, like air, to conduct electricity. Later, scientists found that this process, known as **ionisation**, occurs because the rays can knock an electron completely out of its atom. This turns a neutral atom into an **ion pair**: a negatively charged electron and the positively charged ion it has left behind. The electron and the positive ion are free to move so that the gas becomes an electrical conductor. Radioactive sources can be detected through this effect, for example by causing an electroscope to be discharged (Figure 18).

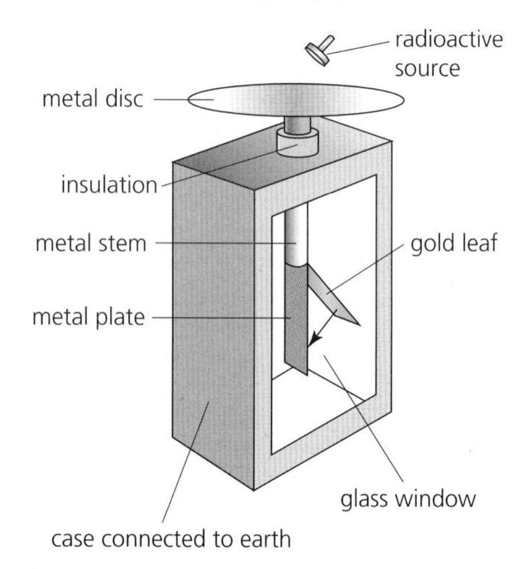


***Figure 18*** *The ions created by the radioactive source discharge a charged electroscope.*

Marie Curie and her husband Pierre took Becquerel's work further. They discovered other materials, thorium and radium, which also gave off ionising rays. In 1903, the Nobel Prize for Physics was awarded jointly to Becquerel and the Curies for the discovery of radioactivity.

### Activity of a radioactive source

Radiation that can ionise gases and cause skin burns must carry a significant amount of energy. This energy comes from changes within the nuclei of the atoms of the radioactive source.

Isotopes, such as radium-226, that emit ionising radiation are **radioisotopes**. Their nuclei are unstable, which means that they are likely to **decay** to a lower energy state at some time in the future. They emit radiation in the process. Sometimes this decay changes the original (parent) nucleus to a different element altogether. This new (daughter) nucleus may be stable, or it may be a new radioisotope – that will in turn decay and emit more radiation.

Radioactive decay is a random event. It is impossible to predict exactly when any particular unstable nucleus will decay, just as when throwing a die you cannot predict when you will get a six. It is possible, though, to calculate the probability that it will decay in a certain time. This will be discussed in Book 2.

The nuclei of some radioisotopes are more likely to decay in a given time than those of other radioisotopes. This means that these radioisotopes emit more radiation in a given time. They are more active. The **activity**, *A*, of a radioisotope is defined as the number of nuclei that decay in one second; this is equal to the number of emissions per second. Activity is measured in the unit becquerel. If a source has an activity of 1 becquerel, 1 Bq, then on average one of its nuclei decays every second. This is an extremely small unit and activities of kilobecquerel, kBq, or megabecquerel, MBq, are much more likely.

Although the becquerel is the SI unit of activity, an older unit, still commonly used, is the curie, Ci. One curie is the number of disintegrations per second in one gram of radium – 266 – which is equivalent to $3.7 \times 10^{10}$ Bq.

### Alpha, beta and gamma rays

It quickly became clear that the rays emanating from radioactive elements were not all the same. Rutherford realised that there were at least two different sorts of rays. One type, which he called alpha (α) radiation, was easily absorbed by materials in its path. The other type, which was more penetrating, he named beta (β) radiation. A third type of very penetrating radiation became known as gamma (γ) rays.

Alpha, beta and gamma rays all cause ionisation. This is how they are detected. Photographic film, Geiger counters and cloud chambers all detect the ionisation caused by radiation. Cloud chambers (Figure 19) reveal the tracks left by radiation (Figure 20) and were very important in the early days of particle physics. Cloud chambers were replaced by bubble chambers and later spark chambers.

A cloud chamber contains a gas, often air, and a highly saturated vapour. Ionising particles such as beta particles ionise the air as they pass through the chamber. The ions produced act as sites for the vapour to condense around. The liquid droplets that form along the trails of ionised air are just big enough to see and photograph.

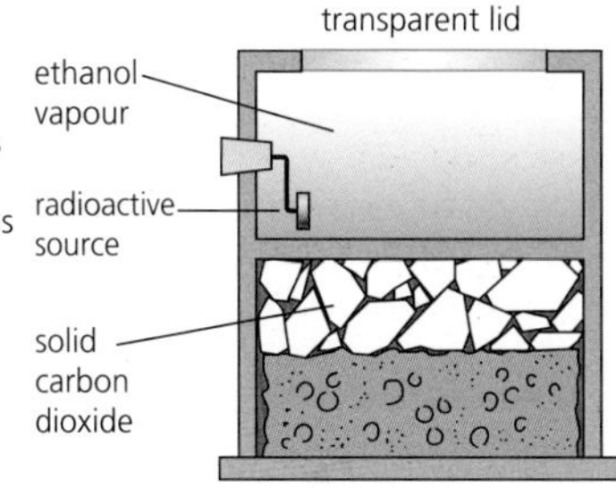


***Figure 19*** *Radiation is detected by the ionisation that it causes.*

### Alpha radiation

Rutherford passed alpha radiation through strong electric and magnetic fields. He showed that it consists of particles that carry a positive charge equal in magnitude to $2e$, where $e$ is the charge on the electron. He also found that the alpha particles have a short range in air, only a few centimetres. Thin sheets of paper could also stop alpha particles. It took an ingenious experiment, completed in 1908, to show what an alpha particle actually is (see Figure 21).

Alpha particles are actually helium nuclei. They are a tightly bound group of two protons and two neutrons, with a charge of $+2e$ and a mass of about 8000 times that of the electron. This combination of relatively large mass and strong electric field makes them highly ionising. When alpha particles pass through a material, they have frequent collisions with atoms. The alpha's large kinetic energy means that it can easily knock an atomic electron out of its orbit. The alpha's large momentum means that it is hardly deflected by the collision. An alpha particle will

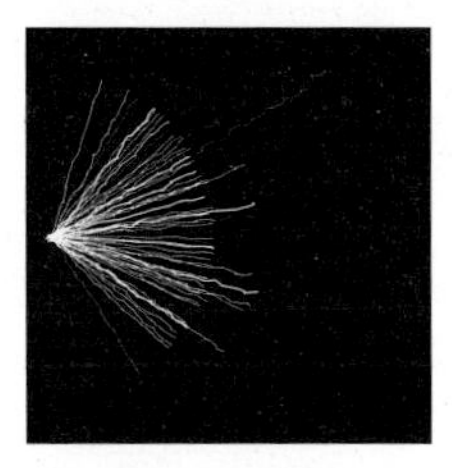

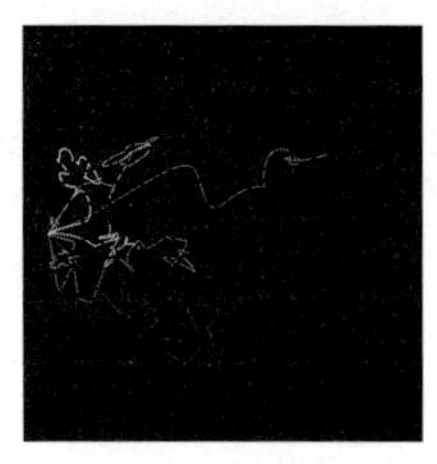

***Figure 20*** *Tracks of alpha radiation (left) and beta radiation (right) in a cloud chamber*

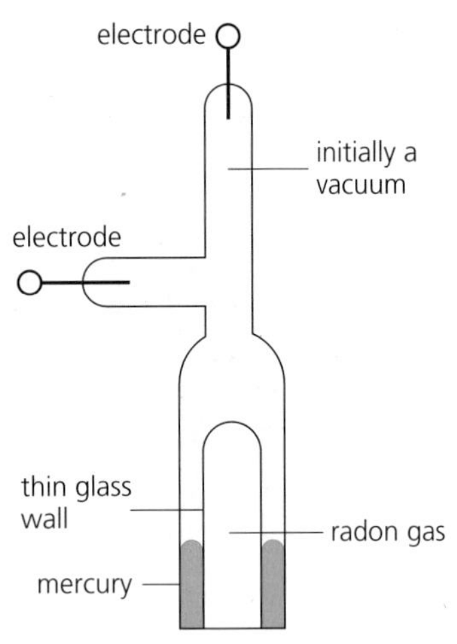


***Figure 21*** *Ernest Rutherford and Thomas Royds sealed some radon gas in a thin-walled glass tube. The glass was thin enough to allow alpha particles to pass through it, but strong enough to withstand atmospheric pressure. The outer tube was evacuated. After a few days, they raised the level of mercury, to compress any gas that might have collected in the tube. They passed a spark between the electrodes and observed the spectrum. This showed that there was helium gas in the outer tube. Rutherford and Royds concluded from their experiment that alpha particles were doubly charged helium atoms.*

undergo thousands of collisions in a short distance until it comes to a halt. After some time it will collect two electrons to become a neutral helium atom.

Decay by alpha emission tends to occur in large unstable nuclei. Inside the nucleus, a stable group of two neutrons and two protons forms. This is expelled from the nucleus at high speed. The emitting, parent, nucleus is transformed into a new daughter nucleus. An example is shown in Figure 22.

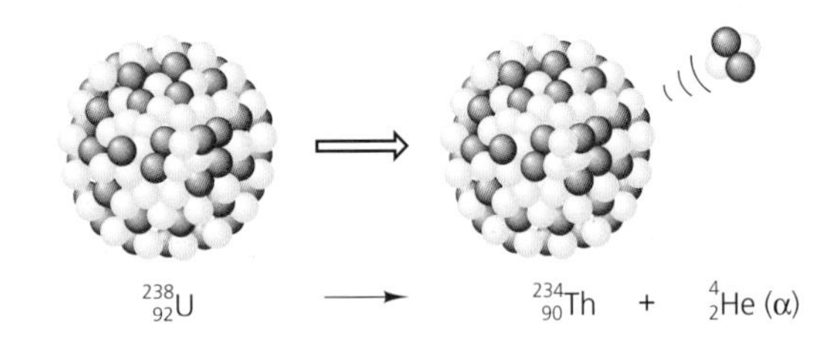


A uranium-238 nucleus is made up of 92 protons and 146 neutrons. It decays to thorium-234 (90 protons and 144 neutrons) by emitting an alpha particle (2 protons and 2 neutrons).

***Figure 22*** *Alpha emission from uranium-238*

The total number of nucleons (protons and neutrons) is not changed by a radioactive decay. The total charge also stays unchanged, so the sum of the proton numbers must be the same before and after any emission.

In alpha decay, the nucleon number, $A$, is reduced by 4 and the proton number, $Z$, is reduced by 2. In a nuclear decay equation, using X as the symbol for the parent radioisotope and Y for the daughter, the alpha decay can be written:

$$^{A}_{Z}X \rightarrow {}^{A-4}_{Z-2}Y + {}^{4}_{2}He$$

An example of an alpha emitter is radium-226. Radium-226 decays into radon, Rn, gas. This decay can be written:

$$^{226}_{88}Ra \rightarrow {}^{222}_{86}Rn + {}^{4}_{2}He$$

## QUESTIONS

**10.** When scientists first discovered alpha rays, they thought that they had no charge: the rays did not appear to be deflected by electric or magnetic fields. Explain what you think the problem might have been with these early experiments.

**11.** An alpha particle travelling through air leaves a trail of ionised atoms and molecules in its wake. Calculate the specific charge of a doubly ionised oxygen atom – that is, an $^{16}_{8}O$ atom that has lost *two* electrons. You may assume that the mass of the orbital electrons is negligible compared with the mass of the nucleus.

**12.** The electronvolt (eV) is a unit of energy (see Chapter 3). An alpha particle is emitted with an energy of around 5 MeV. It takes 10 eV, roughly, to ionise an atom. Estimate how many ion pairs an alpha particle will create before it comes to a stop.

**13.** Radioisotopes that emit only alpha particles are considered not hazardous, as long as they are kept at least five centimetres from the body. However, they are extremely dangerous if they come into contact with the body, for example if someone swallows or breathes in the isotope. Explain why this is.

**14.** The radioisotope americium-241 ($^{241}_{95}Am$) is used in smoke detectors (Figure 23). It decays by alpha emission to neptunium (Np). Write an equation describing this nuclear decay.

***Figure 23*** *There is a small amount of a radioactive isotope in a smoke detector. This causes the air inside to be ionised. A small electric current flows across an air gap. When large smoke particles enter, they interfere with the current, causing the alarm to sound.*

## Beta radiation

Beta particles are much more penetrating than alpha particles. They have a range in air of up to several metres and can pass through thin sheets of an absorber such as plastic or paper. A magnetic field will deflect beta particles much more easily than alpha particles and in the opposite direction. Beta particles are also deflected by an electric field and in a way that shows them to be negatively charged (Figure 24).

Experiments using magnetic fields showed that beta particles have exactly the same charge-to-mass ratio as electrons. In fact, beta particles *are* electrons, even though they come from the nucleus. Beta decay occurs in an unstable nucleus when one of the neutrons decays into a proton and an electron. The proton remains in the nucleus but the electron is emitted at very high speed, often more than 90% of the speed of light. Beta decay leaves the nucleon number of the radioisotope unchanged: there is now a proton instead of a neutron. This extra proton makes the proton number increase by one. Beta decay could therefore be written:

$$^{A}_{Z}X \rightarrow {}^{A}_{Z+1}Y + {}^{0}_{-1}e$$

However, we now know that this decay equation is incomplete. A particle called an antineutrino is also emitted. This will be discussed in Chapter 3.

Beta particles cause ionisation by colliding with atomic electrons. Because the beta particle and the electron have the same mass, the beta particle may be widely deflected by a collision. Not all collisions will cause ionisation; some may 'excite' the atomic electron to a higher energy level. Other collisions may just deflect the beta particle with no change in its kinetic energy. The beta particle has a much less densely ionising track than an alpha particle and its path will be more tortuous (see Figure 20), especially as it begins to slow down.

Some nuclei decay by emitting positively charged beta particles. These have exactly the same mass and other properties as a negative beta particle, but their charge is positive. There is more about these positive electrons (positrons) in Chapter 3.

(a)

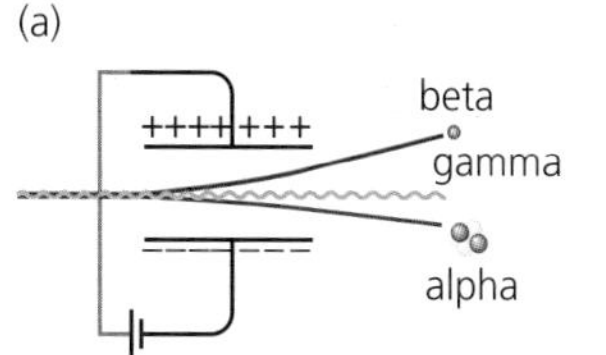


In an electric field, α particles are deflected towards the negatively charged plate; β particles are deflected towards the positively charged plate; gamma radiation is undeflected.

(b)

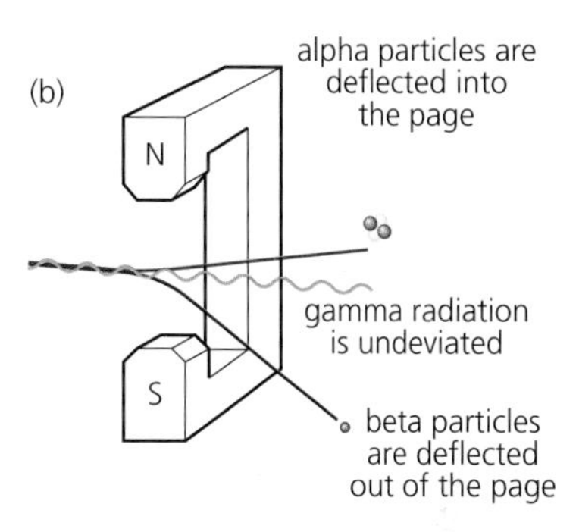


Charged α and β particles are deflected at right angles to the magnetic field; gamma radiation is undeflected.

***Figure 24*** *The effect of electric and magnetic fields on alpha, beta (β⁻) and gamma rays*

### QUESTIONS

**15.** The isotope carbon-14 is a beta emitter that decays into nitrogen, N. Write an equation to represent the decay.

**16.** The cloud chamber pictures in Figure 20 show the tracks left by alpha particles and beta particles. Explain the difference in the tracks.

**17.** The isotope magnesium-23 ($^{23}_{12}Mg$) decays by positive beta emission to sodium (Na). Write an equation to represent the decay. State the difference between the parent nuclei and the daughter nuclei.

### Gamma radiation

Soon after the discovery of radioactivity, it was realised that at least part of the radiation was very penetrating indeed. In 1900, Paul Villard discovered that some rays could pass through thick sheets of metal and still have the ability to blacken photographic plates. He also discovered that electric or magnetic fields could not deflect the radiation (Figure 24). Villard had discovered gamma radiation.

Gamma radiation is high-energy electromagnetic radiation. It has no charge and no mass. Gamma rays can cause ionisation (Figure 25). However, the probability of interaction with an electron is lower than that for alpha or beta particles and so gamma radiation is much less densely ionising.

Gamma emission changes the energy of the parent nucleus, but does not change its nuclear composition.

***Figure 25*** *This cloud chamber photograph shows the ionisation caused by a gamma ray. Most of the tracks are made by secondary electrons, knocked out of their atoms by the gamma ray.*

#### KEY IDEAS

- Some nuclei are unstable. These nuclei decay, emitting ionising radiation in the process. Radioactive decay may change the nucleus to one of a different isotope.
- Unstable nuclei can emit alpha, beta or gamma rays, or a combination of these.
- An alpha particle is a helium nucleus. An alpha decay can be written:

$$^{A}_{Z}X \rightarrow ^{A-4}_{Z-2}Y + ^{4}_{2}He$$

- A beta particle is a high-energy electron emitted from the nucleus. A beta decay can be written:

$$^{A}_{Z}X \rightarrow {}_{Z+1}^{\ \ A}Y + ^{\ 0}_{-1}e$$

plus an antineutrino (see Chapter 3).

- Gamma radiation is high-energy, penetrating electromagnetic radiation.
- During gamma decay the nucleus loses energy, but does not change to a different isotope.

## 2.5 FUNDAMENTAL INTERACTIONS

***Figure 26*** *Although gravity is the dominant force on the scale of planets and stars, within the nucleus it is the weakest of the four fundamental forces.*

The search inside the atom revealed two new forces. The strong nuclear force, which acts to hold nuclei together, has been discussed earlier (section 2.3). A second new force, known as the weak interaction or weak nuclear force, was needed to explain some kinds of radioactive decay.

Scientists now believe that there are only four fundamental forces (Table 4 and Figure 26).

| Force | Relative strength (within the nucleus) | Range |
|---|---|---|
| Strong nuclear | 1 | $10^{-15}$ m |
| Weak interaction | $10^{-5}$ | $10^{-18}$ m |
| Electromagnetic | $10^{-2}$ | infinite |
| Gravity | $10^{-39}$ | infinite |

***Table 4*** *The four fundamental forces*

The **electromagnetic force** holds atoms together. It ties electrons into orbits and binds atoms into molecules. This force acts between all charged particles. Electromagnetic forces have an infinite range, though the strength of the interaction decreases with distance. Because there are two charges, positive and negative, the electromagnetic force can be either attractive or repulsive. Two similar charges repel each other. Two particles carrying opposite charges will attract each other.

Electromagnetism is responsible for the everyday forces between objects. Contact forces, friction, air resistance and tension are all electromagnetic in origin (Figure 27).

***Figure 27*** *Most of the forces acting on the water-skier are electromagnetic. The tension in the tow rope, the drag from the air and water and the buoyancy from the water are all due to the interaction between charged particles.*

**Gravity** is the other force that has a noticeable impact on our lives. Every mass in the Universe attracts every other mass, because gravity is always an attractive force and it has infinite range (Figure 28). The strength of the attraction between two objects depends on their masses and on the distance between them. On the atomic and nuclear scale, the effect of gravity is negligible.

The **strong nuclear force** acts inside protons and neutrons and holds them together. The attraction also acts between nucleons and it is this force that holds the nucleus together. Although it is the strongest of the four forces, it has no effect outside the nucleus. This is because its range is so short, $\approx 10^{-15}$ fm.

The **weak interaction** exerts its influence on all particles. Despite its name, the weak interaction is significantly stronger than gravity, but it only acts over an extremely short range, $\approx 10^{-18}$ m. The weak interaction is responsible for beta radiation and plays an important role in nuclear reactions.

***Figure 28*** *Because there is no negative mass, nothing can be shielded from the effects of gravity. The only way to leave Earth is by expending a lot of energy.*

Without the weak interaction there would be no nuclear fusion, the process that powers the Sun (Figure 29). Nuclear fusion begins when two protons join together (fuse) to form a deuterium nucleus, a positron (a positive electron), $e^+$, and an electron-neutrino, $\nu_e$ (see Chapter 3).

$$^1_1\text{p} + {}^1_1\text{p} \rightarrow {}^2_1\text{H} + {}^{\ 0}_{+1}\text{e} + \nu_e$$

***Figure 29*** *Deep inside the Sun, nuclear fusion is gradually converting hydrogen to helium.*

The probability of this reaction occurring is incredibly small; only 1 in every $10^{31}$ collisions between protons will result in fusion. This makes winning the National Lottery, with odds of around 1 in $10^7$, look like a racing certainty! Even though a typical proton in the Sun undergoes $10^{14}$ collisions every second, it will take an average of 10 billion years before it eventually reacts in this way with another proton. Without this highly improbable reaction, there would be no sunshine and no life on Earth.

## QUESTIONS

**18.** Which forces act on a proton in a nucleus of carbon? Which forces act on a neutron in the same nucleus?

**19.** Uranium atoms, which have 92 protons, are the largest that occur naturally. There are no naturally occurring elements with larger atoms and more protons. Why not?

**20.** Is the weak interaction badly named? Explain your answer.

**21.** Speculate on what atoms, nuclei and even the Universe would be like, if:

**a.** gravity was much stronger

**b.** the strong force was not as strong

**c.** the weak interaction was much weaker... or much stronger and longer range?

## KEY IDEAS

- There are four fundamental interactions between particles: electromagnetic, gravitational, strong and weak.
- The strong interaction is short range (about $10^{-15}$ m) and acts inside the nucleus, holding nucleons together.
- The weak interaction is even shorter range (about $10^{-18}$ m) and is responsible for some forms of radioactivity.
- The electromagnetic force acts on charged particles and is responsible for holding atoms and molecules together.
- Gravity is also a fundamental force, but it is insignificant within the atom.

### Moving on: Grand Unification Theories

In 1850, the renowned British scientist Michael Faraday carried out experiments to prove that the forces of magnetism and electricity were different aspects of the same force (Figure 30). More than a hundred years later, in 1967, Pakistani physicist Abdus Salaam and Americans Steven Weinberg and Sheldon Glashow put forward a theory that linked the weak interaction with the electromagnetic force. The theory predicted the existence of three new particles, the $W^+$, $W^-$ and $Z^0$ particles. In 1983, these particles were discovered by Carlo Rubbia's team at CERN in Geneva. The results were in such good agreement with the theory that the 'electroweak theory' is now widely accepted.

***Figure 30*** *Michael Faraday felt that the different forces could be unified. After linking electricity and magnetism, Faraday tried to link gravity into the scheme but was unsuccessful. Nobody has succeeded in this since.*

Grand Unification Theories (GUTs) link the electroweak and strong interactions together. Most of these predict that the proton is not stable at all, but will decay with a mean lifetime of $10^{30}$ years. These decays are difficult to detect because the lifetime of the Universe so far is only $10^{10}$ years. Physicists are looking in huge tanks containing about 8000 tonnes of water, trying to catch the 1 proton per day that should decay. The results so far are inconclusive.

## PRACTICE QUESTIONS

**1. a.** The nucleus of a particular atom has a nucleon number of 14 and a proton number of 6.

**i.** State what is meant by nucleon number and proton number.

**ii.** Calculate the number of neutrons in the nucleus of this atom.

**iii.** Calculate the specific charge of the nucleus.

**b.** The specific charge of the nucleus of another isotope of the element is $4.8 \times 10^{7}$ C kg$^{-1}$.

**i.** State what is meant by an isotope.

**ii.** Calculate the number of neutrons in this isotope.

*AQA June 2012 Unit 1 Q2*

**2.** An atom of calcium, $^{48}_{20}Ca$, is ionised by removing two electrons.

**a.** State the number of protons, neutrons and electrons in the ion formed.

**b.** Calculate the charge of the ion.

**c.** Calculate the specific charge of the ion.

*AQA June 2013 Unit 1 Q1*

**3.** Tritium and deuterium are isotopes of hydrogen.

**a.** State one difference between an atom of tritium and an atom of deuterium.

**b.** State one similarity between an atom of tritium and an atom of deuterium.

**4. a.** An atom of the radioisotope radium-226 is represented by the symbol $^{226}_{88}Ra$. Determine the number of protons, the number of neutrons and the number of electrons in a neutral atom of radium.

**b.** Radium-226 emits ionising radiation. Explain what *ionising* means.

**c.** Radium-226 is an alpha particle emitter; it decays to radon, Rn. Write out and complete this equation that describes the decay:

$$^{226}_{88}Ra \rightarrow {}^{?}_{?}Rn + ?$$

**5.** At room temperature, radium is a silvery, solid, metal. Radon is a gas. Both are alpha particle emitters. Explain why radon usually presents more of a hazard to human health.

**6.** Describe how you could demonstrate in the lab that alpha particles have a short range in air.

**7.** Rutherford was awarded the Nobel Prize for Chemistry in 1908 for his work on 'The chemical nature of the alpha particles from radioactive substances'. Read this extract from his Nobel acceptance speech and answer the following questions.

*"Shortly after his discovery of the radiating power of uranium by the photographic method, Becquerel showed that the radiation from uranium...possessed the property of discharging an electrified body. In a detailed investigation of this property, I examined the effect on the rate of discharge by placing successive layers of thin aluminium foil over the surface of a layer of uranium oxide and was led to the conclusion that two types of radiation of very different penetrating power were present. The conclusions at that period were summed up as follows:*

*"These experiments show that the uranium, radiation is complex and that there are present at least two distinct types of radiation – one that is very readily absorbed, which will be termed for convenience the α-radiation, and the other of a more penetrative character, which will be termed the β-radiation. When other radioactive substances were discovered, it was seen that the types of radiation present were analogous to the β and α-rays of uranium and when a still more penetrating type of radiation from radium was discovered by Villard, the term γ-rays was applied to them."*

a. In his experiments described in this extract, Rutherford could only identify two distinct types of radiation, although he was aware that there could be more. Suggest a reason why he was only able to identify two types.

b. Later in the speech, Rutherford describes how he showed that alpha and beta radiation carry opposite charges. Explain, briefly, how you would do that.

c. How could you show that $\alpha$ rays from one substance were the same as $\alpha$ rays from another?

d. The bulk of Rutherford's paper describes how he determined the nature of alpha particles. Read his concluding paragraph, which follows here. Explain what Rutherford got right, and what we now know is wrong, with his ideas.

"*Considering the evidence together, we conclude that the α-particle is a projected atom of helium, which has, or in some way during its flight acquires, two unit charges of positive electricity. It is somewhat unexpected that the atom of a monatomic gas like helium should carry a double charge. It must not however be forgotten that the α-particle is released at a high speed as a result of an intense atomic explosion, and plunges through the molecules of matter in its path. Such conditions are exceptionably favourable to the release of loosely attached electrons from the atomic system. If the α-particle can lose two electrons in this way, the double positive charge is explained.*"

**8.** Describe an experiment that you could carry out in the lab to demonstrate that the radiation from a sample of uranium contained three distinct types of radiation. Identify the potential hazards in carrying out the experiment and explain what steps you would take to reduce these hazards.

**9.** Select the correct description. Thomson is credited with the discovery of the electron because he showed that cathode rays were a stream of particles:

**A** that were all negatively charged

**B** that all had the same mass

**C** that all had the same charge

**D** that all had the same specific charge.

**10.** Rutherford's scattering experiment used alpha particles to probe thin gold foil. Which of these conclusions did he make?

**A** Gold atoms are positively charged.

**B** The mass and positive charge of the atom are concentrated in a small region of the atom.

**C** The mass and the positive charge of a gold atom are uniformly distributed.

**D** There are large gaps between atoms.

**11.** Which of these statements about heavy nuclei is **false**?

**A** Very heavy nuclei have more neutrons than protons.

**B** Very heavy nuclei are denser than lighter nuclei.

**C** Very heavy nuclei are more likely to be alpha emitters than lighter nuclei.

**D** Very heavy nuclei are larger than light nuclei.

**12.** The fundamental forces operate at different ranges. Which of these represents the correct order, from shortest range to longest range?

**A** weak interaction – strong interaction – electromagnetic interaction

**B** strong interaction – electromagnetic interaction – weak interaction

**C** electromagnetic interaction – weak interaction – strong interaction

**D** weak interaction – gravity – strong interaction

13. The specific charge on a particle is measured as $+48 \times 10^6$ C kg$^{-1}$. Which of the following is it likely to be?

**A** a carbon-12 ion

**B** an electron

**C** an alpha particle

**D** a beta particle

14. Which of these statements about the strong nuclear force is always false?

**A** It acts between a neutron and a proton.

**B** It acts between two protons.

**C** It holds the nucleus of an atom together.

**D** It holds electrons in their orbits in atoms.

## Stretch and challenge

15. The radius of a nucleus, $r$, is related to its nucleon number, $A$, by

$$r = r_0 A^{\frac{1}{3}}$$

where $r_0$ is a constant.

Table Q1 gives the values of the nuclear radius for some isotopes.

| Isotope | Radius $r$ / $10^{-15}$ m |
|---|---|
| carbon-12 | 2.66 |
| silicon-28 | 3.43 |
| iron-56 | 4.35 |
| tin-120 | 5.49 |
| lead-208 | 6.66 |

*Table Q1*

**a.** Plot a straight-line graph to confirm that

$$r \propto A^{\frac{1}{3}}$$

using the data in Table Q1. From your graph, obtain a value for $r_0$ and estimate the uncertainty in this value.

**b.** Taking the mass of one nucleon as $1.67 \times 10^{-27}$ kg, estimate the density of nuclear matter. State any assumption that you have made.

# 3 ANTIMATTER AND NEUTRINOS

## PRIOR KNOWLEDGE

It would help if you can recall the principle of the conservation of energy and Fleming's left hand motor rule. You should be familiar with the units for charge (coulomb) and potential difference (volt).

## LEARNING OBJECTIVES

In this chapter the unit of energy called the electronvolt (or electron volt) is introduced. You will learn about the discovery and properties of antimatter. The chapter builds on what you have already learned about beta decay in Chapter 2 to include the discovery and properties of neutrinos. The family of fundamental particles known as leptons will be described.

**(Specification 3.2.1.2 part, 3.2.1.3, 3.2.1.5 part)**

The Big Bang theory says that when matter was created in the very early Universe, an equal amount of antimatter was also created. Later, matter and antimatter should have recombined, annihilating each other in a process of mutual destruction that left only radiation. Though the Universe is dominated by radiation, there is obviously some matter left over. There is very little sign of any antimatter. An initial imbalance, just 1 in 10 billion, must have tipped the scales in favour of matter, without which we would not be here.

In theory, antimatter is a perfectly symmetrical reflection of matter. Each ordinary matter particle has its antimatter equivalent, with identical mass but other properties, such as charge, reversed. But the symmetry cannot be perfect; there must be an asymmetry that explains why the Universe is made of matter, rather than antimatter or just radiation. Particles called neutrinos may help us to solve this puzzle. Neutrinos come in three different types or 'flavours', and can change from one flavour to another and back again. This oscillation seems to happen at a different rate in neutrinos compared to antineutrinos.

Unfortunately, neutrinos are difficult to study. They are the most abundant particle in the Universe, and around 100 trillion (that is $10^{14}$) pass through you each second, but they rarely interact with anything. They interact so rarely, in fact, that if you live to be 100, only one neutrino, on average, will interact with an atom in your body in that time. That interaction might cause a tiny flash of light. That is why neutrino 'telescopes' are in very dark places, well away from other sources of light or radiation. The detectors at the Ice Cube Laboratory at the South Pole (Figure 1) are buried between 1 and 2.5 km deep in the ice sheet, while they watch for the rare events that might explain how the Universe evolved.

***Figure 1** (background) A neutrino detector being lowered into the ice at the South Pole. Neutrinos may help us to answer one of the biggest questions in physics – "How did we come to be here at all?"*

## 3.1 MASS AND ENERGY

The creation of particles and antiparticles was first observed in 1933 in photographs of the tracks left by **cosmic rays**, high-energy particles from space. We now know that matter–antimatter pairs of particles can be created when the energy of radiation is high enough (Figure 2). This conversion of energy into mass (and back again) was predicted by Albert Einstein.

The principle of conservation of energy states that the total energy of a closed system is constant; in other words, energy is not created or destroyed. Einstein showed that this idea had to be broadened to include mass. The combined mass and energy in any system is conserved, but energy and mass may be converted from one to the other. This conversion of mass to energy powers radioactivity and nuclear fission. Einstein's theory gave us the physics equation that nearly everyone knows, $E = mc^2$, or in words:

energy transferred = mass difference × (speed of light)$^2$

where the unit of mass is kilogram (kg), the unit of speed of light is metre per second ($\text{m s}^{-1}$) and the unit of energy is joule (J).

*Figure 2* This photograph from a particle detector shows the creation of matter and antimatter. An electron and its antiparticle, a positron, are created from gamma radiation, and then spiral away from each other in a magnetic field.

Einstein's equation applies to all mass and energy, but it has particularly important consequences in nuclear physics, such as the fusion reaction that was described at the end of Chapter 2:

$$^1_1\text{p} + {^1_1}\text{p} \rightarrow {^2_1}\text{H} + {^0_{+1}}\text{e} + \nu_e$$

If we measured the combined mass of the particles on the right-hand side of that equation, we would find that it is less than that on the left-hand side. In other words there is less mass after the reaction than there was before. This mass difference is the $m$ in $E = mc^2$. It has been transferred to other forms of energy. Since $c$, the speed of light, is $3.0 \times 10^8\ \text{m s}^{-1}$, $c^2 = 9.0 \times 10^{16}\ \text{m}^2\ \text{s}^{-2}$. A small mass difference, in kg, leads to a large amount of energy in J.

On the atomic scale, the joule and the kilogram are rather large units to use. In the reaction above, the mass difference is $7.47 \times 10^{-31}$ kg and the energy released is $6.72 \times 10^{-14}$ J, very small numbers that are awkward to work with. So we measure mass in terms of the atomic mass unit, u, which was defined in section 2.3 of Chapter 2 ($1\ \text{u} = 1.661 \times 10^{-27}$ kg). We also need to define an energy unit that is more appropriate for this scale.

### The electronvolt

The unit of energy used in atomic and nuclear physics is the **electronvolt** (eV). One electronvolt (1 eV) is the amount of energy gained by an electron as it accelerates through a potential difference of one volt (1 V). See Figure 3. One joule (1 J) of energy would be transferred by one coulomb (1 C) of charge moving through a potential difference of one volt (1 V). So (1 eV)/(1 J) is in the same ratio as (the charge of one electron)/(1 C), which is $1.60 \times 10^{-19}$, so

$$1\ \text{eV} = 1.60 \times 10^{-19}\ \text{J}$$

Conversely

$$1\ \text{J} = \frac{1}{(1.60 \times 10^{-19}\ \text{J})} = 6.25 \times 10^{18}\ \text{eV}$$

The energy acquired by particles in accelerators is usually given in electronvolts. The cathode ray tubes

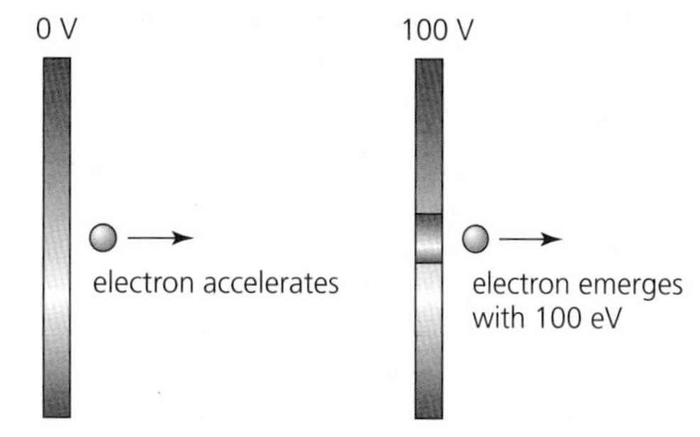


*Figure 3* Accelerating an electron

*Figure 4* The Large Hadron Collider at CERN

used by Thomson achieved electron energies of around 1 keV (one kiloelectronvolt, or $1 \times 10^3$ eV). The Large Hadron Collider (LHC) at CERN in Geneva (Figure 4) was refitted in 2015 and achieved energies of 13 TeV. 1 TeV is 1 (teraelectronvolt, or $1 \times 10^{12}$ eV). Even though this is a very large energy for a particle, it is still only a small fraction of a joule:

$$1\,\text{TeV} = 1 \times 10^{12}\,\text{eV}$$
$$= 1 \times 10^{12} \times 1.6 \times 10^{-19}\,\text{J}$$
$$= 1.6 \times 10^{-7}\,\text{J}$$

So 13 TeV = $13 \times 1.6 \times 10^{-7}$ J, which is just over 2 μJ, about enough energy to lift an apple by 2 μm. That is not much on an everyday scale, but is a huge amount of energy for a single particle collision.

## Measuring mass in terms of the electronvolt

Because mass can be transferred into other forms of energy, we can use the unit of energy to measure mass. Since

$$E = \text{mass} \times c^2$$

then

$$\text{mass} = \frac{E}{c^2}$$

We can use the unit of MeV/$c^2$ to measure the mass of small particles. On this scale the mass of an electron is 0.511 MeV/$c^2$ and one atomic mass unit (1 u) is 931.5 MeV/$c^2$.

### QUESTIONS

1. An electron that is accelerated by a particular X-ray tube gains an energy of 100 keV. Convert this energy to joules.

2. The energy released by a nuclear fission reaction is around 200 MeV. The average household in the USA uses 11 600 kWh of electricity per year (1 kWh means 1000 J of energy every second for an hour).

   a. How many nuclear reactions are needed to supply that amount of electricity?

   The mass of uranium used in each reaction is approximately 236 u.

   b. What mass of uranium is needed, in kg, to fuel the average American family for one year?

   c. Why is that an underestimate?

3. The most energetic cosmic ray ever detected (probably a proton travelling at almost the speed of light) was recorded in Utah in 1991. It came as a shock to physicists, who christened it the OMG particle. Its energy was estimated to be approximately $3 \times 10^{20}$ eV.

   a. How much energy in joules is that?

   b. Compare it to an everyday energy, like the kinetic energy of a moving object, or the work done in lifting or pushing something. (Kinetic energy = $\frac{1}{2}mv^2$ and work done = force × distance moved.)

   c. Why did the OMG particle cause such a shock?

### Stretch and challenge

4. A muon is a particle similar to the electron, but heavier. Its mass is 106 MeV/$c^2$.

   a. What is this in kilograms?

   b. What is this in terms of the atomic mass unit?

5. The reaction between hydrogen and oxygen is one of the most energetic chemical reactions there is, releasing about $0.5 \times 10^{-18}$ J every time one oxygen atom (O) reacts with a hydrogen molecule ($H_2$) to form a water molecule.

   a. Compare $0.5 \times 10^{-18}$ J to the energy released by the fission of uranium ($^{235}U + n \rightarrow {}^{93}Nb + {}^{141}Pr + 2n$), which is about 200 MeV per reaction.

**b.** Compare $0.5 \times 10^{-18}$ J to the energy released by the nuclear fusion of hydrogen to helium,

$$4(^1_1\text{H}) \rightarrow {}^4_2\text{He} + 2\text{e}^+ + 2\text{ neutrinos} + \text{energy (26 000 000 eV)}$$

**c.** If this is to be a fair comparison, we ought to take into account the mass of each fuel. Using the following data, find the energy released per kilogram of fuel used.

Mass of a hydrogen molecule $= 3.35 \times 10^{-27}$ kg

Mass of a hydrogen atom $= 1.67 \times 10^{-27}$ kg

Mass of an oxygen atom = 16.0 u

Mass of a neutron = 940 MeV/$c^2$

Mass of a uranium-235 atom = 235 u

1 u = one atomic mass unit = 931.5 MeV/$c^2$

**KEY IDEAS**

- Mass and energy can be converted to each other according to the formula $E = \Delta mc^2$.
- The energy in atomic and subatomic processes is measured in the unit electronvolt: $1\text{ eV} = 1.60 \times 10^{-19}$ J.
- The mass of subatomic particles can be expressed in the unit MeV/$c^2$.

## 3.2 ANTIMATTER

It was the British theoretical physicist, Paul Dirac (Figure 5), who first predicted the existence of **antimatter**. He realised, in 1928, that his equation describing the behaviour of an electron would work equally well if some quantities, such as charge, had the opposite sign. He predicted the existence of a particle with exactly the same mass as the electron, but with a positive charge, now called the positron. Indeed, he suggested that *all* particles must have a 'mirror image' twin – a particle of identical mass but opposite charge. Dirac called these **antiparticles**. Dirac was awarded the Nobel Prize for Physics at the age of 31 for his work on relativistic quantum physics, which predicted the existence of antimatter.

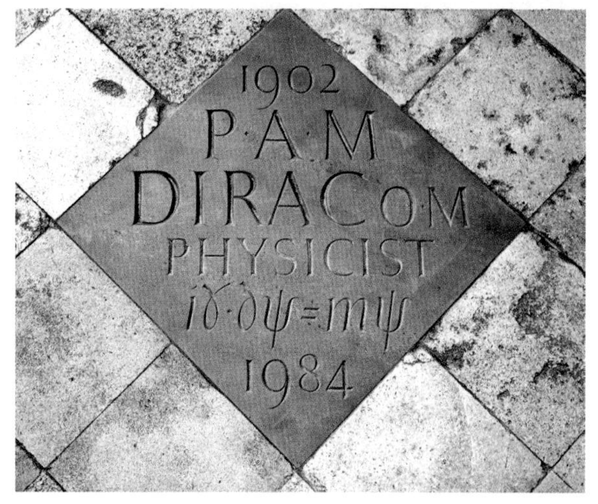


***Figure 5*** *Dirac was born in Bristol in 1902. A plaque in Westminster Abbey carries his 'beautiful' equation.*

Dirac's prediction was based on the underlying symmetry of his mathematics. Predicting the existence of antimatter was a bold step, but Dirac was justified when the positron, a positive electron, was discovered shortly afterwards by Carl Anderson.

In 1932 Anderson was observing tracks made by cosmic rays, high-energy particles from space. He was able to see the tracks left by high-energy electrons using a cloud chamber (Figure 6). Anderson used a strong magnetic field to curve the path of these electrons. He detected tracks that were identical to those made by an electron moving down through the atmosphere, except that they curved in the opposite direction. These were either electrons moving upwards through the apparatus or antielectrons, of equal mass but carrying a positive charge, moving downwards through the apparatus. Anderson put a thin piece of lead across the middle of his apparatus. This would show which way these particles were moving, since a particle would slow down after passing through the lead and the curvature of the track would be greater

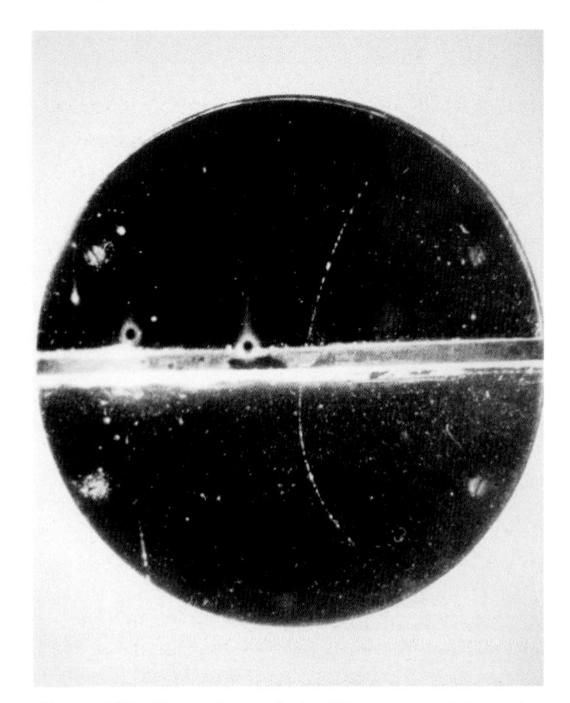

***Figure 6*** *The first evidence of a 'positively charged electron' (the track moves down the picture and curves to the right)*

(Figure 6). Anderson found that the particles were indeed moving down through the chamber. He had the first proof of the existence of the positron.

Antimatter makes only rare and rather fleeting appearances in our Universe. It can occur where there is sufficient radiation energy, perhaps as a result of cosmic ray collisions, or even as result of electric discharges in thunderstorms. Positrons can be emitted by some radioactive isotopes, many of which are created artificially in nuclear reactors. These isotopes, which emit positive beta particles, $\beta^+$, can be useful in medicine (see Assignment 1). Other examples of antimatter, such as antiprotons and antineutrons, have been observed. Although 'antimatter atoms' of hydrogen and helium have been created at CERN, none have been found to occur naturally, in cosmic rays for example.

**QUESTIONS**

6. Sketch a circle to represent the cloud chamber shown in Figure 6. Suppose there is a magnetic field acting into the plane of your drawing. Sketch the path that would be taken in each case by:
   a. a positron moving from left to right (label this A)
   b. an electron moving from left to right (label this B)
   c. a proton moving from left to right (label this C)
   d. a neutron moving from left to right (label this D).

   In each case add labelled arrows to show the direction of the force on the particle. See Figure 7 for a reminder of Fleming's left-hand rule.

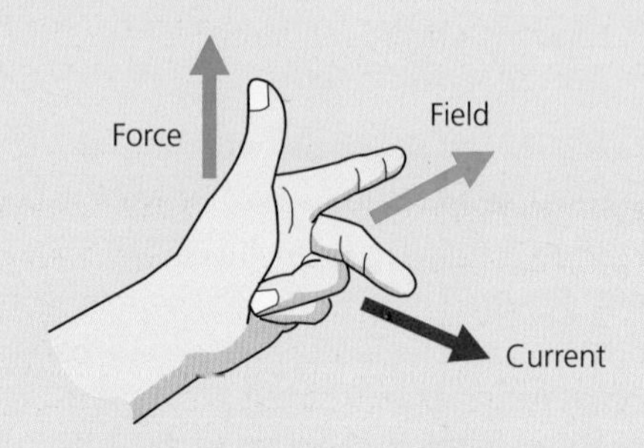


***Figure 7*** *Fleming's left-hand rule can be used to determine the way that charged particles move in a magnetic field. If your First finger (magnetic Flux or Field) points from magnetic North to South, and your seCond finger (Current) points in the direction that a positive charge is moving, then your thumb will show the direction of the force, and so the deflection, of the charged particle. You have to point your second finger in the opposite direction if it is a negatively charged particle.*

7. In 2011 the antimatter team at CERN reported keeping an antihydrogen atom for 1000 seconds. Why is this a difficult thing to do?

## 3.3 ANNIHILATION AND PHOTONS

When a particle meets its own antiparticle, they annihilate each other. The mass of the two particles is entirely transferred to electromagnetic radiation (Figure 8).

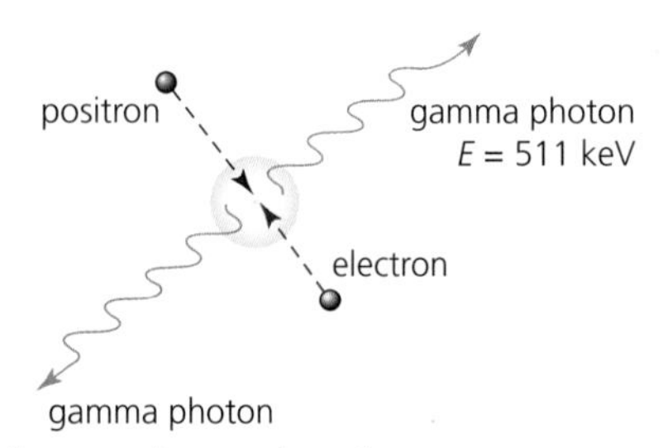


When a positron and an electron meet, they annihilate each other. Two identical gamma photons of energy 511 keV are emitted in opposite directions.

***Figure 8*** *Positron–electron annihilation*

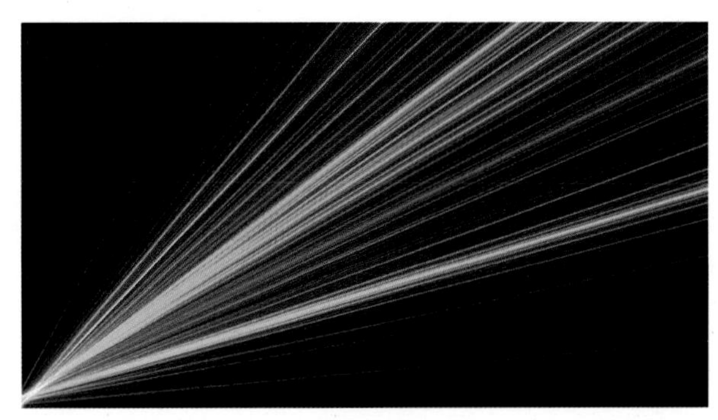

*Figure 9* Sometimes it is more helpful to picture light as a stream of particles, called photons, rather than as a continuous wave. A brighter light means more photons per second, rather than a larger-amplitude wave.

The electromagnetic radiation is emitted as two or more 'packets' of electromagnetic energy known as **photons**. The idea that electromagnetic radiation can be emitted and absorbed in discrete 'packets' of energy (Figure 9), rather than as a continuous wave, was developed in the early years of the 20th century to explain phenomena such as the photoelectric effect (as you will see in Chapter 8).

A photon can be pictured as a particle. Although it has no mass, it has energy and momentum. The energy of the photon, $E$, is proportional to the frequency, $f$, of the radiation:

$$E \propto f \qquad \text{or} \qquad E = hf$$

The constant $h$ is known as the **Planck constant**, equal to $6.63 \times 10^{-34}$ J s (to three significant figures).

In everyday terms the amount of energy transferred by a single photon is small, ranging from a few neV for a photon from the radio region of the spectrum, to MeV for a gamma photon.

Using the wave formula, $c = f\lambda$, so that $f = c/\lambda$, the energy of a photon can be written as

$$E = \frac{hc}{\lambda}$$

### Worked example 1

The energy of a photon of microwave radiation of frequency $f$ = 100 GHz is

$$E = 6.63 \times 10^{-34} \times 100 \times 10^{9} = 6.63 \times 10^{-23}\ \text{J}$$

which is equal to $\frac{6.63 \times 10^{-23}}{1.60 \times 10^{-19}} = 0.414$ meV

### Worked example 2

The shortest wavelength that can be directly observed by the human eye is about 400 nm. A photon from this part of the spectrum, violet visible light, has an energy of

$$E = \frac{hc}{\lambda}$$

$$= \frac{6.63 \times 10^{-34} \times 3 \times 10^{8}}{400 \times 10^{-9}}$$

$$= 4.97 \times 10^{-19}\ \text{J}$$

$$= 3.1\ \text{eV (to 2 s.f.)}$$

The energy of the photons that arise from the annihilation of an electron and a positron is very much greater than a few electronvolts. If the electron and positron are moving slowly relative to each other when they meet, there is no significant kinetic energy to take into account. The amount of energy released by the annihilation can be calculated from $E = mc^2$, where $m$ = the change in mass, which is twice the mass of an electron $= 2 \times 9.1 \times 10^{-31}$ kg. This gives

$$E = mc^2 = 1.64 \times 10^{-13}\ \text{J} = 1.02\ \text{MeV}$$

A photon carries momentum as well as energy. Assuming the electron and positron have negligible momentum before the reaction, there must be no overall momentum after the annihilation (by the principle of conservation of momentum; see section 11.2 of Chapter 11). The energy is shared exactly between two photons, which are emitted in opposite directions (see Figure 8). Each photon has an energy of 511 keV. These are gamma photons.

If the electron and positron meet at high speed, the extra kinetic energy can cause other particles to be created. The tunnel at CERN that now houses the Large Hadron Collider (LHC) previously held the Large Electron Positron collider (LEP). (The 'Large' refers to the collider, not the electron or the hadron!) Electrons and positrons were accelerated in opposite directions until they were made to collide. At these energies, when the electron and positron annihilate each other, many particles are created. In the example in Figure 10, jets of short-lived charged particles are created as well as a gamma ray with enough energy to produce another electron–positron pair. This last process is known as **pair production** (see section 3.4).

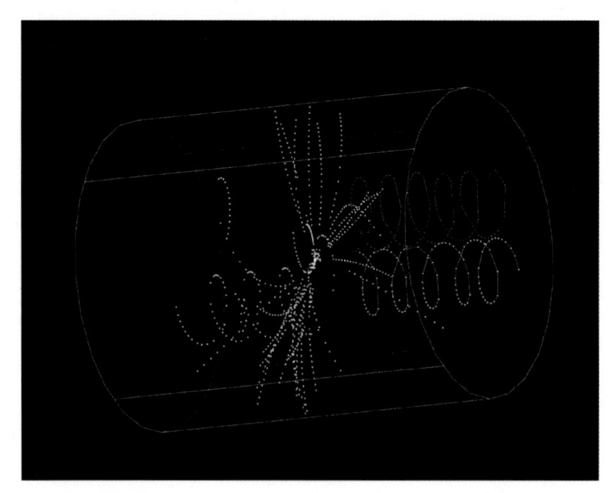

***Figure 10*** *A high-energy collision, in CERN's LEP collider. The electron and positron entered too fast to leave visible tracks before annihilating in the centre of the detector. The collision produced sprays of charged particles (blue dotted tracks), an unseen gamma ray photon, which leads to pair production in the form of an electron (yellow) and positron (red). Because they are of lower energy, their spiral tracks are clearly visible in the detector's magnetic field.*

## QUESTIONS

**8.** Calculate the energy of a photon of blue light with a wavelength of 480 nm.

**9.** When an electron and a positron annihilate each other, the photons each have energy of 511 keV.

**a.** What is the frequency of these photons?

**b.** What is their wavelength?

**10.** Small, high powered laser 'pointers' are now available, some with a range of as much as a hundred miles. Because these laser beams do not diverge much, they are potentially damaging to the eye, which is particularly sensitive to green light.

One of these green lasers emits light of wavelength 532 nm, and can achieve a beam intensity (power per unit area) of 50 μW $cm^{-2}$ at a distance of 100 m.

**a.** What is the frequency of green laser light?

**b.** What is the energy of a photon of this light? Give your answer in joule and in electronvolt.

### Stretch and challenge

**c.** How many photons per second will pass through 1 $cm^2$ at a distance of 100 m?

## KEY IDEAS

- Every particle has an antiparticle – its antimatter equivalent with equal mass but opposite charge.
- Electromagnetic radiation has a particle nature. The 'particles' are called photons. They have zero mass.
- The energy, *E*, of a photon is equal to *hf*, where *h* is the Planck constant and *f* is the frequency of the radiation.
- When a particle meets its antiparticle, they annihilate each other, releasing energy as two (or more) photons of gamma radiation.

## ASSIGNMENT 1: INVESTIGATING POSITRON EMISSION TOMOGRAPHY SCANNING

Matter–antimatter annihilation (see Figure 8) is used in the medical imaging technique of positron emission tomography (PET) scanning. A radioactive isotope, such as carbon-11, which emits positrons, is injected into the body. When the positron is emitted it only travels about 1 mm in the body before it meets with an electron and is annihilated. Because two gamma rays are emitted in opposite directions, it is relatively easy to locate the exact position of the radioactive isotope.

This PET technique, using carbon-11 as a positron emitter, is used in brain imaging (Figure A1). The patient inhales a small dose of carbon monoxide 'labelled' with carbon-11. Carbon monoxide molecules attach themselves to the haemoglobin in red blood cells and are transported around the body. When the carbon-11 decays it reveals areas of high blood flow, which correspond to active regions in the brain. PET scans (Figure A2) can show which areas are busy when the patient is reading, listening or just sitting with their eyes closed.

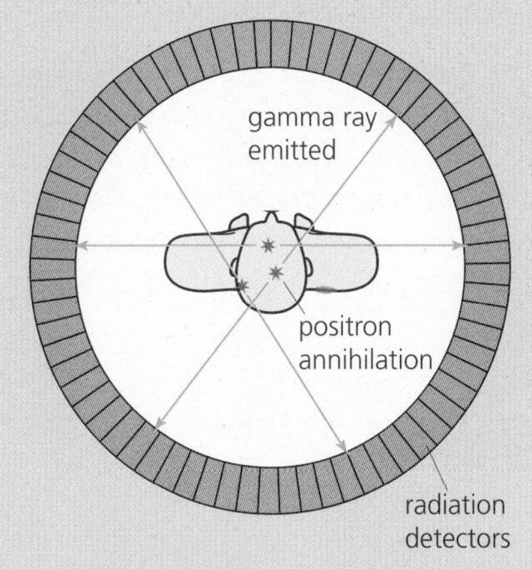


***Figure A1*** *Positron emission tomography (PET)*

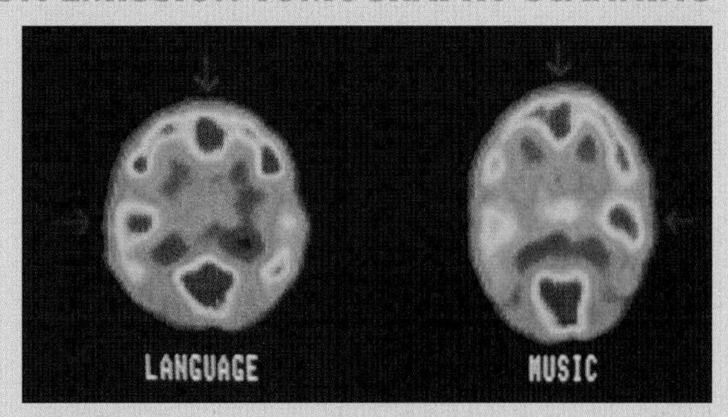


***Figure A2*** *PET scanning is now used for research in neuroscience and psychology.*

### Questions

**A1** Find an interesting piece of research on the Internet involving the use of PET scanning in medicine. (Searching for terms such as 'PET', 'positron emission', 'psychology', and 'neuroscience research' will turn up a lot of possibilities.)

**A2** Prepare a short report to explain the research. You should cover the following points in your report.

- **a** Describe the aim and the method of the research.
- **b** Explain how and why PET scanning is used.
- **c** Discuss the conclusions.

# 3.4 PAIR PRODUCTION

Annihilation is the conversion of matter to radiation. The opposite process, where matter is 'created' from radiation, is referred to as **pair production** (Figure 11). In this process electromagnetic energy in the form of a gamma photon is converted into a particle–antiparticle pair. There are always two particles created: one is conventional matter and the other is its antimatter opposite. This means that charge is 'conserved', since before the event there is only radiation, which carries no charge, and after the pair production there are always two particles of opposite charge, so the total charge is still zero.

A gamma ray photon has to have a minimum energy of 1.02 MeV before it can create an electron–positron pair. This is because the mass of the pair has an energy equivalence of 1.02 MeV. If the radiation had more energy than this, the surplus energy would appear as kinetic energy carried by the positron and

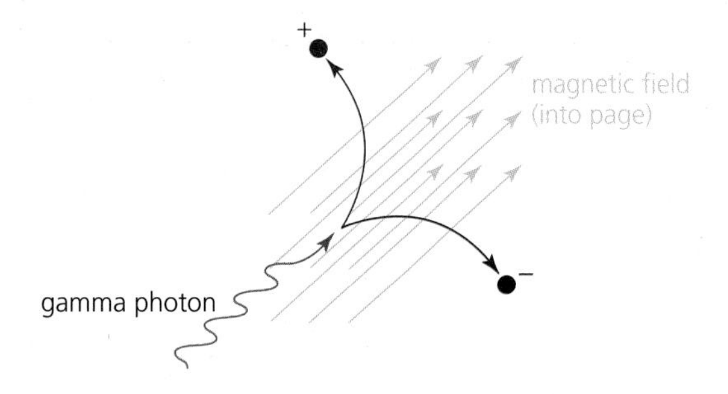


A magnetic field can be used to separate a particle from its antiparticle after pair production.

***Figure 11*** *Pair production*

the electron. Particle detector tracks of electron–positron pairs with different amounts of kinetic energy are shown in the upper and lower parts of Figure 2 in section 3.1.

Matter–antimatter pairs can also be produced in high-energy particle collisions. In 1955 the first antinucleon was discovered at the Berkeley accelerator, which could accelerate protons up to energies of 6 GeV and smash them into protons in a fixed target. The collision produced antiprotons (denoted by $\bar{p}$) in the reaction

$$p + p \rightarrow p + p + p + \bar{p}$$

This really is a remarkable reaction. By colliding two protons together we have produced an extra proton and an antiproton. If this happened when two snooker balls collided, the table would very soon fill up with extra balls! The extra mass needed to create the proton–antiproton pair has come from the kinetic energy of the initial protons. The minimum energy required to do this is

$$2 \times \text{mass of the proton} \times (\text{speed of light})^2$$

that is

$$\text{minimum energy} = 2m_p c^2 = 3.0066\times10^{-10}\ \text{J} = 1.9\,\text{GeV}$$

A year later the antineutron was produced by antiproton–proton collision:

$$\bar{p} + p \rightarrow n + \bar{n}$$

The mass of a particle is always identical to the mass of its antiparticle, but other properties, such as charge, are reversed in sign. A neutron has no charge, so it might seem that the neutron and antineutron are the same. This is *not* the case. Even though a neutron has zero charge, it does have magnetic properties. These are equal in magnitude but opposite in direction to those of an antineutron. The neutron has an internal structure, as you will see in Chapter 4 – it is made up of three quarks. The antineutron is composed of three antiquarks.

### QUESTIONS

**11.** The minimum gamma ray photon energy to create an electron–positron pair is 1.02 MeV. What is the frequency of such a photon?

**12.** Particles are always created in pairs, for example an electron and positron. The creation of a single electron has never been seen even though this would take only half as much energy. Why is this?

### KEY IDEAS

- A particle–antiparticle pair may form from radiation if the energy of a photon is high enough, greater than $2mc^2$.

## 3.5 NEUTRINOS

Neutrinos are the most numerous particles in the Universe. They outnumber the protons and neutrons of ordinary matter by a factor of 1000 000 000. Neutrinos created at the time of the Big Bang still permeate the Universe. There are about 100 or so of them in each cubic centimetre of space. Neutrinos are also emitted by radioactive nuclei and from nuclear reactions. The Earth is bathed in neutrinos from the Sun. Every second about 60 thousand million solar neutrinos pass through every square centimetre of the Earth's surface.

Despite this, neutrinos are extremely difficult to detect. They are not charged, so they do not feel electrostatic attraction or repulsion. Their mass is extremely small, so they are little affected by gravity. They are also unaffected by the strong nuclear force, so interact with other particles only through the weak interaction. But this acts only over a very short range, about $10^{-18}$ m (see section 2.5 of Chapter 2), so neutrinos rarely interact with other matter at all. The vast majority of neutrinos that strike the Earth pass straight through with no deviation.

Experiments that are looking for neutrinos often use large tanks of water, usually placed deep underground, surrounded by sensitive light detectors. They are looking for the very occasional flash of light that signifies, for example, that a neutrino has interacted with a neutron.

The existence of the neutrino was first predicted by Wolfgang Pauli in 1930. At the time, physicists were struggling with a problem in understanding beta radiation (see section 2.4 of Chapter 2). Beta particles are fast-moving electrons emitted by the nuclei of some radioactive atoms. Unlike alpha particles, which are emitted with a well-defined energy, betas are emitted with a range of energies (Figure 12).

Beta particle emission seemed to contravene the principle of the conservation of energy. If a certain amount of energy is transferred by each radioactive decay, why did the emitted beta particle have a range of possible energies? Pauli suggested that another particle, the neutrino, is also emitted in beta decay. The neutrino carries away the balance of the energy, so that the total energy of the decay is always constant.

The neutrino is represented by the symbol $\nu_e$. The subscript 'e' stands for 'electron' and these neutrinos are more properly referred to as electron neutrinos (see section 3.6). The neutrino also has its antiparticle, the **antineutrino**, which is written as $\bar{\nu}_e$. It is this antineutrino that is emitted during beta decay:

$$ {}^{A}_{Z}X \rightarrow {}^{A}_{Z+1}Y + {}^{0}_{-1}e + \bar{\nu}_e $$

Neutrinos and antineutrinos differ in a property called 'spin' – their spins are equal but in opposite directions.

Alpha particles from a given unstable nucleus are emitted with one of a small number of discrete energy values. The lower-energy alphas are accompanied by a gamma photon.

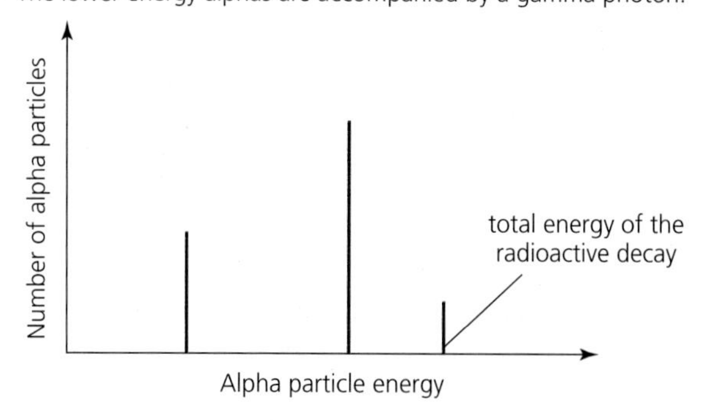


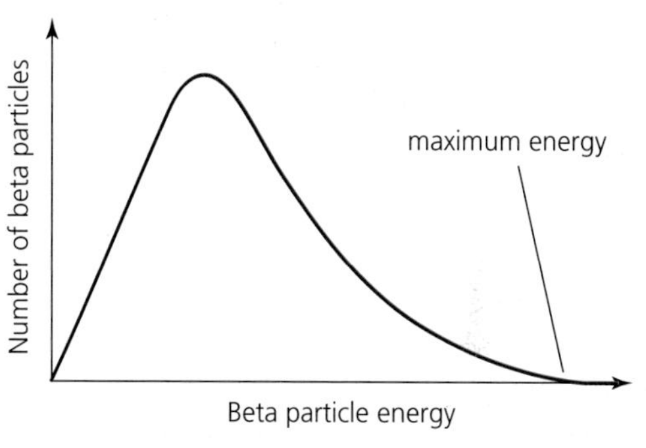


***Figure 12*** *Energy of emitted alpha and beta particles*

**QUESTIONS**

**13.** Read Pauli's letter in Figure 13. Which of Pauli's predictions about the neutrino proved to be correct? Where do we now believe he was mistaken?

**14.** It is possible that the neutrino is its own antiparticle. Why is that possible for a neutrino and not for an electron?

## Experimental evidence for neutrinos

It was 26 years before Pauli's courageous proposition could be confirmed. The problem is that neutrinos are so difficult to detect. On average a neutrino emitted from a beta decay will pass through a light-year's thickness of lead and only interact once. The best way of finding a few of these elusive particles is to look in a place where there is a huge number of them. In 1956 Clyde Cowan and Frederick Reines set up their apparatus next to the Savannah River nuclear reactor in the USA. The neutrino flux was expected to be $10^{13}$ neutrinos per second through each square centimetre ($cm^2$).

*Zurich*

*December 4, 1930*

*Dear radioactive ladies and gentlemen,*

*I beg you to listen most favourably to the carrier of this letter. He will tell you that, in view of the ... continuous beta spectrum, I have hit upon a desperate remedy to save the law of conservation of energy. This is the possibility that electrically neutral particles exist which I will call neutrinos,*[1] *which exist in nuclei ... and which differ from the photons also in that they do not move with the speed of light. The mass of the neutrinos should be of the same order as that of the electrons and should in no case exceed 0.01 proton masses. The continuous beta spectrum would then be understandable if one assumes that during beta decay a neutrino is emitted with each electron in such a way that the sum of the energies of the neutrino and electron is constant ....*

*I admit that my remedy may look very unlikely, because one would have seen these neutrinos long ago if they really were to exist. But only he who dares wins .... Hence one should seriously discuss every possible path to rescue ....*

*Your most obedient servant,*

*W. Pauli*

**Figure 13** *Pauli was unable to attend a nuclear physics conference, as he had to go to a ball, so he sent a letter proposing the existence of a new particle. At the time only two subatomic particles were known to exist – the electron and the proton – so predicting a new particle was quite courageous.*

[1] Pauli actually called his new particle the *neutron*; but the particle we now call the neutron was discovered two years later. Pauli's predicted particle was re-christened by the Italian Enrico Fermi; it became known as the *neutrino* ... the little neutral one!

Cowan and Reines were actually looking for the antineutrinos, $\bar{\nu}_e$, that are emitted from beta-minus decay. In this decay, a neutron, n, decays into a proton, p, and an electron, $e^-$. An antineutrino is also emitted. The decay can be written as

$$n \rightarrow p + e^- + \bar{\nu}_e$$

The experiment was designed to look for the subsequent reaction of the antineutrino with a proton:

$$p + \bar{\nu}_e \rightarrow n + e^+$$

where $e^+$ is a positron. Cowan and Reines used large tanks of liquid scintillator with a high hydrogen content. The liquid also contained a cadmium compound. The idea was that the positron would meet an electron and be annihilated almost immediately, releasing two gamma rays, which would cause a prompt flash of light in the scintillator. A short, but predictable, time later the neutron would be absorbed by the cadmium and would emit a gamma ray, which would lead to another flash of light. The neutrino would reveal itself by the right coincidences between the flashes of light. After three years of experiments

### QUESTIONS

**15.** Why did the liquid scintillators in the Cowan and Reines experiment need a high hydrogen content?

**16.** Why was the experiment installed next to a nuclear reactor?

Cowan and Reines were able to telegraph Pauli to say that they had found the neutrino.

## The search goes on

One current neutrino-detection experiment is the Super-Kamiokande, 1 km below ground, in Japan (Figure 14). It is designed to detect neutrinos from the Sun. We know that there should be a neutrino flux from the Sun, caused by nuclear reactions deep within the solar core. These reactions are well understood and the theory is in excellent agreement with measurements, except for one thing. The problem was that more than half of the solar neutrinos seemed to disappear on their way to Earth. The detector consists of 50 000 tonnes of ultrapure water, surrounded by 11 000 highly sensitive light detectors. These are looking for the flash of light that comes when an electron hurtles away at high speed after being hit by a neutrino. Even in such a large tank of water this is a rare event.

The latest results from the Super-Kamiokande and elsewhere confirm that the solar neutrinos do not go missing. Rather, they change from one flavour of neutrino to another *on the way* to Earth. This oscillation implies that neutrinos have some mass, however small.

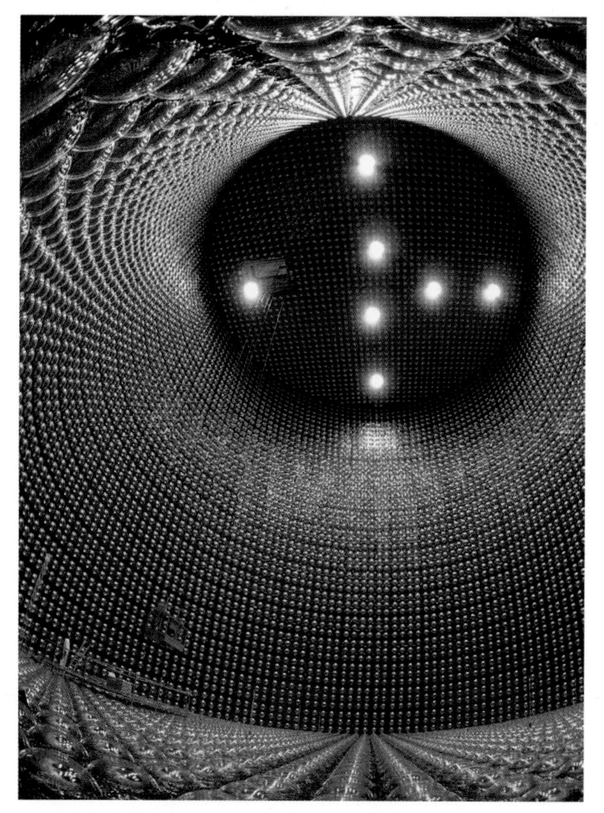

***Figure 14*** *Light detectors in the Super-Kamiokande*

**KEY IDEAS**

- The neutrino and its antiparticle, the antineutrino, are particles that carry no charge and have very small mass.
- Neutrinos interact with other particles only very rarely, via the weak force.
- Pauli predicted the existence of the neutrino by studying the energy released in beta decay.

# 3.6 THE LEPTON FAMILY

Pauli's prediction of the neutrino and its subsequent discovery meant that there were now four subatomic particles known to physics, the other three being the electron, the proton and the neutron. Dirac's work on antimatter instantly doubled this to eight. This was enough to explain the atomic and nuclear phenomena that were known at the time. However, many more new particles were found in the next few years.

## Muons

In the days before the big accelerators, such as the LHC at CERN, the way to study particle physics was to fly a balloon (Figure 15). High in the atmosphere there is a greater flux of cosmic rays. High-energy particles, such as protons and penetrating gamma rays, collide with atoms of hydrogen and oxygen in the highest reaches of the atmosphere, sending a shower of particles towards the Earth (Figure 16).

The **muon** was detected in these cosmic ray showers. Anderson photographed some unusual tracks in his cloud chambers and calculated the charge-to-mass ratio as $8.8 \times 10^8$ C kg$^{-1}$. This was smaller than the

***Figure 15*** *Cosmic ray studies often involved a large balloon filled with hydrogen.*

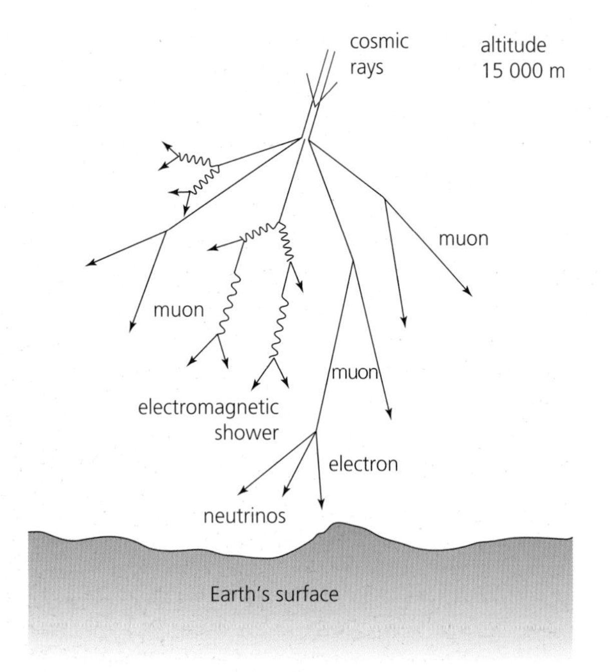


***Figure 16*** *Muons are created at the edge of the atmosphere, about 15 km above the Earth's surface. A muon travelling at 95% of the speed of light would take about 52 μs to reach the surface. But that short time is more than 20 muon lifetimes, so most of the muons should have decayed before they ever reach the surface of the Earth. The fact that they have not is experimental evidence for Einstein's special theory of relativity, which states that moving clocks run slow. The muons are moving so near to the speed of light that time runs much more slowly for them.*

same ratio for the electron, and very much higher than the value for the proton. In fact the muon turned out to have *exactly* the same charge as an electron, $-1.60 \times 10^{-19}$ C, but its mass was 200 times greater. This came as a shock to physicists, who could not see why the electron needed a 'big brother'. Nuclear physicist Isador Rabi was reputedly in a restaurant when he heard of the muon's discovery. "Who ordered that?" he demanded.

The muon is a heavier version of the electron. It has exactly the same charge and, like the electron, it is a **fundamental particle**, meaning it is without any internal structure. However, as well as its mass of 200 times the electron's mass, there is another difference. The muon is unstable, with an average lifetime of $2.2 \times 10^{-6}$ s or 2.2 μs.

A muon decays quickly (Figure 16), typically into an electron and two neutrinos:

$$\mu^- \rightarrow e^- + \bar{\nu}_e + \nu_\mu$$

The first neutrino, $\bar{\nu}_e$, in the above reaction is an electron antineutrino, whereas the second one, $\nu_\mu$, is a muon neutrino.

The electron neutrino and the muon neutrino are not the same: they certainly have different masses. The muon neutrino appears in reactions involving muons.

## QUESTIONS

**17.** This reaction shows the decay of an antimuon. State in words what it decays into.

$$\mu^+ \rightarrow e^+ + \nu_e + \bar{\nu}_\mu$$

## More leptons

In 1978 yet another member of the lepton family was discovered. The tau-minus, $\tau^-$, particle was observed by a team working on electron–positron collisions at Stanford in the USA. The Stanford team collided electrons and positrons together to produce a tau-minus and its antiparticle, the tau-plus:

$$e^+ + e^- \rightarrow \tau^+ + \tau^-$$

For several months these results were not generally accepted, until a team working in Germany replicated the results.

The tau particle has the same charge as the electron and the muon but is much more massive, around 3500 times the mass of the electron. It seems logical to suppose that this new, heavier version of the electron has its own type of neutrino and antineutrino, the $\nu_\tau$ and the $\bar{\nu}_\tau$. These were finally detected in July 2000 by a team at Fermilab, USA.

The lepton family now has three groups within it: the electron and its neutrino, the muon and its neutrino, and finally the tau lepton and its neutrino. When all the antiparticles are taken into account, there are 12 leptons and antileptons. The leptons have all these things in common:

- They are not affected by the strong nuclear force.
- They are affected by the weak interaction, and charged leptons are affected by electromagnetic force.
- They are all fundamental particles, which means that they are not constructed from other smaller particles.
- All three lepton 'generations' are fermions, that is, they have spin value ½.

Recent work has suggested that there may not be any other members of the lepton family, though no-one has yet been able to explain fully why there are three types of lepton and why they should have the masses that they have. The Higgs boson (see Chapter 4) gives the electron its mass, but may not be the cause of the neutrino's mass, which is very difficult to measure. The leptons and antileptons are summarised in Table 1.

**KEY IDEAS**

- The leptons, and their antiparticles, antileptons, are all fundamental particles.
- Each of the three charged leptons (electron, muon, tau) has an associated neutrino, which is also a lepton.
- Each lepton has an antiparticle that has an identical mass but whose other properties, charge for example, are opposite.
- Leptons are affected by the weak interaction.

## Moving on: particles within particles

Experiments at CERN and elsewhere now suggest that there only three generations of lepton. There may be no more elementary particles beyond the tau for us to discover. But what about the proton and the neutron? Experiments and theory were pointing to the fact that they were not fundamental particles – they had structure. The quest to reveal that structure is described in the next chapter.

| | Symbol | Charge (in terms of proton charge) | Mass (in terms of electron mass) | Mass / MeV/$c^2$ |
|---|---|---|---|---|
| **Leptons** | | | | |
| electron | $e^-$ | −1 | 1 | 0.511 |
| electron neutrino | $\nu_e$ | 0 | ≈0 | $<2 \times 10^{-6}$ |
| muon | $\mu^-$ | −1 | 207 | 106 |
| muon neutrino | $\nu_\mu$ | 0 | ≈0 | <0.190 |
| tau | $\tau^-$ | −1 | 3500 | 1790 |
| tau neutrino | $\nu_\tau$ | 0 | <35 | <18.2 |
| **Antileptons** | | | | |
| positron | $e^+$ | +1 | 1 | 0.511 |
| electron antineutrino | $\bar{\nu}_e$ | 0 | ≈0 | $<2 \times 10^{-6}$ |
| antimuon | $\mu^+$ | +1 | 207 | 106 |
| muon antineutrino | $\bar{\nu}_\mu$ | 0 | ≈0 | <0.190 |
| antitau | $\tau^+$ | +1 | 3500 | 1790 |
| tau antineutrino | $\bar{\nu}_\tau$ | 0 | <35 | <18.2 |

***Table 1*** *A summary of leptons and antileptons*

## ASSIGNMENT 2: MAKING A CASE IN DISCUSSION

### Money to burn?

When physicists started research into the basic building blocks of matter, they had no clear idea of what, if any, useful applications would spring from their work. It is called 'blue sky' research, which is research done for the sake of knowledge, rather than applied research, which aims to improve a particular technology. CERN in Switzerland is the home of much blue sky research and is supported by funds from 21 European governments. This kind of research is among the most expensive in the world. But do we need this kind of research? Should the funds be used for things that are more immediately useful?

Prepare to take part in a discussion about the sort of research that governments should support. Do you want to fund CERN's hyper-expensive acceleration projects?

### Preparing for your discussion

When you prepare for a discussion you will need to consider the following:

- What are your starting points? Note down the attitudes and ideas you already have on a piece of paper.
- Now take a long look at your assumptions. Try to sort them into statements of fact and statements of opinion.
- Now look at the statements of fact. Do any need to be checked? Can you find the evidence to back up your facts? If you cannot, are you sure your 'fact' is correct?
- Then look at your opinions. Can you unpick the reasons behind them? Since they are opinions, not facts, you will not be able to find something that will prove them to be correct. But you should be able to find some evidence to suggest that they are reasonable.
- Decide which of your opinions you feel strongest about. What are the things you are extremely unlikely to change? This is not always a bad thing – to believe in something very strongly is not wrong. But do recognise that a strongly held belief is not as strong in an argument as a simple fact.
- Finally, as well as deciding on your strongest opinions, look for areas where you are uncertain or willing to be convinced. Where do you feel able to compromise?

### Holding a successful discussion

After you have assembled your thoughts and some evidence, you are ready to participate in a discussion. A few simple rules will help the discussion to be productive:

- Make sure you listen – and check that you understand what people are saying. Ask questions like 'So are you saying ...?' and try to paraphrase what someone has said. They can correct you if you have misunderstood them.
- Hear what people are saying – people who *listen carefully* learn much more from discussions than the people who *talk continually*.
- Remember that some opinions will annoy and irritate you. Do not let your feelings interfere with your thinking!
- Be prepared to learn and change your position on some issues. And when you agree with people – tell them.
- Try not to tell people they are stupid – even if they are! Insults rarely work as a way of convincing someone.

Remember that discussions should be win–win situations. Even if you end up disagreeing with someone, you can learn something by understanding their point of view. Not every discussion is an attempt to convince someone that you are completely correct in every detail and that they need to change their ideas to match yours.

### Rounding off your discussion

At the end of your discussion, try to agree a statement that summarises the opinions of the whole group. This might contain two lists of statements:

- Statements everyone can support.
- Statements most of the group can support.

## PRACTICE QUESTIONS

**1.** **a.** Every type of particle has a corresponding antiparticle. Give **one** example of a particle and its corresponding antiparticle.

**b.** State **one** difference between this particle and its antiparticle.

*AQA June 2013 Unit 1 Q3b*

**2.** **a.** Pair production can occur when a photon interacts with matter. Explain the process of pair production.

**b.** Explain why pair production cannot take place if the frequency of the photon is below a certain value.

**c.** Energy and momentum are conserved during pair production. State **two** other quantities that must also be conserved.

*AQA Jan 2012 Unit 1 Q3*

**3.** Uranium-238 is a radioactive isotope. The U-238 nucleus decays by emitting an alpha particle:

$$^{238}_{92}\text{U} \rightarrow {}^{234}_{90}\text{Th} + \alpha$$

The energy of the emitted alpha particle is 4.04 MeV (in < 1% of decays), or 4.15 MeV (in ≈23%) or 4.20 MeV (in ≈77%). When the alpha has energy 4.04 or 4.15 MeV, a gamma photon is also emitted.

**a.** How much energy is released by the decay of U-238?

**b.** When an alpha is emitted with energy 4.04 MeV, what is the energy of the emitted gamma photon?

**c.** Beta decay is very different. Beta particles can be emitted with any energy within a continuous range. Explain why this discovery presented a problem (or two!) for physics, and how this was resolved by Pauli.

**4.** Pauli predicted the existence of the neutrino because of a problem with our understanding of beta decay. Which of the following four statements is **incorrect**?

**A** Beta decay appeared to violate the conservation of charge.

**B** Beta decay appeared to violate the conservation of energy.

**C** Beta decay appeared to violate the conservation of momentum.

**D** Beta particles are emitted from a radioisotope with a range of energies.

**5.** Every matter particle has an antimatter equivalent. Which of the following statements is **incorrect**?

**A** The positron is the antimatter equivalent of the electron.

**B** When a positron meets an electron, they annihilate each other.

**C** The mass of a particle is always identical to the mass of its antiparticle.

**D** When an electron and proton annihilate each other, a single gamma ray is released.

**6.** Which of the following statements about neutrinos is correct?

**A** Neutrinos are definitely massless.

**B** Neutrinos interact with other particles only through the strong interaction.

**C** Neutrinos carry no charge.

**D** Neutrinos do not have an antiparticle.

**7.** A gamma ray can create an electron and a positron through pair production. For this to happen, the gamma ray's energy must be:

**A** exactly 0.511 MeV

**B** at least 1.02 MeV

**C** at least 2.4 MeV

**D** less than 1.02 MeV

**8.** The particle detector tracks in Figure Q1 show the production of matter and antimatter from a single gamma ray photon of energy about 2 MeV. The particles produced initially move from top to bottom, and there is a magnetic field directed into the page.

*Figure Q1*

Which is the correct identification of the particles in Figure Q1?

**A** green track proton, red track antiproton

**B** green track antiproton, red track proton

**C** green track positron, red track electron

**D** green track electron, red track positron

**9.** Which row in Table Q1 most accurately describes the properties of leptons?

| | Affected by the strong force | Fundamental | Affected by the weak force |
|---|---|---|---|
| **A** | yes | yes | no |
| **B** | yes | no | yes |
| **C** | no | yes | yes |
| **D** | no | no | no |

*Table Q1*

## Stretch and challenge

**10.** This extract from a website for hospital patients is designed to explain what a PET scan is for, and how it works. Read it carefully and critically, and then answer questions **a** to **d**.

**a.** The extract says, "Positrons are tiny particles which are made as the radio-tracer is broken down inside your body." Explain why this is misleading.

**b.** Positrons are fundamental particles, yet the extract describes them as being "broken down" to create gamma rays. Explain how the gamma rays actually originate.

**c.** The extract explains that "Different levels of positrons are shown as different colours" on the scan. What do the different colours on the scan actually show?

**d.** It is important to explain a PET scan accurately and clearly to a patient who may have little or no physics knowledge. Terms like 'antimatter' and 'radioactivity' are potentially alarming. Write a paragraph explaining the physics of a PET scan. Concentrate on making it accurate, but try to reassure the patient. (You will be assessed on the quality of your written communication.)

**How does a PET scan work?**

Before a PET scan is carried out, a radioactive medicine is produced in a machine called a cyclotron. The medicine is then 'tagged' to a natural chemical such as glucose, water or ammonia. This makes what is called a radio-tracer. Once the radio-tracer is inside the body, it goes to parts of the body that use the chemical it has been tagged to. For example, a radioactive drug called fluorodeoxyglucose (FDG) is tagged to glucose to make a radio-tracer. When inside the body it goes to the tissues that use glucose for energy.

A PET scan works by detecting the energy released by positrons. Positrons are tiny particles which are made as the radio-tracer is broken down inside your body. As positrons are broken down they create gamma rays. These gamma rays are detected by the scanner, which creates a three-dimensional image. The image can show how parts of your body work, by the way in which it breaks down the radio-tracer.

Different levels of positrons are shown as different colours and brightness on a PET image. Some parts of the body break down natural chemicals such as glucose quicker than others. A PET scan is particularly useful in detecting cancer because most cancers use more glucose than normal tissue. Areas of greater intensity, called 'hot spots', show where large amounts of the radio-tracer have built up. Less intense areas, or 'cold spots', indicate a smaller concentration of radio-tracer.

A radiologist will look at the images that a PET scan produces, and report the results to a doctor. A radiologist is a person who has training in interpreting pictures of inside the body.

# 4 THE STANDARD MODEL

## PRIOR KNOWLEDGE

You will need to know about leptons and their properties, from Chapter 3.

## LEARNING OBJECTIVES

In this chapter you will learn how other particles were discovered and about their properties and interactions. You will see how it all fits together in the standard model of particle physics, and about conservation laws that govern particle interactions.

**(Specification 3.2.1.4, 3.2.1.5, 3.2.1.6, 3.2.1.7)**

***Figure 1*** *Professor Higgs congratulates Fabiola Gianotti on a 'momentous day for science', 4 July 2012, after she presented results confirming the existence of the Higgs boson, 40 years after his prediction of it.*

The discovery of the Higgs boson (Figure 1) was the most important discovery in physics for many years. Peter Higgs was understandably emotional as scientists announced that the particle he had predicted 40 years earlier had been found. He had been trying to explain why some particles have mass and some, like the photon, do not; and why similar particles, like the electron, muon and tau leptons, each have a different mass. Higgs proposed that a field, acting throughout all space, was responsible for giving certain particles mass. A particle, known as a gauge boson, would act between the field and all other particles. The Higgs boson, as it became known, is a vital part of the standard model, the currently accepted theory of particle physics. If the Higgs boson could not be found, the standard model might have to be abandoned. Evidence for the Higgs boson would only occur in very high-energy collisions, and so the world's most energetic particle accelerator was constructed.

The Large Hadron Collider (Figure 2) was built in a 27 km circumference circular tunnel at the European Organisation for Nuclear Research (known as CERN), near Geneva. It accelerates protons, and antiprotons, to 99.999 9% of the speed of light in opposite directions, before making them collide inside particle detectors. Scientists found evidence of the Higgs boson in the tracks left by the particles created in these high-energy (7 TeV) collisions.

There is a small chance that the tracks might not have been due to the Higgs boson. However, so many collisions were analysed that scientists were confident in announcing the discovery of 'a Higgs-like boson with a mass equal to 125 ± 0.6 GeV' with only a 0.0001% probability that the observations could be due to chance. Confidence was particularly high because the results had been replicated by two different experiments. Since then, other experiments have corroborated the findings. Peter Higgs' prediction was vindicated, and in 2013 he was awarded a half-share of the Nobel Prize in Physics.

***Figure 2*** *(background) The Large Hadron Collider*

## 4.1 THE PARTICLE EXPLOSION

In the years following the unexpected discovery of the muon (section 3.6 of Chapter 3), many more observations were made of cosmic rays. Cloud chambers and photographic films were used to visualise the tracks left by the rays, mostly high-energy protons, and the particles they create as they crash into atoms in the upper atmosphere. Before long, other new particles were discovered, many of them much heavier than the muon. By the 1960s there were dozens of 'new' particles.

The situation was similar to that in chemistry a hundred years earlier. By 1900 more than 80 chemical elements had been discovered. It did not seem likely that there could be so many different fundamental entities. Scientists felt that there must be an underlying pattern. This proved to be the case, and wc now know that atoms of all the different elements are made from the same three particles – the neutron, the proton and the electron. Perhaps the bewildering array of different subatomic particles (Table 1) was also indicative of a simpler underlying structure.

### Classifying particles

The first steps in making sense of all these particles was to look for patterns, similarities and differences, so that particles could be classified into groups. This was done initially by grouping together particles of similar mass. Two groups of particles, **leptons** (lighter particles) and **hadrons** (heavier particles), were identified (Figure 3). Hadrons were then divided into two subgroups, initially by their relative masses, baryons and mesons – baryons being generally heavier. These and a third group of particles, the gauge bosons, will be discussed later in this chapter.

### Leptons

There are six different members of the lepton family. The electron and the electron neutrino, the muon and its neutrino, and the tau lepton and its neutrino

| Year | Particle | Discovered by (place) | Source of particles | Energy |
|---|---|---|---|---|
| 1897 | Electron | J. J. Thomson (Cambridge) | Cathode ray tube | keV |
| 1911 | Nucleus | Rutherford (Manchester) | Alpha particles | MeV |
| 1931 | Positron | Carl Anderson (CalTech) | Cosmic rays | MeV–GeV |
| 1932 | Neutron | James Chadwick (Manchester) | Alpha particle | MeV |
| 1937 | Muon | Neddermeyer and Anderson | Cosmic rays | GeV |
| 1947 | Pions | Powell (Bristol) | Cosmic rays | GeV |
| 1948– | K mesons $\tau$, $\Xi$, $\Lambda$, $\theta$, $\Sigma$ | (Manchester, CalTech, Bristol, ...) | Cosmic rays | GeV |
| 1955 | Antiproton/Antineutron | (University of California, Berkeley) | Synchrotron | GeV |
| 1957 | Neutrino | Cowan and Reines (California) | Nuclear reactor | GeV |
| 1961 | Muon neutrino | Ledermann (Brookhaven) | Cosmic rays | GeV |
| 1964 | $\Omega^-$ particle | (Brookhaven) | Accelerator | GeV |
| 1967 | Up and down quarks | (Stanford, SLAC) | Linear accelerator | 20 GeV |
| 1974 | J/$\Psi$ (charmed quark) | (SLAC/Brookhaven) | Linac | GeV |
| 1976 | Tau lepton | (SLAC) | Linac | GeV |
| 1977 | Y hadron (bottom quark) | (Fermilab) | Tevatron | 400 GeV |
| 1983 | W, Z bosons | (CERN, Geneva) | Proton synchrotron | 500 GeV |
| 1995 | Top quark | (Fermilab) | Tevatron | 1.8 TeV |
| 2000 | Tau neutrino | (DONUT, Fermilab) | Tevatron | TeV |
| 2012 | Higgs boson | (CERN, Geneva) | LHC | 8 TeV |
| ? | Supersymmetry particles? | ? | ? | 14 MeV? |
| ? | Dark matter? | ? | ? | ? |

***Table 1*** *Timeline of particle discovery*

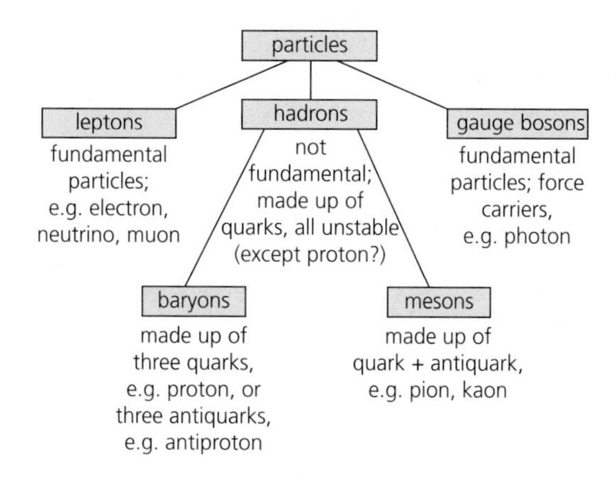


***Figure 3*** *The standard model, the currently accepted theory of particle physics, classifies all particles into one of three families: leptons, hadrons or gauge bosons.*

(see section 3.6 of Chapter 3). Each of these has an antiparticle, so there are also six antileptons. All of these are fundamental particles – they cannot be broken down into other particles. Leptons are not affected by the strong nuclear force.

## Hadrons

Cosmic ray and particle accelerator experiments revealed a large number of other particles in addition to the leptons described above. Pi mesons, kaons, delta mesons and many others were discovered, generally with much larger masses than the leptons (see Table 2). This group of particles is referred to as hadrons (from the Greek, *hadros*, meaning bulky). Hadrons are affected by both the strong interaction and the weak nuclear force.

Some hadrons carry charge; some do not. All the hadrons, with the probable exception of the proton, have been found to be unstable. Like radioactive nuclei, after a certain time they decay into something else. For example, a free neutron (not bound in a nucleus) is unstable. It decays with a half-life of about 11 minutes via the reaction

$$n \rightarrow p + e^- + \bar{\nu}_e$$

(as in beta-minus decay of an unstable nucleus).

The proton seems to be stable, although some theories suggest that it does decay with a lifetime of the order of $10^{32}$ years. As that is much longer than the life of the Universe so far, it is likely to be difficult to detect proton decays. The Super-Kamiokande neutrino detector (see Figure 14 in Chapter 3) has been looking for evidence of proton decays. Despite thousands of tonnes of pure water in a lake surrounded by photo-detectors, there is no evidence yet.

| Name | Symbol | Rest mass (GeV/$c^2$) | *Q* (charge) | *B* (baryon number) | *S* (strangeness) |
|---|---|---|---|---|---|
| K-minus | $K^-$ | 0.4937 | −1 | 0 | −1 |
| K-plus | $K^+$ | 0.4937 | 1 | 0 | 1 |
| K-zero | $K^0$ | 0.4977 | 0 | 0 | 1 |
| K-zero bar | $\bar{K}^0$ | 0.4977 | 0 | 0 | −1 |
| lambda | $\Lambda^0$ | 1.116 | 0 | 1 | −1 |
| neutron | n | 0.9396 | 0 | 1 | 0 |
| phi | $\Phi$ | 1.020 | 0 | 0 | 0 |
| pi-minus | $\pi^-$ | 0.1396 | −1 | 0 | 0 |
| pi-plus | $\pi^+$ | 0.1396 | 1 | 0 | 0 |
| pi-zero | $\pi^0$ | 0.1350 | 0 | 0 | 0 |
| proton | p | 0.9383 | 1 | 1 | 0 |
| sigma-minus | $\Sigma^-$ | 1.197 | −1 | 1 | −1 |
| sigma-plus | $\Sigma^+$ | 1.189 | 1 | 1 | −1 |
| sigma-zero | $\Sigma^0$ | 1.192 | 0 | 1 | −1 |
| xi-minus | $\Xi^-$ | 1.315 | −1 | 1 | −2 |
| xi-zero | $\Xi^0$ | 1.321 | 0 | 1 | −2 |

***Table 2*** *Hadron data*

**KEY IDEAS**

- Leptons and hadrons are two groups of subatomic particles.
- Leptons are fundamental particles. They are not affected by the strong force.
- Hadrons are particles that can interact via the strong nuclear force. They can be classified as baryons or mesons. Hadrons are not fundamental. All of them, except (probably) the proton, are unstable.

# 4.2 HADRON INTERACTIONS AND CONSERVATION LAWS

A large number of hadron interactions and decays have now been studied. Most of the results are from particle accelerators where protons, or other charged particles, are made to collide with a target material at high energies. In the 1950s and 1960s, bubble chambers were used to detect the resulting particles (Figure 4). These are similar to the earlier cloud chambers (see Figure 7 of Chapter 3) but use superheated liquid rather than gas. The complex patterns were analysed and the particles identified. Spark chambers have now replaced bubble chambers as particle detectors.

It soon became apparent that some reactions could, and did, happen, whereas other reactions never actually took place. It seems as if some reactions are forbidden.

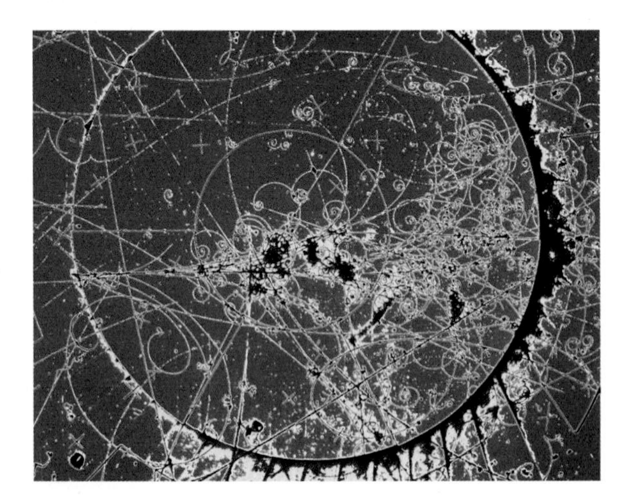

***Figure 4*** *Bubble chamber image of particle tracks. Magnetic fields lead to spiral tracks, from which the particle's mass and charge may be deduced.*

## Mass–energy and momentum conservation

Every interaction between particles has to obey the law of conservation of energy. We have seen, though, that the quantification of energy for subatomic interactions has to include mass. The total **mass–energy** before an interaction or a decay has to equal the total mass–energy afterwards. In the beta-minus decay equation, the mass of the neutron on the left-hand side of the equation is greater than the combined mass of the particles on the right-hand side of the equation. The 'spare' energy is seen as kinetic energy of the decay products.

In beta-plus decay, a proton decays to a neutron, a positron and a neutrino. But a neutron has a greater mass than a proton. It seems as if mass, and therefore energy, has been created by this decay. In fact, the decay only takes place within an 'excited' nucleus, where energy from the nucleus allows the decay to take place. A free proton cannot decay in this way. Indeed, as we have discussed, protons may well be stable.

The total **momentum** of the particles before an interaction, or collision, must also be conserved (Figure 5).

## Charge conservation

Our first ideas about charge came from electrostatic phenomena, such as lightning. We now know that that there are two 'types' of electric charge, which we

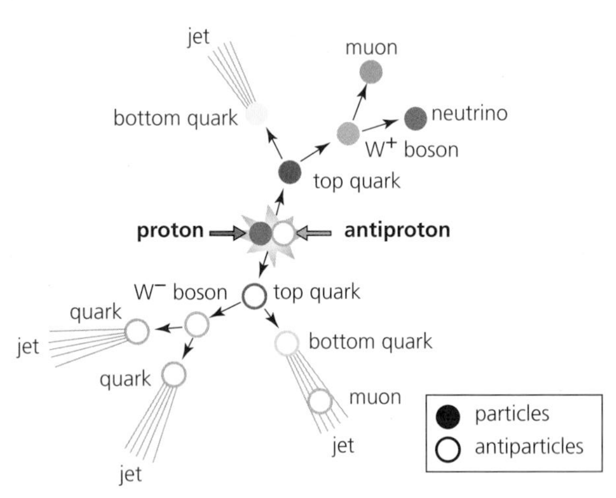


***Figure 5*** *This collision between a proton and an antiproton heralded the discovery of the top quark at Fermilab. A top–antitop pair of quarks are formed. This massive quark–antiquark pair quickly decays by the weak interaction to a bottom–antibottom quark pair. Although the W bosons are not observable, and the neutrinos are almost impossible to detect, the energy and momentum of the jets of particles can be measured. The conservation of momentum states that the total momentum of all the products after the collision must be equal to the total momentum before the collision. The mass and velocity of the undetected particles can therefore be inferred.*

call positive and negative. Positively charged bodies attract negatively charged ones, but repel other positive charges. Similarly, two negatively charged bodies repel each other, so two electrons exert a force on each other that pushes them apart.

Many hadrons, and leptons, carry an electric charge. The charge on the proton is defined to be +1. The charge carried by other particles is related to that. The charge on the electron is equal to −1, the positron charge is +1, and the neutron has a charge of zero. Other particles are denoted using a symbol, often a Greek letter such as p, Δ or Σ. A superscript is used to denote charge; for example, $\pi^+$ carries a single positive charge, $\pi^-$ carries a single negative charge and $\pi^0$ is uncharged. An antiparticle is often denoted by a horizontal line (bar) over the symbol, such as $\bar{p}$, to denote an antiproton. On some occasions, an uncharged particle is written without a superscript; for example, the lambda particle may be written Λ. In these cases you should assume that there is no charge.

One of the rules that govern the interactions between all particles is the **conservation of charge**. In all the charged particle reactions that have been studied, the total charge of the particles has always been conserved.

The total charge after an interaction is always the same as the total charge before the interaction.

Table 3 shows two examples of particle interactions, one allowed by charge conservation, the other not allowed.

### Conservation of baryon number

There are some interactions involving hadrons that are allowed by charge conservation, but have never been detected. This is because there are other conservation laws that apply in addition to charge conservation. These place extra restrictions on which reactions can take place. One of these is the conservation of **baryon number**, *B*.

## QUESTIONS

1. Which of these reactions is forbidden because it contravenes the conservation of charge? You will need to refer to Table 2, two pages back.
   a. $\Sigma^+ \rightarrow \Lambda + \pi^-$
   b. $\pi^- + p \rightarrow \Sigma^- + K^+$
   c. $\pi^- + p \rightarrow \Sigma^- + K^+ + \pi^-$

The hadrons can be divided into two groups, the **mesons** and the **baryons**. Baryons (from the Greek *barus*, or heavy) were thought to be heavier than mesons. We now know that this is not always true. Hadrons are not fundamental particles – they have internal structure, and the underlying difference between mesons and baryons is in their structure. This is discussed in section 4.3.

Particles classified as baryons are given a baryon number, *B*, of 1. Every baryon has its antiparticle with identical mass, but with opposite values of some other properties, like charge. Antibaryons have a baryon number $B = -1$. Mesons and leptons have a baryon number of zero, $B = 0$ (Table 4). Baryon number can only have a value of +1, −1 or 0. It is one of several **quantum numbers** that define a system in particle physics. Quantum numbers describe quantities that are conserved and that can only take certain discrete values. For example, charge is a quantum number because it is always conserved and it is only observed in discrete values, specifically integer multiples of the proton charge.

Interactions between hadrons can only occur if the baryon number is conserved.

The total baryon number before an interaction has to be the same as the total baryon number afterwards.

| | Before | | After | |
|---|---|---|---|---|
| Reaction 1 | p + p | → | $p + p + \pi^- + \pi^+$ | Allowed |
| *Q* | 1 + 1 = 2 | → | 1 + 1 + (−1) + 1 = 2 | |
| Reaction 2 | $p + \pi^-$ | → | $p + \pi^+$ | Not allowed |
| *Q* | 1 + (−1) = 0 | → | 1 + 1 = 2 | |

***Table 3*** *Particle interactions and conservation of charge, Q*

| Baryons | Antibaryons | Mesons |
|---|---|---|
| Baryon number $B = 1$ | Baryon number $B = -1$ | Baryon number $B = 0$ |
| Proton, p, and neutron, n | Antiproton, $\bar{p}$ | Pi mesons (pions), $\pi^+$, $\pi^-$, $\pi^0$ |
| Sigma particles, $\Sigma^0$, $\Sigma^-$, $\Sigma^+$ | Antineutron, $\bar{n}$ | K mesons (kaons), $K^+$, $K^-$, $K^0$ |

*Table 4* Hadrons: some examples of baryons, antibaryons and mesons

| | Before | | After | |
|---|---|---|---|---|
| Interaction 1 | p + p | → | p + p + n | Not allowed |
| $B$ | 1 + 1 = 2 | → | 1 + 1 + 1 = 3 | |
| Interaction 2 | p + p | → | p + p + $\pi^0$ | Allowed |
| $B$ | 1 + 1 = 2 | → | 1 + 1 + 0 = 2 | |
| Interaction 3 | n | → | p + e + $\bar{\nu}$ | Allowed |
| $B$ | 1 | → | 1 + 0 + 0 = 1 | |
| Interaction 4 | n | → | $\pi^0 + \pi^0$ | Not allowed |
| $B$ | 1 | → | 0 + 0 = 0 | |

*Table 5* Particle interactions and conservation of baryon number, B

Table 5 shows four examples of particle interactions and decays.

**QUESTIONS**

2. Are any of these reactions forbidden by the conservation of baryon number?
   a. $\Sigma^+ \rightarrow p + \pi^0$
   b. $p + p \rightarrow p + p + p + \bar{p}$
   c. $p + p \rightarrow p + p + \pi^0 + \pi^0$

## Strangeness

Both conservation of charge and conservation of baryon number have to be satisfied if a reaction is to take place. But some reactions that satisfy these two rules have never been observed. Reactions involving K mesons (also known as kaons) caused particular problems to early particle physicists. Kaons appear as the decay products of some neutral particles, but almost always in pairs, rarely singly, although there was nothing in either charge or baryon number conservation to prevent this. The kaons also have an unusually long lifetime, $10^{-10}$ s, compared to other, apparently similar, hadrons, which have typical lifetimes of the order of $10^{-23}$ s.

Murray Gell-Mann, who won the 1969 Nobel Prize in Physics for his work on classifying particles, suggested that there was yet another property, like charge and baryon number, that had to be conserved in hadron reactions. He called this property **strangeness**. All hadrons are given a strangeness number, $S$ (Table 6). Those with a non-zero value of strangeness are referred to as 'strange particles'. The strangeness quantum number can take values from −3 through to +3. Strangeness is a property that applies only to hadrons, so all leptons have zero strangeness.

| Strangeness $S = -2$ | Strangeness $S = -1$ | Strangeness $S = 0$ | Strangeness $S = +1$ |
|---|---|---|---|
| $\Xi^-$ (XI-minus) | $\Lambda^0$ (lambda) | p (proton) | $K^+$ and $K^0$ (kaons) |
| $\Xi^0$ (XI-zero) | $K^-$ (K-minus) | n (neutron) | |
| | $\Sigma^+$, $\Sigma^-$, $\Sigma^0$ (sigma particles) | $\pi^+$, $\pi^-$, $\pi^0$ (pions) | |
| | | $\Phi$ (phi meson) | |

*Table 6* Strangeness, S, for some hadrons

| | Before | | After | |
|---|---|---|---|---|
| Interaction 1 | $p + \pi^-$ | $\rightarrow$ | $K^0 + \Lambda^0$ | |
| $S$ | $0 + 0 = 0$ | $\rightarrow$ | $1 + (-1) = 0$ | Allowed |
| Interaction 2 | $p + \pi^-$ | $\rightarrow$ | $K^- + \Sigma^+$ | |
| $S$ | $0 + 0 = 0$ | $\rightarrow$ | $(-1) + (-1) = -2$ | Not allowed |

***Table 7*** *Particle interactions and conservation of strangeness, S*

In 'production' reactions, where two or more particles combine, strangeness is always conserved.

Table 7 shows two examples of such interactions, one allowed by strangeness conservation, the other not allowed.

However, when strange particles decay via the weak interaction, their strangeness can change by $\pm 1$. This helps to explain the relatively long decay times for some mesons.

For example, $\Lambda^0 \rightarrow p + \pi^-$ is a decay that is allowed, since charge and baryon number are conserved ($\Lambda^0$ is a baryon, $B = 1$). Similar decays that take place by the strong interaction lead to typical particle lifetimes of the order of $10^{-23}$ s. But the $\Lambda^0$ has a lifetime of around $10^{-10}$ s. This is because $\Lambda^0$ is a strange particle, and the decay products, p and $\pi$, are not. The decay must therefore take place via the weak interaction, which leads to a much longer lifetime.

## QUESTIONS

**3.** Which of these interactions is forbidden because it contravenes the conservation of strangeness?

**a.** $p + \pi^- \rightarrow K^- + \Sigma^+$

**b.** $\pi^- + p \rightarrow K^+ + \Sigma^-$

**c.** $K^+ + p \rightarrow \pi^+ + \Sigma^+$

**d.** $\Phi \rightarrow K^+ + K^-$

**e.** $\Phi \rightarrow \pi^+ + K^-$

**4.** The positive kaon decays to two pions with a mean lifetime of about $1.28 \times 10^{-8}$ s. The decay is written $K^+ \rightarrow \pi^+ + \pi^0$. A similar decay, $\rho^0 \rightarrow \pi^+ + \pi^-$, has a mean lifetime of $4 \times 10^{-24}$ s. What difference in the decay mechanism might explain the huge difference in lifetime? (Hint: the $\rho^0$ is a meson with zero strangeness.)

**5.** For a reaction to take place it must satisfy conservation of charge, baryon number and strangeness. Which of these reactions are forbidden? Give the reason for each of your answers. (You will need to refer to Table 2.)

**a.** $\pi^+ + n \rightarrow \pi^+ + \Lambda$

**b.** $n \rightarrow \pi^+ + e^-$

**c.** $K^+ + K^- \rightarrow \pi^0$

**d.** $\pi^+ + p \rightarrow \Sigma^+ + K^+$

**e.** $p + \pi^- \rightarrow K^+ + K^-$

**f.** $p + \pi^- \rightarrow K^0$

## KEY IDEAS

- Hadrons each have a fixed value for charge, *Q*, baryon number, *B*, and strangeness, *S*. Strong interactions involving hadrons can only take place if the interaction conserves charge, *Q*, baryon number, *B*, and strangeness, *S*.
- All interactions must also conserve mass–energy and momentum.
- Decays of strange particles that take place by the weak interaction may result in a change of strangeness of $\pm 1$.

# 4.3 LEPTON CONSERVATION

Particle interactions obey a number of other conservation laws. One of these applies to lepton number, *L*. In any particle interaction, the number of leptons is always conserved. So the total number of leptons after any interaction must be the same as the number of leptons before the reaction. In fact, the conservation law is even more constraining. Each generation of the lepton family, for example the electron and the muon, have their own **lepton number**, $L_e$ and $L_\mu$ (see Table 8), which must be conserved separately. Notice that the quantum numbers in Table 8 for an antiparticle have the opposite sign to those of the corresponding particle.

| Particle | | Rest mass (MeV/$c^2$) | Charge | $L_e$ | $L_\mu$ | Antiparticle | Charge | $L_e$ | $L_\mu$ |
|---|---|---|---|---|---|---|---|---|---|
| Electron | $e^-$ | 0.511 | −1 | 1 | 0 | $e^+$ | 1 | −1 | 0 |
| Electron neutrino | $\nu_e$ | Small, non-zero | 0 | 1 | 0 | $\bar{\nu}_e$ | 0 | −1 | 0 |
| Muon* | $\mu^-$ | 105.7 | −1 | 0 | 1 | $\mu^+$ | 1 | 0 | −1 |
| Muon neutrino | $\nu_\mu$ | Small, non-zero | 0 | 0 | 1 | $\bar{\nu}_\mu$ | 0 | 0 | −1 |
| Tau* | $\tau^-$ | 1777 | −1 | 0 | 0 | $\tau^+$ | 1 | 0 | 0 |
| Tau neutrino | $\nu_\tau$ | Small, non-zero | 0 | 0 | 0 | $\bar{\nu}_\tau$ | 0 | 0 | 0 |

*These are unstable, with a mean lifetime of 2.2 μs (muon) and $2.9 \times 10^{-13}$ s (tau).

***Table 8*** *Summary of lepton properties*

Consider the lepton numbers in the decay of a muon:

$$\mu^- \rightarrow e^- + \nu_\mu + \bar{\nu}_e$$

Before the reaction: $L_e = 0$

$L_\mu = 1$

After the reaction: $L_e = 1 + 0 + (-1) = 0$

$L_\mu = 1$

So lepton numbers are conserved.

### QUESTIONS

6. Which of these reactions and decays are allowed and which are forbidden?

   **a.** $\bar{\nu}_\mu + p^+ \rightarrow e^+ + n$

   **b.** $\bar{\nu}_\mu + p^+ \rightarrow \mu^+ + n$

   **c.** $\mu^- \rightarrow e^- + \bar{\nu}_e + \nu_\mu$

7. With a partner, construct a quiz of at least five questions. Each question should be a particle collision or decay similar to the ones in questions 5 or 6. Make sure that some of them are possible, and others are forbidden (you should know which conservation law would be contravened). Challenge another pair to see who can get the most questions right.

### KEY IDEAS

- In particle interactions there must be conservation of lepton number for electron leptons and for muon leptons (and also for tau leptons).

## 4.4 THE QUARK MODEL

The number and properties of all the different hadrons became rather confusing. The crucial questions for physics in the 1960s were: Are all these different particles actually fundamental? Or is there a simpler, underlying structure that would explain the composition and properties of these particles? At the turn of the 20th century, when scientists were faced with almost a hundred different types of atom, Rutherford's scattering experiments had helped to show that there was an underlying structure that explained the variety of atoms. In the 1960s it was another scattering experiment, this time carried out at Stanford in the USA, that revealed the structure inside hadrons (Figure 6).

The Stanford Linear Accelerator Center (SLAC) in California accelerated electrons to an energy of around 20 GeV, high enough to probe inside protons and neutrons. The SLAC experiment found that a significant proportion of the high-energy electrons were scattered through a large angle. These results indicated that the neutrons and protons are not particles of uniform density, but have point-like charged particles within them.

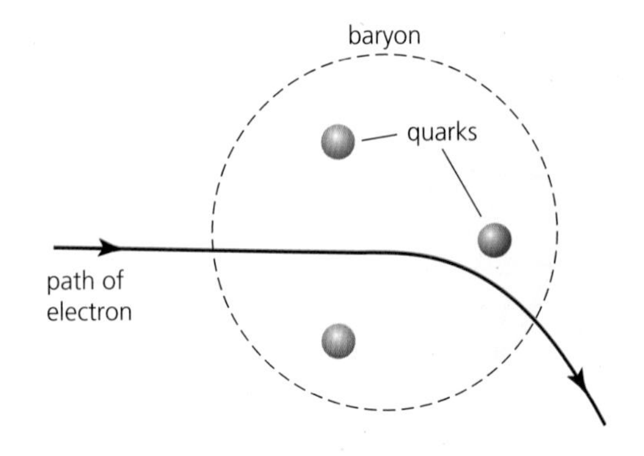


***Figure 6*** *Electron scattering by quarks within a baryon*

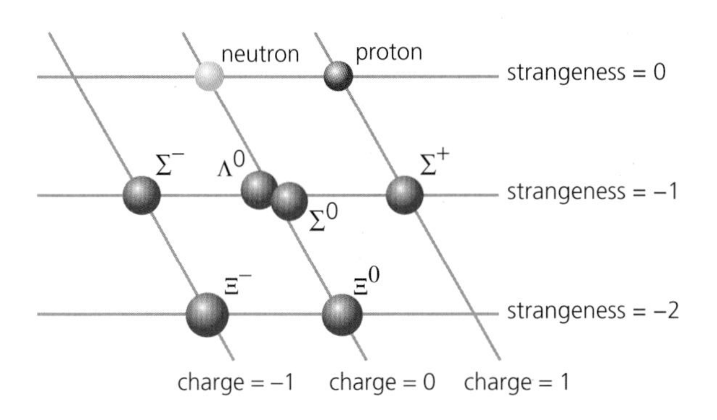


***Figure 7*** *The 'eightfold way' for a family of baryons*

The SLAC results confirmed the theory put forward by Murray Gell-Mann and George Zweig a few years earlier. Gell-Mann had grouped the hadrons together in families (Figure 7), rather as Mendeleev had grouped the elements together in the periodic table about 100 years before. The patterns that he and others created could be explained by supposing that all hadrons were composed of smaller constituents, which Gell-Mann called **quarks**. (He took the term from "Three quarks for Muster Mark", a quotation from *Finnegans Wake* by James Joyce.) The SLAC experiments confirmed that hadrons are not fundamental particles, but are composed of combinations of different types of quark.

Gell-Mann initially suggested that there were three different quarks, which he labelled up, u, down, d, and strange, s. These are sometimes referred to as the 'flavour' of the quark. Each of these quarks has a different mass and specific values for charge, baryon number and strangeness (Table 9). Each quark has a corresponding antiquark of exactly equal mass but opposite values for charge, baryon number and strangeness.

Note that the strange quark, s, is defined to have a strangeness of −1. Its antiparticle, the anti-strange quark, $\bar{s}$, has a strangeness of +1.

## Hadron structure

Using the quark model, it is possible to describe all hadrons in terms of combinations of quarks and antiquarks.

- Baryons are always combinations of three quarks.
- Antibaryons are combinations of three antiquarks.
- Mesons and antimesons are always composed of a quark and an antiquark.

A proton, for example, is composed of two up quarks and a down quark (see Table 10 and Figure 8).

The properties of each hadron can now be explained in terms of the quarks from which it is made. The total charge of the hadron is the sum of the quark charges. The same can be said for the total baryon number and total strangeness. The quark model is surprisingly simple. Only two types of quark, the up and the down, are needed to account for the properties of neutrons and protons, which together make up almost all of everyday, observable matter. The two antiquarks, $\bar{u}$ and $\bar{d}$, are needed to explain the existence of the antiproton and the antineutron.

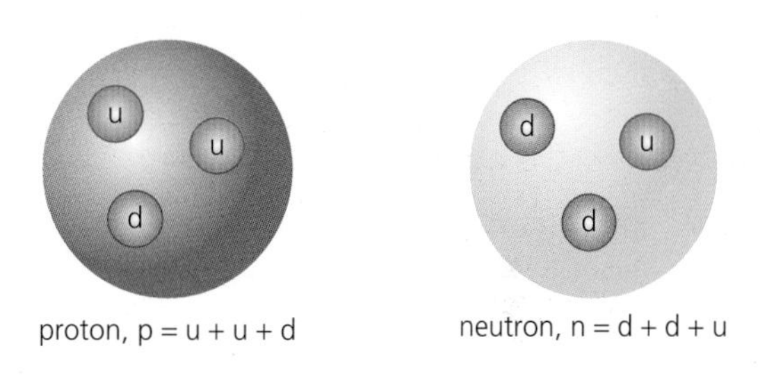


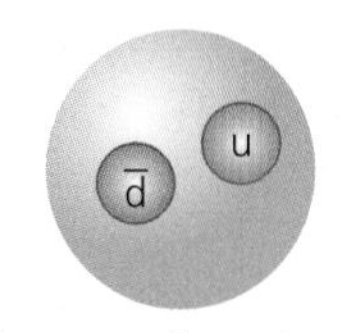


***Figure 8*** *Proton, neutron and pi meson (pion) showing quark structure*

| Quark | Baryon number *B* | Charge *Q* | Strangeness *S* | Antiquark | Baryon number *B* | Charge *Q* | Strangeness *S* | Mass (GeV/$c^2$) |
|---|---|---|---|---|---|---|---|---|
| up, u | 1/3 | 2/3 | 0 | $\bar{u}$ | −1/3 | −2/3 | 0 | 0.005 |
| down, d | 1/3 | −1/3 | 0 | $\bar{d}$ | −1/3 | 1/3 | 0 | 0.01 |
| strange, s | 1/3 | −1/3 | −1 | $\bar{s}$ | −1/3 | 1/3 | +1 | 0.2 |

***Table 9*** *Properties of quarks and antiquarks*

| Proton: | $p = u + u + d$ | | | |
|---|---|---|---|---|
| | **Up, u** | **Up, u** | **Down, d** | **Proton, p** |
| Charge, $Q$ | 2/3 | 2/3 | −1/3 | 1 |
| Baryon number, $B$ | 1/3 | 1/3 | 1/3 | 1 |
| Strangeness, $S$ | 0 | 0 | 0 | 0 |
| **Antiproton:** | $\bar{p} = \bar{u} + \bar{u} + \bar{d}$ | | | |
| | **Anti-up, $\bar{u}$** | **Anti-up, $\bar{u}$** | **Anti-down, $\bar{d}$** | **Antiproton, $\bar{p}$** |
| Charge, $Q$ | −2/3 | −2/3 | 1/3 | −1 |
| Baryon number, $B$ | −1/3 | −1/3 | −1/3 | −1 |
| Strangeness, $S$ | 0 | 0 | 0 | 0 |

***Table 10*** *Quark structure of the proton and the antiproton*

## QUESTIONS

**8.** A neutron is composed of two down quarks and an up quark, $n = u + d + d$. An antineutron is an anti-up and two anti-down quarks, $\bar{n} = \bar{u} + \bar{d} + \bar{d}$. Construct tables (as in Table 10) for the neutron and the antineutron that show how their properties are the sum of the quark properties.

Mesons and antimesons are always quark–antiquark pairs. A meson's antiparticle is the opposite combination of quark–antiquark. For example, the pi-plus meson is formed from the two quarks, $u\bar{d}$, while its antiparticle, the pi-minus, is formed from the opposite combination, $\bar{u}d$ (Table 11). The properties of a meson are the sum of the properties of its constituent quarks (Table 12).

| Meson | Quark composition |
|---|---|
| $\pi^+$ | $u\bar{d}$ |
| $\pi^-$ | $\bar{u}d$ |
| $\pi^0$ | $u\bar{u}$ or $d\bar{d}$ |
| $K^+$ | $u\bar{s}$ |
| $K^-$ | $\bar{u}s$ |
| $K^0$ | $d\bar{s}$ |
| $\bar{K}^0$ | $\bar{d}s$ |

***Table 11*** *Quark composition for pions and kaons*

| | Up, u | Anti-strange, $\bar{s}$ | $K^+$ |
|---|---|---|---|
| Charge, $Q$ | 2/3 | 1/3 | 1 |
| Baryon number, $B$ | 1/3 | −1/3 | 0 |
| Strangeness, $S$ | 0 | 1 | 1 |

***Table 12*** *Quark structure of a $K^+$ meson ($u\bar{s}$)*

## QUESTIONS

**9.** Why is it that mesons can only have strangeness $\pm 1$ or 0, whereas a baryon may have strangeness of $\pm 3$, $\pm 2$, $\pm 1$ or 0?

### Neutron decay

The quark model has been extraordinarily successful in describing and predicting the properties of hadrons. It can be used to understand what is happening in nuclear reactions or decays, for example during a beta particle emission. In beta-minus emission a neutron decays to a proton (Figure 9), emitting an electron and an antineutrino in the process. The quark theory

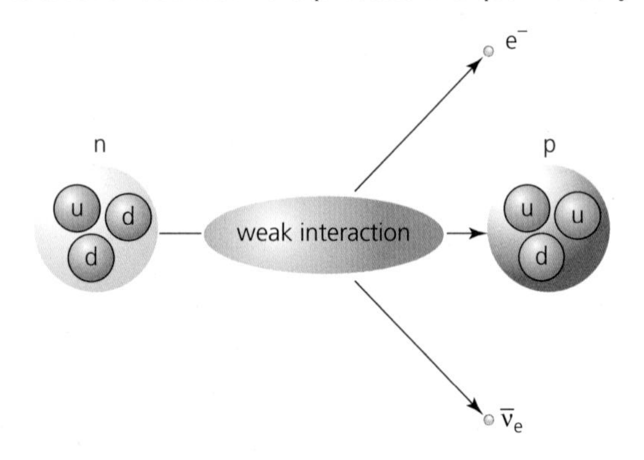


***Figure 9*** *Decay of a neutron into a proton*

tells us that inside the neutron a down quark has changed into an up quark, emitting an electron and an antineutrino.

## Can we observe a quark?

A solitary quark has never been seen. The masses and properties of the quarks have to be inferred from those of the hadrons that they make up. The reason why an individual quark has never been observed is the enormous amount of energy that would be needed to rip it from within a hadron.

Relatively speaking, splitting the atom is easy: it takes only a few electronvolts to tear an electron away and ionise the atom. Breaking up a nucleus requires more energy, in the region of millions of electronvolts. To separate a single quark from a proton would require around $10^{13}$ eV. This amount of energy is more than enough to create a new quark–antiquark pair. Rather than tear the quarks apart, new quarks are created to replace the one you were trying to remove. All that this would achieve would be the creation of a new meson.

It is possible that quarks may be unconfined at enormously high temperatures, around $10^{12}$ K. The Universe was at this temperature a few microseconds after the Big Bang. At these temperatures, neutrons and protons 'melt' into quark matter. It has also been suggested that quark matter may exist in the very high density at the centre of a neutron star. The closest that we can get to these extreme conditions occurs very briefly during the high-energy particle collisions in particle accelerators (Figure 10).

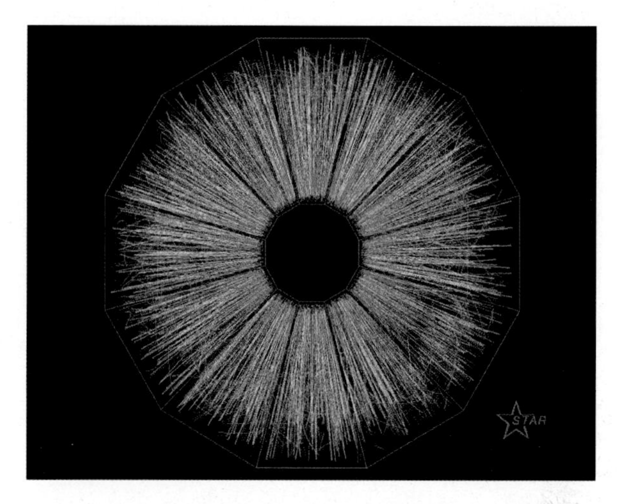


***Figure 10*** *The spray of particles produced at the Brookhaven National Laboratory when two gold nuclei collided at high speed. Quark matter has been created in these collisions, but it lasts for less than* $10^{-25}$ *s.*

## ASSIGNMENT 1: EXTENDING THE THREE-QUARK MODEL

### Stretch and challenge

This is a comprehension exercise. The content here goes beyond what you need to know for A-level, but it should help you to improve your comprehension skills and will go some way towards completing your understanding of the standard model. Read it carefully and then answer the questions that follow.

### The six-quark model

The three-quark model was initially very successful, but several arguments suggested that there should be a fourth quark. One of these arguments was quark–lepton symmetry. In the early 1970s four types of lepton had been discovered (eight counting the antiparticles). These four leptons could be grouped as follows:

| **Leptons** | $e^-$ | $\mu$ | $Q = -1$ |
|---|---|---|---|
| | $\nu_e$ | $\nu_\mu$ | $Q = 0$ |

Only three quarks had been identified. These were grouped as follows:

| **Quarks** | u | ??? | $Q = +2/3$ |
|---|---|---|---|
| | d | s | $Q = -1/3$ |

Symmetry, often a guiding principle in developing physics theory, suggested that there should be another quark with charge +2/3. The quark was expected to have a larger mass than the other three and to carry zero strangeness. It was expected to have a new conserved property of its own. This property became known as 'charm'. In November 1974 two independent teams in the USA discovered a new massive meson, which became known as the J/ψ particle. It had zero strangeness and zero charm, and was eventually identified as being composed of a charm quark and an anticharm quark pair, $c\bar{c}$. Because the particle had no overall charm, it was said to have 'hidden charm'! Later, particles containing a single charm quark together with other quarks were discovered; these particles were said to have 'naked charm'!

With the discovery of the tau lepton (see section 3.6 of Chapter 3) and its neutrino, there were now three families of leptons. To preserve the symmetry, there should be two 'new' quarks, both considerably heavier than those previously discovered. The two new quarks were given the names 'top' and 'bottom'. They were expected to have zero strangeness and zero charm, but they should each have a new conserved property of their own, namely topness and bottomness.

In 1977 a new heavy meson was discovered at Fermilab in Chicago. This new particle, the Y meson, had all the right characteristics to be a bottom–antibottom pair, $b\bar{b}$. In 1995 Fermilab completed the set: a hadron containing the predicted top quark was confirmed, from the pattern of its decay products. The top quark took such a long time to find because it was much more massive than anticipated, more than 40 times heavier than the bottom quark and 35 000 times heavier than the up and down quarks that make up ordinary nuclear matter.

There was now symmetry between the leptons and the quarks in that there are three families of each:

**Leptons** $e, \nu_e$ $\mu, \nu_\mu$ $\tau, \nu_\tau$

**Quarks** u, d s, c t, b

This symmetry is not just satisfying, it is an essential part of quantum chromodynamics (QCD), the theory that explains quark behaviour. The number of pairs of leptons, one with charge −1 and the other with charge 0, must be matched by the number of pairs of quarks, one with charge +2/3, the other with charge −1/3.

| Leptons | | Quarks | |
|---|---|---|---|
| Electron | | Up | |
| | Muon neutrino | | Strange |
| | | | |

*Table A1*

## Questions

**A1** What is the difference between a lepton and a hadron?

**A2** What is the difference between the three lepton families?

**A3** There are three families of leptons and three families of quarks. Copy and complete Table A1 to show this.

**A4** What connects the particles in each column of Table A1? What changes as you go down each column?

**A5** The Y meson had hidden bottom–explain! Suggest the quark structure of a meson with naked bottom. What other properties would your meson have?

**A6** Is it possible to have a particle with hidden charge, or hidden strangeness?

**A7** The properties of the quarks are shown in Table A2. Construct a similar table for leptons.

**A8** At the end of most sections in this book there are some 'Key Ideas'. These are short bullet points designed to remind you of the main points of that section. Write the 'Key Ideas' for the information in this assignment.

| Quark | Charge $Q$ | Baryon number $B$ | Strangeness $S$ | Charm $C$ | Bottomness $B$ | Topness $T$ |
|---|---|---|---|---|---|---|
| Up, u | +2/3 | 1/3 | 0 | 0 | 0 | 0 |
| Down, d | −1/3 | 1/3 | 0 | 0 | 0 | 0 |
| Charm, c | +2/3 | 1/3 | 0 | 1 | 0 | 0 |
| Strange, s | −1/3 | 1/3 | −1 | 0 | 0 | 0 |
| Top, t | +2/3 | 1/3 | 0 | 0 | 0 | 1 |
| Bottom, b | −1/3 | 1/3 | 0 | 0 | −1 | 0 |

***Table A2*** *Properties of the six quarks*

**KEY IDEAS**

- Hadrons are composed of particles called quarks.
- Quarks have antiparticles, known as antiquarks. Antiquarks have the same mass but opposite values of charge, baryon number and strangeness.
- Baryons are composed of three quarks or three antiquarks.
- Mesons are composed of a quark and an antiquark.
- Three types (or flavours) of quark, up, down and strange, were sufficient to describe all the hadrons that were known up to 1970.

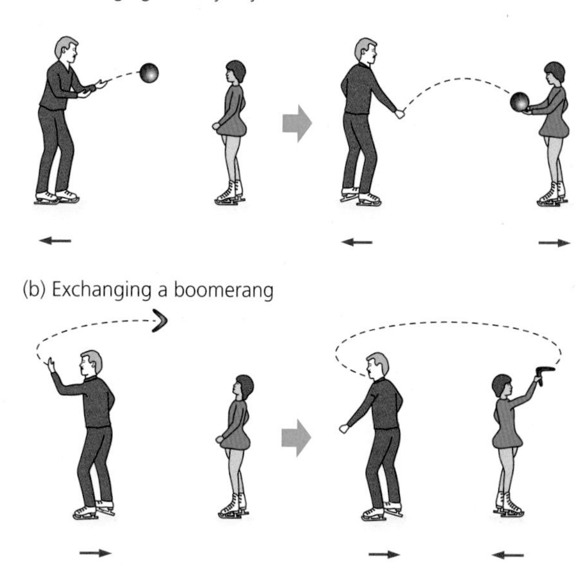


***Figure 11*** *Analogy of particle interactions*

# 4.5 EXCHANGE PARTICLES

The theory of interactions between particles proposed a third group of fundamental particles. On the subatomic scale **exchange particles** are responsible for transmitting the fundamental forces that hold nucleons, nuclei and atoms together. On a larger scale they are responsible for the everyday forces that affect our lives.

## Exchange particles and Feynman diagrams

When two bodies exert a force on each other, perhaps the floor exerting a force on your shoes, what is it that happens between the two bodies? What happens to make the two objects repel each other? The Japanese physicist Hideki Yukawa suggested that when two particles, A and B, exert a force on each other, a **virtual particle** is created. This virtual particle can travel between particles A and B and affect their motion. The virtual particle, which may exist for only a very short time, is referred to as an exchange particle and it is the mediator of the force.

The idea of a force being carried by an exchange particle can be pictured by considering two people on ice skates (Figure 11). If one person throws a ball to the other, both skaters' motion will be affected. In fact, they will be pushed away from each other. We have to stretch the analogy a bit to include an attractive force, but if you imagine a boomerang being thrown, rather than a ball, then the two skaters will be drawn together. The analogy of a ball, or boomerang, being exchanged is not to be taken too literally. In some interactions, it is not, or not just, momentum that is being exchanged, but charge or other properties. Indeed, in some interactions, involving the weak force, this property is being changed.

The exchange particles that are transferred between fundamental particles are known as **gauge bosons** and each fundamental force has its own boson or set of bosons.

## The electromagnetic force

The electromagnetic force is carried between charged particles by the **photon**, $\gamma$. When two charged particles, say two electrons, exert a force on each other, a virtual photon is exchanged between them. We can use simple diagrams to represent what happens (Figure 12). These are called 'Feynman diagrams' after Richard Feynman (Chapter 2), who devised them. The photon is a massless, chargeless particle. In fact it is its own antiparticle: a photon is identical to an antiphoton.

## The strong interaction

Hideki Yukawa was working on the strong nuclear interaction when he first proposed the idea of exchange particles in 1935. He suggested that these exchange particles could be travelling at close to the speed of light across the nucleus. An exchange particle moving at close to the speed of light has to exist for about $10^{-23}$ s if it is to have time to travel across the nucleus. This enabled Yukawa to predict a maximum

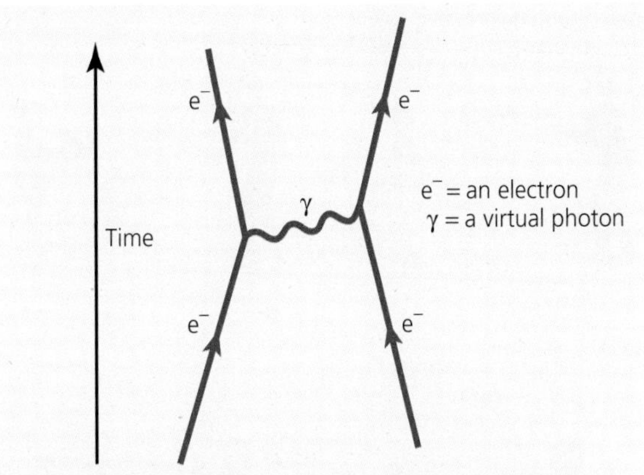


Feynman diagrams represent the events in an interaction. In the diagrams in this book the sequence of time goes up the page. Note that Feynman diagrams tell us nothing about the actual paths of the two particles. The angles in the diagrams have no significance.

***Figure 12*** *Diagram representing two electrons feeling the electric force as a result of a photon being exchanged between them*

mass for the particle, which became known as the pi meson or pion (Figure 13). The pion acts between nucleons to hold the nucleus together (Figure 14).

We now know that, at a deeper level, the strong interaction is mediated by gauge bosons called **gluons** that pass between quarks.

***Figure 13*** *Yukawa's predicted pion was discovered in 1947 in cosmic ray experiments. This cosmic ray collision produces a spray of particles, including 16 pions, shown in yellow.*

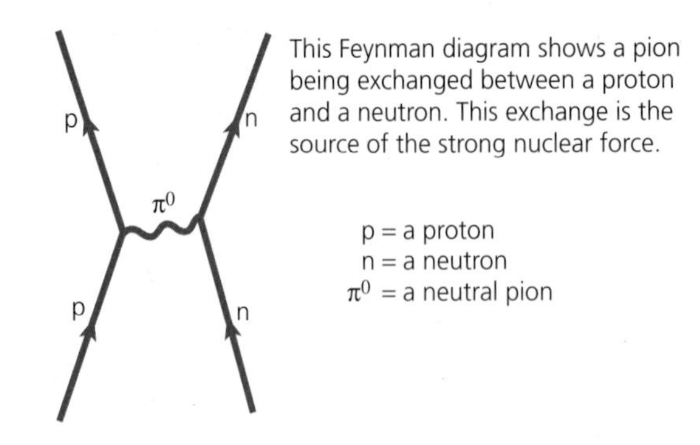


This Feynman diagram shows a pion being exchanged between a proton and a neutron. This exchange is the source of the strong nuclear force.

p = a proton
n = a neutron
$\pi^0$ = a neutral pion

***Figure 14*** *Diagram representing pion exchange*

## The weak interaction

The weak interaction has a very short range. Its gauge bosons are relatively massive, and since a large mass (and therefore energy) means a short lifetime, the exchange particles can only travel a short distance. The weak interaction has three gauge bosons, also known as intermediate vector bosons, $W^+$, $W^-$ and $Z^0$. These bosons were eventually discovered in 1983 at CERN. The weak interaction acts on both leptons and hadrons. It is the only force, other than gravity, that acts on neutrinos. This explains the fact that neutrinos so rarely interact with anything.

The weak interaction can alter the flavour of a quark. Since W bosons carry charge, the weak interaction may lead, for example, to an up quark being changed into a down quark. Look back at Figure 9, which shows the decay of a neutron into a proton (this is what happens in beta-minus decay). Figure 15 shows the Feynman diagram for this interaction, mediated by a $W^-$ boson.

Figure 16 shows the detail of the changes at a fundamental level.

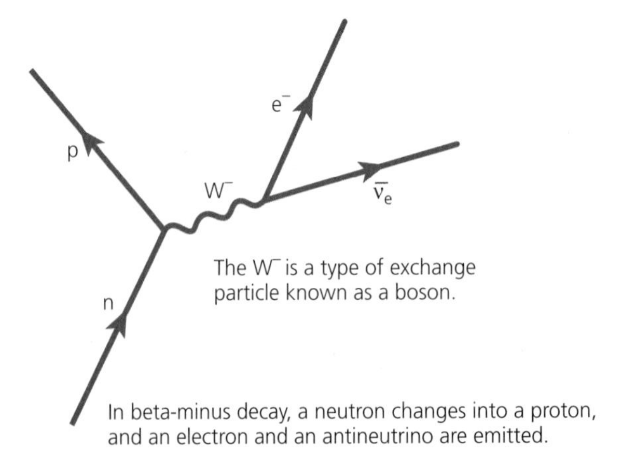


***Figure 15*** *Diagram representing the exchange involved in beta-minus decay*

The weak interaction is also responsible for beta-plus (positron) emission. For example, in the decay of magnesium-23,

$$^{23}_{12}\text{Mg} \rightarrow {}^{23}_{11}\text{Na} + e^+ + \nu_e$$

a proton decays into a neutron. One of the down quarks in one of the protons in the magnesium nucleus has changed to an up quark.

Some other weak interactions are shown in Figure 17. Because the weak interaction can change the flavour of a quark, it may change the strangeness of a hadron, with the result that strangeness is not conserved in weak interactions.

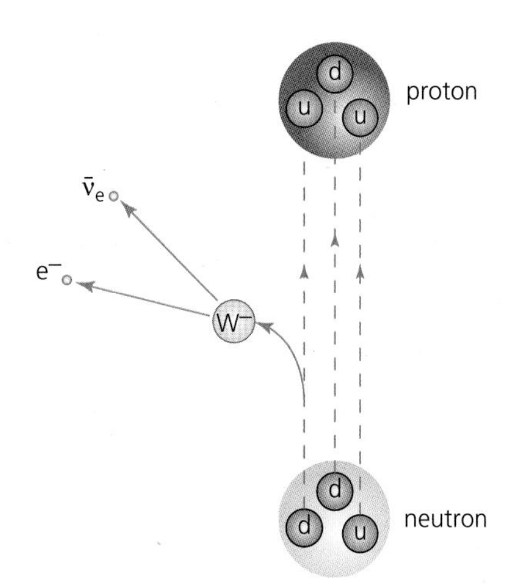


The exchange of a $W^-$ boson changes a neutron into a proton by changing the 'flavour' of a quark. A down quark is transformed into an up quark. The $W^-$ then decays, via the weak force, into an electron and an antineutrino.

***Figure 16*** *Changes in quarks due to the weak interaction in beta-minus decay*

## Gravity

The gauge boson that carries the gravitational force has been named the **graviton**. It is predicted to have zero rest mass and zero charge, but it has never been detected. Because the graviton is massless, it has infinite range. Gravity acts between all particles, but it is the weakest of all the four forces.

Table 13 summarises the forces and exchange particles.

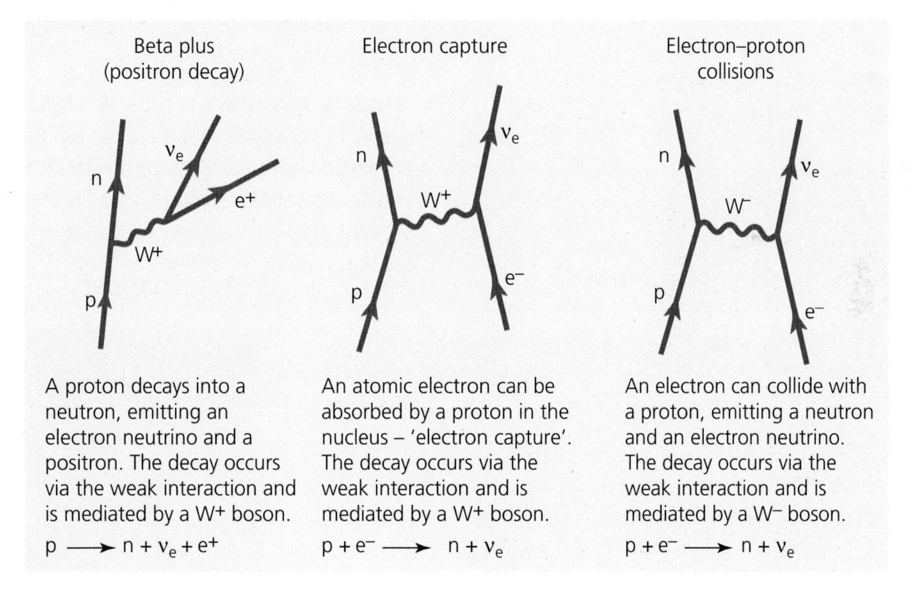


***Figure 17*** *Diagrams representing other weak interactions*

| Force | | Strong | Weak | Electromagnetic | Gravity |
|---|---|---|---|---|---|
| **Exchange particle** | | Gluon | $W^+$, $W^-$ and $Z^0$ bosons | Photon | Graviton |
| **Leptons** | Electrons, muons, taus | ✗ | ✓ | ✓ | ✓ |
| | Neutrinos (e, μ, τ) | ✗ | ✓ | ✗ | ✓ |
| **Hadrons** | Mesons | ✓ | ✓ | ✓ (if charged) | ✓ |
| | Baryons | ✓ | ✓ | ✓ (if charged) | ✓ |

***Table 13*** *Particles and the forces between them*

## QUESTIONS

10. Positron emission happens when a proton is transformed into a neutron. A positron and a neutrino are emitted. Draw a diagram (similar to Figure 16) to show what happens to the quarks.
11. What are the differences and the similarities between gravity and the electromagnetic interaction?
12. Why do neutrinos interact so weakly with other forms of matter?
13. Which of the fundamental forces is responsible for:
    a. binding electrons into atoms
    b. holding the nucleus together
    c. beta decay?

## KEY IDEAS

- Forces are carried by particles known as exchange particles or gauge bosons.
- The gauge bosons for electromagnetic forces and gravity are massless. These forces have infinite range.
- The gauge bosons for the strong and weak interactions have mass. These forces are short range.
- Two of the gauge bosons for the weak interaction, $W^+$ and $W^-$, carry charge. Charge conservation applies at all times in an interaction.
- The strong interaction only acts between hadrons, that is particles made up of quarks.
- The weak interaction acts on all particles.

## ASSIGNMENT 2: PARTICLE PHYSICS – BETTER TOGETHER!

Your task is to write a letter to the government to try to persuade them that it is a good idea to take part in international scientific collaborations such as that at CERN.

Read the following article on the development of knowledge about particle physics. Use this as your starting point, supported by the knowledge you have gained from studying Chapters 3 and 4, and some further research, to explain to the government the advantages of being involved in such collaborative projects, and why it is difficult to pursue such research alone.

(Useful search terms include: international scientific collaboration, CERN, particle physics research.)

Much as we'd like to believe that science can be done by a lone genius, toiling in her basement laboratory, the fact of the matter is that discovering the fundamental secrets of the universe doesn't come cheap!

In 1897, J. J. Thomson discovered the electron, the first fundamental particle, using a cathode ray tube. He designed and carried out the experiments himself. His apparatus, which cost only a few pounds, accelerated electrons to an energy of just less than 500 eV. The energy, complexity and cost of particle physics experiments has risen inexorably since then.

In 1909, Rutherford used alpha particles with energy of a few MeV to probe the atom but higher-energy particles were needed to probe the nucleus. Early in the 1930s, there were two new developments in particle accelerators, one in the USA and the other in England.

**Figure A1** *The first cyclotron accelerated protons to an energy of 1.2 MeV.*

Ernest Lawrence built the first cyclotron in 1931, at the Berkeley Lab in California, using brass, bits of wire and sealing wax (Figure A1). It was only 10 cm across, easily small enough to hold in your hand, and cost about $25. Cyclotrons were used to discover many different isotopes and played an important role in the development of the first nuclear weapons. They are still used today to create artificial radioisotopes and to treat cancer by proton therapy.

***Figure A2*** *This version of the Cockcroft and Walton accelerator was built in 1937 by Phillips in Eindhoven for £6000, and is now in the Science Museum in London.*

In the 1930s, in England, John Cockcroft and Ernest Walton designed an accelerator that split the nucleus for the first time. Their machine, engineered in Holland, could accelerate protons to about 1 MeV (Figure A2). The method they used to multiply the voltage is still used today where high electrostatic voltages are required, in photocopiers for example.

Cockcroft and Walton's machine accelerated particles through a large potential difference. A linear accelerator (linac) achieves higher particle energies by repeating that process several times. Stanford in California is home to SLAC (the Stanford Linear Accelerator Laboratory), the world's largest linac at 2 miles long. SLAC's discoveries include the charm quark, the tau lepton and the quark structure of protons and neutrons. In 2012 it had an annual budget of $350 million.

The most recent discoveries in particle physics have come from the Large Hadron Collider (LHC) at the European Organisation for Nuclear Research (CERN), Geneva. CERN was founded in 1954 with the aim of keeping Europe at the forefront of nuclear research. It has been estimated that the cost of finding the Higgs boson was just over $13 billion. The cost is shared between 21 member states as well as institutions and universities from the USA, Japan and many other parts of the world.

Today, CERN employs 10 times more engineers than research physicists, and much of the engineering is large scale and highly specialist. Complex civil engineering techniques are needed. The LHC runs in a 27 km long circular tunnel and the major experiments are housed in artificial caverns placed around the circumference. For example, the Compact Muon Scattering (CMS) experiment required two shafts to be built, diameter 12 m and 20.5 m, both of which were 100 m deep. The alignment of the main particle accelerator needs to be correct to less than a millimetre – no mean feat over a length of 27 km. Maintaining that precision demands accurate monitoring and a detailed understanding of the region's geology.

The mechanical and electrical engineering challenges are equally demanding. For example, the electromagnets that steer the particle beams have to run at a temperature of 1.9 K (−271.3 °C) and carry an electric current of almost 12 000 A. When the proton beams collide, there are millions of events (collisions and decays) per second. CERN has analysed over 800 trillion collisions while looking for the Higgs boson. The computing power needed to analyse these events is distributed worldwide at 170 computer centres in 35 countries. The data communication requirements led Tim Berners-Lee (Figure A3) to develop the World Wide Web.

***Figure A3*** *In 1989 Tim (now Sir Tim) Berners-Lee invented the World Wide Web to improve communication between scientists involved with research at CERN.*

As scientists strive to understand dark matter, dark energy and the antimatter problem, they need to reach higher and higher energies. The costs will grow – see Figure A4. The LHC has been upgraded to reach 13 TeV, but to reach energies of as much as 100 TeV a new accelerator, in a much larger 100 km circumference tunnel, has been proposed.

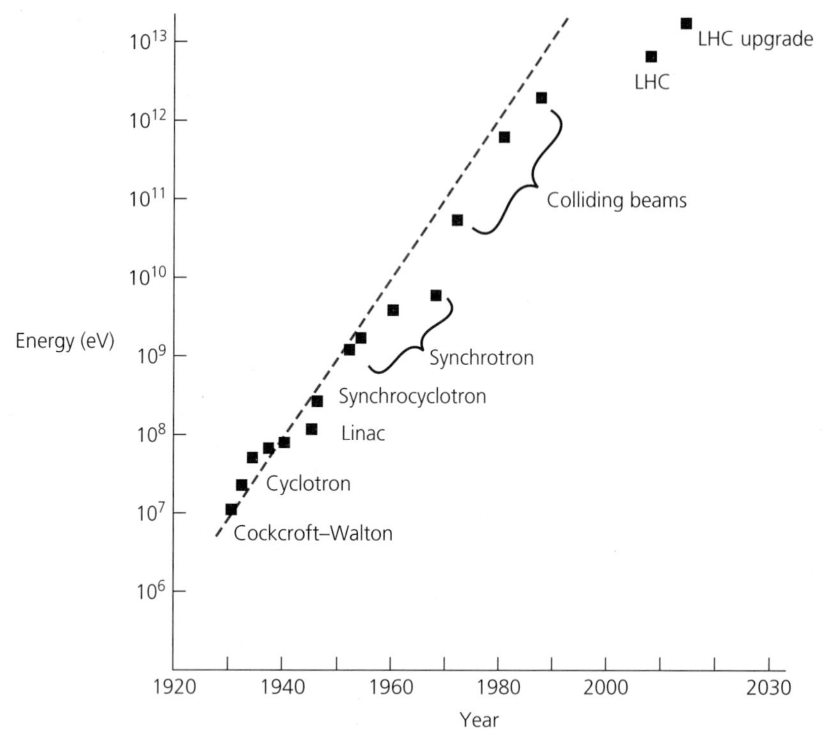


***Figure A4*** *This graph shows how the particle energy produced by accelerators has increased with time. Notice that the scale on the y-axis is logarithmic – it increases in multiples of 10, rather than in equal steps. The log–linear plot is close to a straight line, which shows exponential growth (although the most recent points on the graph, for the LHC, show that we are now falling short of that).*

## The final picture?

The list of fundamental particles now includes 24 'matter' particles (six leptons and six quarks and their antiparticles) and five 'force' particles (gluon, photon, W, Z and graviton) – see Figure 18. Among these, only the graviton has yet to make an appearance. No-one has yet been able to explain why there are three lepton and three quark families, or why our Universe prefers matter to antimatter, or exactly how the Higgs boson gives different masses to different particles, or what dark matter and dark energy really are. There are still a lot of 'big' questions that need answers.

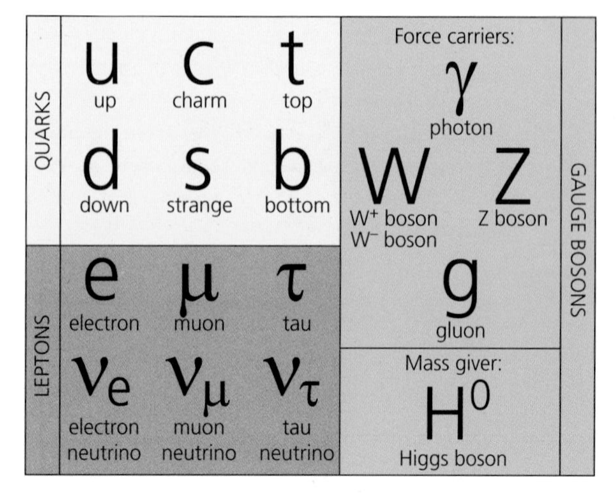


***Figure 18*** *Fundamental particles of the standard model*

## PRACTICE QUESTIONS

1. a. i. Name two baryons.

ii. State the quark structure of the pion, $\pi^+$.

b. i. The $K^+$ kaon is a strange particle. Give **one** characteristic of a strange particle that makes it different from a particle that is not strange.

ii. One of the following equations represents a possible decay of the $K^+$ kaon.

$K^+ \rightarrow \pi^+ + \pi^0$

$K^+ \rightarrow \mu^+ + \bar{\nu}_\mu$

State, with a reason, which one of these decays is not possible.

c. Another strange particle, X, decays in the following way:

$X \rightarrow \pi^- + p$

i. State what interaction is involved in this decay.

ii. Show that X must be a neutral particle.

iii. Deduce whether X is a meson, baryon or lepton, explaining how you arrive at your answer.

iv. Which particle in this interaction is the most stable?

*AQA June 2011 Unit 1 Q1*

2. a. i. State how many quarks there are in a baryon.

ii. Hadrons fall into two groups, baryons being one of them. State the name that is given to the other group of hadrons.

iii. Give **two** properties of hadrons that distinguish them from leptons.

b. The forces between particles can be explained in terms of exchange particles. Copy and complete Table Q1 by identifying an exchange particle involved in the interaction.

| Interaction | Exchange particle |
|---|---|
| electromagnetic | |
| weak | |

*Table Q1*

c. The following equation shows electron capture:

$p + e^- \rightarrow n + \nu_e$

i. Draw a Feynman diagram that represents this interaction.

ii. Explain why, when electron capture occurs, a neutrino rather than an antineutrino is produced.

*AQA Jan 2013 Unit 1 Q3*

3. Hadrons and leptons are two groups of particles. Write an account of how particles are placed into one or other of these two groups. Your account should include the following:

- how the type of interaction is used to classify the particles
- examples of each type of particle
- details of any similarities between the two groups
- details of how one group may be further subdivided.

The quality of your written communication will be assessed in your answer.

*AQA June 2013 Unit 1 Q3a*

4. Which of these is the odd one out?

**A** $W^-$ boson

**B** electron

**C** $\pi^+$

**D** proton

5. The $\Delta^{++}$ baryon is doubly charged. Use the table of quark properties (Table 9 in this chapter) to identify its correct quark composition.

   **A** ddd

   **B** $\overline{d}\overline{d}\overline{d}$

   **C** uuu

   **D** $\overline{u}\overline{u}\overline{u}$

6. Which of the following statements is **not** correct?

   **A** Baryons are always formed from three quarks.

   **B** Mesons are always formed from a quark and an antiquark.

   **C** Leptons are always lighter than baryons.

   **D** All mesons are hadrons.

7. During beta-minus decay, which of the following is correct?

   **A** An up quark is transformed to a down quark by the strong force.

   **B** An up quark is transformed to a down quark by the weak force.

   **C** A neutrino is emitted, so that charge is conserved.

   **D** An antineutrino is emitted to conserve charge.

8. In which row in Table Q2 are all the statements true?

9. In particle decays, which of the following is true?

   **A** Strangeness is never conserved in decays mediated by the strong interaction.

   **B** Strangeness is always conserved in decays mediated by the strong interaction.

   **C** Strangeness is always conserved in decays mediated by the weak interaction.

   **D** Strangeness is conserved in all decays.

10. Conservation laws are very important in physics. They reflect underlying symmetries in nature and are a clear guide as to what is possible. This applies to particle physics. Murray Gell-Mann has said: "Anything that is not expressly forbidden, will happen."

    Read excerpts A and B and then answer the following questions.

***Excerpt A***

'Conservation' (the conservation law) means this … that there is a number, which you can calculate, at one moment – and as nature undergoes its multitude of changes, this number doesn't change. That is, if you calculate again, this quantity, it'll be the same as it was before. An example is the conservation of energy: there's a quantity that you can calculate according to a certain rule, and it comes out the same answer after, no matter what happens, happens.

**— Richard P. Feynman**

*From 'The Great Conservation Principles', The Messenger Series of Lectures, No. 3, Cornell University, 1964*

| | Strong force | Weak interaction | Electromagnetic | Gravity |
|---|---|---|---|---|
| A | Acts only between hadrons | Acts only between leptons | Acts only between charged particles | Acts between masses |
| B | Holds nucleons together | Causes radioactivity | Holds molecules together | Holds galaxies together |
| C | Shortest range | Short range | Infinite range | Infinite range |
| D | Bosons are massless | Bosons are massless | Bosons are massless | Bosons are massless |

*Table Q2*

*Excerpt B*

In all of physics there are only six conservation laws. Each describes a quantity that is conserved, that is, the total amount is the same before and after something occurs. These laws have the restriction that the system is closed, that is, the system is not affected by anything outside it.

| | |
|---|---|
| Conservation of charge | Conservation of angular momentum |
| Conservation of momentum | Conservation of baryons |
| Conservation of mass–energy | Conservation of leptons |

*From 'quarknet': US Department of Energy*

**a.** In excerpt A, Feynman talks about the 'conservation of energy' whereas excerpt B lists 'conservation of mass–energy'. Are the statements contradictory? Explain your answer.

**b.** The properties of some hadrons and leptons are shown in Tables 2 and 8 in this chapter. Use these to help you to explain which of the following reactions is/are allowed and which is/are forbidden.

**i.** $\Lambda^0 \rightarrow p + \pi^-$

**ii.** $K^+ + p \rightarrow \Lambda^0 + p + p$

**iii.** $\nu_\mu + p \rightarrow \mu^+ + n$

Stretch and challenge

**c.** Write a short paragraph explaining how each of the following conservation rules is applied in particle physics. Give some examples.

**i.** Conservation of mass–energy

**ii.** Conservation of charge

**iii.** Conservation of lepton number

**d.** Conservation of strangeness is not mentioned in excerpt B. Should it be included? Explain your reasoning.

# 5 WAVES

## PRIOR KNOWLEDGE

You will remember some basic ideas about waves and their behaviour. You should recall that all waves can be reflected and refracted, and know that sound and electromagnetic radiation (including light) travel as waves.

## LEARNING OBJECTIVES

In this chapter you will build on that basic knowledge. You will learn about the differences between progressive and stationary waves, about the superposition of waves and about polarisation.

**(Specification 3.3.1.1, 3.3.1.2, 3.3.1.3)**

Ten-year-old Tilly Smith was on a beach holiday in Thailand when she saved her family, and around 100 others, from disaster. When the tide rushed out very suddenly, Tilly raised the alarm. She had heard about this 'drawback' in a geography lesson and knew a huge wave would follow. The beach was evacuated safely. It was Boxing Day, 2004. An earthquake centred below the sea near Sumatra had triggered a huge wave, known as a tsunami (Figure 1). The tsunami struck the coasts around the Indian Ocean with enormous force. Together, the earthquake and the tsunami killed in the region of 230 000 people and caused massive destruction in many countries.

The tsunami was caused when the earthquake suddenly lifted a section of the seabed by several metres, displacing around 30 km$^3$ of water. This disturbance sent a series of waves across the ocean. Out in the deep ocean the waves were only 60 cm high, though they were travelling at the speed of a jet aircraft, between 500 and 1000 km h$^{-1}$. As the tsunami reached the shallower water near the coast, it slowed to 20 km h$^{-1}$ and grew rapidly to a height of 20–30 m in some places.

It is possible that an even larger tsunami could occur in the North Atlantic. One side of the Cumbre Vieja volcano on La Palma in the Canary Islands is unstable. If a large part falls into the sea, it could cause an enormous wave. The wave would travel at around 800 km h$^{-1}$ across the Atlantic, spreading out as it went. Some of its energy would be dissipated, but it could still be around 50 m high on striking the coastline of North America.

Unfortunately, not every beach or shoreline has a Tilly Smith to sound the alarm. The first indication of a tsunami may well be the earthquake itself. But an international tsunami warning system now covers the Atlantic, Pacific and Indian Oceans. Scientists monitor seismographs, tidal gauges and tsunami detection buoys and it should be possible to alert people and also avoid false alarms.

***Figure 1*** *An approaching tsunami*

# 5.1 PROGRESSIVE WAVES

The ripples spreading across the surface of water (Figure 2) are a familiar example of wave motion. We see a repeated, regular disturbance of the surface as a wave travels across the water. Each point on the surface of the water is displaced, rising up to a crest and down into a trough. This pattern of displacement is repeated regularly as the crests and troughs pass by. Water molecules gain gravitational potential energy, which is then transferred to kinetic energy and back to potential again as each point on the surface rises, falls and rises again. Oscillating water molecules act on their neighbours so that as one point on the surface goes up towards a crest, an adjacent point also moves upwards but reaches the crest slightly later. The process is repeated across the water surface so that the crest moves across the water.

Water ripples are examples of **progressive** waves. Progressive waves transfer energy but they do not transfer any matter. There is no net movement of water. Similarly, when you shout, the vibrations of your vocal cords cause sound waves in the air, which carry the energy across the room, but the air does not go with it. If you flick the end of a rope, as in Figure 3, the wave carries the energy to the other end, but you still have the rope in your hand. In each case energy is transferred, but there is no net movement of the material (known as the **medium**) that carried the wave.

***Figure 2*** *Ripples on water*

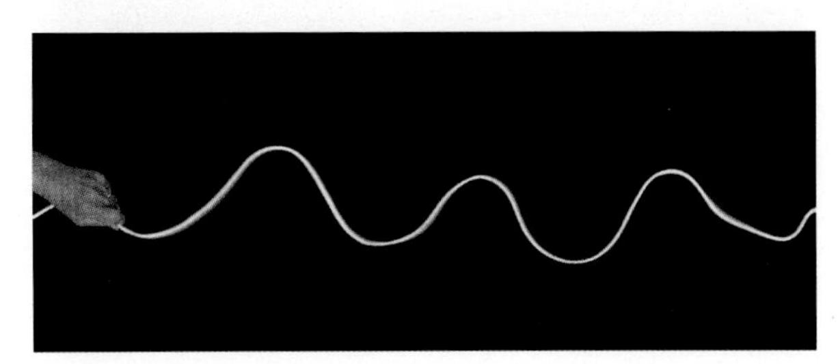

***Figure 3*** *A progressive transverse wave on a rope*

However, the individual particles of the medium do move. Each particle oscillates about its equilibrium position – that is, where it would be with no wave to disturb it. When the wave has passed, the particles return to their original positions so that the passage of a wave leaves the medium itself unchanged.

## QUESTIONS

1. Teacher: "What is a wave?"

   Student: "A wave is a mobile energy store."

   Teacher: "A good answer, but ..."

   In what respect is it a good answer? Why is the teacher not completely happy with the answer?
   Use the example of a wave on a string (Figure 4) to illustrate your suggestions.

### Transverse waves

In Figure 3 the end of a long rope is being wiggled up and down. A progressive wave travels along the rope. Each point on the rope is oscillating up and down, displaced from its equilibrium position. These oscillations are out of step with one another. In Figure 4, as the wave travels from left to right, the motion of point 12 is slightly delayed compared with point 13, the motion of point 11 is sightly delayed compared with point 12, and so on along the rope. In fact the wave is a sequence of oscillating particles, each pulling on adjacent particles, each with a displacement slightly out of step with its neighbours.

These waves are **transverse** waves. This means that the oscillations of the particles are at right angles to the direction in which the wave travels (the direction of propagation). Surface waves on water, waves on stretched strings and electromagnetic waves are all examples of transverse waves.

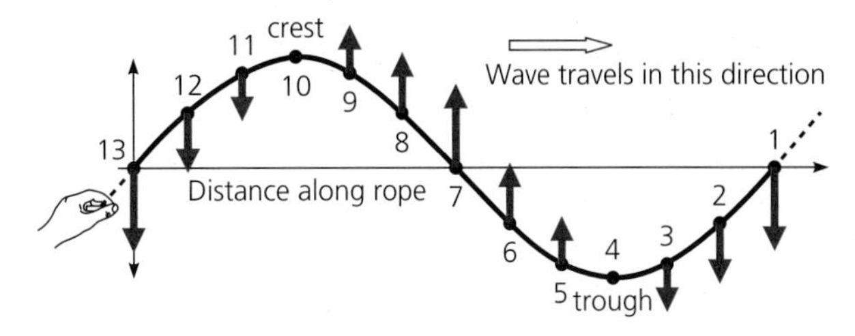


***Figure 4*** *Each particle of the rope is oscillating up and down about its equilibrium position. Each particle's oscillation is slightly out of step with that of adjacent ones – it has a slightly different displacement at any instant of time.*

## ASSIGNMENT 1: ANALYSING ENERGY TRANSFER BY WAVES

**(MS 0.3, MS 3.1, MS 3.3, MS 4.2)**

### Questions

**A1** Look at the photograph of the circular ripples in Figure 2 on the previous page. How do the ripples change as they move across the surface? What happens to the wave's energy?

Imagine you have been given a 'Ripple-o-meter' that can measure the energy of a ripple, *E*. Actually, the 'Ripple-o-meter' cannot measure all the ripples at once. It really measures the energy of part of a ripple, the energy per unit length (*E*/*L*). Suppose that you are lucky enough to have several 'Ripple-o-meters'. You place them at different distances, *r*, from the source of the ripples and take readings of *E*/*L*.

**A2** Describe what you would expect to find.

**A3** Compare the waves in Figure A1 with those in Figure 2. What differences are there? How could these ripples be formed? How could you create sound waves or light waves that are similar to the water waves here?

***Figure A1*** *Straight water ripples*

### Stretch and challenge

**A4** **a.** Explain quantitatively what you would expect to find from your experiment with circular ripples using 'Ripple-o-meters' – that is, suggest a mathematical relationship for the results.

**b.** What graph could you plot to test your relationship?

**c.** How could you use your graph to find the total energy in each ripple?

**d.** What have you assumed?

**e.** How does your answer relate to a different example of wave motion, say the light from a star? What are the similarities and differences?

**A5** **a.** How would your answer to question A4 change if the waves were straight, as in Figure A1, rather than circular ripples?

**b.** How does this relate to a real source of straight-line light or sound waves?

## QUESTIONS

2. Suppose that the wave in Figure 4 is moving from left to right at 1 $cm\,s^{-1}$. Sketch the wave on the rope and add a sketch to show how the rope would look 1 s later.
3. A Mexican wave (Figure 5) looks like a progressive wave – each person is 'oscillating' slightly later than the person before them. Why is it *not* a true progressive wave?

***Figure 5*** *Mexican wave*

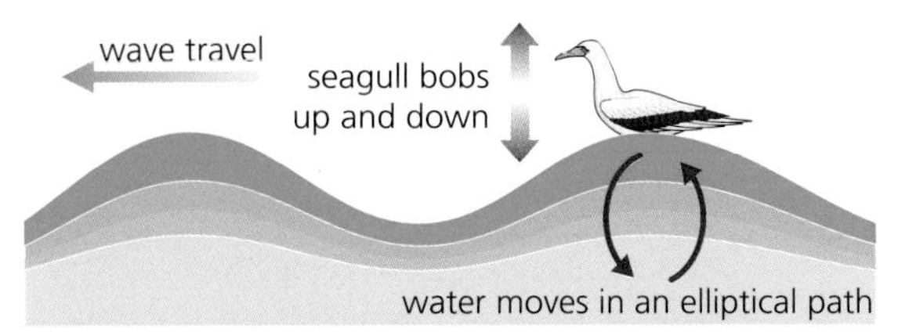


**Figure 6** *A surface water wave is not a true transverse wave. The water particles are not merely displaced up and down by the wave, they move in roughly elliptical paths: in the direction of the wave motion at a crest, and in the opposite direction in a trough.*

Transverse waves can, as we commonly see, travel across a water *surface*, but they cannot propagate through a *body* of water. (A surface water wave is not a perfect transverse wave. The water particles are not merely displaced up and down by the wave, but move in roughly elliptical paths: in the direction of the wave motion at a crest, and in the opposite direction in a trough (Figure 6).)

## Longitudinal waves

Progressive waves can also propagate through a material as a **longitudinal** wave. This is how sound waves travel through air. A vibrating object, such as a tuning fork or a loudspeaker, pushes periodically on the air near it, compressing and expanding it at regular intervals. The result is a pressure wave. Molecules in the air are forced closer together and then further apart in the direction of propagation, resulting in regions of compression and expansion (or 'rarefaction') of the air. This can be modelled using a Slinky spring (Figure 7).

A compression is a region of relatively high pressure and density where the molecules are closer together than their average positions. In a rarefaction the molecules have moved further apart than usual,

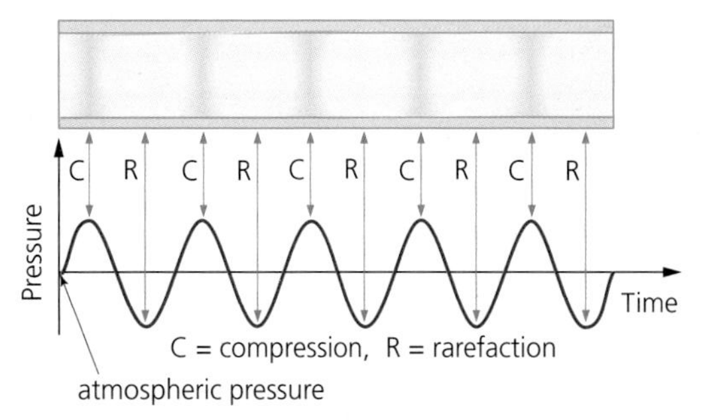


**Figure 8** *Sound is a pressure wave: the air pressure varies as the sound wave passes. The pressure is highest at a compression and lowest at a rarefaction. Even for a very loud sound these changes are small compared to atmospheric pressure, typically 20–30 Pa compared to atmospheric pressure of around 100 000 Pa.*

leading to a region of lower density and pressure. These variations can be plotted on a graph (Figure 8).

## Mechanical waves and electromagnetic waves

All the waves in Table 1 (see the next page) are mechanical waves. They transfer energy through a material (medium) by the oscillations of particles. Mechanical waves may be transverse or longitudinal.

Electromagnetic (EM) waves are *not* mechanical – they do not require a medium in which to travel. Light, infrared (IR) and ultraviolet (UV) radiation from the Sun (Figure 9) would not reach us otherwise. Since all other waves need a medium, one theory suggested that space is not empty at all, but instead is filled with a material known as the 'aether'. The aether was supposed to act as a medium to carry the oscillations that made light waves. But all efforts to detect the aether have failed. Electromagnetic waves can propagate through a vacuum, without any particle oscillation involved. So what is it that oscillates?

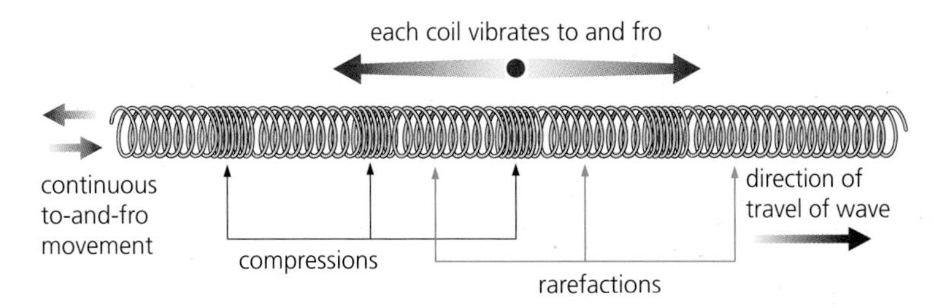


**Figure 7** *A longitudinal wave*

| Wave | Medium | Type | Cause |
|---|---|---|---|
| Sea surface wave | Water | Transverse (approximately) | Wind. Other disturbances to the water surface (from throwing in a pebble, to earthquakes) |
| Waves on strings, for example, in musical instruments | Stretched nylon or metal strings | Transverse | Plucking a string (guitar), scraping a bow across it (violin) or striking a string (piano) |
| Seismic waves | Rocks in Earth's lithosphere and mantle | Transverse and longitudinal | Earthquakes, explosions |
| Sound waves | Often air, but can be liquid or solid | Longitudinal | Vibrations of molecules in the air, caused by a vibrating object |

***Table 1*** *Some examples of mechanical waves*

The answer is that EM waves are linked electric fields and magnetic fields. Such fields are regions of space where an electric charge would feel a force (these ideas will be dealt with more fully in Book 2). As an EM wave passes, the electric and magnetic fields oscillate. As the electric field changes, it induces a change in a magnetic field at right angles to it. Similarly, as the magnetic field varies, it leads to a change in the electric field. This periodic fluctuation propagates through space as a self-perpetuating wave (Figure 10). The directions of the varying electric and magnetic fields are always at right angles to the direction of wave propagation. Hence EM waves are transverse waves.

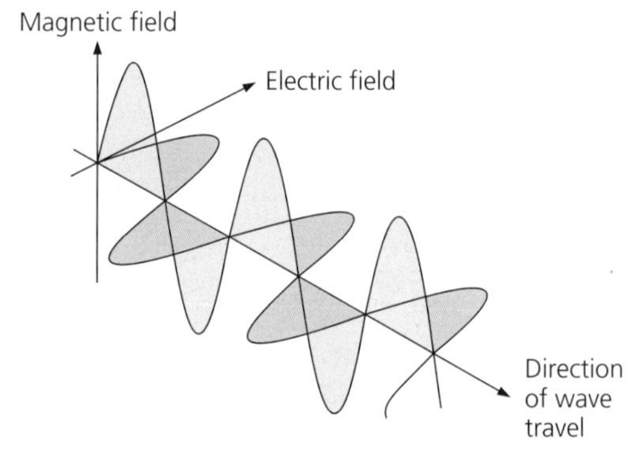


***Figure 10*** *An electromagnetic (EM) wave*

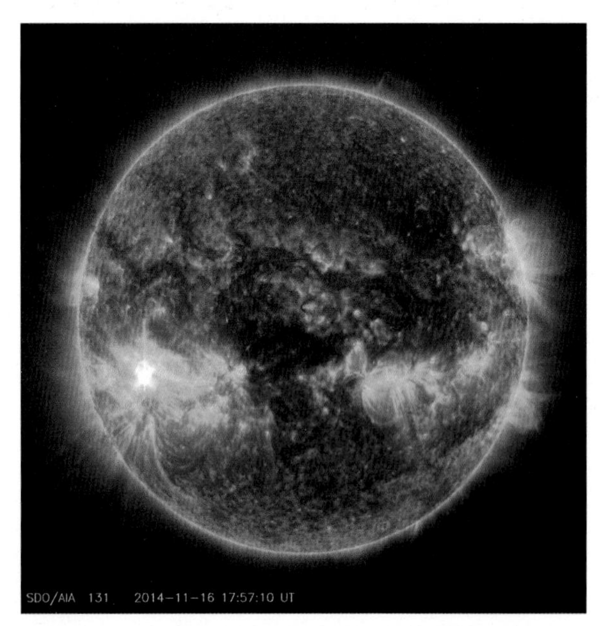


***Figure 9*** *A solar flare photographed at UV wavelength. The Sun emits EM radiation across the whole spectrum, from radio waves to gamma rays. EM waves all travel through a vacuum at the same speed of $2.997 \times 10^8$ m $s^{-1}$. Thankfully the Earth's atmosphere shields us from the most harmful wavelengths.*

## QUESTIONS

4. a. Sound travels as a longitudinal wave. It travels through solids much more quickly than it travels through air. Suggest why.
   b. "In space no-one can hear you scream." Why not? Why is it you could see an alien spaceship, but not hear it?
5. Seismic waves can travel through rock as either transverse or longitudinal waves but transverse waves cannot travel through a body of a liquid. Explain why.
6. Design a brief demonstration to show that sound waves from a loudspeaker cause longitudinal waves in the air.
7. Explain how you would prove to a class of younger students that sound is a mechanical wave and light is not.
8. Gamma rays are electromagnetic radiation of very high frequency. How does their speed in a vacuum compare with that of radio waves?

### Stretch and challenge

9. The graph in Figure 11 shows how, at an instant in time, the displacement of particles from their equilibrium position varies with distance along a *longitudinal* wave.

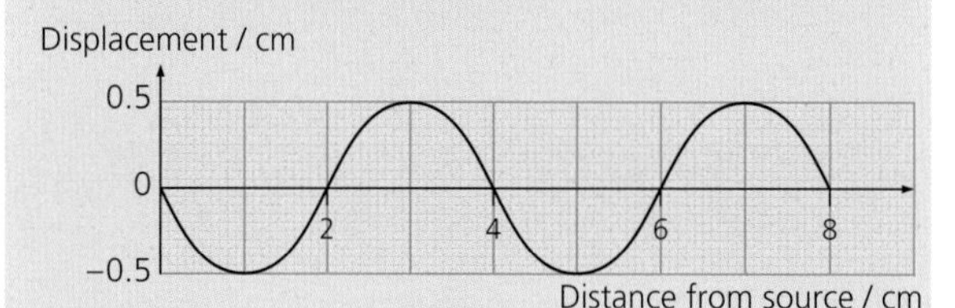


***Figure 11***

a. Copy the diagram and mark with dots the positions of the particles that are at their equilibrium positions.
b. Positive displacement is in the direction of travel of the wave. Which of the particles you have marked are at compressions? Which are at rarefactions? Mark them C and R.
c. Mark with an X the particle whose equilibrium position is 1 cm from the source of the disturbance.
d. In which direction is particle X moving?

### KEY IDEAS

- Progressive waves transfer energy, without causing any permanent displacement of the medium.
- In a transverse wave the oscillations are at right angles to the direction that the wave travels.
- In a longitudinal wave the particles oscillate in the same direction as the wave travels.
- Surface water waves, waves on a string and electromagnetic waves are transverse waves.
- Sound waves are longitudinal waves.
- Electromagnetic waves do not need a medium to travel through. They all travel at the same speed in a vacuum.

## 5.2 LOOKING IN DETAIL AT WAVES

### Displacement–distance graphs, wavelength and phase

Progressive waves can be drawn as they would look at one instant in time (Figure 12). Such a 'snapshot' shows how the displacement of the particles depends on their distance from the wave source, at a certain time.

The displacement is measured from the equilibrium position and may be either positive or negative. The maximum displacement caused by the wave is known as the **amplitude**, $A$ (see Figure 12). The amount of energy transferred by a wave depends on its amplitude.

The distance between any two consecutive points on a wave that have identical displacement and velocity is referred to as the **wavelength**, $\lambda$ (see Figure 12). The wavelength is measured in metres.

### QUESTIONS

10. Two seagulls are 150 m apart on the surface of the sea. They are bobbing up and down as waves pass them. When one of the seagulls is at the crest of a wave, the other is in a trough. When this happens there is one wave crest between the seagulls. What is the wavelength of the waves?

If we observed two points on the wave that are exactly one wavelength apart (such as 'a' and 'b' in Figure 12), we would see that they oscillate in step with each other. These points are said to be **in phase**. Points on the wave that are half a wavelength apart reach the opposite extremes of their oscillation at the same time, like points 'a' and 'c' in Figure 12. These points are in **antiphase**, i.e. completely out of phase with each other. Other points on the wave, like 'a' and 'd', have a **phase difference** that depends on what fraction of a wavelength lies between them.

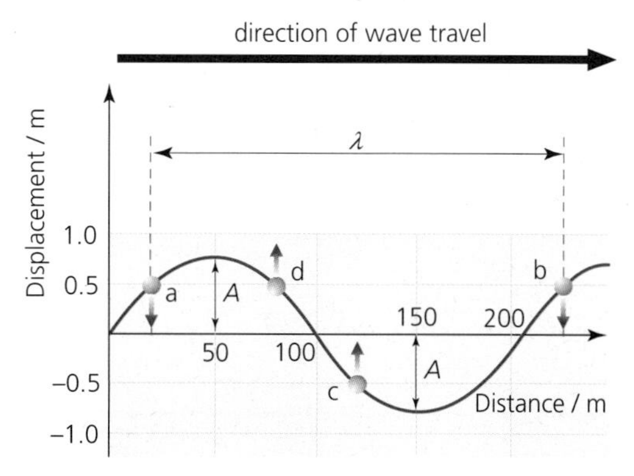


***Figure 12*** *A displacement versus distance graph showing a wave at an instant in time*

The phase difference between two waves, or between two points on a wave, can be expressed as an angle. One whole wave, or one cycle (Figure 13),

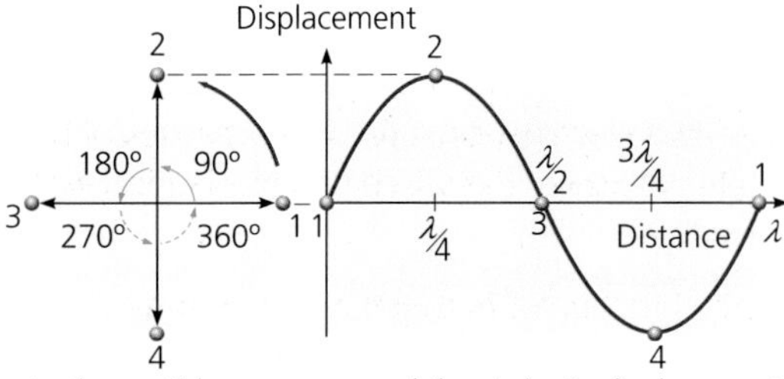


As the particle moves around the circle, its displacement above and below the **x**-ais traaces out a sine wave.

***Figure 13*** *Phase difference as an angle*

represents an angle of 360°. Two points with a phase difference of 360° are in phase because they are one wavelength apart. Points with a phase difference of 0°, 360°, 720°, ..., $n \times 360°$ are in phase (where $n$ is an integer).

Two points with a phase difference of 180° are half a wavelength apart and so exactly out of phase. In fact any two points that are an odd number of half-wavelengths apart will be exactly out of phase.

Phase difference is sometimes expressed in **radians** rather than in degrees. One radian is defined as the angle turned through at the centre of a circle when a distance of one radius is travelled around the circumference (Figure 14). Since the circumference of a circle of radius $r$ is $2\pi r$, there are $2\pi$ radians in a whole circle. So $2\pi$ radians is equivalent to 360°,

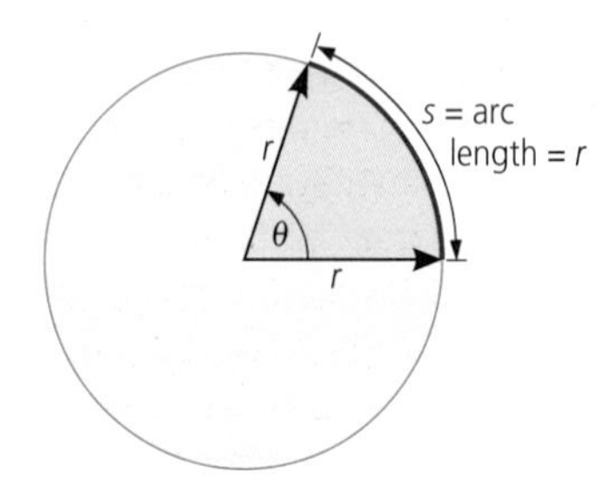


A radian is defined as the angle at the centre of a circle that is subtended by an arc that is the same length as the radius.

The angle $\theta$ is defined as $\theta = \frac{\text{arc length}}{\text{radius}} = \frac{s}{r}$ in radians.

This is often written as $s = r\theta$. When $s = r$, $\theta = 1$ radian.

***Figure 14*** *Definition of a radian*

which means that one radian is an angle equivalent to approximately 57.3°. A phase difference of $2\pi$ is the same as 360°, and it refers to two points that are a whole number of wavelengths apart. Table 2 summarises how phase differences can be expressed.

## QUESTIONS

**11. a.** Sketch a displacement versus distance graph for a transverse wave with a wavelength of 1 m and an amplitude of 1 m. Show three wavelengths.

**b.** Mark a point at the top of the first wave crest and label it A. Now mark, and label, a point that has a phase difference, compared to A, of:

$\pi/2$ (label it B)

$\pi$ (label it C)

$2\pi$ (label it D).

**12.** Two points on a wave have a phase difference of $\pi/6$ radians.

**a.** Express this phase difference in degrees.

**b.** If the wavelength of the wave is 3 m, what is the smallest possible distance between these points?

## Displacement–time graphs and frequency

Rather than sketch a wave at a point in time, we can concentrate on one point and sketch how its displacement varies with time (Figure 15). The graph looks very similar to Figure 12, but now two points on the waveform are separated by a certain time, rather than by a certain distance. The time taken for one complete wave to pass by a point is known as the **period**, $T$, of the wave, and it is often measured in seconds.

The phase difference between points on the wave now depends not on what fraction of a wavelength lies between them, but on what fraction of the period $T$ separates them.

| Phase difference in fractions of a cycle | Phase difference in degrees | Phase difference in radians |
|---|---|---|
| 1, 2, 3, ... | 360°, 720°, 1080°, ... | $2\pi$, $4\pi$, $6\pi$, ... |
| 1/2, 3/2, 5/2, ... | 180°, 540°, 900°, ... | $\pi$, $3\pi$, $5\pi$, ... |
| 1/4, 5/4, 9/4, ... | 90°, 450°, 810°, ... | $\pi/2$, $5\pi/2$, $9\pi/2$, ... |
| 1/3, 4/3, 7/3, ... | 120°, 480°, 840°, ... | $2\pi/3$, $8\pi/3$, $14\pi/3$, ... |

***Table 2*** *Phase differences expressed in different ways*

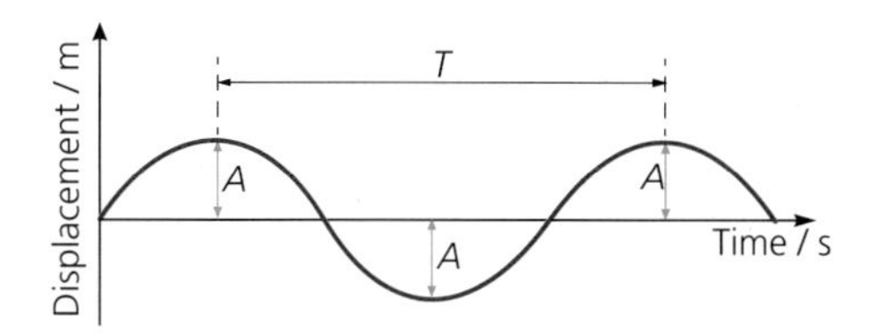


*Figure 15* Displacement versus time graph for a point on a wave

The period of wind-driven ocean surface waves is typically around 10 s. That means that there is one-tenth of a wave every second. The number of waves every second is known as the **frequency** of the waves and it is measured in hertz (Hz). In this case $f = \frac{1}{10} = 0.1$ Hz.

The frequency of a wave depends only on the wave source and is not affected by the medium in which the wave travels. For example, the frequency of the sound wave from a guitar string depends only on the frequency of vibration of the guitar string, and it stays constant as it travels through the air.

The relationship between the period of a wave, $T$, and its frequency, $f$, can be written as

$$f = \frac{1}{T} \quad \text{or} \quad T = \frac{1}{f}$$

## QUESTIONS

**13.** A tuning fork emits a note of frequency 256 Hz. What is the period of these sound waves?

**14.** One of the unusual features of a tsunami is its long period, typically about 1 hour. What is the frequency of such a tsunami?

**15.** What is the phase difference of the two waves shown in Figure 16:

**a.** in fractions of a cycle

**b.** in degrees

**c.** in radians?

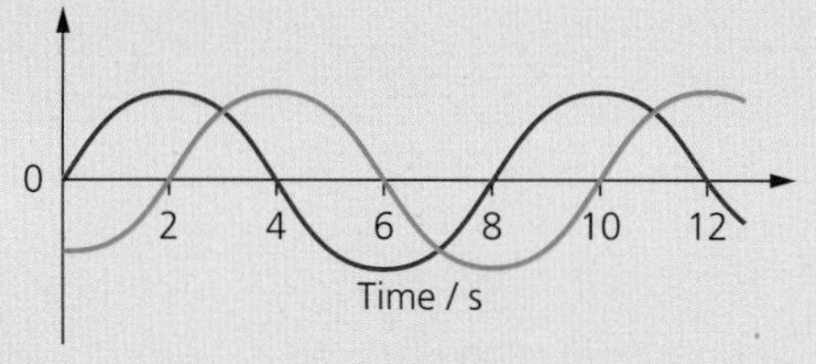


*Figure 16*

# 5.3 WAVE SPEED

The speed of a wave depends on the properties of the medium. In particular, the speed of a mechanical wave depends on:

- the size of the forces between each vibrating particle – the **elasticity** of the medium
- the **inertia** of the vibrating particles – how easy or difficult it is to accelerate each particle.

A material can be thought of as a series of masses connected by springs (Figure 17). As a mechanical wave, such as a sound wave, progresses through the material, the springs stretch and exert forces on the masses. A wave would travel relatively slowly through a system of weak springs and heavy masses.

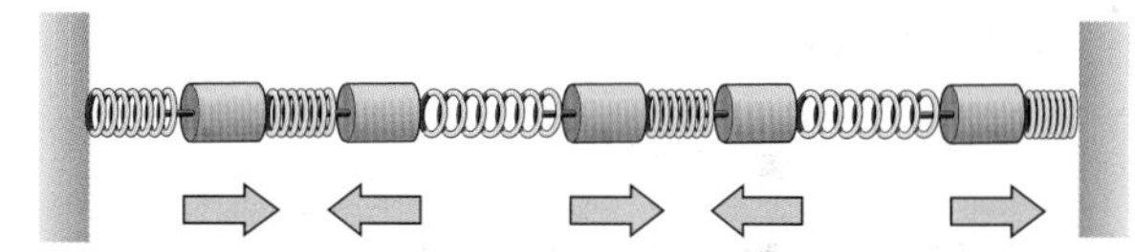

*Figure 17* Analogy of a sound wave passing through a material

Sound travels much more quickly through a solid than through a liquid or a gas (Table 3), because the forces between adjacent particles are so much stronger in a solid.

The speed of a wave, $c$, is connected to its wavelength, $\lambda$, and its frequency, $f$. Think about the ripples that spread out when you dip your finger into and out of the water of a lake (similar to those in Figure 2). If the circular waves that spread out have a speed, $c$, then after 1 s the first wave will have travelled $c$ metres from your finger (Figure 18). If the frequency at which you dip your finger is $f$, then after 1 s there are $f$ waves occupying a distance $c$. The length of each wave, $\lambda$, is therefore

$$\lambda = \frac{c}{f}$$

This relationship is true for *all* waves and is usually written as

$$c = f\lambda$$

This equation can be used to calculate the speed of a wave, if its wavelength and frequency are known.

| Material | Speed of sound / m $s^{-1}$ |
|---|---|
| Iron | 5000 |
| Water | 1500 |
| Air | 330 |

*Table 3* Typical speed of sound

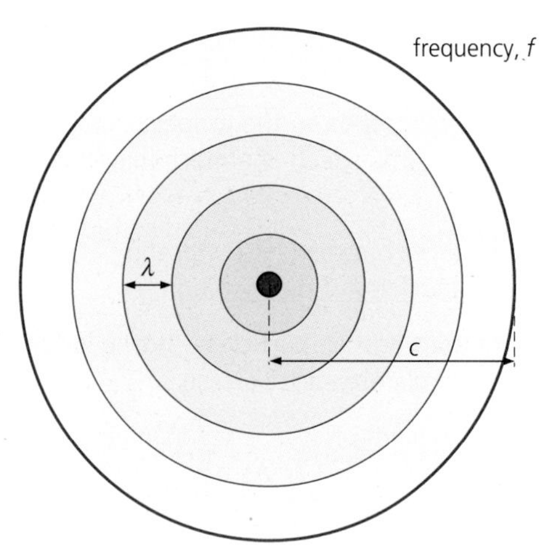


*Figure 18* In one second there will be f waves, and the first wave will have travelled a distance c metres, so $\lambda = c/f$.

### Worked example

A wind-driven ocean wave has a frequency of around 0.1 Hz and a wavelength of around 100 m. This gives a wave speed of

$$c = 100\,\text{m} \times 0.1\,\text{Hz} = 10\,\text{m}\,\text{s}^{-1}$$

Compare this with the speed of a tsunami that has a low frequency of $2.7 \times 10^{-4}$ Hz, but a long wavelength of up to 500 km in the open sea.

The wave speed of the tsunami is

$$c = 2.7 \times 10^{-4}\,\text{Hz} \times 500\,000\,\text{m}$$
$$= 140\,\text{m}\,\text{s}^{-1} \text{ (about } 500\,\text{km}\,\text{h}^{-1}\text{)}$$

### QUESTIONS

**16.** The speed of sound in air is about 330 m $\text{s}^{-1}$. Calculate the wavelength of a sound wave that has a frequency of 256 Hz.

**17.** The BBC transmits Radio 4 on the long-wave band at a wavelength of 1500 m. The speed of radio waves in air is approximately $3 \times 10^8$ m $\text{s}^{-1}$. Calculate the frequency of these radio waves.

**18.** The graph in Figure 19 shows how the displacement of a molecule in the air varies with time as a sound wave passes by.

**a.** Calculate the frequency of the sound wave.

**b.** Use the graph to plot a velocity versus time graph for the air molecule.

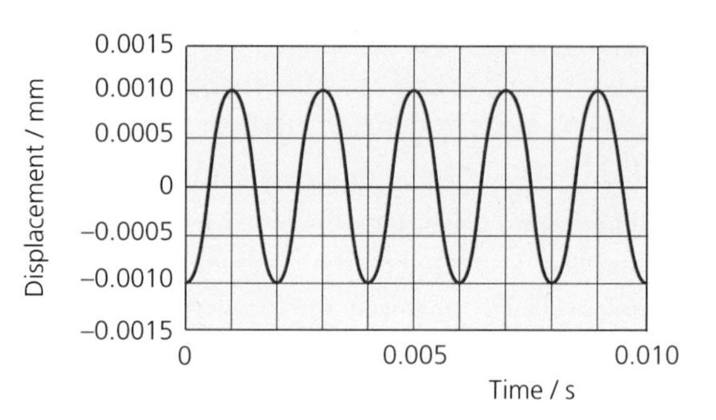


*Figure 19* Displacement versus time for an air molecule in a sound wave

### KEY IDEAS

- The wavelength, $\lambda$, of a progressive wave is the distance between any two consecutive points on a wave that have identical displacement and velocity.
- Two points on a wave that are any whole number of wavelengths apart will have exactly the same displacement and velocity. These points are said to be in phase.
- The phase difference between two waves at any given point, or between two points on a wave, can be expressed as a fraction of a cycle, or as an angle in degrees or radians.
- The frequency $f$ of a wave source is the number of waves per second, measured in hertz (Hz).
- The time taken for one complete wave to pass a point is the period $T$; in seconds, $T = \frac{1}{f}$.
- The wave speed, $c$, is equal to the wavelength multiplied by the frequency; $c = f\lambda$.

## ASSIGNMENT 2: UNDERSTANDING SEISMIC WAVES

**(MS 3.1, MS 3.2)**

### Stretch and challenge

Earthquakes can be very frightening events for the people involved (Figure A1) and can cause a huge amount of damage (Figure A2).

***Figure A1*** *"I was parking my car when the first wave from the earthquake arrived. It felt like a sharp jolt. At the time I was sure that the driver behind had run into me. I got out of the car to look, but there had been no contact. Moments later the ground began to shake and masonry fell from some of the buildings."*

***Figure A2*** *Earthquake damage in San Francisco*

The seismic waves that arise from earthquakes are of two distinct types. The first wave to arrive is known as the primary or P-wave. P-waves are longitudinal waves, which travel relatively quickly, typically $7500\,m\,s^{-1}$. The jolt from the P-wave rarely causes damage to buildings, which are built to be strong in the vertical direction. The P-wave is just the herald of worse problems to come. It is the slightly slower moving, transverse S-waves that cause much of the damage. Unless buildings are built with earthquakes in mind, they cannot withstand the side-to-side shear forces (Figure A2).

The seismogram shown in Figure A3 is a simplification. In reality there are reflections from the boundaries of different layers within the Earth, such as that between the core and the mantle.

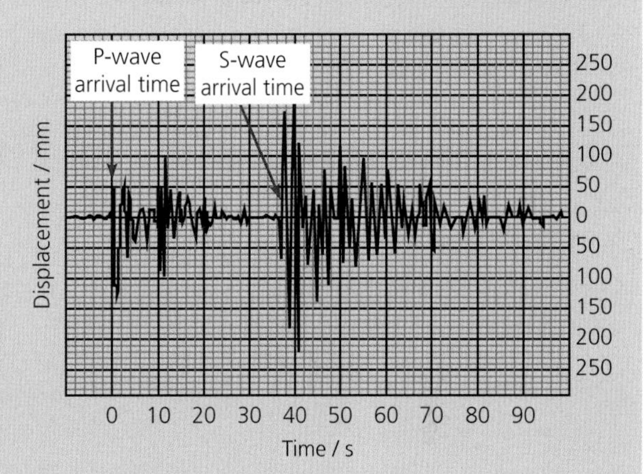


***Figure A3*** *Seismograph trace*

| Magnitude | What it feels like | How often worldwide |
|---|---|---|
| < 2.0 | Normally only recorded by seismographs. Most people cannot feel them | Millions each year |
| 2.0–2.9 | A few people feel them. No building damage | Over 1 million each year |
| 3.0–3.9 | Some people feel them. Objects indoors can be seen shaking | Over 100 000 each year |
| 4.0–4.9 | Most people feel them. Indoor objects shake or fall to floor | 10 000 to 15 000 each year |
| 5.0–5.9 | Can damage or destroy buildings that are not designed to withstand earthquakes. Everyone feels them | 1000 to 1500 each year |
| 6.0–6.9 | Widespread shaking, even far from epicentre. Damages buildings | 100 to 150 each year |
| 7.0–7.9 | Widespread damage | 10 to 20 each year |
| 8.0–8.9 | Widespread damage | One each year on average |
| 9.0–9.9 | Severe damage to most buildings | One in 5–50 years |
| > 10.0 | Never recorded | Never recorded |

***Table A1*** *The Richter scale*

Earthquakes are ranked on the Richter scale (Table A1). An earthquake of less than magnitude 2 would not be noticed by most people; tremors of that size happen almost all the time. An earthquake of 9 on the Richter scale results in virtually total destruction of all buildings and a large loss of life. The earthquake that caused the 2004 tsunami rated more than 9.

It is possible to rate an earthquake on the Richter scale from a seismogram. The amplitude of the S-wave is the key factor. However, we also need to take into account how far the seismograph is from the epicentre. A 'nomogram' is used to find the Richter value. A line is drawn connecting the maximum displacement on the seismogram and the distance to the epicentre, and the value on the Richter scale is the point where the line intercepts it. In Figure A4, a maximum amplitude of 10 mm at a distance of 100 km indicates 4.0 on the Richter scale.

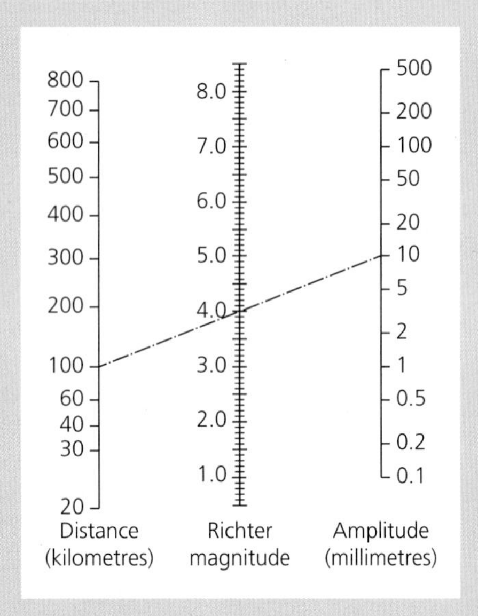


***Figure A4*** *A Richter nomogram*

### Questions

**A1** Figure A3 is a trace from a seismograph showing the delay between the arrival of the P-wave and the S-waves. This P–S interval can be used to estimate the distance from the earthquake's epicentre. If P waves travel at 7500 m s−1 and S waves travel at 4000 m s−1, estimate how far away this seismograph reading was from the epicentre.

**A2** It takes at least three readings from different seismograph stations to identify the position of the earthquake's epicentre. Why is this?

**A3** Use the seismogram (Figure A3) and the nomogram (Figure A4) to estimate where this earthquake fits on the Richter scale.

**A4** Explain why the distance from the epicentre has to be taken into account when rating an earthquake.

**A5** **a.** The Richter scale is not a linear scale. Imagine you are looking at seismograms that are all taken at 100 km from the epicentre of an earthquake. Copy and complete Table A2 to show the maximum amplitude of the S-waves for various magnitudes of earthquake.

| Magnitude on the Richter scale | Maximum amplitude of S-wave / mm |
|---|---|
| 2 | |
| 3 | 1 |
| 4 | |
| 5 | |

*Table A2*

**b.** Plot a graph of the results from your table. Plot the Richter scale value on the *x*-axis.

**c.** What shape of graph is this? What value of amplitude would you expect for a Richter scale value of 6?

## 5.4 POLARISATION

We cannot see directly that light is a wave, but we can see it behave in a similar way to waves on strings or on the surface of the sea. Light can be reflected from a surface, just as a wave on the sea is when it hits a harbour wall. It can also be refracted, just as water waves are on reaching a shallower region (see Chapter 7). Neither reflection nor refraction is convincing proof that light is a wave. After all, a solid object, like a tennis ball, can be 'reflected' from a racket. But light can also be **polarised**, which is a phenomenon that is only shown by transverse waves.

Transverse waves, as we have seen, have oscillations that are perpendicular to the direction of wave propagation. There are an infinite number of ways that this can happen. Figure 20 shows some different possible orientations for the oscillating electric field vector of an electromagnetic wave.

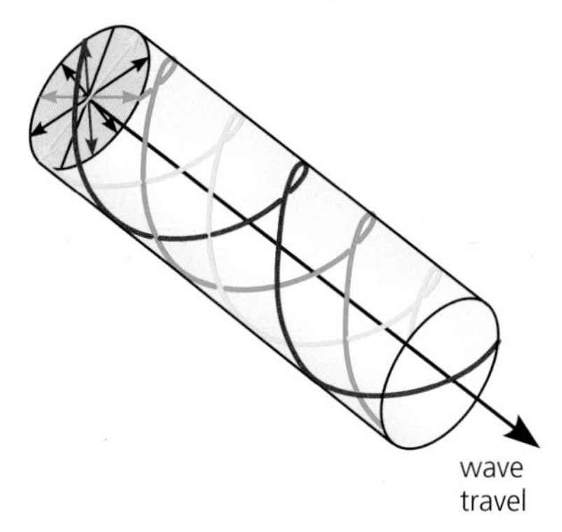


***Figure 20*** *An unpolarised wave has vibrations that move in any direction perpendicular to the wave motion. Ordinary light is unpolarised.*

If the oscillations are confined to just one plane, the wave is said to be **plane polarised**. Because an electromagnetic wave has an oscillating electric field and an oscillating magnetic field, we take the plane of the electric field to be the plane of polarisation.

Polarisation provides the most convincing evidence that light is a transverse wave. Visible light is usually unpolarised, though some sources, such as liquid-crystal displays, emit polarised light. Light can also become polarised by reflection or by transmission through certain materials.

Light can be polarised by transmission through a **polarising filter**. Polaroid is a plastic with long-chain molecules that are aligned in the same direction. A Polaroid filter (Figure 21a) absorbs light that is polarised in a plane parallel to the chain molecules, and allows light that is polarised at 90° to the chains to be transmitted.

Light can also be polarised by reflection from a non-metallic surface. When light strikes a surface, some of it may be absorbed, some may travel through the material, and the rest will be reflected from the surface. Light that is polarised in a plane parallel to the surface is preferentially reflected (Figure 22). The light reflected from the surface of water or glass, for example, is at least partially polarised, so using Polaroid sunglasses can reduce glare (see Figure 23 on the next page).

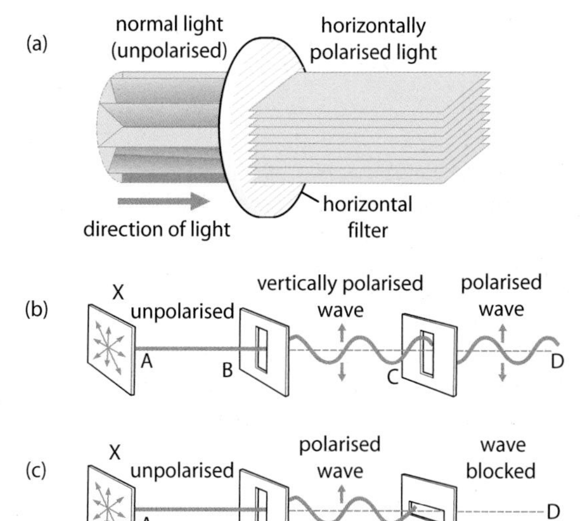


***Figure 21*** *(a) A polarising filter allows waves only with oscillations in a particular plane to be transmitted. (b) Two filters that are parallel to each other allow polarised light to be transmitted. (c) If a second filter is placed at right angles to the first, no light can be transmitted.*

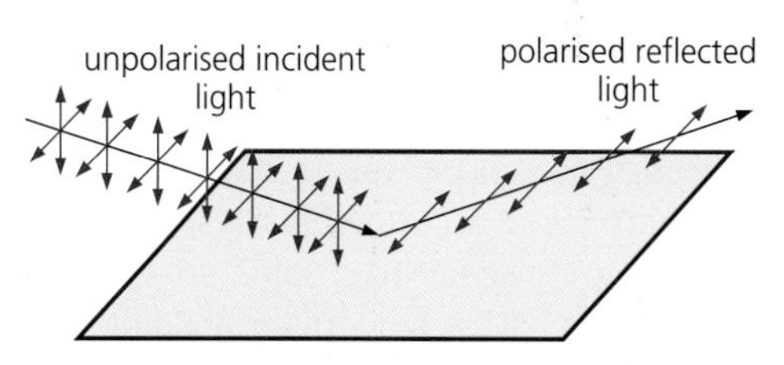


***Figure 22*** *Reflection of light from a non-metallic surface results in a degree of polarisation parallel to the surface.*

When light is reflected in random directions by a rough surface or by small particles, or molecules, in the atmosphere, we say that the light has been scattered. Scattered light from molecules in the atmosphere is partially polarised. The light from the sky is at its most polarised, and most blue (because blue light is scattered more than red light), at 90° from the direction of the Sun. A Polaroid filter on a camera can improve the contrast in a sunny scene.

## Polarisation of radio waves

All electromagnetic waves are transverse and therefore capable of being polarised. The radio waves used to transmit TV signals are polarised. The alignment of the receiving aerial must match the direction of the electric field vector of the transmitted wave, otherwise the received signal will be weak (see Figure 24 on the next page).

For example, the TV transmitter that covers much of Bristol, Bath and north Somerset transmits

(a) Without polarising lenses

(b) With polarising lenses

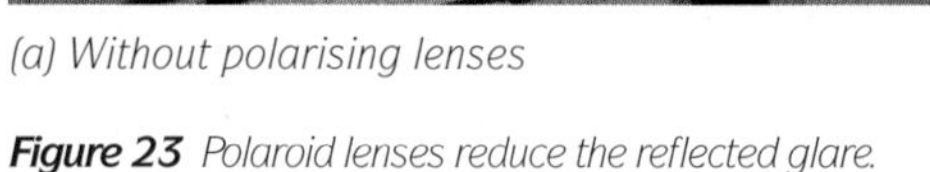

***Figure 23*** *Polaroid lenses reduce the reflected glare.*

horizontally polarised radio waves. The transmitter has an output power of around 250 kW and is mounted on a 300 m high mast on Penhill, itself 300 m above sea level. Despite this, the signal cannot reach all the villages and homes in the area because the Mendip Hills get in the way. The short-wave (UHF) radio waves used for television do not diffract (bend around obstacles) sufficiently to be picked up by an aerial in some areas. Many houses therefore have a second aerial pointing at a different transmitter (Figure 25). Notice how the aerials are differently aligned to receive waves from different transmitters.

***Figure 25*** *Neighbouring TV transmitters transmit radio waves polarised in opposite planes, to prevent signals interfering with each other.*

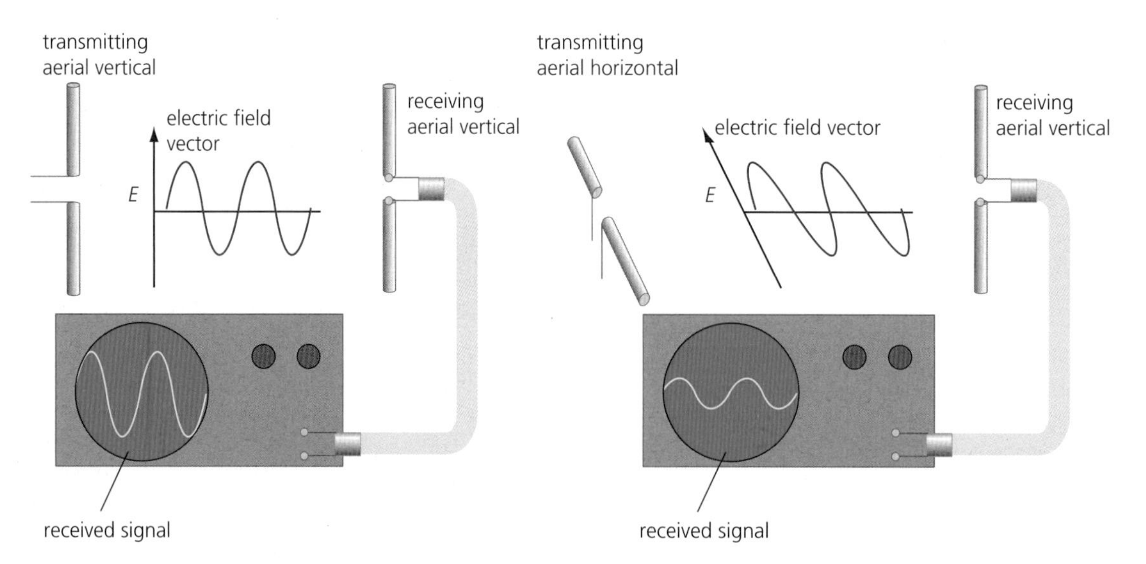


***Figure 24*** *The electrons oscillating in the transmitting aerial cause a radio wave with its electric field polarised in the same plane as the transmitting aerial. The receiving aerial has to be aligned in the same plane to get the maximum signal.*

## Optical activity

'Optically active' materials can rotate the plane of polarisation of light. Sugar solution is able to do this. The more concentrated the solution, the larger the angle through which the polarisation plane is moved. The effect is also used in liquid-crystal displays, where the plane of polarisation can be controlled by an applied voltage. Most modern flat-screen TVs use this effect – try watching TV using Polaroid glasses and then tilting your head!

Some optically active materials are affected by stress (a deforming force). The larger the stress applied to them, the greater the change in the plane of polarisation. The effect can also be dependent on wavelength, which causes coloured patterns when the stressed material is viewed through a polarising filter. This 'photoelasticity' is useful in finding areas of excessive stress (Figure 26).

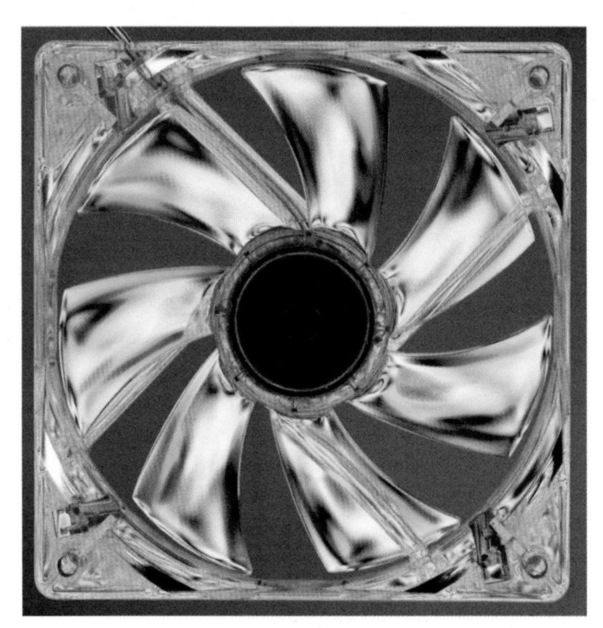

***Figure 26*** *Stress pattern caused by optical activity that is wavelength dependent*

### QUESTIONS

**19.** Is it possible to have polarised sound? Explain your answer.

**20.** "My car has a liquid-crystal display showing the speedometer. If I'm wearing my Polaroid sunglasses and put my head on one side slightly, the display goes black and I can't read the speedo. I've also noticed that I can see coloured patterns in the windscreen. Do I need to seek medical advice or buy a new pair of sunglasses?" Explain.

#### Stretch and challenge

**21.** Wearing Polaroid sunglasses can reduce the glare that you see from reflective surfaces, like the sea. Tilting your head from side to side varies how bright the reflections look. Explain why this happens.

**22.** It has been suggested that if car windscreens were polarising and headlights were beams of polarised light, we could avoid glare without having to dip our headlights. Good idea? Explain.

**23.** "Sugar solution can rotate the plane of polarisation of light. All sugars seem to rotate it in the same way, to the right, say. If you try artificial sugar, which chemically seems to be the same in every way, it does not turn the light. Bacteria eat sugar; if you put bacteria in the artificial sugar water ... and when the bacteria are finished and you pass polarized light through the remaining sugar water you find it turns to the left."

*Richard Feynman*

Suggest an explanation for Feynman's observations.

### KEY IDEAS

- The waves in the electromagnetic spectrum are transverse waves.
- These transverse waves can be plane polarised by transmission or by reflection.
- A polarising filter plane-polarises a light wave. A second polarising filter can then vary the amount of light transmitted, depending on its orientation relative to the first filter.
- Radio waves emitted from a transmitting aerial are plane polarised. The receiving aerial needs to be correctly aligned to receive the full signal.

## ASSIGNMENT 3: DEMONSTRATING POLARISATION

**(PS 1.2, PS 2.1, PS 4.1)**

Your task is to prepare a short, say 10–15 minute, lesson for a small group of younger students about polarisation. You should plan to include at least some of the demonstrations listed in Experiment 1, or the demonstration described in Experiment 2, to help you explain what polarisation is.

### Questions

**A1** Begin to write a plan for your lesson. Think about how you will start. For example, it will help if you can find out what the students already understand about waves. What questions could you ask?

**A2** It would be possible to give an initial demonstration of the idea of polarisation using a long spring or even a piece of thick rubber tubing. How would you do this?

**A3** Which experiment (1 or 2) will you use? Read through the details given here and make your decision about what to show. What apparatus will you need? Write down clear steps for yourself of the order in which you will do things, so that your lesson goes smoothly. Include any safety precautions you will need to take. You will need to try out the demonstrations first.

**A4** What will you need to explain to the students? Are there any diagrams that it could be useful to draw and show, perhaps using presentation software?

**A5** If you get a chance to deliver your lesson, take feedback from your audience and then write an evaluation of how you did and how you might improve on this another time.

### Experiment 1. Light polarisation

This experiment consists of a series of demonstrations involving polarised light. Most can be carried out using very simple apparatus, for example two Polaroid filters (a pair of old Polaroid sunglasses would do), a bright torch, a light meter and a sunny day.

- It is important to consider safety. Students should be reminded that looking directly at the Sun for any length of time can damage eyes.
- One polarising filter, or Polaroid sunglasses, can be used to examine the polarisation of light from different parts of the sky.
- A beam of polarised light can be produced by shining a torch, or a ray box, through a Polaroid filter.
- The intensity of light can be measured using a light meter. It may be possible to find a suitable app for a smartphone or tablet.
- A second Polaroid filter can be placed between the detector and the light source. The intensity transmitted through the two Polaroid filters will depend on the relative orientation of the filters. Varying this angle will alter the amount of light transmitted.
- Observing reflected light through a Polaroid filter will reveal that the intensity of light transmitted depends on the angle of the filter. This is because reflected light is partly polarised.
- If a Perspex set square is placed between two polarising filters orientated at right angles, it should be possible to see stress patterns in the Perspex.

### Experiment 2. Microwave polarisation

A microwave receiver and transmitter can be used to demonstrate polarisation of microwaves (Figure A1). The microwave transmitters used in schools transmit polarised waves, which can be detected by a receiver linked to an amplifier. The amplified signal is used to generate an audio output. A louder sound means that more microwave power is being received.

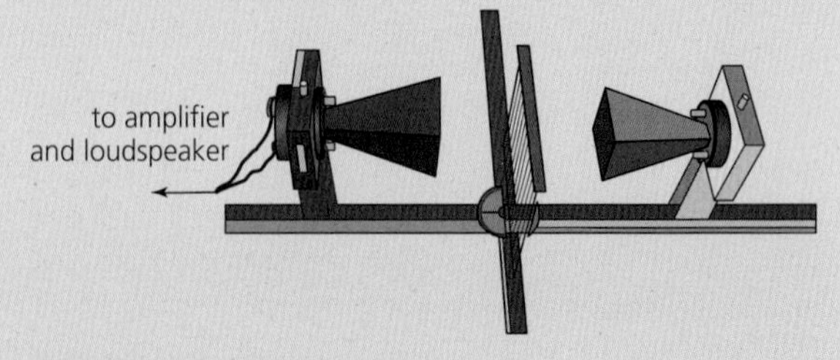


***Figure A1*** *Microwave experiment*

The microwave power emitted by school apparatus is very low and presents no danger. However, it is sensible to warn students that microwave energy can be dangerous and it is good practice to limit their exposure, for example by turning the transmitter off when not in use.

The polarised nature of the waves can be demonstrated by rotating the receiver about a horizontal axis parallel to the direction of propagation. A polarising grid, consisting of a parallel set of conducting wires, can be inserted between the transmitter and receiver. As the grid is rotated, the received power will vary. The grid is acting as an aerial, and it will absorb the most energy when the wires are parallel to the electric field of the microwaves.

## 5.5 SUPERPOSITION OF WAVES

It seems remarkable but it is possible to add two sounds together and produce silence. 'Active noise control' works on this principle (Figure 27). A reverse copy of the noise, 'anti-noise', is added to the original noise to produce a much quieter sound.

*Figure 27* The constant noise from a helicopter rotor is stressful for the pilot and co-pilot and makes communication difficult. They wear 'active headphones' that reduce the noise by playing a reverse copy of the noise back into the ears.

This sound 'cancellation' is possible because of what happens to waves when they are added together. When two similar waves meet, for example, two sound waves or two water waves, the total displacement is the vector sum of the two individual displacements. This is known as the **principle of superposition**. The vector sum means that we need to take account of whether the displacement was positive or negative. For example, if two water waves meet with their crests together, the result will be an extra high crest. If a crest meets a trough of equal size, the resulting displacement of the water will be zero (Figure 28). Superposition can be observed when any two similar waves meet. For example, when the ripples on the surface of water cross each other, they briefly add together and then carry on as before (Figure 29).

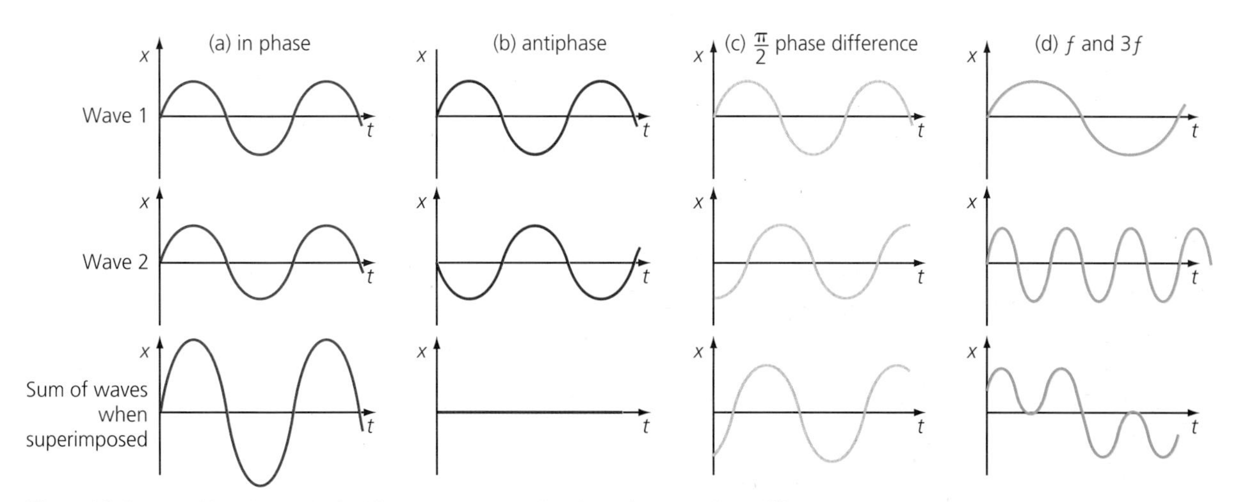


*Figure 28* Superposition: the result of adding two waves together depends on the phase difference between them.

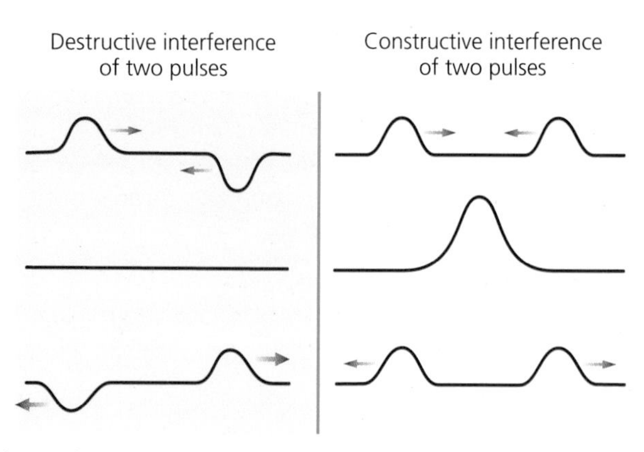


***Figure 29*** *Two waves meet, add together and then pass through each other, undisturbed by the process.*

**QUESTIONS**

**24.** The two wave pulses shown in Figure 30 are heading towards each other on a stretched string. Sketch the displacement of the string after 0.1, 0.2 and 0.3 s.

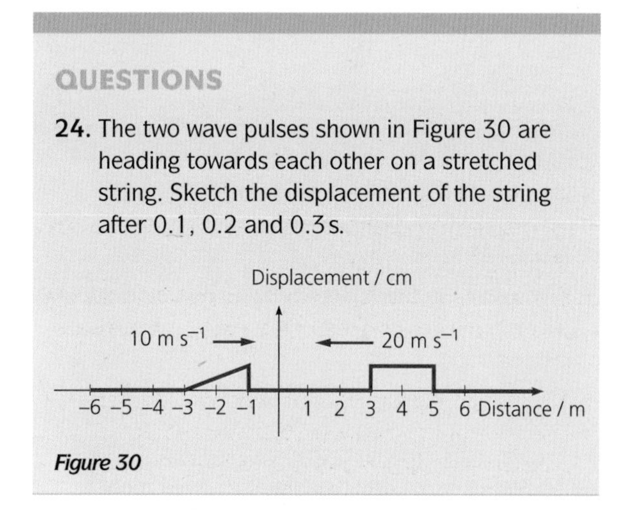


***Figure 30***

## 5.6 STATIONARY WAVES

Musical instruments rely on **stationary waves** (or 'standing' waves) to produce notes (Figure 31). A stationary wave on a violin string, for example, makes the air around it vibrate, causing a progressive sound wave to move through the air.

***Figure 31*** *The strings vibrate as stationary waves.*

Progressive waves carry energy in the direction that the wave travels. But some waves do not appear to propagate at all. These stationary waves are formed by two progressive waves of the same frequency travelling in opposite directions (Figure 32).

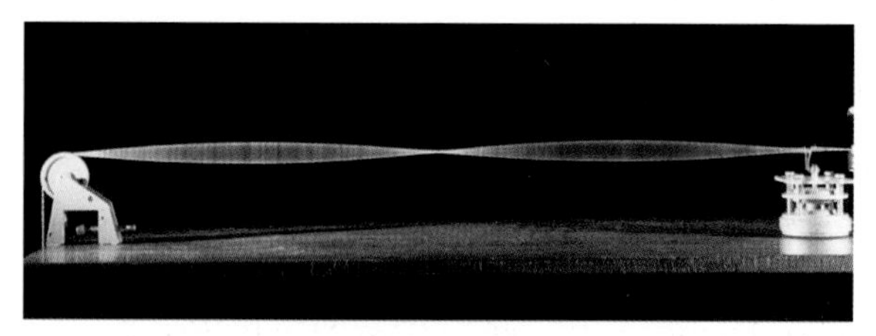

***Figure 32*** *A stationary wave on a string*

Imagine a rope with a person at each end, both wiggling the rope at the same frequency so as to send progressive waves down to the other end. These two waves add together, in accordance with the principle of superposition. At certain frequencies, there are points on the rope where the two waves always meet with a phase difference of 180°. If the two waves have the same amplitude, these points do not oscillate at all; they are stationary points called **nodes**. At other points along the rope there are places where the two waves always meet in phase. At these points the displacement goes from zero to a maximum value of twice the individual amplitudes. These points are called **antinodes**. If the frequency of the wave stays fixed, the positions of the nodes and antinodes do not change (Figure 33). There is always a distance of half a wavelength between two successive nodes, or two successive antinodes.

Progressive waves have crests that move at the speed of the wave. Energy is transferred by the wave. In a stationary wave the nodes and antinodes are both fixed and no energy is transferred along the wave. Stationary waves are formed in systems that have boundaries. Progressive waves are reflected at these

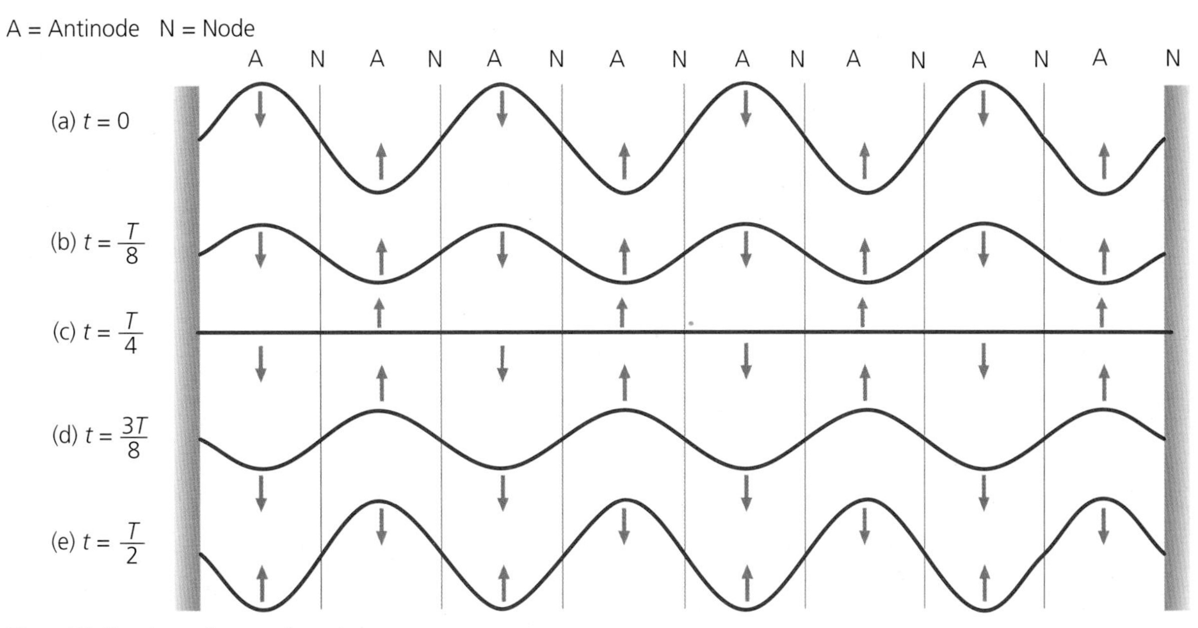


***Figure 33*** *Time-lapse diagram of a stationary wave*

boundaries and stationary waves are caused by the superposition of these reflected waves. The energy is not transferred beyond the boundaries of the system and becomes stored in the stationary wave.

There is another difference between progressive and stationary waves. In a progressive wave there is a phase difference between adjacent points (see section 5.2). In a stationary wave all the points between consecutive nodes are vibrating *in phase*. All these points reach their maximum displacement at the same time, though this is different for each point. All the points on a progressive wave have the same amplitude.

## Waves on strings

The standing wave on the string of a musical instrument is set up because of the superposition of the progressive waves that reflect from the fixed ends of the string. This phenomenon can be investigated using the apparatus shown in Figure 34. The vibration transducer causes a wave to be sent along the string. At certain frequencies a standing wave is set up. These frequencies are known as **harmonics**.

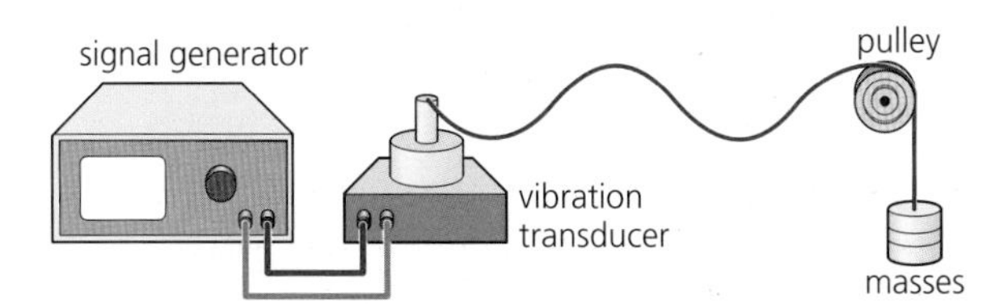


***Figure 34*** *The driving frequency of the input signal is varied until it matches a natural frequency of vibration of the string, called a harmonic. This is an example of resonance.*

The frequencies at which the harmonics occur depend on several factors.

- The tension in the string: in the experiment adding extra masses to the end of the string will increase the tension.
- The mass per unit length of the string: thicker, heavier strings vibrate more slowly and the harmonics occur at lower frequencies.
- The length of the string: a shorter string has shorter-wavelength vibrations, which leads to higher frequencies.

For a string that is fixed at both ends, there can be no vibrations at the ends of the string. These points are always nodes. The lowest frequency at which a standing wave is formed is known as the first harmonic (also known as the fundamental frequency). This standing wave has a single antinode in the centre of the string (as in Figure 32). There is one half of a full wavelength on the string, so for a string of length $l$ the wavelength of the first harmonic is

$$\lambda = 2l$$

and because

$$f = \frac{c}{\lambda}$$

then the frequency of the first harmonic is

$$f_1 = \frac{c}{2l}$$

The second harmonic is formed with a node in the centre of the string. In this case there is a whole wavelength on the string, so

$$\lambda = l$$

and

$$f_2 = \frac{c}{l} = 2f_1$$

The second harmonic has twice the frequency of the first harmonic. A similar argument shows that, for a vibrating string, all the harmonics have frequencies that are integral multiples of the frequency of the first harmonic (Figure 35).

The speed, $c$, at which a wave travels along a stretched string depends on the mass per unit length of the string, $\mu$, and the tension in the string, $T$. In fact

$$c = \sqrt{\frac{T}{\mu}}$$

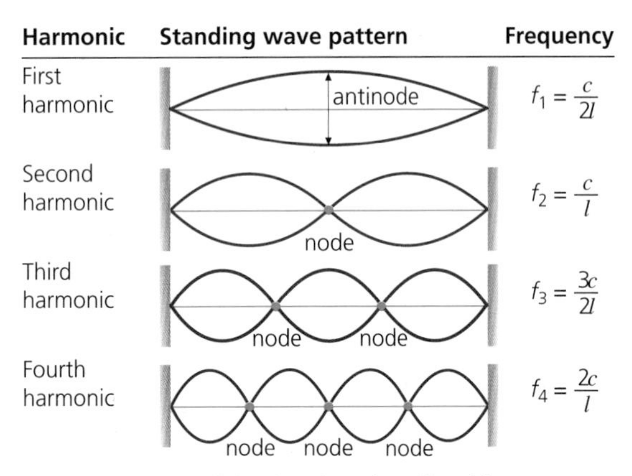


***Figure 35*** *Modes of vibration of a string of length!*

That means the frequency of the first harmonic is

$$f = \frac{c}{\lambda} = \frac{1}{\lambda}\sqrt{\frac{T}{\mu}} = \frac{1}{2l}\sqrt{\frac{T}{\mu}}$$

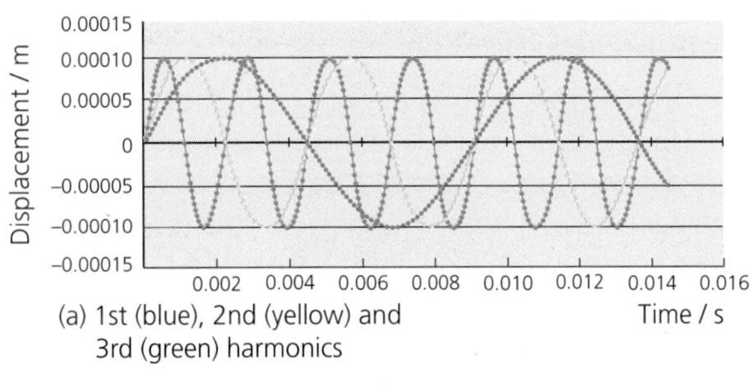


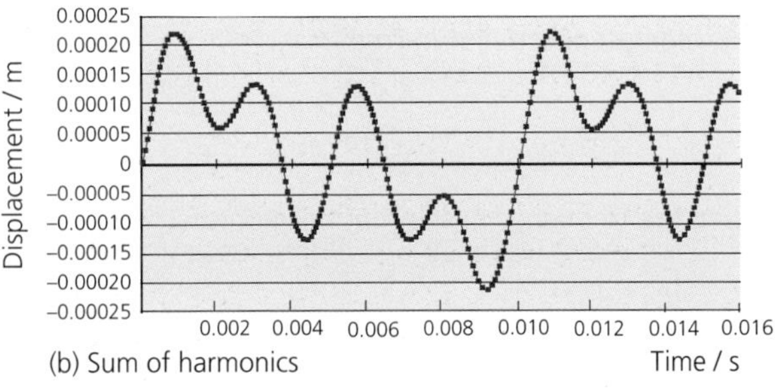


***Figure 36*** *Addition of harmonics to form a complex wave. The 'quality' of a sound wave depends on the mixture of harmonics that are present.*

The vibrations on a real string, on a guitar or violin for example, are much more complex. Many of the harmonics can be present at the same time, though the amplitude of the higher harmonics is usually smaller (Figure 36). As the string vibrates, it causes a sound wave in the air around it. This is amplified by the sound box, which is the body of the instrument (Figure 37). The sound wave will vibrate with the same mixture of frequencies that the string vibrated with, though some of the frequencies are amplified more than others by a sound box.

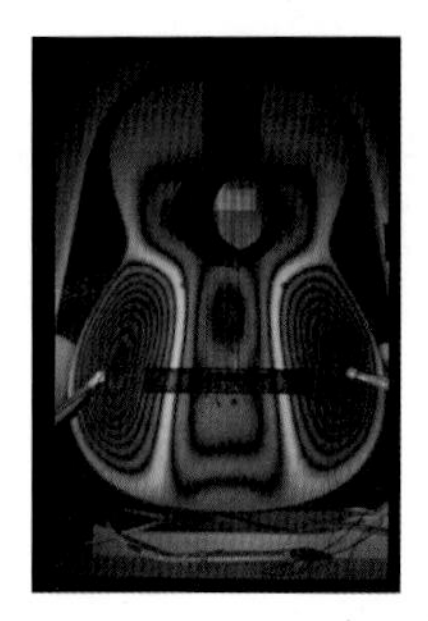

***Figure 37*** *Holograms are used to show two-dimensional stationary waves on the front plate of a guitar. These patterns are used to help in the design of better guitars.*

## QUESTIONS

**25.** Estimate the frequency of the lowest note on a guitar. (The tension in a guitar string is approximately 100 N.)

**26.** Why does a note played on a violin sound different from the same note played on a guitar?

**27.** The G string on a guitar plays a note of frequency 196 Hz when played 'open'. The string can be shortened by pressing on a fret. Estimate the frequency of the highest note that can be played by pressing on a fret.

**28.** If a guitar string is plucked, then touched briefly in the middle of the string, a higher note is heard. How much higher? Why?

**29.** A set of violin strings is replaced with strings of twice the mass. What will the new tension have to be, if the strings are to play the same notes as before?

**30.** A local violin-maker wants to build a violin that is one-quarter of the standard size, but plays the same notes as the standard version. What advice can you give her?

# REQUIRED PRACTICAL: APPARATUS AND TECHNIQUES

## Investigation into the variation of the frequency of stationary waves on a string

The aim of this practical is to verify the relationship

$$f = \frac{1}{2l} \times \sqrt{\frac{T}{\mu}}$$

for the first harmonic of a wave on a stretched string. The practical gives you the opportunity to show that you can:

- use appropriate analogue apparatus to record a range of measurements and to interpolate between scale markings
- use appropriate digital instruments to obtain a range of measurements
- use methods to increase accuracy of measurements
- use a signal generator
- generate and measure waves using a vibration transducer.

### Apparatus

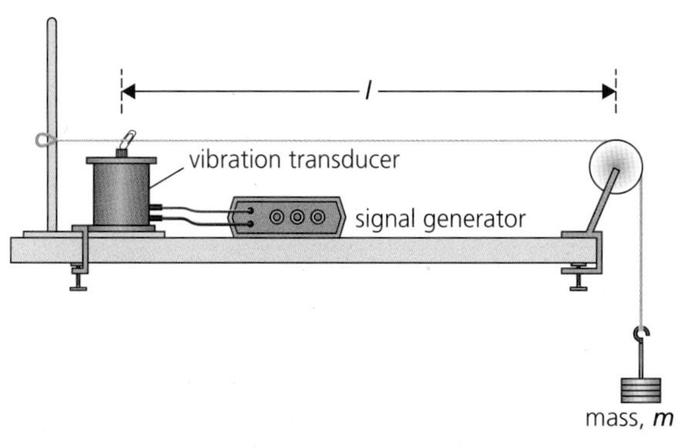


***Figure P1*** *The experimental set-up*

The standard experiment for investigating stationary waves on a string uses a vibration transducer to vibrate the end of the string. A vibration transducer is similar to a loudspeaker: it has a metal post in place of a paper cone.

One end of the string is connected to the post of the vibration transducer and the other end has a loop so that masses can be hung from it via a pulley (see Figure P1). The tension in the string can be varied by changing the mass, $m$.

A signal generator is connected to the vibration transducer so that the frequency of oscillations can be controlled (see Figure P2). The signal generator sometimes has a digital readout of frequency, although this can be inaccurate. Often, the frequency has two controls: one determines the range of frequencies and the other is a fine control. The scale of the fine control is sometimes logarithmic, rather than linear, so that a small movement of the control can make a large change to the output frequency.

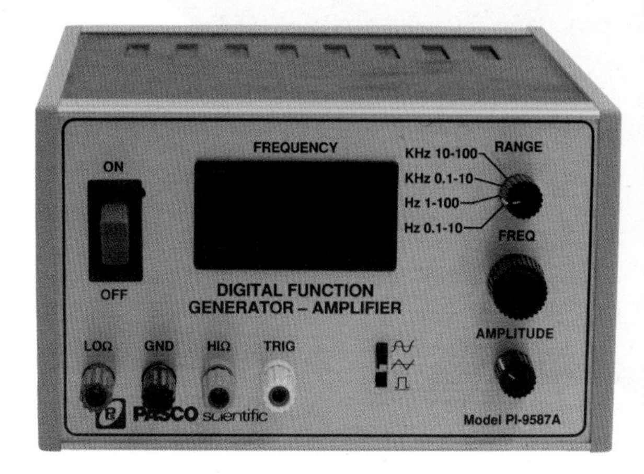


***Figure P2*** *Signal generator*

An alternative means of measuring the frequency is useful, for example an oscilloscope, a frequency meter or a stroboscope with a digital display. Using a stroboscope has the advantage that the stationary wave patterns are observed easily, but care has to be taken because flashing lights can cause problems for some people. Frequencies in the range 7 Hz to 15 Hz should be avoided.

### Techniques

The frequency of the first harmonic for a wave on a string (Figure P3) is given by

$$f = \frac{1}{2l} \times \sqrt{\frac{T}{\mu}}$$

Since only one variable should be changed at a time, several separate experiments are necessary.

1. To investigate the effect of tension, $T$:

   The tension is varied by changing the mass, $m$, on the end of the string. $T = mg$, assuming that the pulley is frictionless. The length, $l$, of the string and its mass per unit length, $\mu$, must be kept constant.

2. To investigate the effect of varying the mass per unit length, $\mu$:

   Mass per unit length is more difficult to change. Several different strings, each of different thickness or density, could be used. It is possible to braid strings together to make strings of $\mu$, $2\mu$, $3\mu$, $4\mu$, etc. The length, $l$, of the string and its tension, $T$, must be kept constant.

   The procedure is similar to that described in step 1.

   A graph of frequency ($y$-axis) against $\sqrt{1/\mu}$ is plotted: a straight line verifies the relationship.

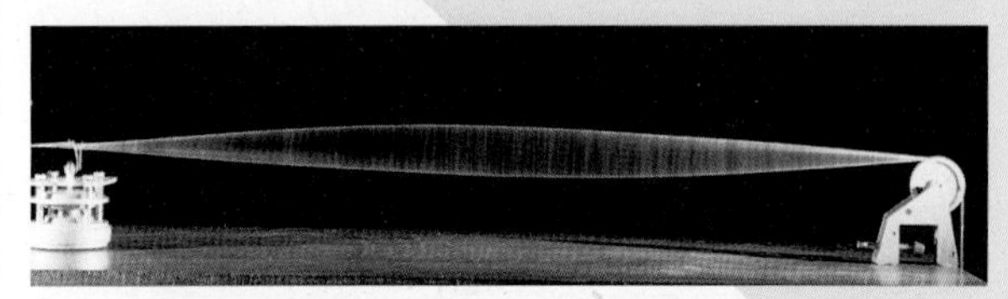

***Figure P3*** *The first harmonic*

3. To investigate the effect of varying the length, $l$, of the string:

   The procedure is similar to that described in step 1.

   A graph of frequency ($y$-axis) against $1/l$ is plotted: a straight line verifies the relationship.

The following techniques are good practice:

- The string should be horizontal. A spirit level or a set square can be used to check this.
- The independent variable in each case should be varied over as wide a range of values as is practicable.
- It can be difficult to find the exact frequency at which the first harmonic occurs. Fine adjustments of the signal generator are needed to get the maximum amplitude on the stationary wave.
- Markers can be used to show the position of the nodes, which may not be exactly at the ends of the string. The length between the markers is measured with a metre ruler.
- There is an element of judgement in deciding at what frequency a standing wave is established. This is a random error, since the frequency may be underestimated or overestimated. Therefore, the uncertainty in the results can be reduced by repeating the readings. It is good practice to take a complete series of readings as the variable (for example, mass) is increased and then take a second complete set of readings as the variable is decreased.

## QUESTIONS

**P1** Explain why the string vibrates strongly only at certain frequencies.

**P2** Suppose that you find that the string vibrates strongly at a frequency of 75 Hz. At this frequency you can see four nodes, one at each end and two on the string.

a. Which order harmonic is this, and how do you know?

b. At what other frequencies would you expect to see large-amplitude vibrations?

**P3** It is a good idea to use markers to mark the nodes. The markers could be large optical pins stuck into a cork. Why will these help your measurement? How would you use them to best effect?

**P4** The vibration generator should not be run at its largest amplitude, as this might affect your results. Suggest what might happen.

**P5** Explain how the stroboscope seems to freeze the motion of the string. At which frequencies will this occur?

P6 Suppose you are investigating the effect of changing the tension in the string. How would you use repeat readings to obtain a more precise answer?

P7 How could you find the velocity of waves on the string?

## Sound waves in pipes

It is possible to set up longitudinal stationary waves, by reflecting a sound wave for example (Figure 38).

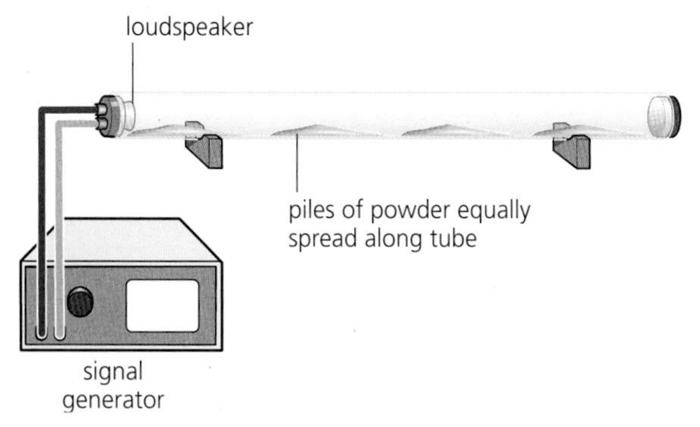


***Figure 38*** *The apparatus uses a loudspeaker at one end of a long transparent tube, and the reflected sound from the other end, to set up a stationary sound wave in the tube. Fine powder is sprinkled in the tube. The powder collects at the points of minimum vibration and so the nodes become visible.*

The air in the tubes and pipes of musical instruments, such as organs, oboes and trumpets (Figure 39), can be made to vibrate as a stationary wave. Sound waves can reflect from the open end of a pipe, as well as the closed end, so that standing waves can be set up in both kinds of pipe (Figure 40). There is always a node at a closed end of the pipe, and an antinode at an open end. That means that different modes of vibration occur in pipes that are closed at one end, like an oboe, or open at both ends, like a flute.

- In an open pipe, all the harmonics can exist, $f_1, 2f_1, 3f_1, 4f_1, ....$
- A closed pipe will only support the odd harmonics, $f_1, 3f_1, 5f_1, ....$

The fundamental note of a closed pipe is always a lower pitch, half the frequency, of an open pipe of the same length (Figure 40).

***Figure 39*** *In a trumpet the length of the vibrating air column can be altered by use of valves, but the real skill of the trumpeter is in blowing in the right way to excite the required harmonics.*

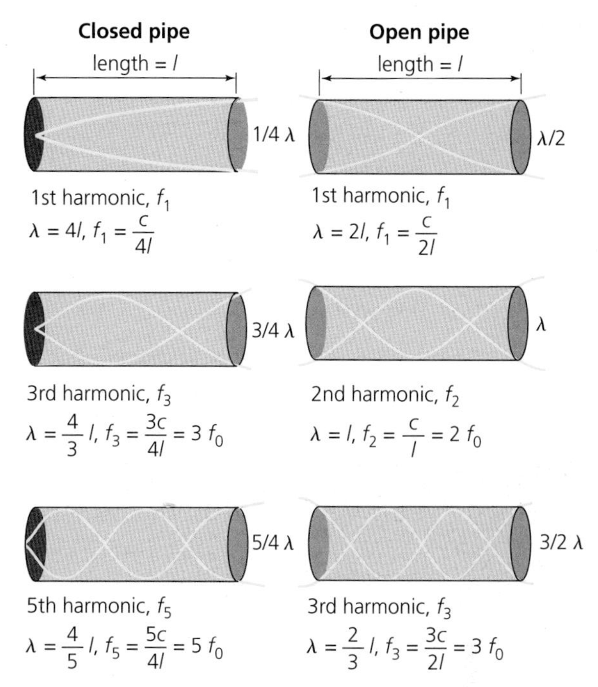


***Figure 40*** *Vibrations in open and closed pipes. Note that although the sound wave is a longitudinal wave, it can be drawn as a transverse wave, effectively a plot of pressure changes (y-axis) against position (x-axis).*

## Microwaves

A stationary wave is set up in a microwave oven. There are nodes and antinodes of microwave intensity. A turntable has to be used to ensure even cooking. If the turntable is removed, and a piece of chocolate is given a very short time in the oven, you can see where the antinodes were (Figure 41).

***Figure 41*** *The chocolate has been melted at antinodes of microwaves.*

You can estimate the speed of light using a microwave oven and a piece of chocolate in this way. See the next worked example.

### Worked example

The frequency of the electromagnetic waves in a microwave oven is around 2.5 GHz. The distance between two successive antinodes shown by the chocolate in Figure 41 is approximately 8 cm, which gives a wavelength of 16 cm.

Using $c = f\lambda$, we get

$$c = 2.5 \times 10^9 \times 0.16 = 4 \times 10^8 \, \text{m s}^{-1}$$

This is a bit high, but the right order of magnitude.

### QUESTIONS

**31. a.** The air in a pipe that is open at each end is made to vibrate. If the pipe is 30 cm long, what frequency would the first harmonic be?

**b.** How would the frequency of the first harmonic change if the pipe were then closed at one end?

**32. a.** Sketch the stationary wave pattern for the fifth harmonic for a wave on a string.

**b.** Sketch the ninth harmonic for a sound wave in a pipe closed at one end.

### KEY IDEAS

- The principle of superposition states that when two waves meet, the resulting displacement at any point is the vector sum of the two individual displacements.
- Stationary waves are formed by the superposition of two progressive waves of the same frequency that are travelling in opposite directions.
- Stationary waves formed by the superposition of waves of equal amplitude have points of zero amplitude called nodes that are half a wavelength apart.
- Stationary waves have points of maximum amplitude called antinodes that are half a wavelength apart.
- Stationary waves do not transfer energy in the direction of propagation as a progressive wave does.
- Every point of a progressive wave eventually reaches maximum displacement (amplitude). This is not true of a stationary wave, where the amplitude depends on position.
- There is a phase difference between any two adjacent points on a progressive wave. This is not the case for a stationary wave, where all the points between any two consecutive nodes are in phase, though there is a phase change of $\pi$ radians (180°) either side of a node.
- The frequencies at which a stationary wave is set up on a string or in a tube are known as harmonics.
- The lowest frequency at which a stationary wave is set up on a string or in a tube is the first harmonic.

### ASSIGNMENT 4: MEASURING THE SPEED OF SOUND IN AIR

**(MS 1.1, MS 1.2, MS 1.5, MS 2.2, PS 1.1, PS 2.1, PS 2.3, PS 3.2, PS 3.3)**

The speed of sound in air is not a constant, but depends on factors such as temperature. In dry air at 0 °C, sound travels at approximately 330 m s$^{-1}$. That is quick enough to present some measurement difficulties for a school physics laboratory. However, there are several different methods, each with its own practical problems and uncertainties.

One simple method is to measure the delay between seeing an event and hearing it. The speed of light in air is about a million times faster than that of sound.

If you are some distance away from an event, a thunderstorm for example, there is a noticeable delay between seeing the lightning and hearing the thunder.

The speed of sound is (distance from the event)/(time delay). To get a precise result, the distance, and therefore the delay, needs to be as large as possible.

An observer 200 m away from an event would record a time of 200/330 = 0.61 s. An uncertainty in timing of 0.1 s leads to a percentage uncertainty of (0.1/0.61) × 100 = 17%. If the distance is measured to the nearest metre, the percentage uncertainty in the distance measurement would be (1/200) × 100 or 0.5%. Adding the percentage uncertainties, this method would give an uncertainty of around 18%. This could be reduced by increasing the distance, but creating an event that can be easily seen and heard at a fairly large, known, distance is a problem.

### Questions

**A1** The delay between seeing and hearing the same event is a problem when timing races at school sports day. Suppose the timekeepers for the 100 m race stand at the finishing line. They are equipped with a stopwatch in each hand, in an attempt to get accurate times for first and second place. They are told to start timing when they see the puff of smoke from the starting gun, not when they hear it, and stop when the winner (right-hand stopwatch) and second place (left-hand stopwatch) cross the line.

Study the results in Table A1. Comment on the accuracy and precision of the results. What random and systematic errors might have occurred here? Are all the timekeepers' results equally valid? Calculate the best value for the mean and estimate the uncertainty of your answers. Suggest what improvements could be made to the method.

| | Timekeeper | | | |
|---|---|---|---|---|
| Position | Mr Smith | Dr Ellis | Bethan R | Eric H |
| Winner / s | 12.211 | 12.17 | 11.91 | 12.25 |
| Runner-up / s | 13.320 | 13.31 | 13.00 | 13.35 |

***Table A1*** *Times recorded by the timekeepers for the 100 m sprint final*

**A2** a. One way to increase precision is to use an echo method (Figure A1). By how much will the uncertainty be reduced?

b. The percentage uncertainty can be reduced still further by timing a number of echoes. Explain how this works. What could be done to make the result as accurate and precise as possible?

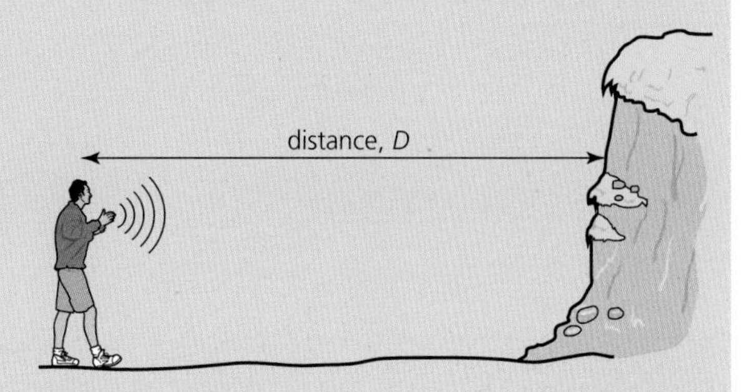


***Figure A1*** *The echo method*

Electronic timers with high precision can be used to measure the passage of sound waves, and hence their speed, but they need an electrical signal to start and stop automatically. The method requires two microphones and a sound-activated timer (Figure A2) or a single-sweep oscilloscope (one that can capture transient phenomena). There are apps for smartphones and tablets that can do all of these things.

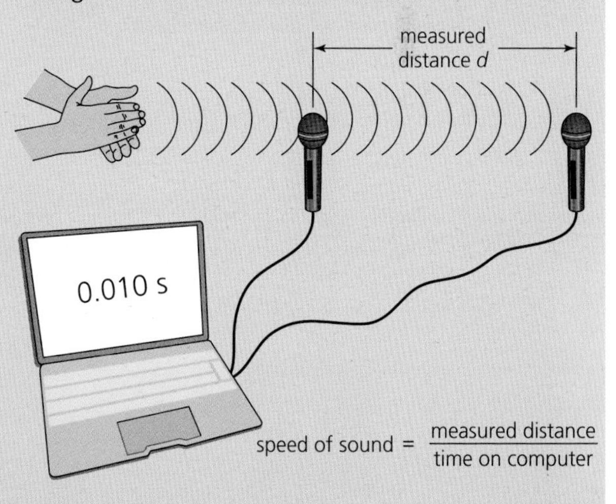


***Figure A2*** *An electronic timing method*

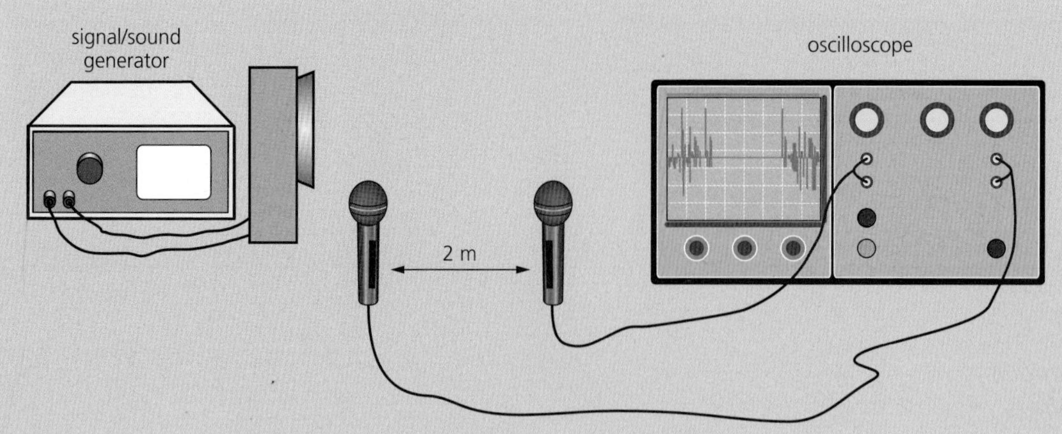


*Figure A3* A method using an oscilloscope to measure the time interval

## Questions

**A3** Can you foresee any problems with the method shown in Figure A3? How could you get round them?

As strange as it might seem, a stationary sound wave can be used to measure the speed of sound. One method uses the equipment shown in Figure A4. A loudspeaker and a reflector are used to set up a stationary sound wave. The frequency of a signal generator is adjusted until a stationary sound wave is set up in the tube. The easiest way to detect this is to listen. When a stationary wave is established, the apparatus will resonate and the loudness will be at a maximum. As the microphone is moved up and down the tube, the oscilloscope will show a maximum signal, followed by a minimum signal, corresponding to nodes and antinodes. A sound level meter can be used instead of an oscilloscope.

**A4** **a.** What is the purpose of the tube in Figure A4?

**b.** Explain how you would use the results of the experiment in Figure A4 to find the speed of sound in air. How could the precision of the results be improved?

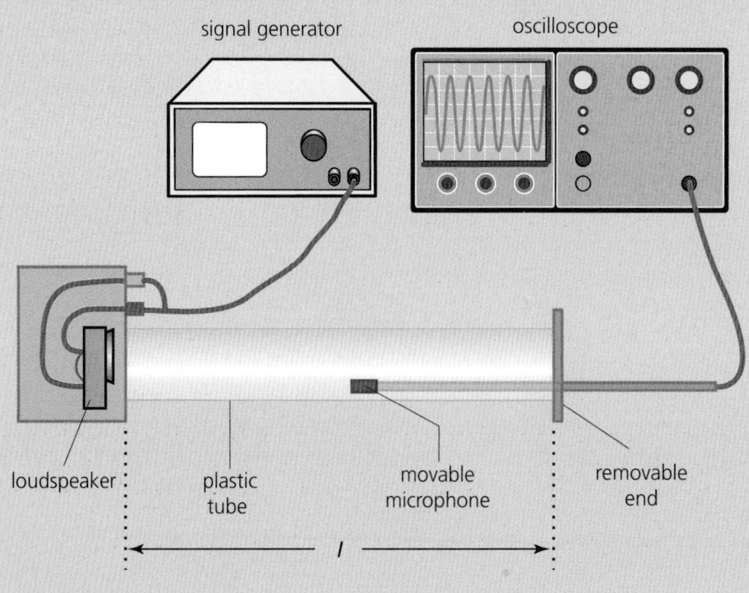


*Figure A4* A stationary wave method

## Stretch and challenge

A second way of determining the speed of sound by using stationary waves is the 'resonance tube' method. If you blow over the neck of a bottle, it will emit a note (Figure A5). This note is its natural frequency and it depends on the length of air in the bottle. This effect can be used to measure the speed of sound.

***Figure A5*** *A stationary sound wave is set up in the air in the bottle.*

A set of tuning forks, or notes played through a loudspeaker, are used to set up a standing wave in the air column (Figure A6). For each tuning fork, the position of the pipe is adjusted until a maximum sound is heard (a sound level meter could be used). The shortest length of air column that gives a maximum sound corresponds to the first harmonic. When this happens, the length of the tube, $l$, is almost equal to one-quarter of a wavelength. The length is not exactly $\lambda/4$ because the antinode occurs just outside the tube. This is called the end correction, $c$. So

$$\lambda/4 = l + c \quad \text{or} \quad \lambda = 4l + 4c$$

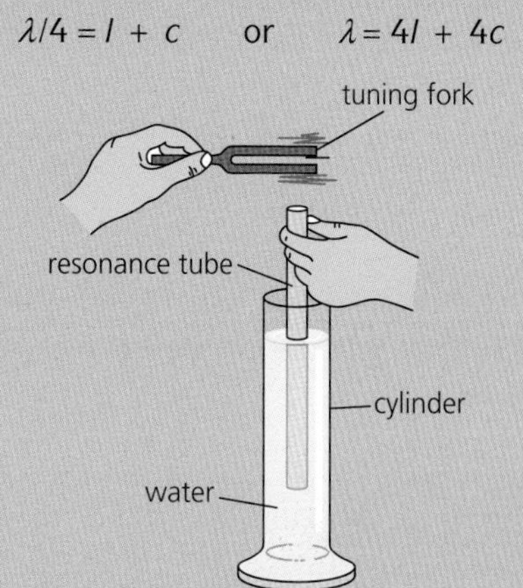


***Figure A6*** *The resonance tube method*

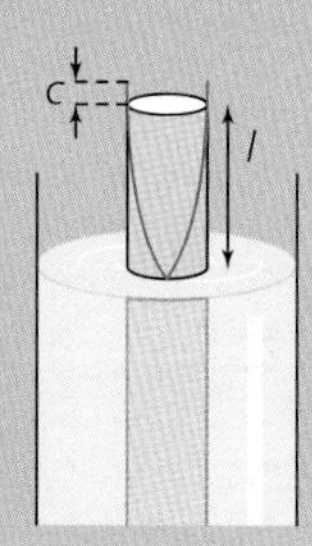


***Figure A7*** *The end correction*

### Questions

**A5** You can get a set of results for notes of different frequencies. How would you use those results to find a value for the speed of sound in air? How would you find the end correction? (You would need to plot a straight-line graph. What would you plot on each axis?) How would you get an estimate of the uncertainty in your answer?

## Applying the properties of waves

It is possible to create stationary light waves by reflecting light, for example by reflecting a laser beam back on itself. This standing wave can be used to direct a stream of atoms to a high degree of precision (Figure 42). Although light can be used to shape a material at a very small scale, there are fundamental properties of waves – in particular, diffraction – that make it impossible to see individual atoms with light. The photograph in Figure 42 is imaged using a stream of atoms, rather than light waves. Diffraction, along with interference, will be discussed in the next chapter.

chromium atoms

laser

laser

deposited lines

$\lambda/2 \approx 213$ nm

standing wave

***Figure 42*** *Chromium atoms are directed down onto a surface, typically silicon, where they are deposited. Just before the atoms hit the surface, they are deflected by the standing wave of laser light rather like raindrops falling onto a corrugated iron roof. The atoms collect into lines and so form an array of narrow lines on the surface. These are shown in the photograph, which is a scanning atomic force microscope image.*

## PRACTICE QUESTIONS

**1.** Earthquakes produce transverse and longitudinal seismic waves that travel through rock. Figure Q1 shows the displacement of the particles of rock at a given instant, for different positions along a transverse wave.

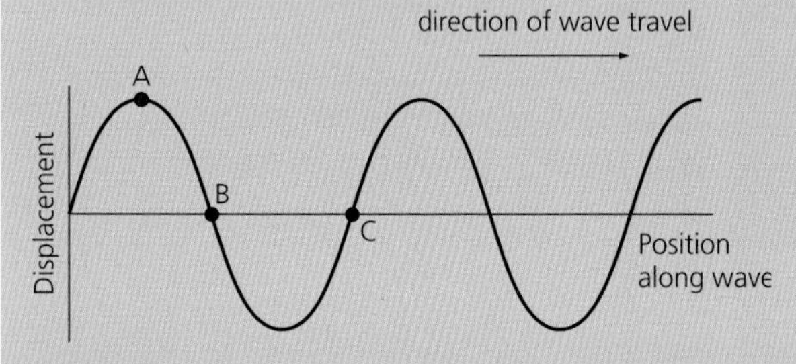


***Figure Q1***

**a.** State the phase difference between

**i.** points A and B on the wave

**ii.** points A and C on the wave.

**b.** Describe the motion of the rock particle at point B during the passage of the next complete cycle.

**c.** A scientist detects a seismic wave that is polarised. State and explain what the scientist can deduce from this information.

**d.** The frequency of the seismic wave is measured to be 6.0 Hz.

**i.** Define the frequency of a progressive wave.

**ii.** Calculate the wavelength of the wave if its speed is $4.5 \times 10^3\,\mathrm{m\,s^{-1}}$.

*AQA Unit 2 June 2013 Q6*

**2.** Figure Q2 shows a continuous progressive wave on a rope. There is a knot in the rope.

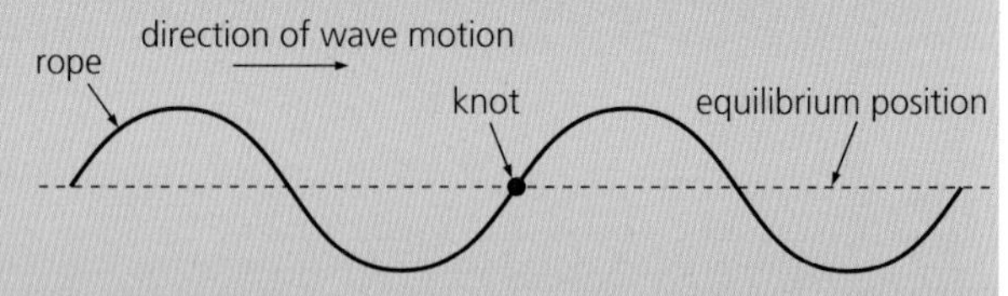


***Figure Q2***

**a.** Define the amplitude of a wave.

**b.** The wave travels to the right. Describe how the **vertical** displacement of the knot varies over the next complete cycle.

c. A continuous wave of the same amplitude and frequency moves along the rope from the right and passes through the first wave. The knot becomes motionless. Explain how this could happen.

*AQA Unit 2 Jan 2012 Q7*

3. When a note is played on a violin, the sound it produces consists of many harmonics.

Figure Q3 shows the shape of the string for a stationary wave that corresponds to one of these harmonics. The positions of maximum and zero displacement for one harmonic are shown. Points A and B are fixed. Points X, Y and Z are points on the string.

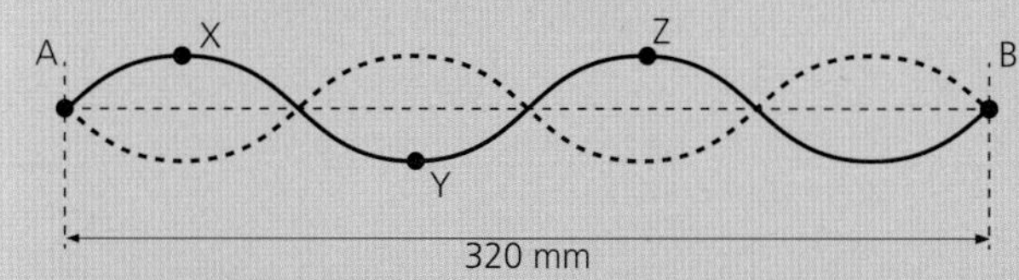


***Figure Q3***

a. Describe the motion of point X.

b. State the phase relationship between

i. X and Y

ii. X and Z.

c. The frequency of this harmonic is 780 Hz.

i. Show that the speed of a progressive wave on this string is about $125\,m\,s^{-1}$.

ii. Calculate the time taken for the string at point Z to move from maximum displacement back to zero displacement.

d. The violinist presses on the string at C to shorten the part of the string that vibrates. Figure Q4 shows the string between C and B vibrating in its first harmonic. The length of the whole string is 320 mm and the distance between C and B is 240 mm.

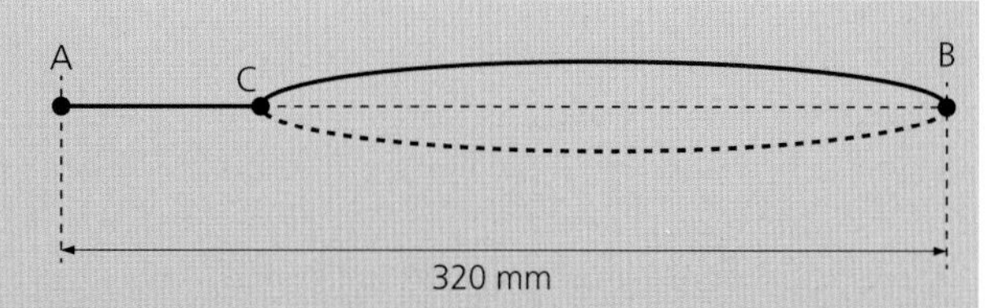


***Figure Q4***

i. State the name given to the point on the wave midway between C and B.

ii. Calculate the wavelength of this stationary wave.

iii. Calculate the frequency of this first harmonic. The speed of the progressive wave remains at $125\,m\,s^{-1}$.

*AQA June 2012 Unit 2 Q6 (adapted)*

4. Sugar solution is optically active. It can rotate the plane of polarisation of light. The amount of rotation depends on the concentration of the sugar solution. This effect is used to measure the concentration of sugar solution. The rotation of the plane of polarisation is measured with a polarimeter (Figure Q5).

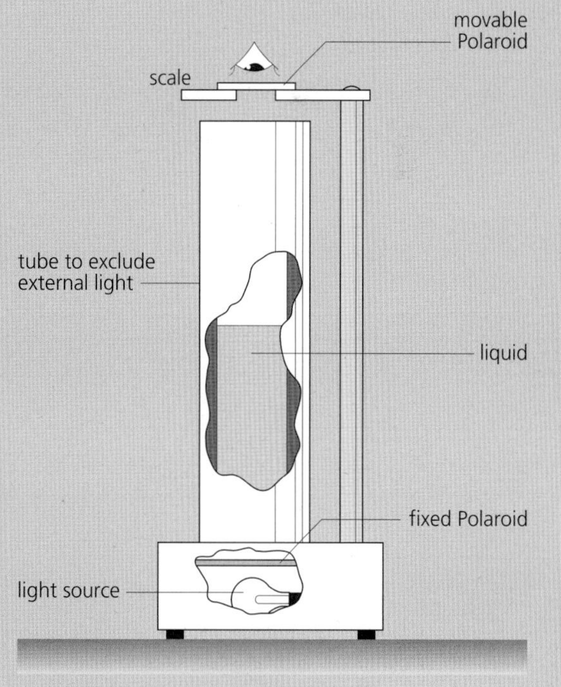


***Figure Q5***

Initially, with no sugar solution in place, the lower Polaroid filter (the polariser) is placed at 90° to the second filter (the analyser). These Polaroids are said to be crossed and there is no transmission of light through the polarimeter. The tube containing sugar solution is then placed between the two crossed Polaroid filters. Light will be observed passing through the analyser. The analyser is then rotated through an angle, $q$, until no light is observed. The angle $q$ is given by

$$q = alc$$

where $a$ is the specific rotation of sugar solution (in degrees $m^2 kg^{-1}$), $l$ = the length of the polarimeter tube (m) and $c$ is the concentration of sugar solution (in kg of sugar per $m^3$ of water).

An experiment is carried out using varying concentrations of sugar and the rotation of the plane of polarisation is measured in each case. The results in Table Q1 were recorded.

| Concentration of sugar / $kg\,m^{-3}$ | Angle of rotation / degrees |
|---|---|
| 50 | 19 |
| 80 | 31 |
| 110 | 44 |
| 140 | 55 |
| 170 | 65 |
| 200 | 74 |
| 230 | 89 |
| 260 | 100 |

*Table Q1*

a. Plot a suitable graph to verify that the angle of rotation is proportional to the concentration of the solution.

b. If the length of the polarimeter tube was 20 cm, find a value for $a$, the specific rotation for sugar.

5. Which of these statements A to D is true?

**A** Longitudinal waves cannot be polarised.

**B** Transverse waves cannot be diffracted.

**C** Transverse waves cannot be polarised.

**D** Longitudinal waves cannot be diffracted.

6. The wavelength of the microwaves generated in a microwave oven is likely to be around:

**A** 12 μm

**B** 12 mm

**C** 12 cm

**D** 12 m

7. Which row of Table Q2 is correct?

| | Stationary wave | Progressive wave |
|---|---|---|
| A | All points on the wave oscillate in phase. | All points on the wave oscillate in phase. |
| B | There is a phase difference of 180° between points on either side of a node. | All points on the wave oscillate in phase. |
| C | There is a phase difference of 360° between points on either side of a node. | There is a phase difference of 180° between points that are one wavelength apart. |
| D | There is a phase difference of 180° between points on either side of a node. | There is a phase difference of 360° between points that are one wavelength apart. |

*Table Q2*

8. Which of the following statements about a stationary wave on a string is **not** true?

**A** The amplitude of the oscillations of a point depends on its position along the string.

**B** There are always at least two points where the amplitude is zero.

**C** The amplitude of the oscillations of a given point varies with time.

**D** The displacement of a point at any time is the sum of the displacements of a wave and its reflection.

9. Which row in Table Q3 correctly describes an experiment that shows that light is a transverse wave?

| | Test | Result |
|---|---|---|
| A | Shine a torch through a Polaroid filter | Light is dimmed |
| B | Shine a torch through a Polaroid filter and rotate the filter | Light is dimmed at a certain angle |
| C | Shine a torch through two Polaroids and rotate one of the filters | Light is dimmed at a certain angle |
| D | Shine a torch through two Polaroids and rotate both of the filters | Light is dimmed at a certain angle |

*Table Q3*

10. A large slab of chocolate is placed in a microwave oven, which has had the rotating plate removed. The microwave oven is switched on for a few seconds and the chocolate is found to have melted in a few spots that are about 6 cm apart. The frequency of the microwaves is approximately:

**A** 2.5 GHz

**B** 5.0 GHz

**C** 2.5 MHz

**D** 5.0 MHz

## Stretch and challenge

11. There are several different types of wave that travel across or through the sea. We are familiar with surface waves, which rise and fall due to wind action and the force of gravity. But there are other types: read the descriptions below.

*Capillary waves*: These are small-amplitude surface waves with a short wavelength. The surface tension of the water, the same force that pulls water into drops, is dominant in these waves, usually referred to as ripples. The maximum wavelength of a capillary wave is about 1.73 cm. The velocity of capillary waves increases with decreasing wavelength, the minimum velocity being 23.1 $cm\,s^{-1}$.

*Internal waves*: These waves, which can have amplitudes as high as several hundred metres, form deep below the surface of the ocean, between layers of water with markedly different densities. They travel at low speed but seem to be responsible for much of the mixing of nutrients (among other things) in the oceans.

*Seismic P-waves*: These are compression waves caused by seismic disturbances, known as P-waves (primary), because they are the first waves to arrive at the surface during an earthquake. They travel at about 1450 $m\,s^{-1}$ in water, with a period of anything between 0.05 s and 50 min.

**a.** From the descriptions state and explain which of these types of wave are longitudinal and which are transverse.

**b.** Copy and complete Table Q4 for each type of wave. Look back through the chapter, do further research and use estimation where necessary. Use a range of values where appropriate.

| | Velocity / $m\,s^{-1}$ (range) | Wavelength / m (range) | Frequency / Hz (range) |
|---|---|---|---|
| Surface wave | | | |
| Capillary wave | | | |
| Internal wave | | | |
| Seismic P-wave | | | |

*Table Q4*

# 6 DIFFRACTION AND INTERFERENCE

## PRIOR KNOWLEDGE

You will be aware of the wave nature of light from Chapter 5, particularly the way that waves add together according to the principle of superposition of waves.

## LEARNING OBJECTIVES

In this chapter you will learn how the wave nature of light causes diffraction and interference effects, and you will calculate the magnitude of these effects.

**(Specification 3.3.2.1, 3.3.2.2)**

The camera on a mobile phone is often its most important selling point. Advertisers tempt us to buy with promises of high-resolution images taken by phones that have 40-megapixel cameras. An image formed from 40 million image points should certainly be high definition. But megapixels are not the whole story when it comes to image quality (Figure 1). A camera phone with 40 megapixels, densely packed on a very small sensor chip, may produce poorer images than a good-quality camera with 20 or 30 megapixels over a much larger sensor. There are other important factors, particularly the size and quality of the lens.

***Figure 1*** *The phone camera is portable and allows instant communication of the images it takes. But it cannot match larger cameras for image quality.*

One measure of image quality is resolution, which describes the size of the smallest detail that can be clearly seen, or resolved, in the photograph. This is partly determined by the number of pixels in the image, but also depends on the diameter of the lens (Figure 2). Light is a wave, and the physics of wave motion sets a limit on the image quality that a small lens can achieve.

***Figure 2*** *When it comes to cameras and telescopes, size is important. The wider the lens, the brighter and more detailed the image.*

Light waves spread out as they pass through an opening (aperture), such as the pupil of your eye or a camera lens. This spreading, or diffraction, limits our ability to distinguish separate objects. The effects of diffraction are reduced with larger apertures; that is why a camera with a large-diameter lens is needed to take high-resolution photographs, particularly of objects in the distance.

# 6.1 INTERFERENCE

When people talk about interference affecting their reception, they usually mean the unwanted noise on a radio receiver, or some crackling on a mobile phone. In physics, the word **interference** has a much more precise meaning. Interference is the effect produced when two sets of waves that have a constant phase difference (see section 5.2 of Chapter 5) are added together. An example of interference of light is shown in Figure 3.

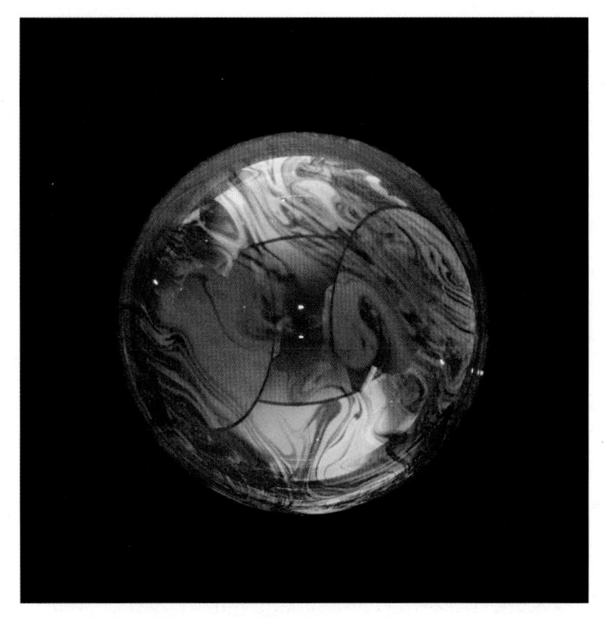

***Figure 3*** *This effect is due to interference of light waves of identical wavelength on reflection from the bubble surface.*

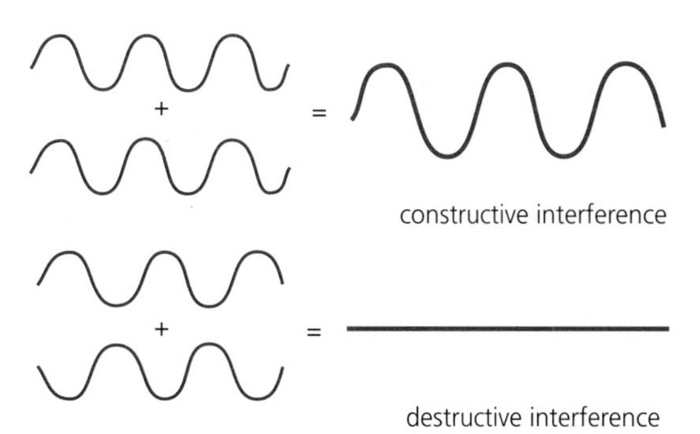


Two similar waves may add together (reinforce) to make a double-height wave, or add together so as to cancel each other out.

***Figure 4*** *The superposition of waves cause interference.*

Interference is a particular example of the principle of superposition of waves (see section 5.5 of Chapter 5). This principle states that the resultant displacement caused when two waves overlap is equal to the vector sum of the displacements from each individual wave (Figure 4). The principle applies to all electromagnetic waves and sound waves.

Electromagnetic waves can be added together at any point in space, even though they may have come from different directions. Superposition will result in a steady interference pattern if the waves have the same frequency and a constant phase difference.

Suppose that your home is between two television transmitters, which are broadcasting in phase on the same frequency (Figure 5). The total signal strength that you receive is the sum of the two separate waves. If the waves arrive in phase, you will receive a stronger signal.

The waves at **A** have travelled equal distances, **AP** and **AQ**, so they must be in phase. They will add together, giving twice the signal amplitude. Waves at **B** have a path difference of **BQ** – **BP** = 1 wavelength. These two waves will also be in phase, so the signal will increase in strength. Waves at **C** do not fit this simple rule. **CP** is 3.5 wavelengths and **CQ** is 4 wavelengths. There is a half-wavelength difference, so the waves will be in antiphase. That means they will cancel out. The antenna at **C** would receive a very weak signal. **A**, **B** and **C** are three points in space. Performing the same calculations for other points shows that there are lines along which the signal is stronger (constructive interference) and lines of almost complete cancellation (destructive interference).

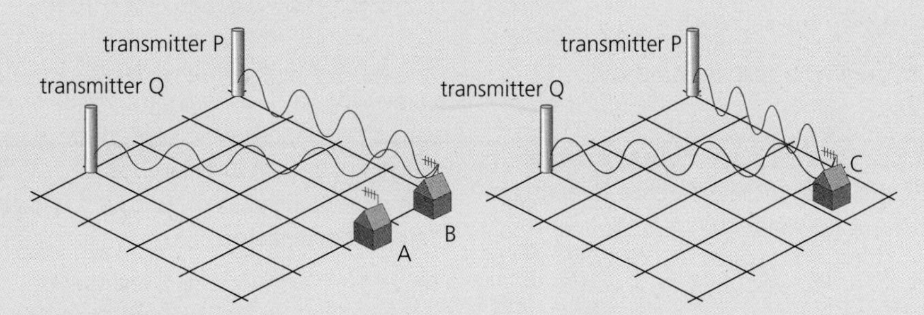


***Figure 5*** *Interference between two radio waves, emitted from transmitters P and Q with zero initial phase difference*

If the waves arrive out of phase, you will receive a weaker signal. Although the waves may be in phase when they leave the transmitter, the phase difference between them when they arrive at your house depends on how far each wave has travelled. In particular, the total signal strength will depend on the **path difference** between the waves, that is, the difference in the distance travelled by each wave (see Figure 5). When the path difference is zero or a whole number of wavelengths, $\lambda$, $2\lambda$, $3\lambda$, $4\lambda$, ..., $n\lambda$, where $n = 0, 1, 2, \ldots$, the waves will be exactly in phase and will add together to give a larger wave. This is **constructive interference**.

When the path difference is an odd number of half-wavelengths, $\frac{1}{2}\lambda$, $\frac{3}{2}\lambda$ (or $1\frac{1}{2}\lambda$), $\frac{5}{2}\lambda$ (or $2\frac{1}{2}\lambda$), ..., $(n + \frac{1}{2})\lambda$, where $n = 0, 1, 2, \ldots$, then **destructive interference** occurs, since the waves will arrive in antiphase. It would be inconvenient to have strips of the country where there was very poor TV or radio reception because of destructive interference. Therefore, adjacent transmitters ensure that the waves cannot interfere by using either slightly different frequencies or different polarisations (see Figure 25 in Chapter 5).

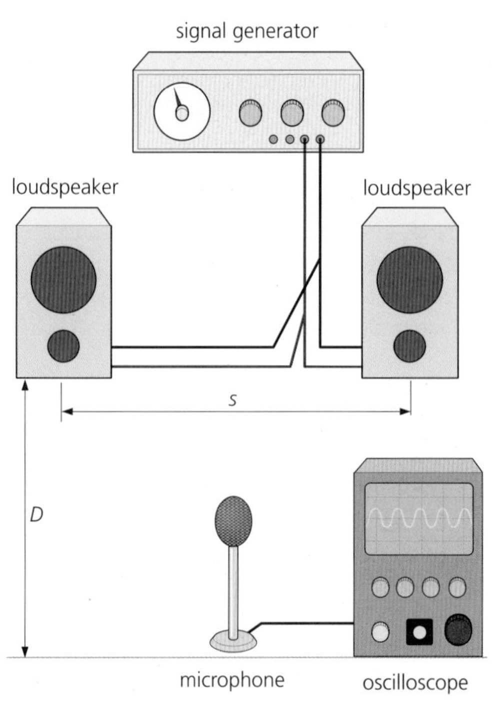


***Figure 6*** *Apparatus for observing interference between sound waves*

### Interference and sound

Sound waves are also added together according to the principle of superposition, but persistent interference patterns are only seen (or heard!) in special circumstances. Two loudspeakers playing an identical sound, a note of the same frequency and about the same amplitude, would allow us to observe regions of silence and of louder sound. Sources that produce waves of the same frequency that maintain a constant phase difference are said to be **coherent**.

The apparatus in Figure 6 will produce an interference pattern that can be observed with a microphone, with regions of silence (interference minima) where destructive interference is cancelling out the waves. The distance between these regions, $w$, will depend on:

- the wavelength of the sound wave, $\lambda$
- the distance between the speakers, $s$
- the distance between the speakers and the microphone, $D$.

It can be shown (see Figure 12 a few pages on) that the separation of two adjacent interference minima (or between two adjacent maxima) is

$$w = \frac{\lambda D}{s}$$

This is often called the **two-source interference formula**.

### Worked example 1

If two loudspeakers were placed 2.0 m apart, each playing a sound wave of frequency 400 Hz, how far apart would the interference maxima be?

The wavelength of the sound waves is

$$\lambda = \frac{c}{f}$$

where $c$ is the speed of sound, say 330 m s$^{-1}$. So

$$\lambda = \frac{330}{400} = 0.83 \text{ m}$$

If we observe the interference pattern at a distance of 5.0 m from the speakers, the separation of adjacent maxima will be

$$w = \frac{\lambda D}{s} = \frac{0.83 \times 5.0}{2.0} = 2.1\text{m}$$

The same analysis applies to any two waves of the same type, such as two light waves or two water waves (see Figure 7), provided the waves are coherent. In light interference the bands of maxima and minima produced are called **fringes**; $w$ is the **fringe separation**.

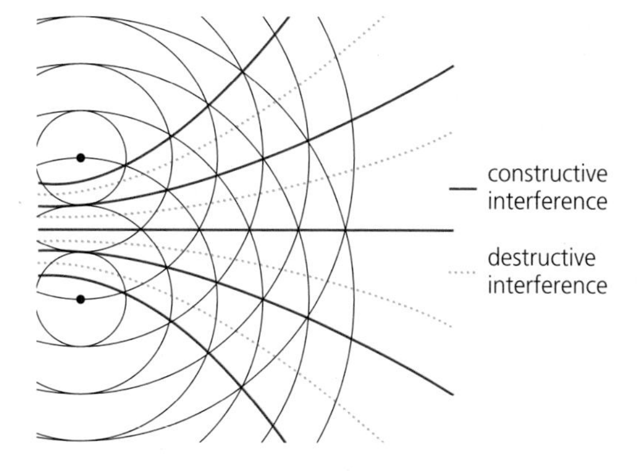


***Figure 7*** *Interference between ripples on a water tank. The diagram shows how this pattern results from constructive and destructive superposition of waves from two point coherent sources.*

## QUESTIONS

1. Two loudspeakers are placed so that they face each other at a distance of 10 m apart. The loudspeakers are playing identical notes of wavelength 1.0 m. A microphone is moved along a line from one loudspeaker to the other. What would the microphone detect? Explain your answer.
2. An interference pattern is created by two loudspeakers placed 3.0 m apart. An observer, 5.0 m away from the loudspeakers (see Figure 8), detects quiet regions that are 50 cm apart.
   a. What frequency note are the loudspeakers playing?
   b. Suggest *three* changes that could be made to the situation that would make the quiet regions further apart.

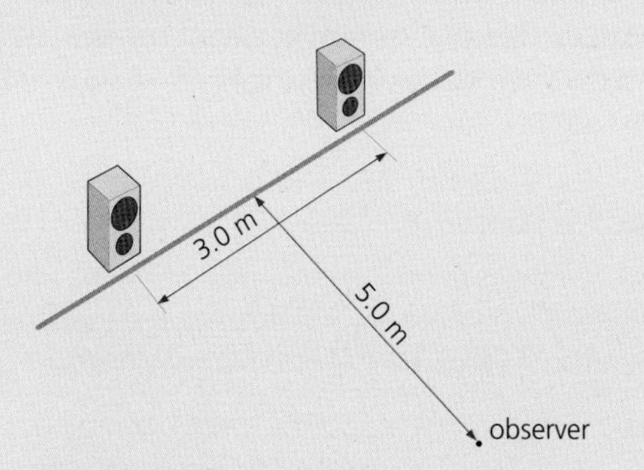


***Figure 8***

3. Why do you not notice interference effects in a room with two stereo speakers playing?
4. There are apps for smartphones and tablets that will produce a constant tone at a frequency that you choose (search for signal generator apps). If you get two phones with this app installed, set them to the same frequency, starting at about 1 kHz, and place them a metre or so apart, and then you should be able to hear quiet and loud patches by moving your head around.
   a. How far apart would you expect the quiet patches to be? (Speed of sound in air ≈ 330 m $s^{-1}$.)
   b. The experiment works better outdoors, well away from buildings. Why?

### Stretch and challenge

   c. If you set the frequencies just slightly different, 1.00 kHz and 1.01 kHz for example, and keep your head still this time, what would you hear? Explain this.

## Interference and the nature of light

It seems strange to suggest that adding two beams of light together can lead to darkness, and yet under some circumstances this is possible. The interference of light is one of the main pieces of evidence for the wave theory of light. This theory has not always been accepted. Isaac Newton suggested that light was made up of tiny solid particles, which he called corpuscles. His theory was an educated guess, and one that he did not defend strongly, although he used it successfully to explain the reflection and refraction of light.

The case for the wave nature of light was first put convincingly by Thomas Young (1773–1829), a polymath who made important advances in the study of light, materials, mathematics, music theory and languages. Young demonstrated that light could form an interference pattern by putting hairs or silk threads in front of an illuminated slit (Figure 9). Supporters of the corpuscle theory could not explain this interference pattern. Young's double-slits experiment is still used to demonstrate the wave nature of light. The experiment also allows us to calculate the wavelength of the light (see Assignment 1).

It is normally very difficult to see interference effects with white light, for the following reasons.

- The wavelengths are so small (about 500 nm on average). This means that the interference fringes are usually too close together to be discernible.
- The range of colours in white light blurs the interference effects. Destructive interference for one colour is masked by the presence of other colours.
- It is difficult to get white light sources that maintain a constant phase difference. Light is emitted in very short bursts, which have random phase and random polarisation.

To show any stable and detectable interference effect, the sources of light waves need to be **coherent**. That is, they need to be:

- the same frequency
- of constant phase difference
- polarised in the same plane.

Coherence can be achieved by using monochromatic light, that is, light of a single frequency. The constant phase is achieved by using two different parts of the same wave, possibly by allowing the wave to pass through two slits (see Figure 10). The two waves will spread out (diffract) as they pass through the narrow gaps, and so overlap.

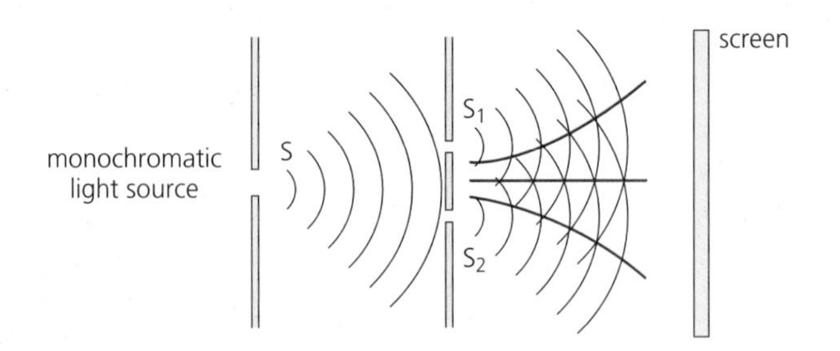


Waves emerging from $S_1$ and $S_2$ are coherent, so produce observable interference.

***Figure 10*** *Two-slits interference of monochromatic light*

Each slit acts as a point source of waves. The resulting interference system is the same as that for two point sources in a ripple tank of water, or the sound waves from two loudspeakers. You cannot see the ripples in an electromagnetic field, but you can detect the effect when the waves hit a screen or a detector. Constructive interference produces a bright fringe, while destructive interference produces a dark fringe (Figure 11).

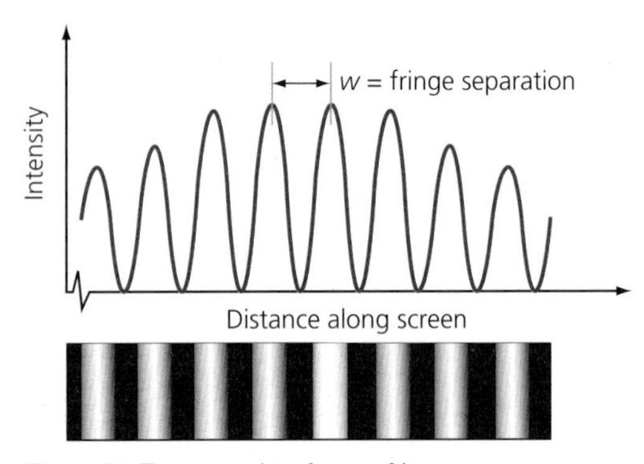


***Figure 11*** *Two-source interference fringes*

The formula that relates wavelength to the spacing of the interference fringes is just the same as for sound waves. Figure 12 shows the derivation of this formula.

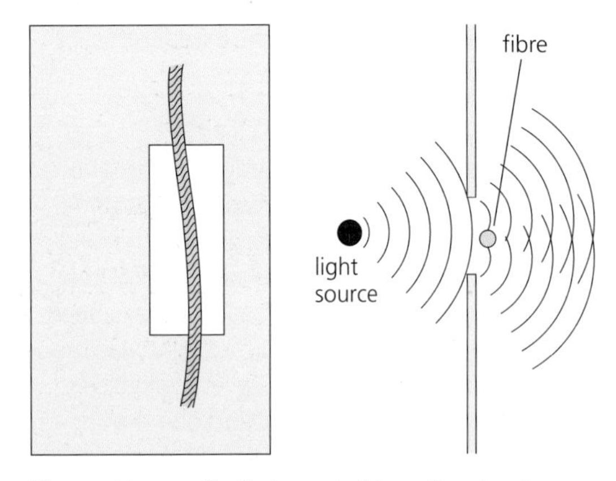


Thomas Young effectively created two slits when he stretched a fibre in front of an illuminated slit. Waves from each side of the fibre diffracted, overlapped and caused interference.

***Figure 9*** *Young's 'double-slits' experiment*

### Worked example 2

How many fringes per millimetre would be visible with the screen at a distance $D = 1.0$ m, using a slit separation of 1.0 mm and yellow light of 500 nm wavelength?

$$w = \frac{\lambda D}{s}$$

$$= \frac{500 \times 10^{-9}\,\text{m} \times 1.0\,\text{m}}{1.0 \times 10^{-3}\,\text{m}}$$

$$= 500 \times 10^{-6}\,\text{m} = 0.5\,\text{mm}$$

So there would be two fringes per millimetre.

## QUESTIONS

5. All electromagnetic waves show the same sort of pattern of two-source interference (Figure 11), but the spacing of fringes varies enormously. Why?
6. Suggest two ways of increasing fringe separation for the same colour of light in a two-slits experiment.
7. Two radio transmitters are 1.0 km apart. Both transmit a 3.0 MHz radio wave. The waves are coherent.
   a. If a car moves along a road parallel to the line joining the transmitters, how will the received signal change?
   b. If the road is 10 km from the transmitters, how far apart will positions of maximum reception be?

### Stretch and challenge

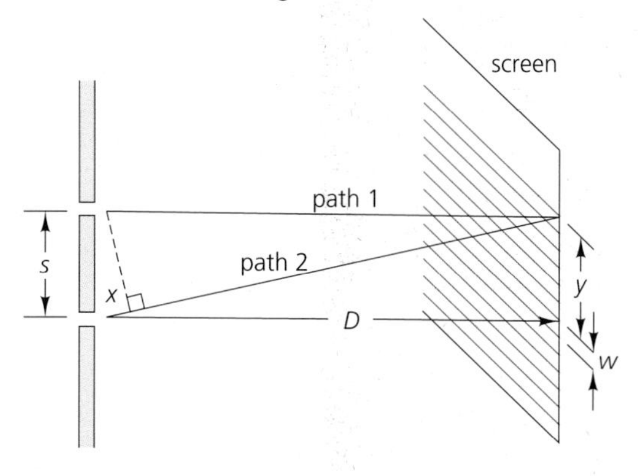


Paths 1 and 2 differ by a few wavelengths of light in a total distance $D$ of around a metre.

We can say to within less than a thousandth percentage uncertainty that: path 1 = path 2 = $D$.

The tiny difference between path 1 and path 2 (shown as $x$ on the diagram) must be a whole number of wavelengths for constructive interference to occur.

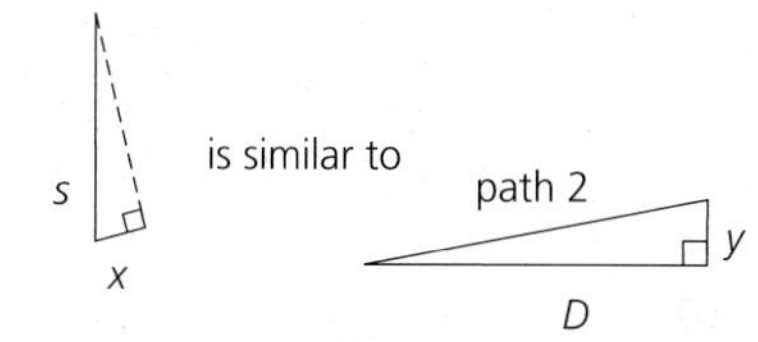


There are two similar triangles on the diagram, as shown above. For similar triangles, the ratio of sides is the same, so:

$$\frac{x}{s} = \frac{y}{\text{path 2}} \approx \frac{y}{D}$$

If we now move up to the next bright fringe, $x$ increases by one wavelength, $\lambda$, and $y$ increases by the fringe separation, $w$.

Therefore $\frac{x}{s} = \frac{y}{D}$

gives: $\frac{(x+\lambda)}{s} = \frac{(y+w)}{D}$

and so: $\frac{\lambda}{s} = \frac{w}{D}$ or $w = \frac{\lambda D}{s}$

The equation is valid whenever there is two-source interference with $w \gg \lambda$.

*Figure 12 Deriving the two-source interference formula*

## ASSIGNMENT 1: INVESTIGATING TWO-SOURCE INTERFERENCE IN WATER WAVES, MICROWAVES AND SOUND

**(MS 0.1, MS 2.3, PS 1.1, PS 1.2, PS 3.2, PS 3.3, PS 4.1)**

The interference pattern caused by two coherent sources is the same for all types of waves. In this assignment you will investigate this interference pattern in three different types of wave, and apply the two-source wave formula, $w = \frac{\lambda D}{s}$, in each case.

### Part 1: Water waves

Water waves can be made to interfere in a ripple tank. Two dippers fixed to an oscillating beam act as two sources of circular waves. The waves can best be observed by projecting the pattern onto a whiteboard, using an overhead projector (Figure A1).

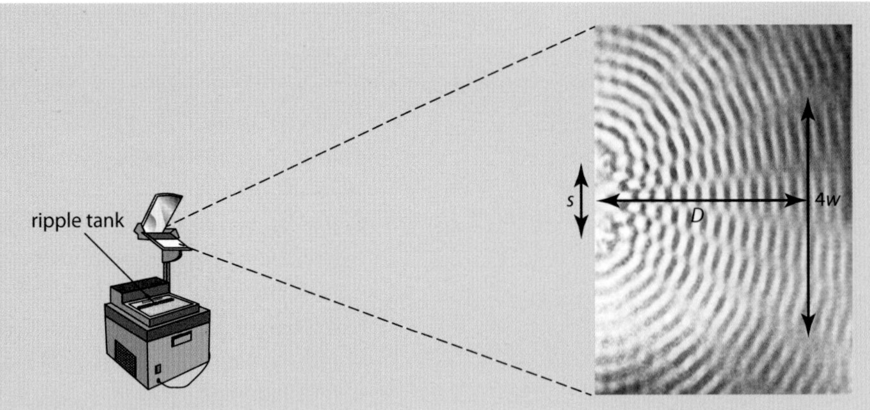


*Figure A1* *Set-up for observing the interference of water waves*

The interference pattern in the ripple tank is projected onto the whiteboard, where the readings can be taken. Lines where the waves cancel each other out can clearly be seen. The vertical line drawn on the whiteboard represents the 'screen' where the fringes are observed. The wavelength can be adjusted by changing the speed of the motor that drives the oscillating beam.

### Questions

**A1** Describe how the interference pattern changes when the wavelength is reduced.

**A2** Describe how moving the dippers closer together affects the interference pattern.

**A3** You could use readings taken from the screen, to calculate the wavelength, using $\lambda = \frac{ws}{D}$.

- **a.** Suppose $D = 2.00$ m, $s = 30$ cm and $w = 0.50$ m. What would the wavelength be?
- **b.** How could you check your answer using a hand-held mechanical stroboscope?

### Part 2: Sound waves

Sound waves can be made to interfere using two loudspeakers as shown in Figure A3. You can use a sound level meter (or a suitable app on a smartphone) to find the areas of maximum or minimum sound. Alternatively, you could use your ear. It makes it easier if you put your finger (gently) in the other ear.

### Questions

**A4** Describe how the frequency of the note played by the loudspeakers affects the spacing of the quiet patches.

**A5** What measurements would you need to take to be able to calculate the speed of sound in air from this experiment? How could you reduce the uncertainty in your measurements?

**A6** In one run of the experiment the speakers were placed 2.0 m apart, and the signal generator was set at 1.0 kHz. A sound level meter placed 6 m away from the speakers detected quiet patches (interference minima) about 1 m apart. Use these figures to calculate the speed of sound in air.

### Part 3: Microwaves

Microwaves can be used to produce a two-source interference pattern (Figure A4).

The transmitter needs to be far enough away to illuminate both slits. The receiver is usually connected to an amplifier with a loudspeaker, so that the maxima can be heard. An oscilloscope can be used instead of a loudspeaker so that the maxima can be seen. The receiver is moved along a line parallel to the metal barriers. As it is moved, a series of interference maxima and minima should be detected.

### Questions

**A7** Describe how you would use the apparatus in Figure A4 to find the wavelength of the microwaves.

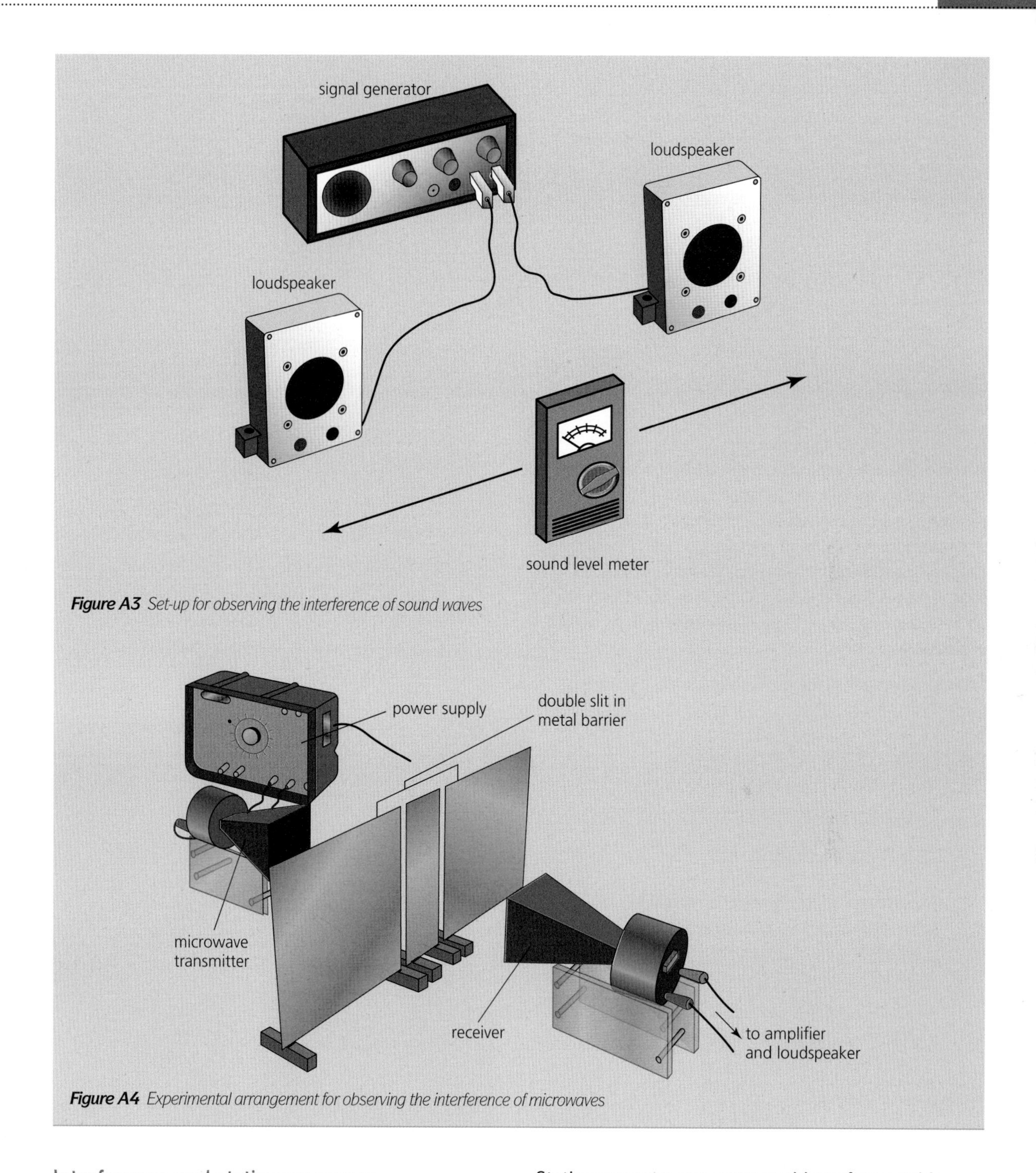

*Figure A3* Set-up for observing the interference of sound waves


*Figure A4* Experimental arrangement for observing the interference of microwaves

## Interference and stationary waves

The stationary waves that were discussed in Chapter 5 are an example of two-source interference. In a stationary wave, the two waves are travelling in opposite directions, often caused by one wave and its reflection. When the incoming wave and the reflected wave travel along the same line, the interference can set up a stationary wave (Figure 13).

Stationary waves can cause problems for portable TVs, radios and phones. The signal can change when people walk around the room or cars drive past. These problems occur because of interference between the incoming wave and its reflections from people or objects. The problem is much greater when there is a metal surface nearby, such as a car body, which acts as a good reflector.

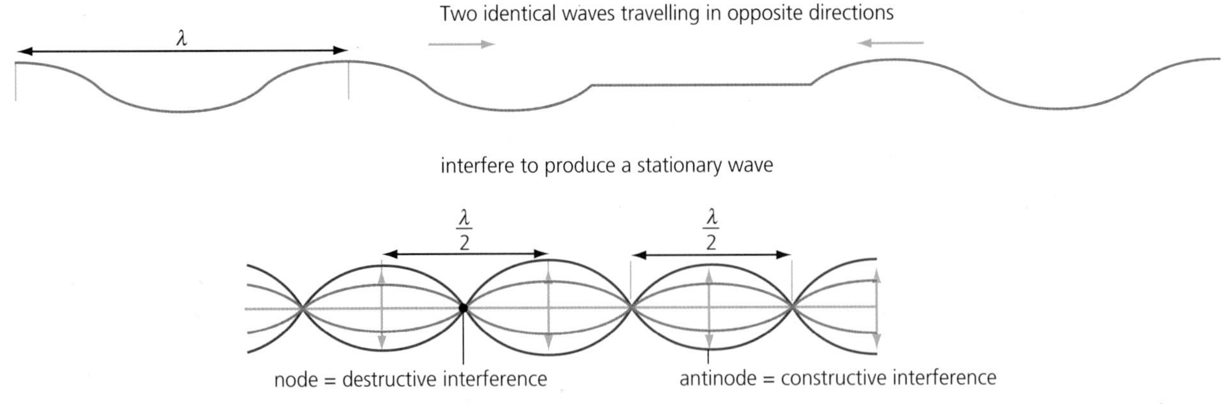


***Figure 13*** *Superposition of two waves travelling in opposite directions creates a stationary wave*

**Antinodes** are positions of constructive interference, separated by a distance $\lambda/2$ (see Figure 13). **Nodes** are positions of destructive interference, also separated by $\lambda/2$. Mobile phones operate on a wide range of wavelengths, but a typical value is 30 cm. If there was a stationary wave caused by reflections in a room, moving just 7.5 cm could move you from an antinode to a node.

More problematic is the wi-fi reception from a wireless router. Fortunately, there is now an app to map the wi-fi signal in your house, and it should be possible to avoid the nodes!

## QUESTIONS

8. Explain why a mobile phone conversation sometimes fades as you walk around the house.
9. Television aerials have a reflector a short distance from their active aerial element (Figure 14). Suggest two reasons for this.

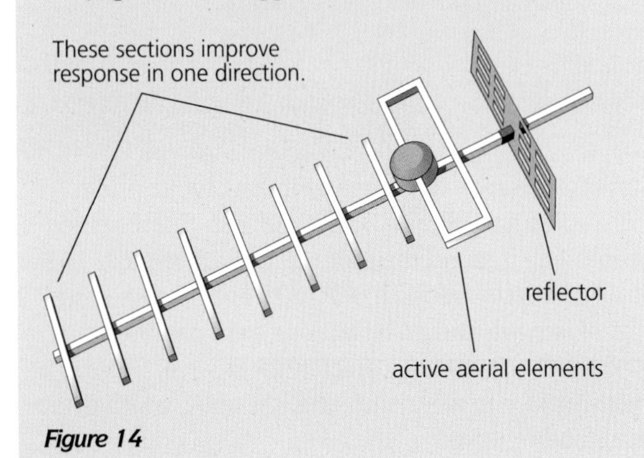


***Figure 14***

10. A television with an outdoor aerial is found to suffer from a fluctuating signal when an aircraft flies over. Explain why this happens.
11. Wi-fi usually operates at a frequency of 2.4 GHz. Interference between reflected waves could lead to standing waves in a room. How far apart would any nodes be?

## KEY IDEAS

- Superposition is the vector addition of waves.
- Coherent wave sources are needed for superposition to result in an observable interference pattern.
- Coherent waves are waves of the same frequency that have a constant phase difference. For complete cancellation, the waves also need to have the same amplitude and, in the case of transverse waves, the same polarisation.
- Two sources with the same phase will interfere constructively where the path difference is a whole number of wavelengths. Destructive interference will occur for an odd number of half-wavelengths.
- The fringe separation in two-source interference (for example, Young's double-slits experiment) is given by the equation $w = \frac{\lambda D}{s}$.

## ASSIGNMENT 2: UNDERSTANDING INTERFERENCE IN THIN FILMS

### Stretch and challenge

About 4% of the light that hits a transparent glass surface at 90° to the surface is reflected. This can be reduced by coating the lens with a thin layer of a material such as magnesium fluoride or a metal oxide (Figure A1). This gives rise to a second reflection from the lens, one from the coating and one from the glass. It is destructive interference between these two reflected waves that reduces the amount of reflected light.

The coating has to have a thickness equal to one-quarter of a wavelength of light, so that the light reflected from the glass surface has travelled half a wavelength further than that reflected from the surface of the coating (Figure A2). It is not possible to make a coating that is the correct thickness for all wavelengths of light. If we choose to make the lens non-reflective in the middle of the optical range, say 550 nm (yellow), it will only work exactly for that particular wavelength, and some wavelengths will still be reflected. If the incident light is 'white', the reflected light will be deficient in yellow and will be richer in red and blue light. The lens will look purple.

***Figure A1*** *Lenses can be coated with an anti-reflective coating that prevents light being reflected from the lens.*

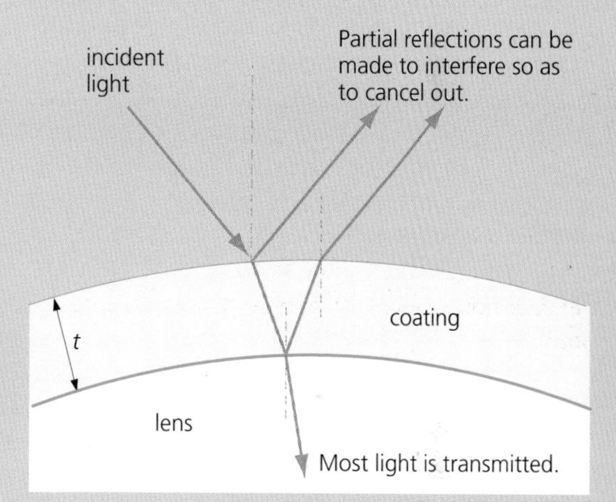


The path difference between two reflected rays is approximately $2t$. If $2t$ is equal to half a wavelength, then the two waves will be out of phase and the waves will interfere destructively so that no light is reflected.

***Figure A2*** *Thin-film interference: anti-reflective coatings prevent light being reflected from the lens. Since less light is reflected, more light is transmitted through the lens.*

#### Questions

**A1** Oil films on water appear coloured when viewed in 'white' light. This is due to destructive interference. Explain, with the help of a diagram, why the coloured fringes appear.

**A2** Research other examples of this thin-film interference. Write a brief article describing your chosen example(s), making sure you explain why the effects occur. (Useful search phrases: interference effects in ... soap bubbles / oil films on water / tempering metals / beetles; Newton's rings.)

## 6.2 DIFFRACTION

We use an enormous range of radio wavelengths in our modern communication systems. The wavelength of the VLF (very low-frequency) radio waves that are used for maritime distress signals and AM radio can be as long as 100 km. The wavelength of the EHF (extremely high-frequency) radio waves used for high-speed digital communications links with satellites is only 1 mm (Figure 15).

The longest waves are not very directional: the waves spread out in all directions and they are able to pass quite effectively around obstacles, such as hills and buildings, because of their long wavelength. This spreading of waves is known as **diffraction** and it is more pronounced for longer electromagnetic waves (see Figure 16). It is diffraction that allows mobile telephones to work even when there is no 'line-of-sight' transmission path between the mobile phone and the base station. The short wavelength – microwave – end of the radio spectrum is less prone to diffraction. High-frequency microwave links, used by TV companies for outside broadcasts, for example, have to have 'line of sight' between transmitter and receiver.

A way of explaining the diffraction of waves is known as **Huygens' construction**. Consider a plane wave approaching a gap (an aperture). The line, or surface, along which the wave disturbance has the same phase at all points we call a 'wave front' (Figure 17). Christiaan Huygens suggested that each point on a wave front could be thought of as a point source of a new wave.

***Figure 15*** *A microwave transmitter used for relaying television and mobile phone signals*

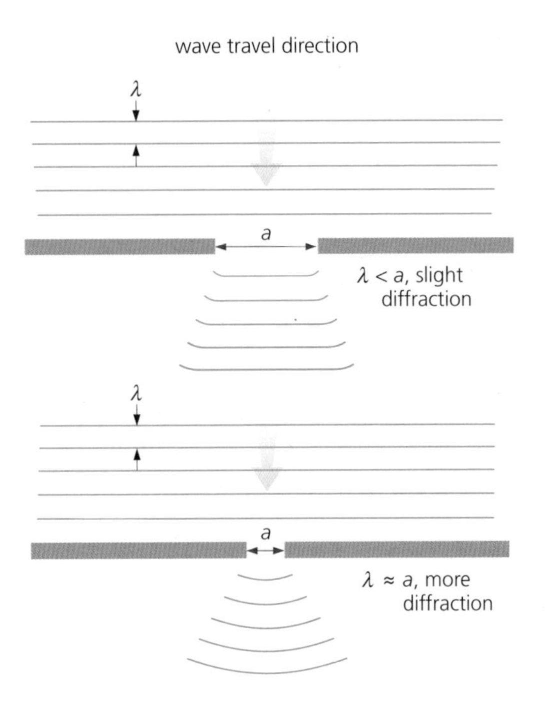


***Figure 16*** *Diffraction of waves through a gap*

Each of these new waves, called 'secondary wavelets', spreads out and overlaps with other secondary wavelets. The new wave front is formed by the constructive interference of all the secondary wavelets. As a plane wave passes through an aperture, the spreading of new wavelets causes diffraction.

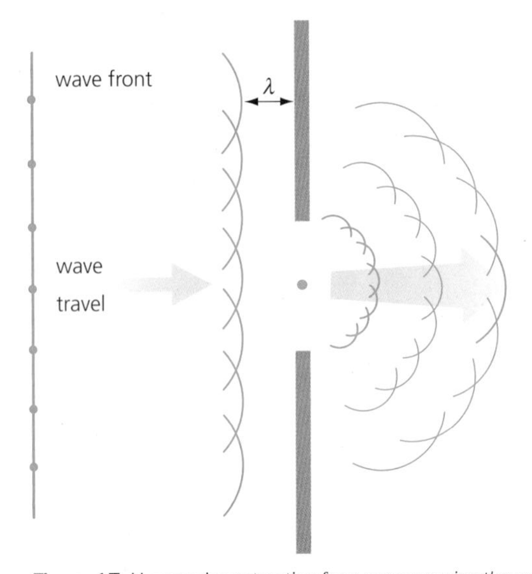


***Figure 17*** *Huygens' construction for a wave passing through a gap*

## QUESTIONS

**12.** Diffraction occurs with longitudinal waves as well as with transverse waves. One of the ways in which people can work out the direction of a sound wave is because of the shadowing effect of the head. The ear that is further from the sound is effectively blocked by the head. This method is much more successful at high frequencies. People find it more difficult to locate the source of low-frequency sounds. Explain why this is.

**13.** Why is it that houses in the shadow of a hill get poor TV reception but can still get good radio reception? (TV transmission is typically at 600 MHz, while medium-wave radio is transmitted at a frequency of around 1000 kHz.)

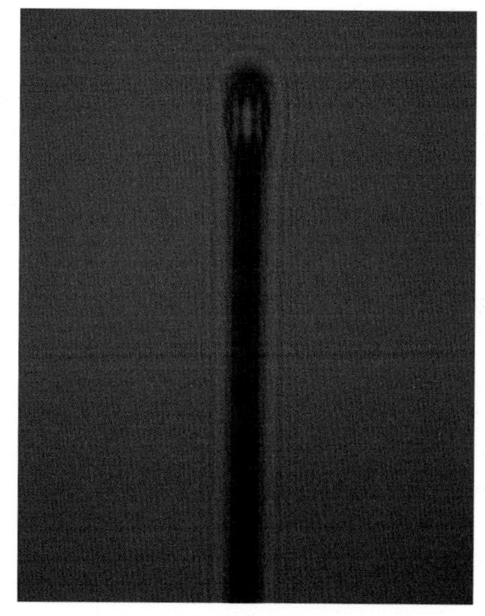

***Figure 18*** *Light waves can be diffracted so that light passes into the region of geometric shadow. This is only noticeable for very small obstacles or apertures like this needle eye.*

### Diffraction of light through a narrow slit

Light waves can also be diffracted, yet the effects are not noticeable in everyday situations because the wavelength of light is so short, between 400 and 700 nm. It is only when the obstacles or apertures are almost as small as the wavelength that diffraction effects become apparent (Figure 18).

When light is allowed to pass through a very small slit, we can see the light spreading out into the region where we would expect to see only shadow. But this is not the only effect. A series of light and dark fringes becomes visible (Figure 19). These are similar to the interference fringes from two slits, but here there is only one source of light. What is causing the interference?

The answer is that secondary wavelets from different parts of the slit are interfering with each other. In some directions they add together constructively, and a bright fringe is formed. In other directions the secondary wavelets are out of phase and the net effect is a dark fringe (see Figure 20).

In the direction given by the angle $\theta$, there is a path difference of $\left(\frac{a}{2}\right)\sin\theta$ between the wavelet that leaves from the edge of the slit (1) and the wavelet that leaves from half-way across the slit (3). If this path difference is equal to half a wavelength, the wavelets will cancel out and there will be an interference minimum (here that means darkness). The wavelets from just above 1 will cancel with those just above 3, and so on.

The same thing will happen when the path difference is $\frac{3}{2}\lambda$, $\frac{5}{2}\lambda$, and so on. When the path difference is equal to a whole number of wavelengths, a maximum (a bright fringe) will appear. This gives rise to the diffraction pattern for a single slit (see Figure 19).

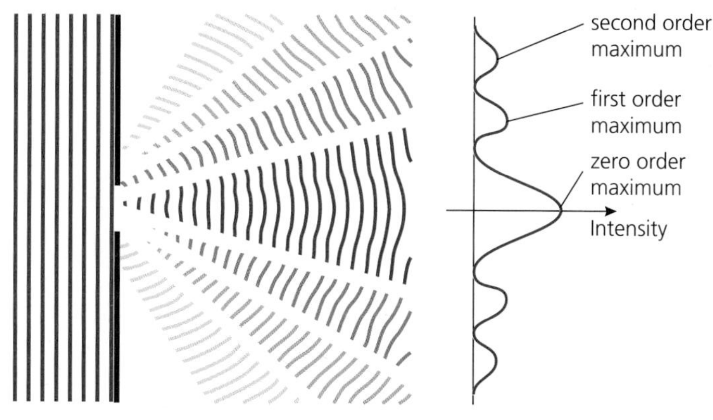


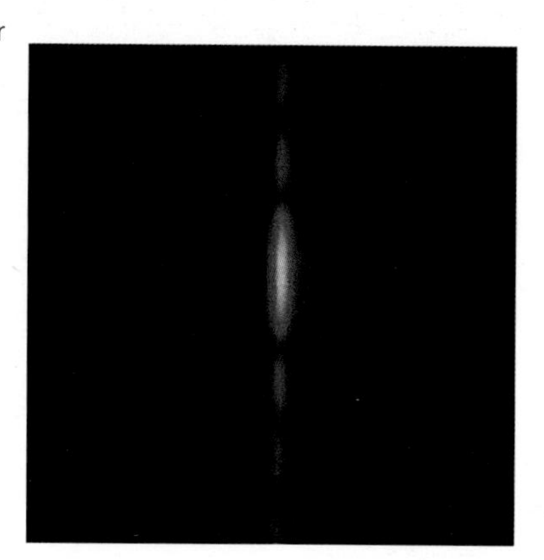

***Figure 19*** *Diffraction of monochromatic light at a slit*

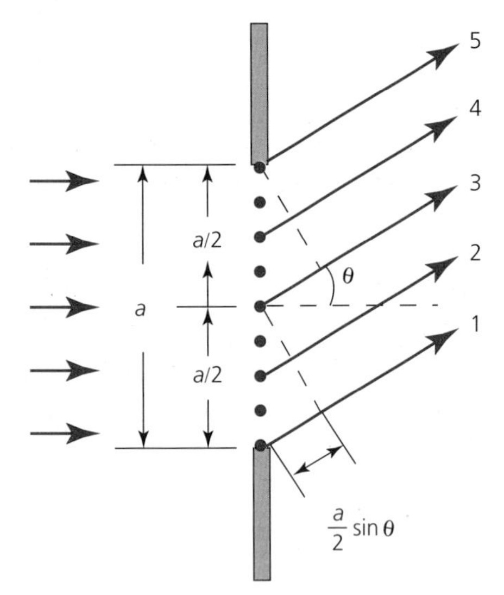


***Figure 20*** *In certain directions, wavelets from the top half of the gap cancel those from the bottom half.*

When a single slit is illuminated with white light, that is light that contains many different wavelengths, the diffraction pattern is more complex. The central (zero-order) maximum appears white, since there is no phase difference between wavelets from different parts of the slit. However, the first secondary maximum occurs when the path difference between wavelets from the edge and middle of the slit is equal to a whole wavelength. This will be different for different colours of light. Violet is the shortest wavelength, and so constructive interference will happen at a shorter path difference (and hence smaller angle of diffraction) than for the other colours. This will be followed by blue and so on until the path difference is large enough for constructive interference to be seen in red light (Figure 21).

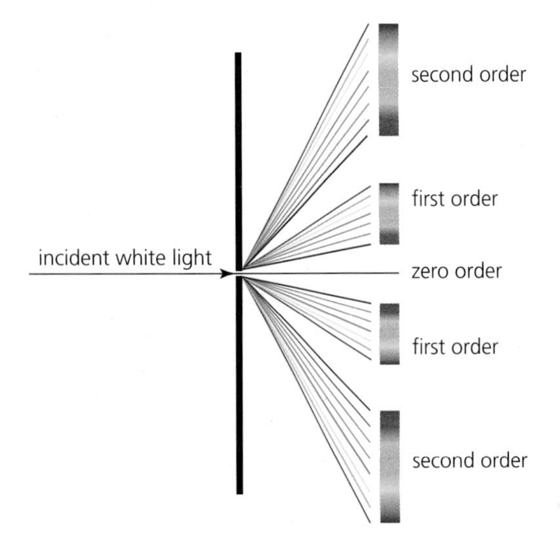


***Figure 21*** *Diffraction of white light at a slit*

Diffraction effects are always greater when the size of the obstacle or aperture is of the same order of magnitude as the wavelength. Hence the narrower the slit through which light passes, the greater the diffraction. This can be seen in Figure 22.

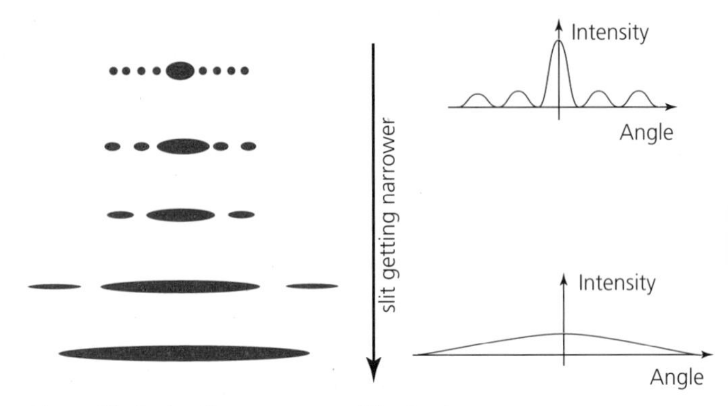


***Figure 22*** *As the slit gets narrower, diffraction becomes more noticeable and the central maximum gets wider.*

## QUESTIONS

**14.** How would the diffraction pattern in Figure 19 differ if blue light were directed at the same slit?

**15.** In the apparatus shown in Figure 23, a microwave transmitter operating at a frequency of 10 GHz is aimed at a gap between two metal plates. The plates are initially 30 cm apart.

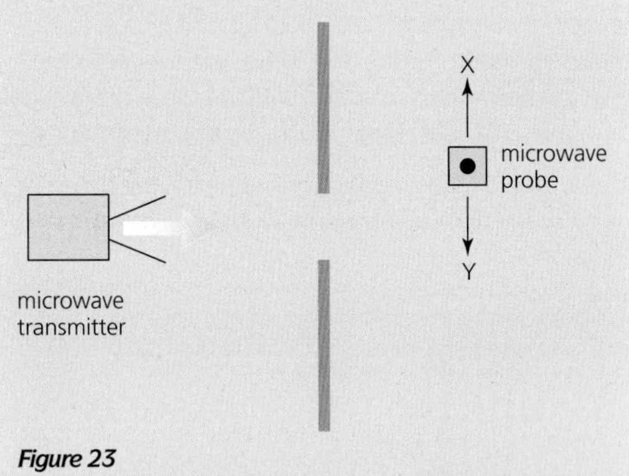


***Figure 23***

**a.** Describe what you would detect with a microwave probe moved along the line XY.

**b.** The plates are moved closer together until the gap is only 3 cm. What would the microwave probe detect now?

### Stretch and challenge

**16.** Look at the interference pattern from two slits in Figure 11. Interference between the two slits should produce a set of equally spaced, equally bright fringes. But this is not the case. The bright fringes are not the same intensity, and one or two seem to be missing altogether. Can you explain why the fringes of the two-slits interference pattern are not equally bright? (*Hint*: Look at the diffraction pattern for the single slit in Figure 19.)

## Diffraction and resolution

Diffraction of light passing through an aperture has important consequences. It sets a limit on the amount of detail that can be seen in an image. When light passes through the pupil of your eye, it is diffracted and forms a diffraction pattern on your retina. If you are looking at two light sources that are sufficiently close together, their diffraction patterns will overlap on your retina (Figure 24). It is not possible for you to tell them apart, and we say that the objects are not resolved. The way to see greater detail – achieve greater **resolution** – is to use a camera or a telescope with a larger aperture, so that diffraction is less of a problem.

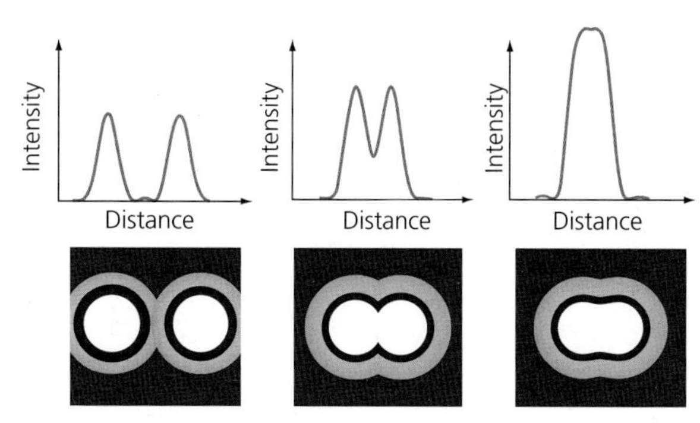


***Figure 24*** *Diffraction through an aperture causes the image of a point source of light, such as a star, to spread into a disc, with fringes around it. If two such images overlap, it may be impossible to distinguish the individual images; in other words we cannot resolve them. The problem is worse with smaller apertures.*

## Diffraction gratings

Much of our understanding of the composition and structure of stars and galaxies is based on measurements of the light that they emit. Different atoms and molecules emit and absorb different frequencies of light. By studying the light emitted from a star, we can compare its spectrum (Figure 26) with the spectrum from laboratory light sources. This tells us which elements are present in the star. The strength of different spectral lines gives us information on the star's temperature. If the spectral lines are slightly shifted compared to laboratory light sources, we can infer that the star is moving relative to the Earth, and we can work out its velocity. All this is done by using a **diffraction grating** to produce a spectrum.

A diffraction grating is simply a set of many parallel thin slits, achieved by etching parallel opaque lines

### QUESTIONS

**17.** Why do radio telescopes (Figure 25) need to be so much larger than optical (light) telescopes?

***Figure 25*** *Radio telescopes*

***Figure 26*** *Spectra from ten stars (of decreasing surface temperature going downwards), each produced by the star's light being passed through a diffraction grating*

on a sheet of transparent material, usually glass. As light waves pass through the grating, each slit causes diffraction. The waves from each slit overlap and interfere to give areas of constructive and destructive interference. In most directions there is complete destructive interference and no light is transmitted. Constructive interference takes place in a few directions. In these directions diffracted light beams occur. As in the diffraction pattern from a single slit, these diffracted beams are called the 'orders' of diffraction (Figure 27).

The direction of the diffracted beams is related to the spacing of the slits and the wavelength. This can be shown using Huygens' construction (Figure 28).

We can use a simplified form of the diagram in Figure 28 to derive an expression that will allow us to calculate the angles at which the bright maxima will appear. This is shown in Figure 29.

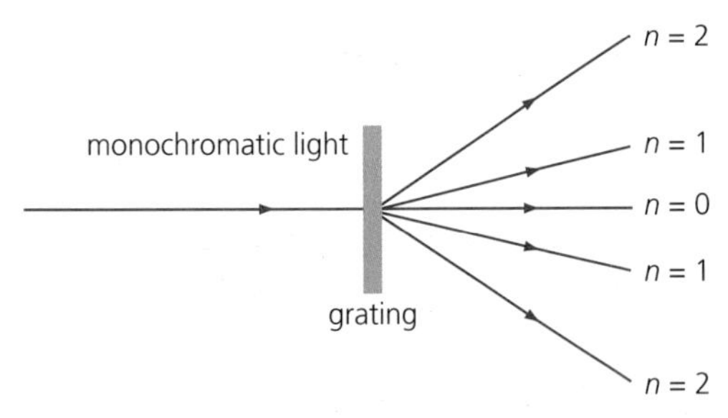


***Figure 27*** *Orders of diffraction with a diffraction grating. The central beam is the zero-order maximum (n = 0). Adjacent beams on either side are first-order maxima (n = 1), and subsequent beams are numbered accordingly.*

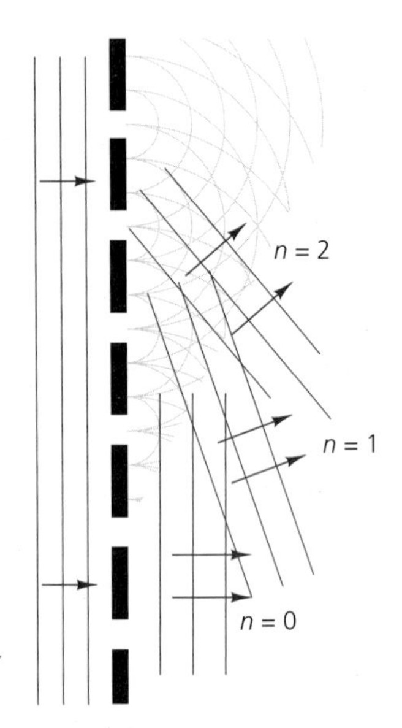


***Figure 28*** *Huygens' construction for light passing through a diffraction grating. The red lines on the right of the grating show directions of constructive interference.*

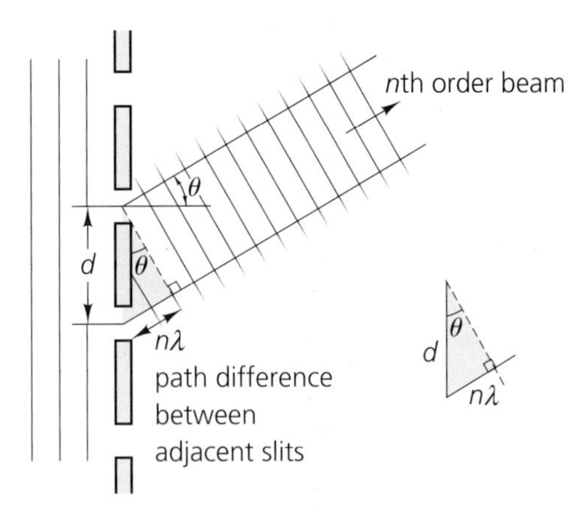


***Figure 29*** *Deriving the diffraction grating formula*

The straight-through beam, $\theta = 0°$, is formed as light from each slit arrives exactly in phase. There is no path difference at all and every wavelength will arrive in phase.

The first-order beam is formed in the direction in which adjacent slits have a path difference of a whole wavelength, $\lambda$. These waves are therefore in phase. The second-order beam is in a direction in which light from adjacent slits is two wavelengths out of phase, that is, the path difference is $2\lambda$.

In general, the *n*th-order beam is at an angle $\theta$ to the original wavefront direction. Along this beam, waves from adjacent slits have a path difference of $n\lambda$.

We can relate this path difference to the distance between each slit (the grating spacing), *d*, by looking at the central (shaded) triangle of Figure 29. For that triangle,

$$\sin\theta = \frac{\text{opposite}}{\text{hypotenuse}} = \frac{\lambda n}{d}$$

which can be written as

$$d\sin\theta = n\lambda$$

This is known as the **diffraction grating formula**.

### Worked example

Suppose that we use a diffraction grating with 400 000 lines per metre. We can use the expression $n\lambda = d\sin\theta$ to calculate the wavelength of light that gives a second-order beam at an angle of 25°.

The grating spacing, *d*, is found from the reciprocal of the number of lines per metre. In this case

$$d = \frac{1}{400\,000\ \text{m}^{-1}} = 2.5 \times 10^{-6}\ \text{m}$$

Using $n\lambda = d \sin \theta$ we find

$$2\lambda = 2.5 \times 10^{-6} \times \sin 25°$$

This gives

$$\lambda = 5.3 \times 10^{-7}\ \text{m} = 530\ \text{nm}$$

We can also use the diffraction grating formula to calculate how many diffracted beams would be visible. The value of $\sin \theta$ cannot be bigger than 1, so the maximum value of $n$ is given by $n\lambda = d$:

$$n \leq \frac{\lambda}{d} = \frac{(2.5 \times 10^{-6}\ \text{m})}{(5.28 \times 10^{-7}\ \text{m})} = 4.73$$

The fourth-order beam will be the last possible beam. The total number of visible beams is therefore $4 + 4 + 1 = 9$ (that is, four orders on either side of the centre, plus the central maximum).

---

When diffraction gratings are illuminated with light that is not monochromatic, white light for example, the constituent wavelengths (colours) are dispersed. Spectra can be seen on either side of the central maximum, which itself is the same mix of wavelengths as the source. The spectra arise because the condition for constructive interference is that the path difference must equal an integral number of wavelengths, $\lambda$, $2\lambda$, $3\lambda$, …, $n\lambda$. Since this is equal to $d \sin \theta$, each wavelength will show constructive interference at specific values of $\theta$ (Figure 30).

Diffraction gratings are used in spectrometers to produce spectra in chemical analysis, for example (see section 8.1 of Chapter 8), and in astrophysics to investigate both the movement and the composition of stars and galaxies (Figure 31). They are also used in photography to create special effects.

Gratings designed for analysing ultraviolet (UV) and infrared (IR) radiation are made from silicon rather than glass, since glass absorbs some wavelengths in the UV and IR regions of the spectrum.

***Figure 30*** *The number of spectra produced depends on the spacing of the grating. The higher orders are less bright and spread over a wider angle, often overlapping other orders.*

***Figure 31*** *This image of Saturn's rings from the NASA spacecraft, Cassini, was taken with an ultraviolet spectrometer, UVIS, which also provides information on the composition of Saturn's moons.*

The diffraction gratings considered so far are 'transmission gratings' – the light diffracts as it passes through them. It is also possible to construct 'reflection gratings' (Figure 32), which work on the same principle, of constructive and destructive interference of diffracted reflected beams.

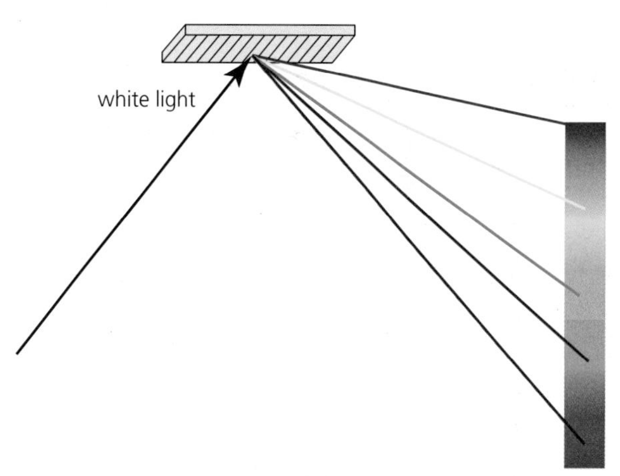


***Figure 32*** *A finely grooved metal grating acts as a reflection diffraction grating.*

## QUESTIONS

**18.** When white light passes through a diffraction grating, most of the orders of diffraction are dispersed into different colours, forming spectra. The central, zero-order beam is always white. Explain why this is so.

**19.** A diffraction grating has 200 000 lines per metre. How many diffracted beams will be visible if red light, $\lambda = 600$ nm, is used to illuminate the grating?

**20. a.** A CD can create a spectrum when it reflects white light (Figure 33). Explain how this happens.

**Figure 33**

Stretch and challenge

**b.** Look at a CD creating a spectrum. Assuming the CD is acting as a diffraction grating, estimate the number of lines per millimetre.

**c.** Try a DVD instead, and comment on the difference.

**21. a.** It should be possible to make a diffraction grating for sound waves. How would you go about this? What dimensions would such a grating have?

**b.** Can you think of any applications for such a device? What effect would it have on a sound wave?

## X-ray diffraction

Beams of X-rays reflecting off planes of atoms in a crystal behave in the same way as light at a diffraction grating. X-ray diffraction patterns are used to give information about the structure of crystalline materials. Dorothy Hodgkin (Figure 34) extended the technique to organic molecules and her work revealed the structure of insulin, vitamin B12 and penicillin. It was later used by Rosalind Franklin in work that led to the discovery of the structure of DNA (Figure 35).

***Figure 34*** *Dorothy Hodgkin made use of X-ray diffraction in her research.*

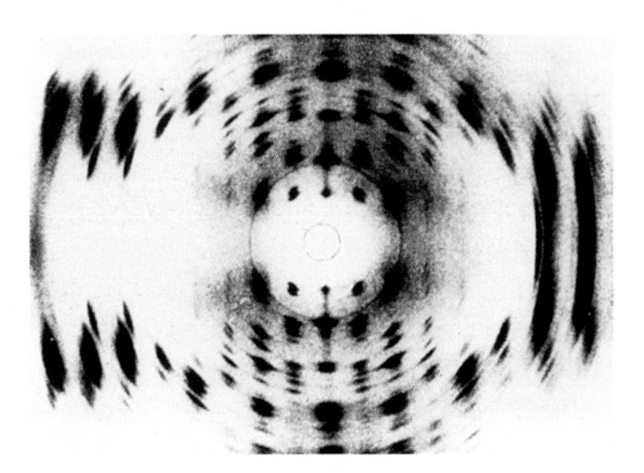

***Figure 35*** *X-ray diffraction pictures gave vital information in the determination of the structure of DNA.*

### KEY IDEAS

- Diffraction is the spreading of waves, after passing an obstacle or an aperture of a similar size to the wavelength, into the region of 'geometrical shadow'.
- The diffraction of light through a small gap produces several intensity maxima: the central maximum is twice as wide and many times more intense than the others.
- The smaller the gap, the wider the central maximum and the more spread out the diffraction pattern.
- The longer the wavelength of the waves incident on a gap, the greater the angle at which constructive interference occurs.
- A diffraction grating consists of many slits separated by a distance $d$: it gives rise to several orders of diffraction ($n = 0, \pm1, \pm2, \pm3, \ldots$), which are at angles $\theta$, given by $d \sin\theta = n\lambda$.

# REQUIRED PRACTICAL: APPARATUS AND TECHNIQUES

## Investigation of interference effects – Young's double-slits experiment and interference by a diffraction grating

The aim of these experiments is:

1 to observe interference fringes from two light sources, and use the relationship $w = \frac{\lambda D}{s}$

2 to investigate the effect of multi-slit interference, including the diffraction grating.

**The practicals give you the opportunity to show that you can:**

- use appropriate analogue apparatus to record a range of measurements and to interpolate between scale markings
- use methods to increase accuracy of measurements
- use callipers to measure small distances, including using vernier scales
- use a laser to investigate characteristics of light, including interference and diffraction.

## Practical 1: Young's double-slits experiment: two-source interference effects

### Apparatus

Thomas Young managed to demonstrate these effects using a candle. Fortunately, we now have lasers. Laser light is monochromatic and differs from other light sources in that photons are released by a process of 'stimulated emission'. This means that the monochromatic light from the laser is effectively one long wave train, with constant phase. Lasers are therefore ideal for showing interference effects with visible light.

Laser light does present some safety issues, however, especially to eyesight. A laser beam diverges very little, which means that the intensity of the beam does not decrease much with distance. (Intensity is the energy per second per unit cross-sectional area of the beam, measured in W $m^{-2}$.) A high-intensity laser beam entering the eye, which tends to focus the beam to an even smaller area, will damage the retina, possibly creating blind spots.

Lasers are classified according to the risk they present. Class 1 lasers are inherently safe, because they are either low-power or enclosed, such as the one used in a DVD player.

Lasers in a school laboratory are class 2. These are low-power (< 1 mW) visible lasers. They are not intrinsically safe, so care needs to be taken to prevent the laser beam entering the eye either directly or by reflection. Class 2 lasers need to carry the warnings shown in Figure P1.

LASER RADIATION
DO NOT STARE INTO BEAM
CLASS 2 LASER PRODUCT

***Figure P1*** *Hazard warnings for a class 2 laser*

High-power lasers need to be used in controlled rooms and protective goggles need to be worn. These are not permitted in schools.

### QUESTIONS

**P1** Laser beams are used to monitor the position of the Lunar Orbiter. Each beam diverges to a diameter of 3 km by the time it gets to the Moon, 400 000 km away. Calculate the angle of divergence of the laser beam, assuming that it starts from a point. Give your answer in radians and in degrees.

**P2** Infrared lasers may be more damaging to your eye than visible lasers. Suggest why this might be.

**P3** When using a laser in an interference experiment, why is it a good idea to use a matt screen to form your interference fringes on?

### Techniques

Two coherent sources of light can be formed by shining laser light through double slits. It is important to make sure that both slits are illuminated. The resulting interference pattern is a set of equally spaced bright fringes, as shown in Figure P2. The fringe separation, $w$, can be found by measuring the width of several fringes. Vernier callipers can be used to measure $w$. The distance from the slits to the screen, $D$, can be measured with a metre ruler. The separation of the slits, $s$, is usually given, but could be measured using a travelling microscope.

The equation $w = \lambda D/s$ can be used to find the wavelength of the laser light, where $w$ = separation of fringes, $\lambda$ = wavelength of laser light, $D$ = distance from double slits to screen and $s$ = separation of slits.

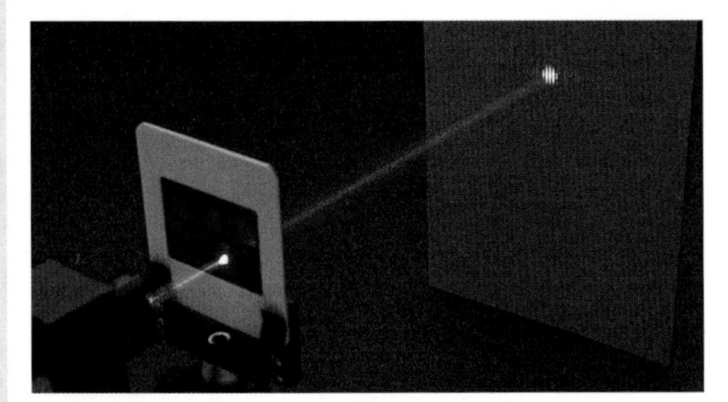

**Figure P2** *Set-up for the experiment, showing the two-slits interference pattern on the screen*

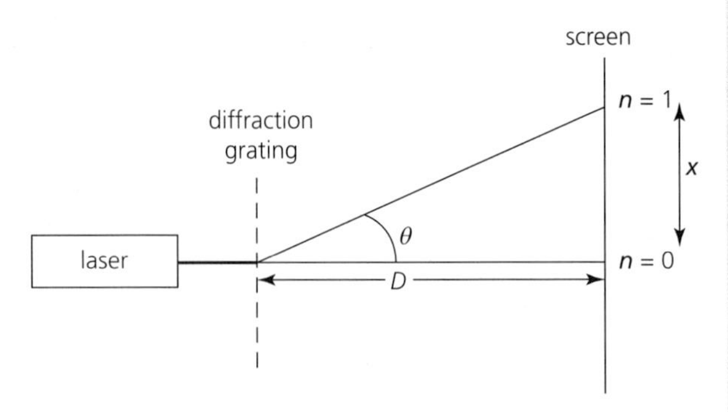


**Figure P3** *Geometry for diffraction grating experiment (Not to scale. $D \approx 3$ m would give $x \approx 0 \rightarrow 1$ m for a coarse grating with 100 lines per mm.)*

## QUESTIONS

**P4** If the slits are separated by 0.25 mm, and the screen is 2.00 m away, the fringes would be about 5 mm apart.

a. Calculate the wavelength of the laser light.

b. Estimate the uncertainty associated with this answer.

c. How could you reduce the uncertainty?

## Practical 2: Interference by a diffraction grating

### Apparatus

If the number of slits is increased from two to three, then from three to four, and so on, but the slit separation and the width of each slit is kept constant, the interference maxima become brighter and sharper. A 'coarse' transmission diffraction grating has up to about 100 lines per millimetre and the resulting interference pattern is a series of bright lines or dots. 'Fine' diffraction gratings could have as many as 6000 lines per millimetre; the maxima (bright dots) will be much further apart than those produced by the coarse grating.

### Techniques

The interference pattern from a diffraction grating can also be used to measure the wavelength of light. Rearranging the diffraction grating equation $n\lambda = d \sin\theta$ gives

$$\sin\theta = n\left(\frac{\lambda}{d}\right)$$

The angle $\theta$ can be found for each bright maximum and then $\sin\theta$ plotted against $n$, the order of the maximum. A graph of $\sin\theta$ ($y$-axis) against $n$ ($x$-axis) should be a straight line, with a gradient equal to $\left(\frac{\lambda}{d}\right)$. The best way to find $\theta$ is to find $\tan\theta$, which can be measured as $\left(\frac{x}{D}\right)$, where $x$ is the distance along the screen from the central maximum ($n = 0$) to the maximum, and $D$ is the distance from the grating to the screen (Figure P3). Then $\theta = \tan^{-1}\left(\frac{x}{D}\right)$.

## QUESTIONS

**P5** The experiment described above is carried out with $D = 3.0$ m and a grating with 100 lines per millimetre. The results in Table P1 were obtained.

| $n$ | $x$ / m |
|---|---|
| 0 | 0.00 |
| 1 | 0.20 |
| 2 | 0.40 |
| 3 | 0.60 |
| 4 | 0.77 |
| 5 | 1.03 |
| 6 | 1.20 |

*Table P1*

a. Using the results in Table P1, calculate $\tan\theta$, $\theta$ and $\sin\theta$ for each order, $n$.

b. Plot a graph of $\sin\theta$ against $n$.

c. Use your graph to find the wavelength of the laser light.

d. The wavelength of the laser light is actually 633 nm. How accurate was the experiment?

### Stretch and challenge

**P6** The experiment was repeated with a much finer grating, 600 lines per millimetre. The results in Table P2 were obtained.

| $n$ | $x$ / m |
|---|---|
| 0 | 0.00 |
| 1 | 1.03 |
| 2 | 1.93 |
| 3 | 3.25 |
| 4 | no max. |
| 5 | no max. |
| 6 | no max. |

*Table P2*

Only four measurements of $x$ could be obtained. Explain why this was the case.

## PRACTICE QUESTIONS

1. a. A laser emits *monochromatic light*. Explain the meaning of the term *monochromatic light*.

   b. Figure Q1 shows a laser emitting blue light directed at a single slit, where the slit width is greater than the wavelength of the light. The intensity graph for the diffracted blue light is shown.

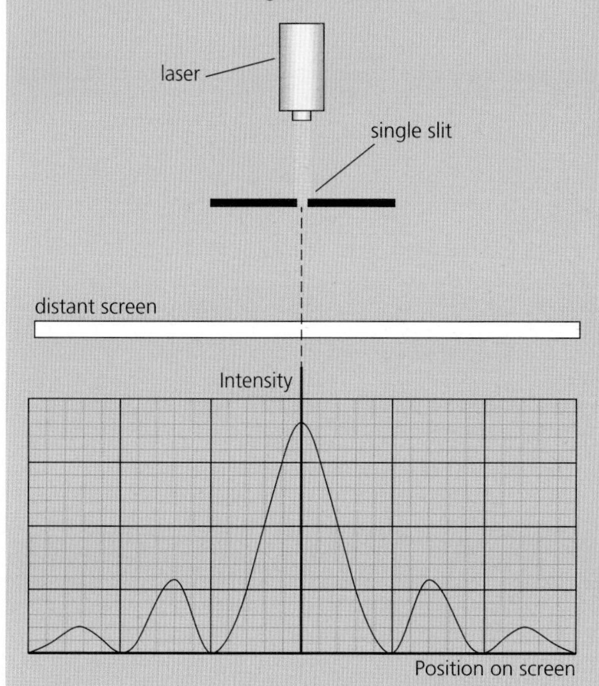


*Figure Q1*

The laser is replaced by a laser emitting red light. Make a sketch copy of Figure Q1 and then on the same axes sketch the intensity graph for a laser emitting red light.

   c. State and explain **one** precaution that should be taken when using laser light.

   d. The red laser light is replaced by a non-laser source emitting white light. Describe how the appearance of the pattern would change.

*AQA June 2013 Unit 2 Q7*

2. Figure Q2 shows the paths of microwaves from two narrow slits, acting as coherent sources, through a vacuum to a detector.

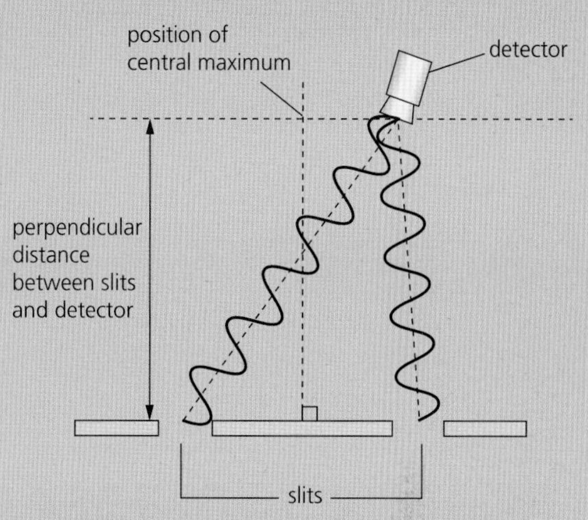


*Figure Q2*

   a. Explain what is meant by *coherent sources*.

   b. i. The frequency of the microwaves is 9.4 GHz. Calculate the wavelength of the waves.

      ii. Using Figure Q2 and your answer to part **b i**, calculate the path difference between the two waves arriving at the detector.

   c. State and explain whether a maximum or minimum is detected at the position shown in Figure Q2.

   d. The experiment is now rearranged so that the perpendicular distance from the slits to the detector is 0.42 m. The interference fringe spacing changes to 0.11 m. Calculate the slit separation. Give your answer to an appropriate number of significant figures.

   e. With the detector at the position of a maximum, the frequency of the microwaves is now doubled. State and explain what would now be detected by the detector in the same position.

*AQA Jan 2013 Unit 2 Q7*

3. a. In an experiment, a narrow beam of white light from a filament lamp is directed at normal incidence at a diffraction grating. Copy and complete the diagram in Figure Q3 to show the light beams transmitted by the grating, showing the zero-order beam and the first-order beams.

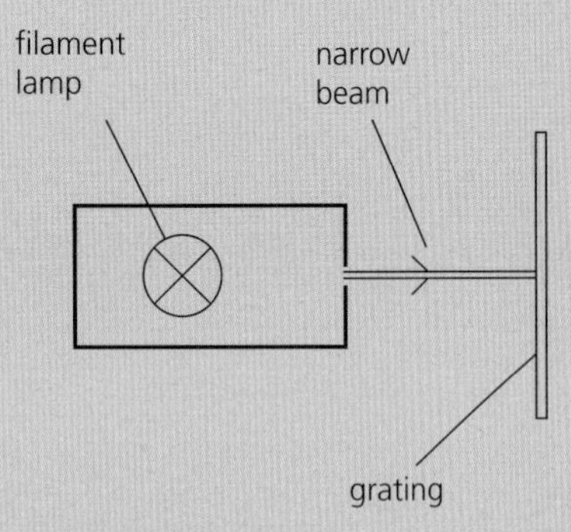


*Figure Q3*

b. Light from a star is passed through the grating. Explain how the appearance of the first-order beam can be used to deduce **one** piece of information about the gases that make up the outer layers of the star.

c. In an experiment, a laser is used with a diffraction grating of known number of lines per mm to measure the wavelength of the laser light.

i. Draw a labelled diagram of a suitable arrangement to carry out this experiment.

ii. Describe the necessary procedure in order to obtain an accurate and reliable value for the wavelength of the laser light. Your answer should include details of all the measurements and necessary calculations. The quality of your written communication will be assessed in your answer.

*AQA Jan 2012 Unit 2 Q5*

4. Two identical loudspeakers are playing the same frequency sound, 680 Hz, from a signal generator. A microphone is moved along the line ABC. See Figure Q4.

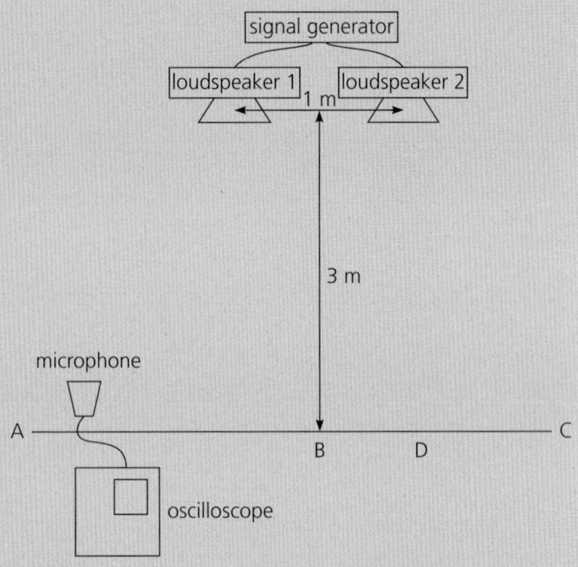


*Figure Q4*

a. Describe how the strength of the signal from the microphone will change as it is moved along the line ABC.

b. Explain why the signal changes.

c. How would your answer to part **a** be changed if the signal generator was adjusted to a higher frequency?

d. A 'quiet' region is observed at point D. If this is the first quiet region observed after point B, calculate the distance BD. Take the speed of sound as $c = 340\ \text{m s}^{-1}$.

5. Diffraction gratings are used in a spectrometer to measure the wavelength of light.

a. Describe and explain what you would see if you illuminated a diffraction grating with white light.

A diffraction grating with 300 lines per millimetre is used to measure the wavelength of sodium light, $\lambda = 589.00$ nm.

b. At what angle would you expect the first diffraction maximum to appear?

c. How many diffraction maxima could you see with this grating?

d. The characteristic yellow light from sodium lamps actually comprises two wavelengths of light, $\lambda = 589.00$ nm and $\lambda = 589.59$ nm. With this diffraction grating the two lines are too close together to resolve. How could you improve the visibility of the two lines?

6. The waves from two radio aerials meet at a point such that they completely cancel each other out. Which of these statements about the waves is **false**? They must have

**A** the same polarisation.

**B** the same frequency.

**C** exactly the same phase.

**D** exactly the same amplitude.

7. Two narrow slits, 2 mm apart, are illuminated by a laser beam. An interference pattern is formed on a screen 3 m away. The distance between five fringes is measured as 4.8 mm. The wavelength of the laser was

**A** 640 µm

**B** 640 nm

**C** 128 nm

**D** 128 µm

8. An interference pattern in sound is created by playing the same frequency note through two loudspeakers. The regions of quiet and loud are found to be very close together. Which row of Table Q1 shows the most effective way of increasing that spacing?

| | | |
|---|---|---|
| A | Move the speakers further apart | Increase the frequency of the sound |
| B | Move the speakers further apart | Decrease the frequency of the sound |
| C | Move the speakers closer together | Increase the frequency of the sound |
| D | Move the speakers closer together | Decrease the frequency of the sound |

*Table Q1*

9. A diffraction grating, with $N$ lines per metre, is used in an experiment to measure the wavelength of the light from a laser. Several orders, $n$, of diffraction maxima are visible. The angle, $\theta$, at which each maximum occurs is measured. Which row of Table Q2 shows the correct way to plot the results?

| | $y$-axis | $x$-axis | Gradient |
|---|---|---|---|
| A | $\theta$ | $n$ | $\lambda N$ |
| B | $\sin\theta$ | $n$ | $\lambda/N$ |
| C | $\sin\theta$ | $n$ | $\lambda N$ |
| D | $\theta$ | $N$ | $\lambda/N$ |

*Table Q2*

10. A diffraction grating with 1000 lines per millimetre is illuminated with monochromatic red light of wavelength 640 nm. How many diffraction maxima will be visible?

**A** 3

**B** 2

**C** 1560

**D** none

11. Jonathan looks at a star. When he observes the star through a small telescope it is just possible to see that it is a binary star – there are two stars with separate images. Which of the following is correct *and* is sufficient to explain Jonathan's observations?

**A** The light from the star is diffracted less by the telescope because the telescope has a wider aperture than Jonathan's eye.

**B** The image is brought into sharper focus by the telescope because the focal length of the telescope can be adjusted.

**C** The images produced by the telescope are brighter because the telescope has a wider aperture than Jonathan's eye.

**D** The images produced by the telescope are magnified, and magnifying an image always improves its resolution.

## Stretch and challenge

**12.** Read the passage about interference and diffraction effects carefully and then answer the questions that follow.

The Morpho butterfly (Figure Q5) is one of the world's largest. The beautiful, blue colour of its wings is not due to pigment but to their structure. The wings are actually a colourless, translucent membrane covered by a layer of scales. Each scale is about 100 μm long and 50 μm wide. They form a set of grooves that reflect light. The light waves reflected from each groove overlap and this leads to interference effects.

***Figure Q5*** *The Morpho butterfly*

The exoskeleton in some beetles has a similar repeating grooved structure that acts as a reflection diffraction grating (Figure Q6).

***Figure Q6*** *Beetle with a grooved exoskeleton*

Some other beetles get their coloured sheen from another interference effect, thin-film interference (Figure Q7). The beetle's exoskeleton is made up of thin closely spaced layers that partially reflect the light.

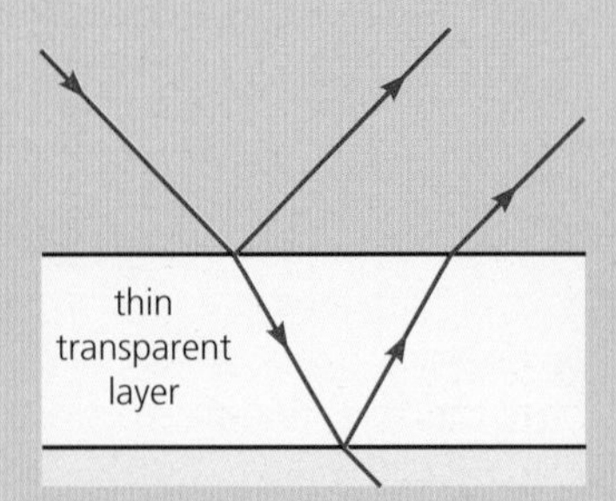


***Figure Q7*** *Thin-film interference*

**a.** The passage says that the exoskeleton of some beetles 'acts as a reflection diffraction grating'.

  **i.** Describe what you see when a *transmission* diffraction grating is illuminated with white light.

  **ii.** Explain how this is caused.

  **iii.** How does a reflection diffraction grating produce a similar effect?

  **iv.** The beetle in Figure Q6 appears to have coloured regions on it. If you looked at the real beetle, these colours would move if you shifted your point of view. Why does this happen?

**b.** The 'grooves' on a Morpho butterfly's wings are about 200 nm apart. Suggest why this helps to make the wings look blue.

**c.** In order to see interference effects, there must be at least two coherent sources of light. Explain why the reflected beams in Figure Q7 are coherent.

**d.** The colourful plumage of some birds, peacocks for example, is at least partly due to interference effects while in other birds the colour is simply due to pigment. Suggest **two** tests you could do to distinguish between the two.

# 7 REFRACTION AND OPTICAL FIBRES

## PRIOR KNOWLEDGE

You will have met the laws of reflection of light and should be able to describe the principle of refraction.

## LEARNING OBJECTIVES

In this chapter you will learn about refractive index and how to calculate angles of refraction. You will learn about the phenomenon of total internal reflection, apply this knowledge to optical fibres and find out how they are used.

**(Specification 3.3.2.3)**

***Figure 1*** *Computer graphic of the global Internet network, produced by the Cooperative Association for Internet Data Analysis at the University of California. Pink areas are US Internet locations; UK ones are dark blue.*

It is difficult to say where and when the first connection on the Internet was made. It may have happened at the University of California, in 1969, when the cryptic message 'LO' was sent from a computer in Los Angeles to another at Stanford, a few hundred miles away. The full message, 'LOGIN', was lost due to a system crash. From this inauspicious start, the Internet has conquered the world. By 2015 there were more than three billion Internet users, about 40% of the world's population (Figure 1).

Physics and physicists have made crucial contributions to this remarkable growth. In the early 1970s, Donald Davies, working at the National Physical Laboratory in Middlesex, developed 'packet switching', a method of routing data through a network. In 1990, physicist Tim Berners-Lee designed the World Wide Web, and a year later CERN put the first website online.

Physicists now focus on making the Internet faster. Early Internet traffic was mainly text, emails and data. It ran on telephone systems using electrical signals in copper wires. Now high-resolution images and video streaming need much higher data transmission rates. According to 'Nielsen's law', data transmission rates grow by 50% every year. Light, rather than electricity, carries the signals. Optical fibres now form the backbone of the UK's communications network, carrying data as light pulses at hundreds of megabits per second (Mb $s^{-1}$). The main data bottlenecks are the links to each house or office and the wi-fi inside. Light might help there, too. LED lights can be made to flash at millions of times per second, far faster than the eye can detect, and much faster than wi-fi.

## 7.1 REFLECTION OF LIGHT

In Chapter 6 you saw that when light passes through a very small aperture, diffraction effects become noticeable. You also saw that light can show interference effects, when there are two or more coherent sources. These phenomena are evidence for the wave nature of light. But under most everyday conditions, diffraction and interference effects are small and can be neglected, and we can treat light as if it travelled in a straight line. This simplifies matters enormously. We can show the direction of a light wave as a **ray** – that is, a straight line with an arrow in the direction of propagation (Figure 2).

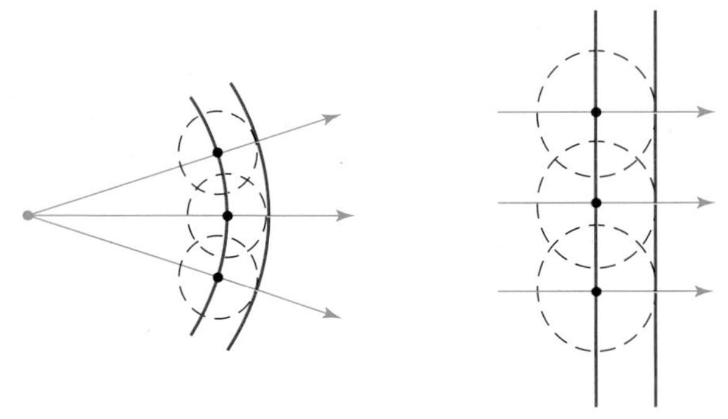

***Figure 2*** *A ray diagram shows the direction of travel of light wave fronts. Each ray (blue) is drawn perpendicular to the wave fronts (red) it represents. The dotted lines show the secondary wavelets (using Huygens' construction; see section 6.2 of Chapter 6).*

***Figure 3*** *The angle of reflection is equal to the angle of incidence.*

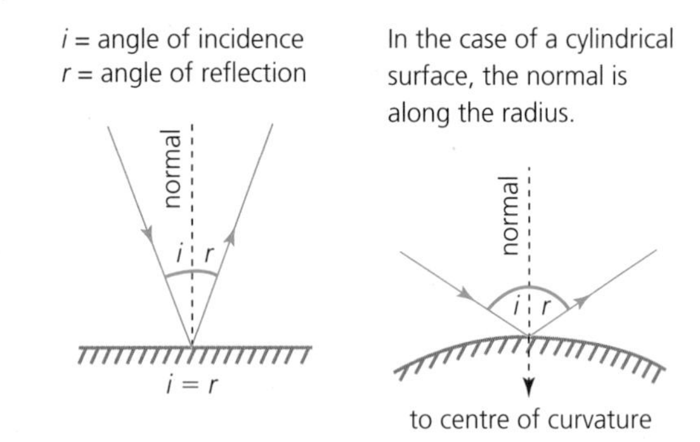


***Figure 4*** *Reflection at mirror surfaces*

### Reflection at a plane mirror

Optical fibres are used in our communications network. They carry signals as pulses of light. The inner walls of the glass fibre behave as perfect mirrors. Light repeatedly reflects from them, as it does from a plane mirror (Figure 3), at the same angle at which it approached – known as the **angle of incidence**.

The angle of incidence and the angle of reflection are measured from the **normal**. This is a line drawn at right angles to the surface of the mirror at the point where the incident ray meets it (Figure 4).

The **laws of reflection** are:

- For any reflecting surface, the angle of incidence ($i$) and the angle of reflection ($r$), both measured from the normal, are equal.
- The incident ray, the reflected ray and the normal lie in the same plane.

In the case of a plane mirror, we can apply the law of reflection to several rays from an object to construct a theoretical ray diagram (Figure 5). From that, we can predict the type and position of the image seen in the mirror (Figure 6).

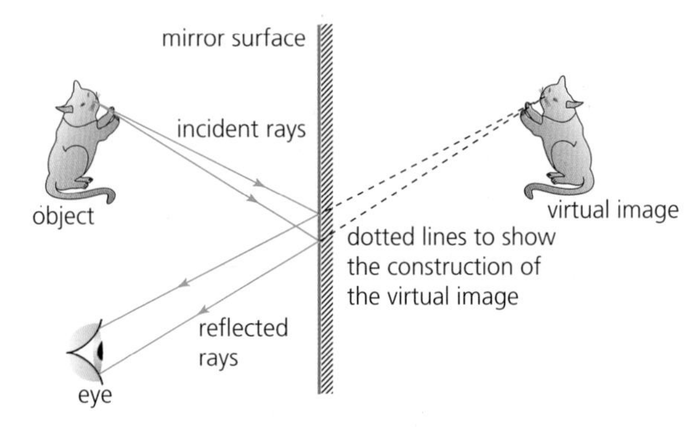


***Figure 5*** *Producing a virtual image*

*Figure 6* The virtual cat is the same size, and the same way up, as the real cat, but left and right are reversed (the image is laterally inverted).

The reflected rays diverge from the mirror surface (see Figure 5). The brain interprets the image as a set of rays diverging from a point behind the mirror surface. An image that can be projected onto a screen, as in a slide projector, is called a **real image**. However, no light rays actually pass through the position of the image in a mirror (Figure 5), so it is called a **virtual image**.

The position of the virtual image formed by a plane mirror can be found using the **no-parallax** technique (Figure 7). The experimental results confirm the predictions of the ray diagrams:

- The image is erect (upright).
- The image appears laterally inverted (left to right).
- The image is the same distance from the reflecting surface as the object is.

If pin B is correctly positioned at at the image point, the image of pin A and the top of pin B will stay together even if you change your viewing position.

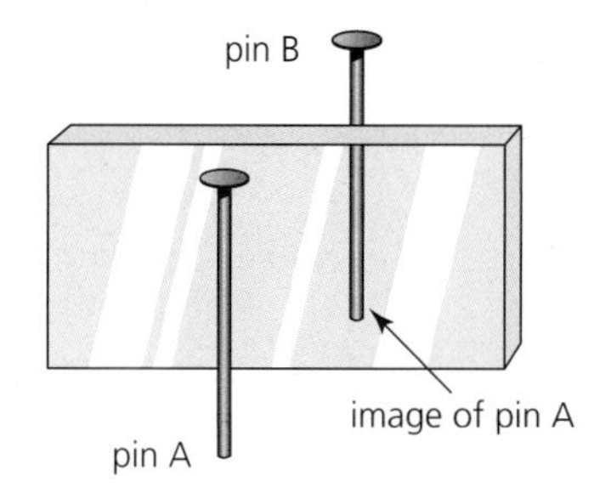


*Figure 7* Using the 'no-parallax' method to locate an image in a plane mirror

## QUESTIONS

1. Light reflects from rough surfaces, too, like this page. Explain, using a diagram, why you cannot see an image of your face in the page.
2. The image in a plane mirror appears laterally inverted (left and right are swapped). Explain why the image does not appear to be inverted top to bottom.

### Stretch and challenge

3. The reflection of light is not always total. Look carefully at Figure 3 and you will see that some light is transmitted. Suggest the factors that might affect the amount of light that is transmitted compared to the amount that is reflected.
4. The illusion known as Pepper's ghost is used in the theatre (Figure 8) to present ghostly images on stage. Make a sketch copy of the diagram in Figure 8 and add light rays to your drawing to show how it works.

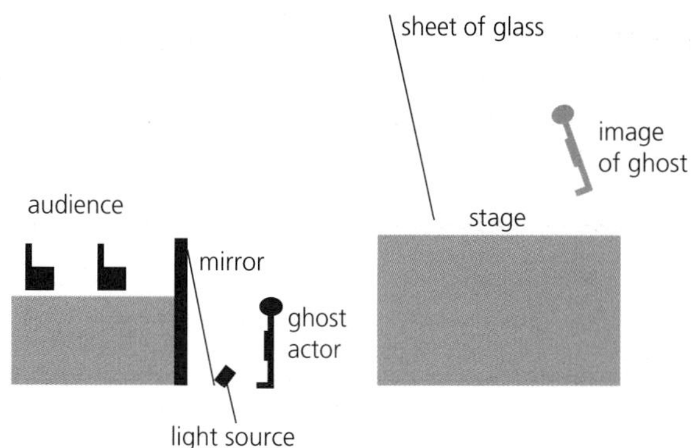


*Figure 8* 'Pepper's ghost'

**KEY IDEAS**

- For reflection of light at a plane mirror, the angle of incidence is equal to the angle of reflection. The incident ray, the reflected ray and the normal lie in the same plane.
- The image and the object are the same distance from the reflecting surface and lie on a line normal to the surface. The image is virtual and erect and appears laterally inverted.

## 7.2 REFRACTION AT A PLANE SURFACE

When a wave meets a boundary between two materials, in general some of its energy is reflected and the remainder is transmitted into the new medium (Figure 10).

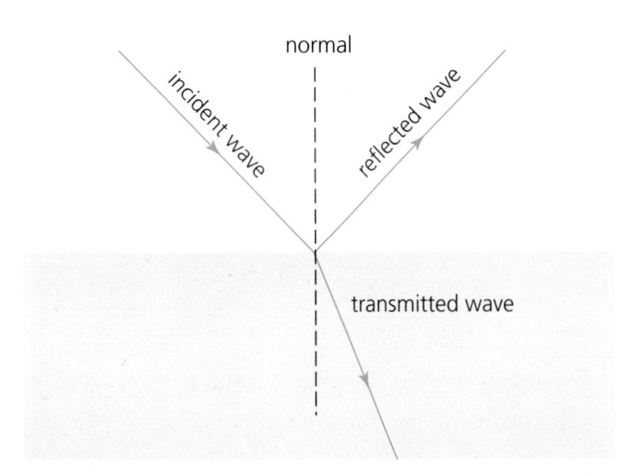


*Figure 9 The principle of conservation of energy means that, at the boundary between two materials, the energy of the reflected wave added to the energy of the transmitted wave must be equal to the energy of the incident wave. In the case of a perfect reflector, no energy is transmitted.*

Wave speed depends on some of the properties of the medium, so, as a wave crosses a boundary between two different materials, it will speed up or slow down. For example, light travels at a higher speed in air than it does in water or glass, and water waves slow down as they move from deep to shallow water. **Refraction** is the change in the direction of a wave that occurs due to the change in wave speed as it travels from one medium to another. Refraction of light through glass and water can lead to some unusual effects (Figure 10).

Refraction affects both transverse waves and longitudinal waves. Sound waves are refracted as they

*Figure 10 Refraction of light can produce an image that is displaced from the object.*

cross a boundary between media in which the speed of sound is different. For example, sound travels faster in warmer air than in cold air. At night a 'temperature inversion' may occur, where the air near the ground is warmer than that higher up. Sound waves will travel faster in the warmer air and so will tend to curve back towards the ground at night. This means that you can often hear things at night from a greater distance.

When a wave slows down, the wave fronts (see section 6.2 of Chapter 6) get closer together. This is to be expected, as wave speed = frequency × wavelength ($c = f\lambda$). The frequency is determined by the source and remains constant. So if the speed decreases, the wavelength must also decrease. Figure 11 shows water waves travelling from a deep region to a shallower region. When waves strike the boundary at an angle, part of the wave front (the left-hand part of the wave front in Figure 11) reaches the new medium before the rest of the wave. That part will therefore

*Figure 11 Waves travelling from deep water through an area of shallow water*

change speed first; in Figure 11 it slows down before the rest of the wave front. As the rest of the wave front enters the medium, and so slows down, the wave front changes direction. The wave has been refracted.

A ray, showing the direction of the waves, moves towards the normal to the boundary when the wave slows down, and away from the normal when the wave speeds up. The deflection of the ray depends on the ratio of the wave speed in each medium: a bigger change in speed will result in a bigger change in angle. One way of thinking about this is shown in Figure 12.

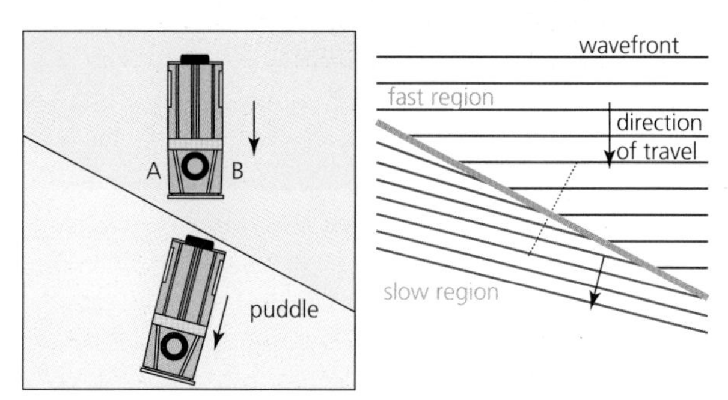


*Figure 12* A car approaching a puddle is an analogy for the change in direction of wave fronts as they hit a slower region at an angle. When the car reaches the puddle, wheel A hits the water first and slows down. This causes the car to swerve into the puddle.

Figure 13 shows rays of light hitting a glass block at an angle to the boundary, refracting towards the normal as the light waves slow down in the glass, and then refracting away from the normal on exit from the block.

The refraction of light can lead to misleading images (see Figures 10 and 14). Figure 15 shows the ray diagram that explains why the straw appears to be bent at the water surface.

*Figure 13* The light rays follow a zig-zag path as the waves first slow down and then speed up as they re-enter the first medium.

*Figure 14* The straw looks bent, but this is because the direction of the light ray has been changed by refraction.

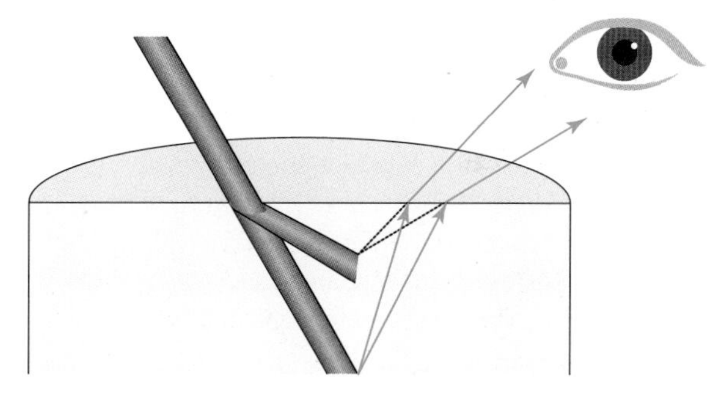

*Figure 15* The underwater image of the straw is a virtual image.

## QUESTIONS

5. Figure 16 shows a series of plane waves reaching a medium where the wave speed is slower. Copy and complete the diagram to show what happens to the wave fronts as they pass through the medium and emerge from the other side.

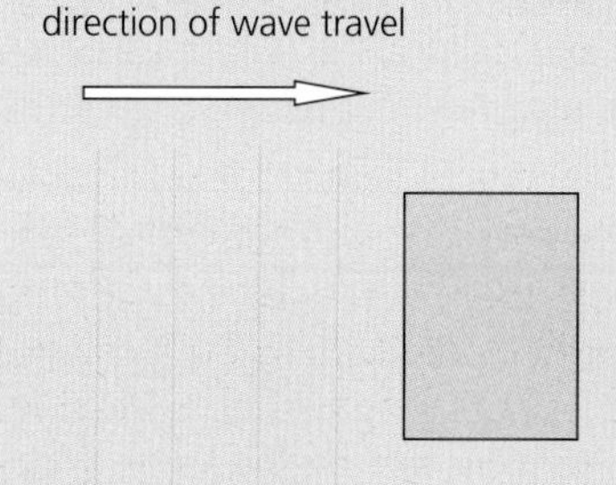


*Figure 16*

6. Light travels more slowly in water than in air. Sketch the wave fronts from an underwater light source, as they reach the water surface at an angle and refract into the air.

7. The refraction of light makes a swimming pool look less deep than it actually is. Explain this phenomenon with the aid of a diagram.

Stretch and challenge

8. Explain why surface waves on the sea tend to strike a beach at, or near, normal incidence – that is, the wave direction is parallel to the normal at the boundary between water and shore.
9. Consider the equation:

   wave speed = frequency × wavelength

   When the wave speed decreases, for example when light enters glass, the wavelength decreases. Why is it that the frequency remains fixed?

## 7.3 REFRACTIVE INDEX

Light waves travel at a constant speed, $c$, in a vacuum ($c = 299792458 \text{ m s}^{-1}$). When they pass through matter, the waves slow down. We define the **refractive index** of a transparent substance, $n$, as the ratio of the wave's speed in a vacuum, $c$, to the speed in the substance, $c_s$:

$$n = \frac{c}{c_s}$$

It is a ratio and so has no unit. Since the speed of light in a vacuum is higher than in any material, the refractive index of a substance is always greater than 1 (Table 1). The higher its value, the more the rays are refracted. The refractive index of air is around 1.0003, so we often take it as 1. The precise value depends on the density of the air, which varies with temperature. Convection currents in the atmosphere lead to fluctuations in density, which alter the refraction of light (Figure 17). These variations make stars seem to twinkle when viewed from Earth.

| Substance | Refractive index |
|---|---|
| Diamond | 2.42 |
| Glass | 1.40 to 2.00 |
| Perspex | 1.50 |
| Water | 1.33 |
| Sea water | 1.34 |
| Ice | 1.31 |
| Air | 1.00 |

*Table 1* Refractive index of different substances

In general, denser substances tend to have a higher refractive index, though there is no strict relationship. Materials with a high refractive index are often referred to as **optically dense**.

*Figure 17* The latest observatories use lasers to measure the variations in refractive index in the atmosphere and feed them back so that a 'twinkle-free', high-quality image may be formed.

### Snell's law

When light is refracted at the boundary between two materials, the amount of refraction depends on the ratio of the refractive indices of the two materials (Figure 18).

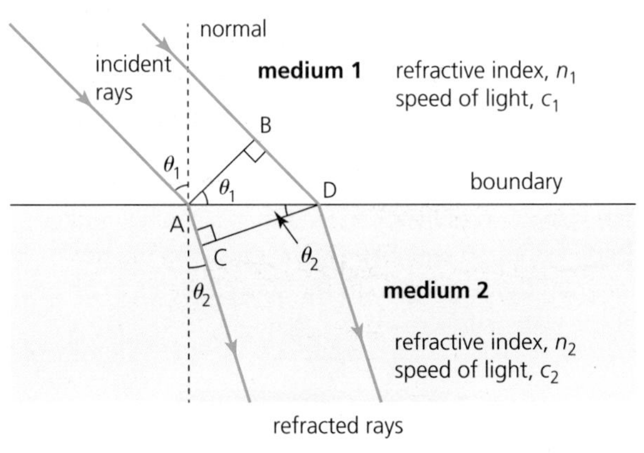


*Figure 18* This shows two rays representing the direction of a set of plane wave fronts. Notice that the rays are parallel before, and after, the movement across the interface.

**Snell's law** links the angles of incidence and refraction with the speed of light in each medium.

$$n_1 \sin \theta_1 = n_2 \sin \theta_2$$

The law is derived from Fermat's principle, which says that light will take the path that minimises its journey time. Look at Figure 18. It shows a wave front AB moving across the boundary between two media, 1 and 2, which could be air and water, for example. In the next interval of time, $t$, point B will travel to D, but point

A will only travel as far as C, because it has moved into a medium where the speed of light is slower.

We are aiming to derive an expression that links the geometry, in particular the angles of incidence and refraction, to the speed of light in each medium.

From Figure 18, $\sin\theta_1 = \frac{BD}{AD}$ and $\sin\theta_2 = \frac{AC}{AD}$, so dividing these gives

$$\frac{\sin\theta_1}{\sin\theta_2} = \frac{BD}{AC}$$

But we can express the lengths BD and AC in terms of the speed of light and time:

$$BD = \text{speed} \times \text{time} = c_1 \times t$$

$$AC = c_2 \times t$$

Since the times to travel AC and BD are the same, we can cancel $t$ from those equations:

$$\frac{BD}{AC} = \frac{c_1}{c_2}$$

So

$$\frac{\sin\theta_1}{\sin\theta_2} = \frac{c_1}{c_2}$$

This is one form of Snell's law. It states that, for a given wavelength of light, the ratio of the speed of light in each medium is equal to the ratio of the sines of the angle that the ray makes with the normal in each medium. Therefore, the angle is always smaller in the more optically dense medium. Note also that the incident and refracted rays and the normal all lie in the same plane.

Further, since $n_1 = \frac{c}{c_1}$ and $n_2 = \frac{c}{c_2}$, we obtain

$$\frac{\sin\theta_1}{\sin\theta_2} = \frac{n_2}{n_1}$$

This can be rearranged as

$$n_1 \sin\theta_1 = n_2 \sin\theta_2$$

## Worked example

A ray of red light strikes an edge of an equilateral triangular prism at an angle of incidence of 30°.

a. Calculate the angle of refraction. The refractive index of the glass is 1.45.

b. Sketch the path of the ray of light until it leaves the prism, marking in all the angles of incidence and refraction.

c. Explain what difference it would make if the ray of light had been white.

a. $n_1 \sin\theta_1 = n_2 \sin\theta_2$
Assume $n_1 = 1$. Then

$$\sin\theta_2 = \frac{1}{1.45}\sin 30° = 0.345$$

$$\theta_2 = 20.2°$$

b. See Figure 19.

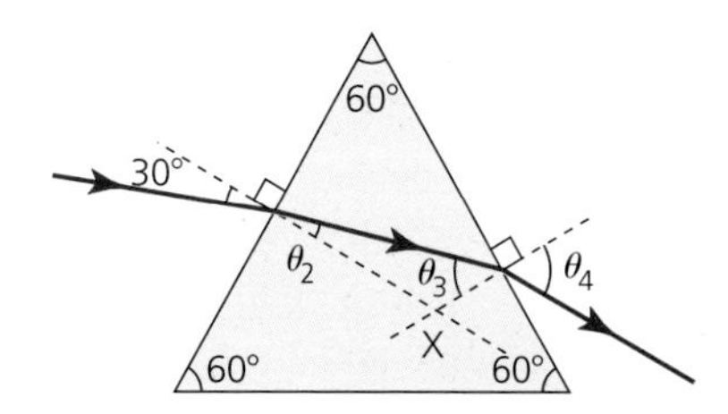


*Figure 19*

The angle at X must be 360° − 90° − 90° − 60° = 120°. The angle of incidence $\theta_3$ is therefore 180° − 120° − 20.2° = 39.8°.

Using Snell's law again for the emerging ray,

$$n_2 \sin\theta_3 = n_1 \sin\theta_4$$

$$\sin\theta_4 = \frac{n_2}{n_1}\sin\theta_3 = \frac{1.45}{1}\sin 39.8° = 0.928$$

$$\theta_4 = 68.1°$$

c. If the incoming light had been white, the shorter wavelengths (the violet end of the spectrum) would have been refracted further towards the normal when the light first met the prism, and further away from the normal as the light left the prism, thereby creating a spectrum.

## QUESTIONS

10. A ray of light strikes a double-glazed window at an angle of 35° to the normal, as in Figure 20. The panes of glass are separated by an air gap.

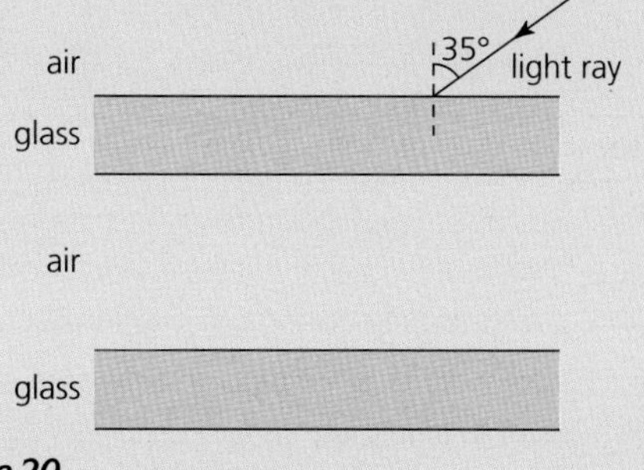


*Figure 20*

a. Copy Figure 20 and on it sketch the path of the light as it passes through the first pane, the gap and the second pane of the window.

b. Calculate the angle of refraction in the first pane of glass. (Take the refractive index of the window glass to be 1.4.)

c. Hence label on your sketch all the angles of incidence and refraction with their values.

d. There will be some partial reflection at each interface. Include these reflected rays in your diagram and explain what effect it may have on an object viewed through the window.

11. A convex lens converges light and can be used to bring light rays to a focus.

a. Parallel rays of red light are incident on a convex lens as shown in Figure 21. On a copy of the diagram, sketch the path of the rays through and emerging from the lens.

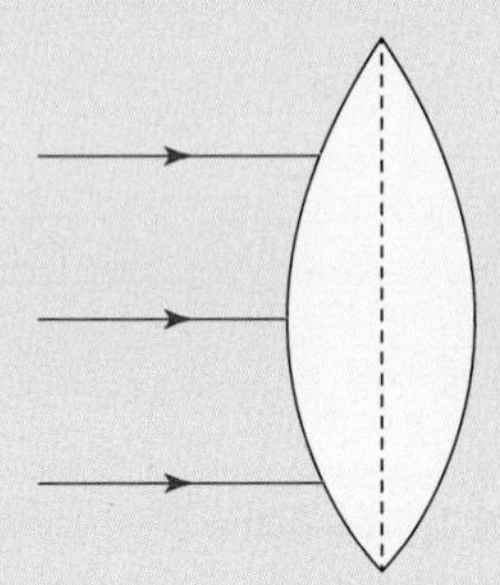

*Figure 21*

b. Explain why the lens brings the red light rays to a focus.

c. What difference would it make to the action of the lens if the glass had a higher value of refractive index?

d. What difference would it make to the action of the lens if the light were blue rather than red? (*Hint*: The glass will have a higher refractive index for blue light than for red.)

e. Considering your answers to parts **c** and **d**, describe what effect the lens would have on white light. How would this affect an image produced by the lens?

f. Why do large telescopes use mirrors, rather than lenses?

12. When you see an object, light from it has travelled through your cornea, the aqueous humour, the lens and the vitreous humour (Figure 22). At each boundary the light is refracted, so that it finally comes to a focus on the light-sensitive cells in the retina at the back of your eye.

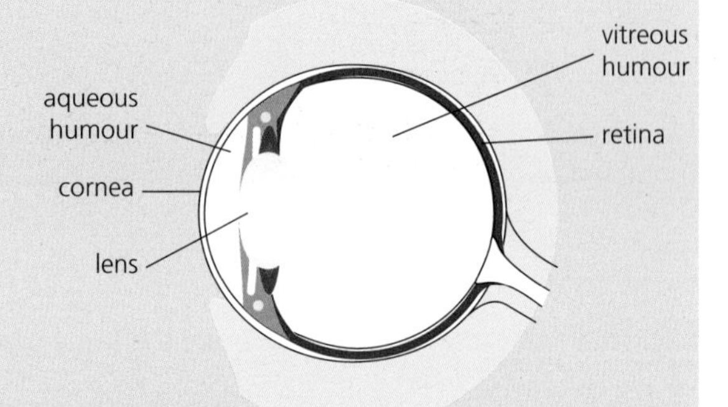


***Figure 22*** *Cross-section through the eye*

Table 2 lists values for the refractive index for the various parts of the eye.

| Medium for light path through the eye | Refractive index |
|---|---|
| Air | 1.0003 |
| Cornea | 1.376 |
| Aqueous humour | 1.336 |
| Lens (outer layer) | 1.406 |
| Vitreous humour | 1.337 |

*Table 2*

a. Using the information in Table 2, calculate the ratios of the refractive indices at each boundary.

b. Where does most of the refraction take place?

c. The refractive index of water is 1.33. Explain, using a diagram and a simple calculation, why it is difficult to see when you are swimming underwater. How do goggles help?

## KEY IDEAS

- Rays of light refract towards the normal when they enter a material that is optically denser (that is, where light travels more slowly).
- Rays of light refract away from the normal when they enter a material that is optically less dense (that is, where light travels faster).
- The refractive index of a substance is given by

$$n = \frac{\text{speed of light in a vacuum}}{\text{speed of light in substance}} = \frac{c}{c_s}$$

- The greater the refractive index of a material, the more optically dense it is.
- The refractive index of air is very nearly 1.
- Snell's law states that, at a boundary between media 1 and 2,

$$n_1 \sin\theta_1 = n_2 \sin\theta_2$$

where $\theta$ is the angle with the normal to the boundary.

## ASSIGNMENT 1: USING SNELL'S LAW TO FIND THE REFRACTIVE INDEX OF GLASS

**(MS 0.6, MS 1.5, MS 3.2, MS 3.3, MS 3.4, PS 1.1, PS 2.1, PS 2.2, PS 2.3, PS 3.1, PS 3.3)**

We can verify Snell's law and find the refractive index of glass by measuring the refraction of a light ray as it passes through a rectangular block of glass.

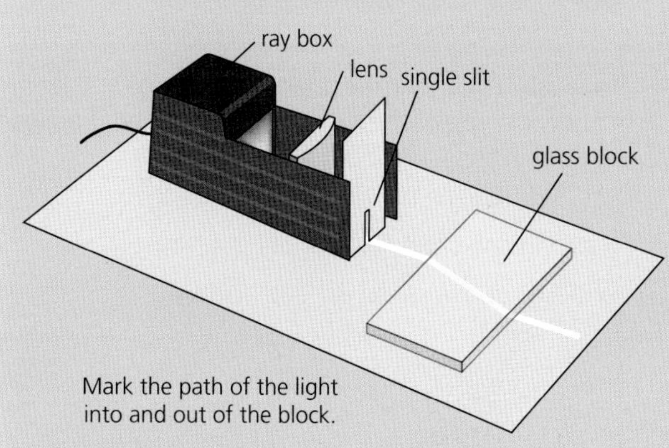


***Figure A1*** *The set-up*

Using the apparatus in Figure A1, the angle of refraction can be measured for a range of angles of incidence, say 10° to 80°. The glass block has to be removed to measure the angle of refraction, so for each angle of incidence the light path has to be marked, with a sharp pencil, as it enters and leaves the block. Remember that angles of incidence and refraction are always measured *from the normal*, not from the boundary between the two media.

A typical set of results is shown in Table A1.

| Angle of incidence, $\theta_1$ / degree | Angle of refraction, $\theta_2$ / degree |
|---|---|
| 10 | 7 |
| 20 | 15 |
| 30 | 22 |
| 40 | 29 |
| 50 | 32 |
| 60 | 40 |
| 70 | 41 |
| 80 | 45 |

***Table A1***

The equation $n_1 \sin\theta_1 = n_2 \sin\theta_2$ can be written in the form

$$\sin\theta_1 = (n_2/n_1) \times \sin\theta_2$$

Comparing this with $y = mx + c$, a graph of $\sin\theta_1$ ($y$-axis) against $\sin\theta_2$ ($x$-axis) should be a straight line, with a gradient of $n_2/n_1$. As the experiment is done in air, $n_1 \approx 1$, so the gradient will give $n$, the refractive index of the glass.

### Questions

**A1** Use the results in Table A1 to plot an appropriate graph to find the refractive index of glass.

**A2** Identify the sources of error in the experiment.

**A3** Estimate the uncertainty in the readings, and in your answer for the refractive index.

**A4** Suggest how the uncertainty in the answer could be reduced.

**A5** Suppose that you now wished to investigate the refraction between glass and water. Design an experiment to find the ratio of the refractive indices of glass and water.

## 7.4 LIGHT IN AN OPTICAL FIBRE

Optical fibres are long, solid cylinders of glass or transparent polymer, with a very small diameter – a single fibre has a diameter around 15 μm, about the same as a human hair. A cable used for signal transmission is made of many such fibres. The type of fibre used in communication networks is made up of two layers of glass, of slightly different refractive indices. These are called **step-index fibres**. Light (usually infrared) is transmitted along the inner layer, which is made from very pure glass; this is called the **core**. The outer layer is called the **cladding**. This is less pure glass with a lower refractive index.

The laws of reflection and refraction govern the path of light through an optical fibre. We can use Snell's law to work out what happens to rays of light when they enter the fibre.

Suppose the rays of light (in air) from a light-emitting diode (LED) strike the end of an optical fibre at angles of incidence up to 30°. If the core of the fibre is made of glass of refractive index 1.60, we can calculate what will happen to a ray of light as it travels down the fibre.

If the end is flat and at right angles to the fibre, the angle of incidence for a ray from the edge of the beam, $\theta_1 = 30°$. Assuming $n_{air} = 1$, then

$$n = \frac{\sin\theta_1}{\sin\theta_2}$$

Therefore

$$\sin\theta_2 = \frac{\sin 30°}{1.6} = \frac{0.5}{1.6}$$

$$\theta_2 = 18.2°$$

From the right-angled triangle (Figure 23):

$$90° + \theta_3 + \theta_2 = 180°$$

$$\theta_3 = 90° - \theta_2 = 71.8°$$

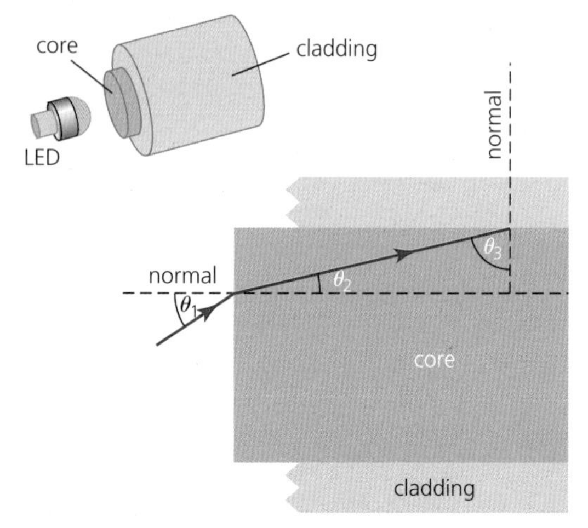


***Figure 23*** *A light ray entering an optical fibre*

This angle, $\theta_3$, is now the angle of incidence at the boundary between the core and the cladding. Suppose the cladding has a refractive index of 1.50.

Using Snell's law, with the angle of refraction at the core–cladding boundary $\theta_4$, we have

$$1.6 \sin 71.8° = 1.50 \sin\theta_4$$

giving

$$\sin\theta_4 = 1.013$$

If you try to find the inverse sine of 1.013 on your calculator, it will indicate that you have made an error. This does not mean that the calculation is wrong, only that the mathematics has not kept up with the physics of refraction. Snell's law only works if the ray actually crosses the boundary. At high angles of incidence the rays cannot cross over from a material of higher refractive index to one of lower refractive index – they only reflect.

QUESTIONS

13. To reduce light loss, an LED is stuck onto the end of an optical fibre with a transparent glue. The refractive indices of the glue and the core of the fibre are 1.40 and 1.60.

    a. Calculate the ratio: $\frac{\text{refractive index of glue}}{\text{refractive index of core}}$

    b. If a ray of light from the LED hits the end of the fibre core at 50° to the normal, at what angle will it strike the wall of the core?

## 7.5 TOTAL INTERNAL REFLECTION

When a light ray travels from one medium to a less optically dense medium, it is refracted away from the normal. The angle of refraction is larger than the angle of incidence. With increasing angle of incidence, there comes a point when the light is refracted along the boundary. If the angle of incidence is increased beyond that point, no light escapes from the first medium at all (Figure 24).

In Figure 24 a light ray is passing through a semicircular glass block. At low angles of incidence (as in A), light refracts as expected. As the angle of incidence in the glass block is increased, more of the light is reflected (as in B). Eventually, an angle of incidence is reached (as in C). At higher angles, reflection increases until the internal surface acts as a perfect mirror (as in D). This is called **total internal reflection**, because all of the incident light energy is reflected.

The change in reflection is not sudden; there is partial reflection inside the block for all angles of incidence. The reflected ray gets stronger as the angle of incidence rises. At one particular angle, the **critical angle**, $\theta_c$, the refracted ray lies along the boundary between the media (angle of refraction 90°). Beyond the critical angle ($i > \theta_c$), the internal reflection is total; there is no refracted ray.

The critical angle depends on the refractive indices of the media. At the critical angle, the ray travels at 90° from the normal in the less dense medium 2, so

$$n_1 \sin\theta_c = n_2 \sin 90°$$

So

$$\sin\theta_c = \frac{n_2}{n_1}$$

In the case of rays travelling in a medium of refractive index $n$ where the less dense medium is air, we have $n_2 \approx 1$, so

$$\sin\theta_c = \frac{1}{n}$$

### Total internal reflection in an optical fibre

Total internal reflection allows a light signal to pass along a very long optical fibre. Provided the incident angle at the core–cladding interface is always greater than the critical angle, the light rays undergo repeated reflections at that boundary and are trapped in the core. A single pure glass fibre (that is, without the cladding) could be used to carry light in this way, but its surface would be easily scratched, allowing some light to leak out (Figure 25). What effect does the cladding have on total internal reflection?

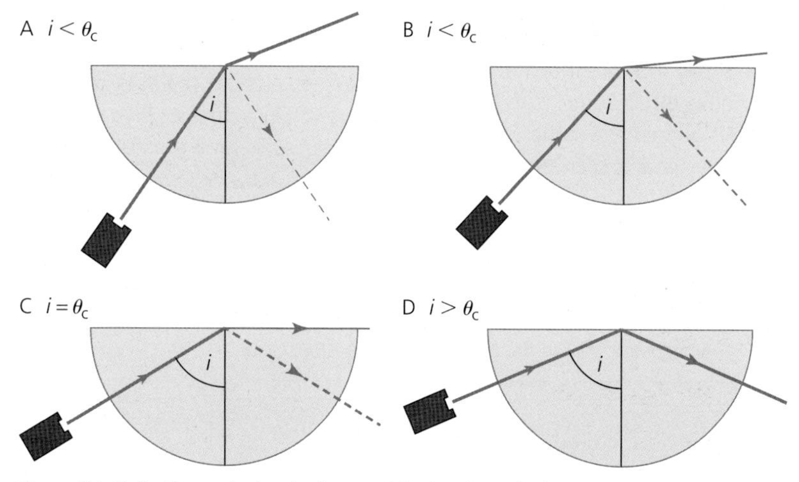


***Figure 24*** *Reflection and refraction in a semicircular glass block*

Surface scratches lead to light leakage in a single unclad fibre.

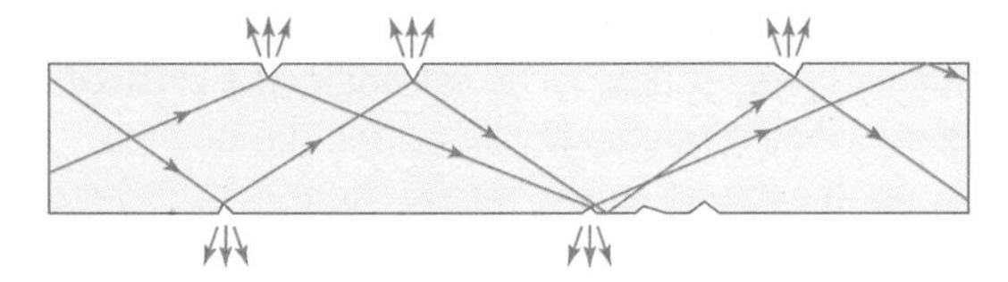

Scratches in the outer cladding do not matter because it does not carry a light signal.

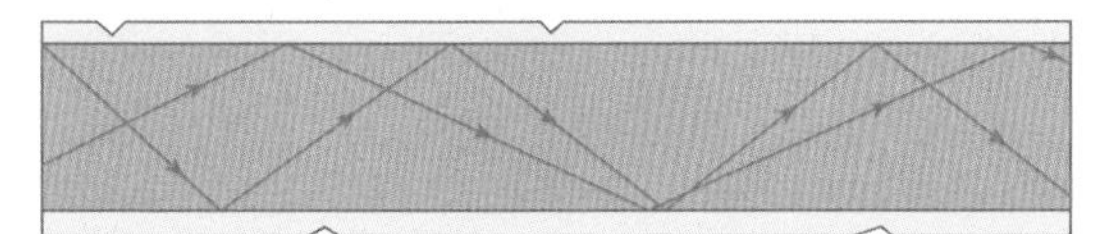

***Figure 25*** *Light leakage from an optical fibre is much reduced by cladding.*

Suppose the core of an optical fibre has a refractive index of 1.60. In air, the critical angle of the glass is given by

$$\sin\theta_c = \frac{1}{n} = \frac{1}{1.60} = 0.625$$
$$\theta_c = 38.7°$$

If the cladding is made of less optically dense glass, the critical angle becomes much larger. Suppose the cladding has a refractive index of 1.50. The critical angle is now given by

$$\sin\theta_c = \frac{n_2}{n_1} = \frac{1.50}{1.60} = 0.9375$$
$$\theta_c = 69.6°$$

The protected fibre reflects only light rays with a very high angle of incidence. This increase of critical angle for the cladded fibre might seem to be a disadvantage, because less of the light will reflect inside. However, it turns out to be an advantage in signal transmission. When only large angles of incidence are allowed, the rays that are internally reflected will travel by a more direct route (see Figure 26, where the zig-zag ray is almost the same length as the direct ray) and so the signal it carries will suffer a smaller relative time delay.

Ray B has the longest possible path with an angle of incidence equal to the critical angle. The ratio of the length of ray B's journey to the length of ray A's journey is $\frac{PQ}{PR} = \frac{PQ}{\sin\theta_c} = \frac{1}{0.9375} = 1.0667$

Along 1 km of fibre the maximum path is 1.0667 km.

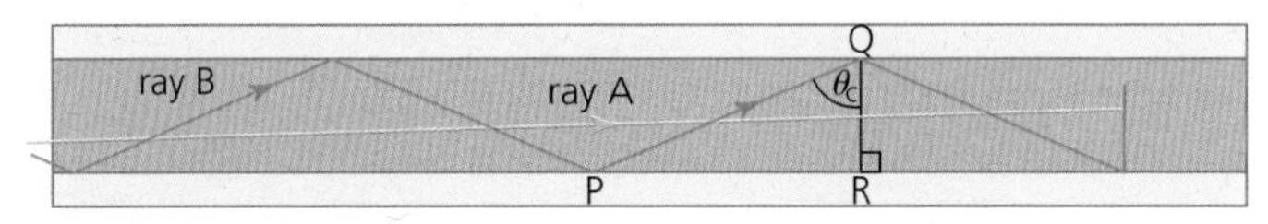


***Figure 26*** *Calculating path difference in an optical fibre*

## QUESTIONS

**14. a.** Calculate the critical angle for diamond (refractive index = 2.42).

**b.** What effect does this have on the appearance of a cut diamond (Figure 27)?

***Figure 27*** *Why does a diamond sparkle?*

**15.** Figure 28 shows the path of a light beam through an equilateral prism. The angle of incidence at the first air–glass interface is 50° and the refractive index of the glass is 1.4.

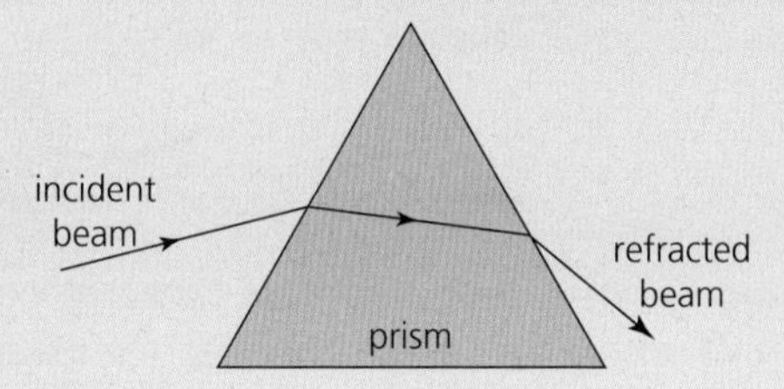


***Figure 28***

**a.** Copy the diagram. On your sketch, draw in the normals where the light rays cross a boundary and label all the angles of incidence and refraction with their calculated values.

**b.** Redraw the diagram with the angle of incidence at the *glass–air* interface as 50°. You will need to work back to find the initial angle of incidence from air to glass.

## ASSIGNMENT 2: EXPLAINING RAINBOWS

**(MS 0.6, MS 2.2, MS 2.3, MS 2.4, MS 4.1, MS 4.5)**

A rainbow is caused by the Sun's light being refracted and then internally reflected in raindrops. Figure A1 shows the path of a ray of red light through a raindrop.

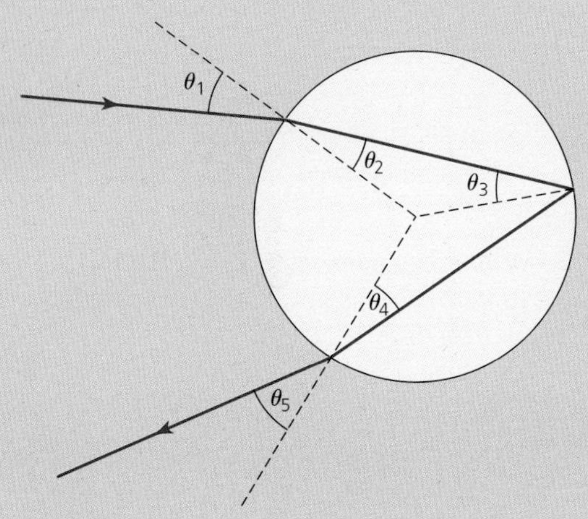


*Figure A1* *Refraction and total internal reflection in a raindrop*

### Questions

**A1** If $\theta_1 = 70°$, calculate $\theta_2$.

**A2** What is the angle of incidence, $\theta_3$, at the back of the raindrop? (*Hint*: Look for an isosceles triangle.)

**A3** Calculate the critical angle for a light ray travelling from air to water and hence explain why the light is reflected at the back of the raindrop.

**A4** What is the angle of incidence, $\theta_4$?

**A5** Calculate the angle of refraction, $\theta_5$, as the ray of light leaves the raindrop.

**A6** Water has a higher refractive index for blue light than for red light. Make a sketch of Figure A1 and on your diagram add in the path that blue light would take.

**A7** Use your sketch to explain how rainbows occur.

### Stretch and challenge

**A8** It is sometimes possible to see a secondary rainbow. This happens when there are two internal reflections inside the raindrop. With the help of a diagram explain the formation of a secondary rainbow.

**A9** The sky is noticeably darker between the primary and secondary rainbows (Figure A2). This is known as Alexander's dark band. Why does it occur? (You may need to do a little research.)

*Figure A2* *A rainbow with a secondary rainbow and Alexander's dark band*

## KEY IDEAS

- Total internal reflection occurs when a light wave reaches a boundary between one medium and another of lower refractive index, at angles of incidence greater than the critical angle.
- In general, the critical angle is given by $\sin\theta_c = \frac{n_2}{n_1}$ where $n_2 < n_1$
- For light travelling from a medium into air, the critical angle is given by

$$\sin\theta_c = \frac{1}{n}$$

- Optical fibres rely on repeated total internal reflection to transmit light pulses.

# 7.6 DISPERSION AND ATTENUATION IN AN OPTICAL FIBRE

## Modal dispersion

Rays taking different paths along an optical fibre will have different journey times to the end of the fibre (look back at Figure 26). This is called **modal dispersion** because the different modes of travel are dispersed in terms of their time of arrival. In communications it is important to keep modal dispersion as small as possible. Information signals are sent through the optical fibre as a series of short pulses of light. With dispersion, short pulses sent from the transmitter spread out by the time they reach the receiver. This is called **pulse broadening** and it reduces the quality of the signal (Figure 29). Severe pulse broadening can cause pulses to overlap and then information is lost. Hence there is a maximum pulse frequency for each type of fibre.

### Worked example

What is the maximum data rate (pulses per second) that can be sent along 10 km of fibre, if the refractive index of the core is 1.55 and that of the cladding is 1.45?

The critical angle between the core and the cladding is given by

$$\sin\theta_c = \frac{n_2}{n_1} = \frac{1.45}{1.55} = 0.935$$

$$\theta_c = 69.3°$$

The ray that strikes the cladding at the critical angle will travel $d/\sin 69.3°$ (see Figure 26), while the straight-through ray travels a distance $d$. So over 10 km, the longest path will be

$$\frac{10}{0.935} = 10.7 \text{ km}$$

so 700 m longer. Because the speed of light in the core is $3.00 \times 10^8/1.55 = 1.935 \times 10^8$ m s$^{-1}$, this is equivalent to a time delay of

$$\frac{700}{1.935 \times 10^8} = 3.62 \times 10^{-6} \text{s}$$

If we assume that another pulse must not arrive in this time, the maximum pulse frequency is $1/(3.62 \times 10^{-6}) = 0.28 \times 10^6$ pulses per second.

One way of reducing modal dispersion is to use core and cladding of almost identical refractive indices. Another way is to keep the core of the fibre as narrow as possible. This is what is done in a **monomode** fibre, as opposed to a standard **multimode** fibre. The core of such a fibre is only a few micrometres in diameter – about 100 times thinner than a human hair.

### QUESTIONS

**16.** Explain how using core and cladding materials of almost identical refractive index would reduce modal dispersion to a minimum.

**17.** What is the time delay between the straight-through path and the critical-angle path for 1 km of the optical fibre with refractive indices of 1.50 and 1.60? (*Hint:* Refer back to Figure 26.)

**18.** A data signal is sent along the optical fibre with refractive indices as in question 17. What would be the maximum frequency of transmission of pulses along 10 km of this cable?

## Material dispersion

The refractive index of a material varies with wavelength. The values quoted in Table 1 in section 7.3 are for a 'standard wavelength' of light – the 589 nm bright orange light that you might have seen in sodium street lamps. For almost all materials, refractive index increases with frequency, so blue light is refracted more than red light. For flint glass, for example, the typical refractive indices are 1.640 for red light, 1.646 for yellow and 1.660 for blue. This variation of refractive index is the reason that white light beams are split up into different

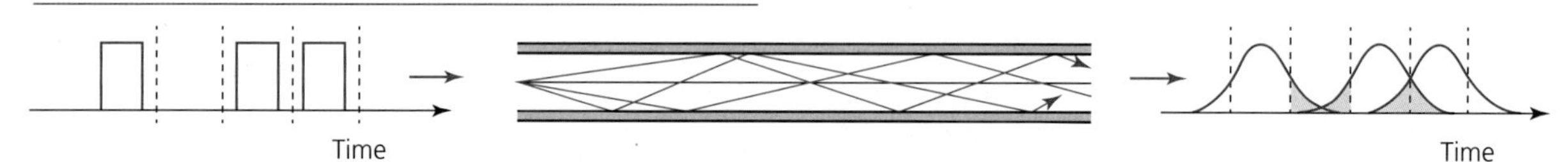


***Figure 29*** *Short pulses are sent into the step-index fibre, but modal dispersion means that each output pulse spreads over a longer time, merging with other pulses and limiting the maximum pulse frequency that can be transmitted by the fibre.*

colours by prisms and raindrops. The separation of different colours is called **chromatic dispersion**, or **material dispersion** because it is caused by properties of the material.

As with modal dispersion, it is important to avoid material dispersion in the optical fibres used in communications, because otherwise a range of wavelengths transmitted as a sharp pulse would spread out as it travelled along the fibre (Figure 30).

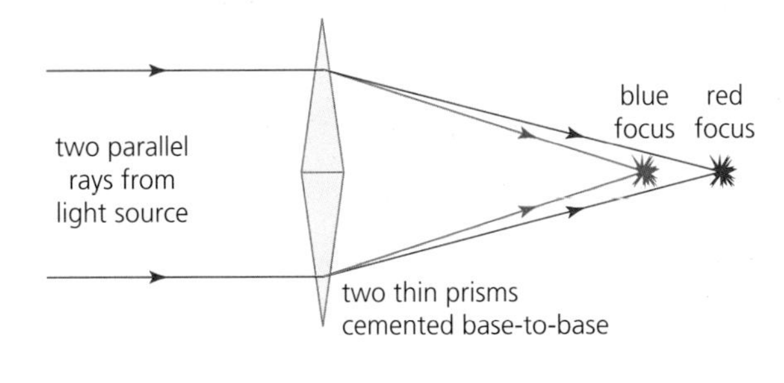


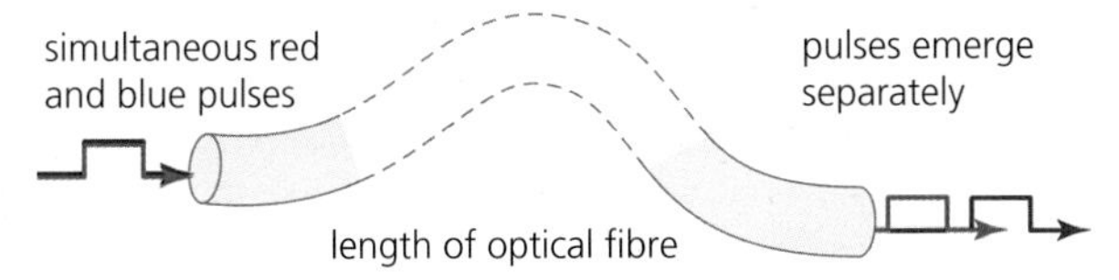


***Figure 30*** *Chromatic or material dispersion, and its effect on pulses sent along an optical fibre*

It is difficult to make glass that has no chromatic dispersion. Instead, light with a very narrow wavelength band is used. Lasers make ideal light sources for optical fibres because they are monochromatic: their light covers an extremely narrow wavelength band. Other sources include light-emitting diodes (LEDs). These are much cheaper but they transmit light over a wider band of wavelengths. The amount of chromatic dispersion also depends on the wavelength itself; for glass, there is very little chromatic dispersion at wavelengths in the infrared at around 1300 nm. Most telecommunications systems use infrared transmitters.

## Attenuation

The glass used to make optical fibres is incredibly transparent – if the oceans were made of it, you would be able to see to the bottom of the deepest part. However, there are some losses that weaken the signal. This 'weakening' is called **attenuation** and the unit for its measurement is the **decibel** (dB). Optical fibres have an attenuation coefficient, which is measured in decibels per kilometre.

The attenuation of a light signal in an optical fibre is due to several different processes:

- **Scattering.** A small variation in the composition of the glass, or its structure, leads to scattering. Some scattered light rays then hit the cladding at less than the critical angle and are refracted out of the core.
- **Absorption.** A photon of light may interact with an electron in an atom in the glass and be absorbed. Later the electron will lose energy and re-emit a photon but not necessarily of the same wavelength (see Figure 31) and not in the same direction.

***Figure 31*** *Iron impurities in the glass absorb red light and scatter green light. Iron is sometimes deliberately added to the glass for car windscreens as it also absorbs infrared and so helps to keep the car cooler.*

- **Bending losses.** If an optical fibre is bent too sharply, some light will strike the cladding at less than the critical angle and will not be reflected back into the core.

The overall attenuation depends on the composition of the fibre and the wavelength of the light (Figure 32).

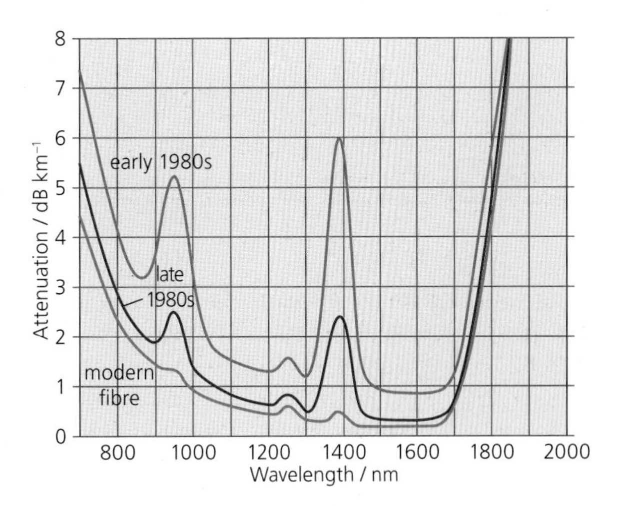


***Figure 32*** *Modern optical fibres suffer less attenuation because of advances in materials technology.*

## QUESTIONS

**19.** A semiconductor light transmitter has a wavelength range from 750 nm to 790 nm. At these wavelengths, the glass of an optical fibre has refractive index 1.6404 and 1.6400, respectively. The light transmitter is switched on for 100 ns. How long would it take for light at the extremes of the wavelength range to travel 30 km along the optical fibre? What is the time difference (in nanoseconds) between the arrival of the two colours? (Take $c = 3.00 \times 10^8\ \mathrm{m\,s^{-}}$.)

**20.** There are three wavelength ranges (known as 'windows') used to carry signals down an optical fibre. Look carefully at Figure 32 and suggest the wavelength ranges used.

**21.** Why does the maximum data transmission rate in an optical fibre depend on its length?

## ASSIGNMENT 3: UNDERSTANDING THE DECIBEL

**(MS 0.1, MS 0.3, MS 0.5, MS 2.5, MS 3.1, MS 3.2, MS 3.4, MS 3.10)**

### Stretch and challenge

A decibel is a unit of measurement that is the logarithm of a ratio. The ratio can be of any two physical values (of the same variable), but the decibel is often used to compare two values of power or intensity (power through an area). The decibel is particularly associated with sound levels, usually with reference to the quietest sound that can be heard.

A healthy human ear can detect an enormous range of sound intensity. The faintest audible sound has an intensity of $1 \times 10^{-12}\ \mathrm{W\,m^{-2}}$, referred to as $I_0$. This is equivalent to a pressure change of less than one billionth of an atmosphere or a movement in the air of less than the diameter of an atom. The loudest sound that can be heard, without permanent damage to the ear occurring, is an intensity of $10\ \mathrm{W\,m^{-2}}$. This is 10 000 000 000 000 times more than the quietest sound. To cover such an enormous range of values, we need to use a logarithmic scale, like the decibel scale shown here.

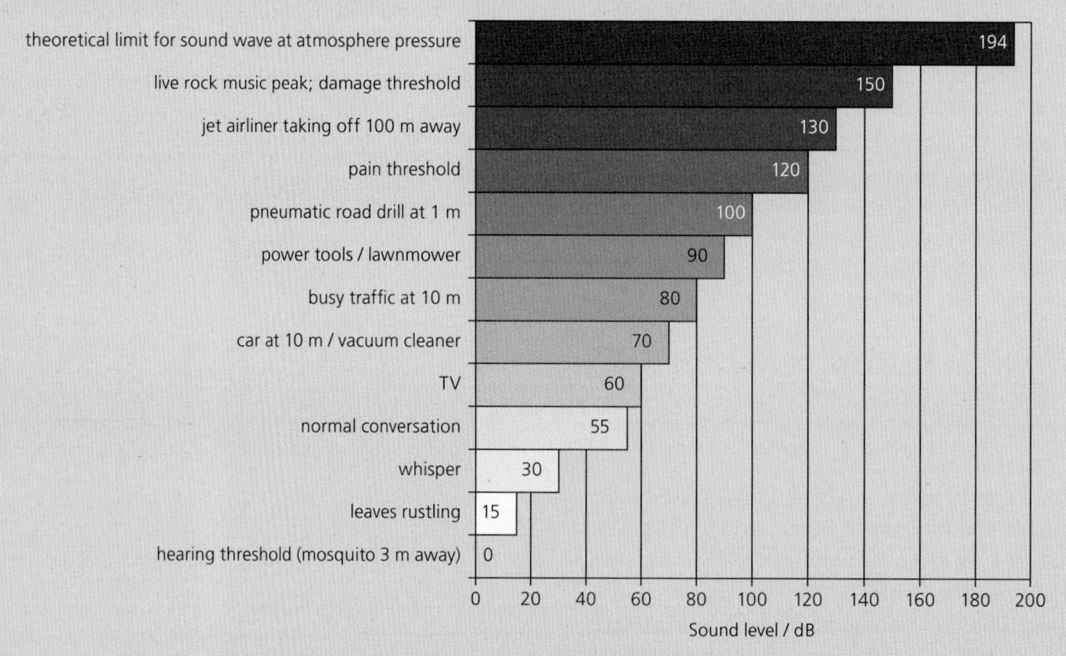


*The sound-level decibel scale. Each 10 dB step means 10 times the intensity.*

The sound intensity level, $I$, in decibel (dB), is

$$I\ (\text{dB}) = 10\log_{10}\left(\frac{I}{I_0}\right)$$

If the sound rose to 10 times the threshold of hearing, the sound level in dB would be

$$10\log_{10}(10) = 10\ \text{dB}$$

But 20 dB is not 20 times the intensity of $I_0$, as the following shows:

$$20 = 10\log_{10}\left(\frac{I}{I_0}\right)$$

$$2 = \log_{10}\left(\frac{I}{I_0}\right)$$

Take inverse logarithms ($10^x$) of both sides of the equation:

$$10^2 = \frac{I}{I_0} = 100$$

So 10 dB is an increase by a factor of 10, but 20 dB is an increase by a factor of $10^2$ (100). A value of 130 dB would be an increase by a factor of $10^{13}$.

## Questions

**A1** An amplifier boosts the power of a signal by 3 dB. If the input power is 1 mW, what would the output power be?

**A2** An amplifier increases the electrical power of a signal by 65 dB. The power of the output signal is measured at 2.50 mW. What was the input power?

**A3** A signal of intensity 30 mW $m^{-2}$ passes through three consecutive amplifiers, which each boost the signal intensity by 20 dB. What will the output power be?

**A4** An electrical signal of power 5 mW is sent along a cable. A repeater station is needed when the signal drops to 1 mW, to boost the signal. If the loss down the cable is 5 dB $km^{-1}$, what will be the maximum distance between repeater stations?

**A5** A signal is transmitted along an optical fibre and its power is measured at different distances from the input. The readings are shown in Table A1. Plot a graph of attenuation (in dB) versus distance and find the attenuation coefficient (in dB $km^{-1}$).

| Power / mW | Distance / km |
|---|---|
| 15.02 | 0.00 |
| 14.11 | 1.20 |
| 11.12 | 2.60 |
| 10.68 | 3.20 |
| 8.41 | 4.40 |
| 7.52 | 5.90 |

*Table A1*

**A6** A student used a sound pressure level meter to monitor the sound levels of a teacher talking in the classroom. The results are shown in Table A2.

| Distance from front of class / m | Sound level of teacher's voice / dB | | |
|---|---|---|---|
| | Normal | Very loud | Shouting |
| 0.3 | 70 | 82 | 88 |
| 0.9 | 60 | 72 | 78 |
| 1.8 | 54 | 66 | 72 |
| 3.7 | 48 | 60 | 66 |
| 7.3 | 42 | 54 | 60 |

*Table A2*

**a.** How much does the intensity of sound increase when the teacher shouts compared to when she is using her normal voice? Give your answer in dB and as a ratio of sound intensities.

**b.** Is this the same factor for a student sitting at the front of the class, compared to a student at the back of the class?

**c.** What is the average variation of sound level with distance? Give your answer in dB $m^{-1}$.

**d.** It has been suggested that the following relationship holds:

$$\left(\frac{\text{distance to 1}}{\text{distance to 2}}\right)^2 = \frac{\text{intensity at 2}}{\text{intensity at 1}}$$

Is this supported by the data?

**e.** How far away from the teacher can the students sit and still be able to hear her voice?

## KEY IDEAS

- Dispersion in an optical fibre refers to a spread in time, so that a sharp input pulse will broaden due to dispersion.
- Modal dispersion occurs due to the different path lengths taken by light rays travelling down a fibre. It can occur even with monochromatic light.
- Material, or chromatic, dispersion is due to a frequency-dependent refractive index, that is, different frequencies of wave travel at different speeds.
- Dispersion broadens a pulse and therefore limits data transmission rates.
- The weakening of a signal with distance is known as attenuation and is measured in decibel (dB), or dB $km^{-1}$ for an optical fibre.
- The attenuation in an optical fibre is due to absorption and scattering in the fibre.

## ASSIGNMENT 4: ANALYSING DATA TRANSMISSION

**(MS 0.6, MS 1.5, MS 2.3, MS 2.4)**

### Stretch and challenge

In this assignment you will calculate the data transmission rate that is required for real-time transmission of high-definition video.

The image on a computer or digital TV screen is made up of pixels. A **pixel**, or picture element, is the smallest area in an image whose colour can be controlled. Each pixel in a digital image is coded. The code determines what colour that pixel will be. The code is expressed as a binary number. Each digit, or **bit**, in the binary code can only be 0 or 1. For example, the four-bit number representing the number 13 is written 1101. Another example is shown in Table A1. The more bits that are used for the code, the more shades of colour (or grey) can be represented in the image and the more accurate the image will be, as shown in the figure.

| Binary place value | $2^7$ | $2^6$ | $2^5$ | $2^4$ | $2^3$ | $2^2$ | $2^1$ | $2^0$ |
|---|---|---|---|---|---|---|---|---|
| Decimal equivalent | 128 | 64 | 32 | 16 | 8 | 4 | 2 | 1 |
| Example | 1 | 0 | 0 | 1 | 0 | 1 | 1 | 1 |

***Table A1*** *In a binary number, each digit (bit) represents a power of 2. The decimal equivalent of the number shown here, 10010111, is 151 (that is,* $1 \times 128 + 0 \times 64 + 0 \times 32 + 1 \times 16 + 0 \times 8 + 1 \times 4 + 1 \times 2 + 1 \times 1$*).*

Binary signals are carried in an optical fibre by pulses of light. Either the light is on, which represents a 1, or the light is off, which represents a 0.

To get a rough estimate of how many pulses of light per second (bits of information) we need for high-definition real-time video streaming, we need to know:

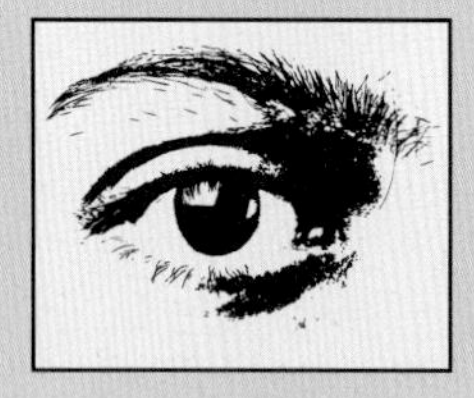

1 bit
2 possible values (light or dark)

2 bits
4 possible values

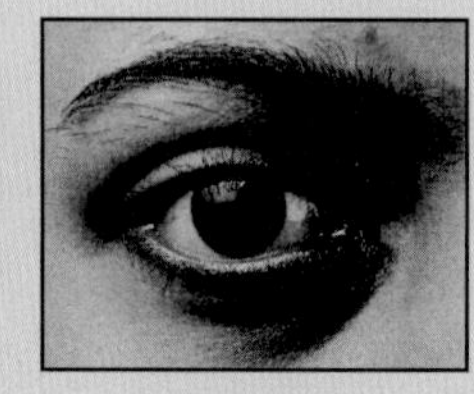

4 bits
16 possible values

*More bits for each pixel means a higher-definition image.*

- how many bits are needed for each pixel
- how many pixels are needed for each frame (one whole-screen image)
- how many frames are shown per second.

The number of bits per pixel determines how many colours can be shown in the pixel. Modern systems use 24 bits per pixel, which allows $2^{24}$ – more than 16 million – different colours. High-definition (HD) screens have 1920 (horizontal) × 1080 (vertical) pixels. HD TV typically uses 60 frames per second. Therefore, the total number of bits required is given by

$$24 \times 1920 \times 1080 \times 60 \approx 3 \times 10^9 \text{ bits per second} = 3 \text{ gigabits/second}$$

In practice, sophisticated data-compression software is used and the required data rate drops to between 5 and 25 megabits/second.

Your task in this assignment is to imagine that a school has installed an optical fibre system to distribute video to its classrooms.

Read the following information and then answer questions **A1** to **A8**. Multimode step-index optical fibre was used since it is significantly cheaper than monomode fibre. The fibre has a core diameter of 62.5 ± 3 μm, which is much larger than monomode fibre but allows lower-cost LEDs or lasers to be used. The cladding has a diameter of 125 ± 2 μm. The fibre-optic cable was designed to work at 850 nm. At that wavelength, the attenuation is 3 dB $km^{-1}$. The refractive index of the core is 1.496 and the refractive index of the cladding is 1.466.

### Questions

**A1** Explain what is meant by multimode optical fibre.

**A2** Explain what is meant by step-index optical fibre.

**A3** Is the attenuation in the optical fibre likely to be a problem for the school?

**A4** Calculate the critical angle between the core and the cladding.

**A5** Use your answer to question **A4** to calculate the time delay between the straight-through path and the critical-angle path for 1 km of the optical fibre. What is the uncertainty in this calculation?

**A6** Use your answer to question **A5** to calculate the maximum bit rate that can be transmitted over 1 km of fibre.

**A7** Compare your answer to question **A6** with the requirements for high-definition video and comment on the school's choice of system.

**A8** It is suggested to the school that 'graded-index' fibre, in which the refractive index varies gradually between the core and the cladding, might be better. Do some research and then explain what advantage it could offer.

## ASSIGNMENT 5: LEARNING ABOUT USES OF OPTICAL FIBRES IN ENGINEERING AND MEDICINE

Read the following and then answer the questions at the end.

### Aircraft safety

The stresses and strains on aircraft components need to be continuously monitored so that any structural problems can be detected in advance. One way of achieving this is to use optical fibres that have integrated reflection diffraction gratings (Figure A1). The 'diffraction grating' is actually a region of the core where the refractive index varies periodically. Some of the light that is travelling along the fibre is reflected at each boundary between regions of different refractive index. Each of these reflected light rays has a path difference, compared to the adjacent reflection, of twice the grating separation (that is 2 × $d$ in Figure A1). These light waves will only be in phase for a particular wavelength, so the grating has the effect of reflecting one wavelength and transmitting the rest of the spectrum. The wavelength that is reflected is dependent on the spacing of the grating. If the fibre is bonded to a critical part of the aircraft, it will monitor any strains (extensions) that occur, with a sensitivity of around one part in one million.

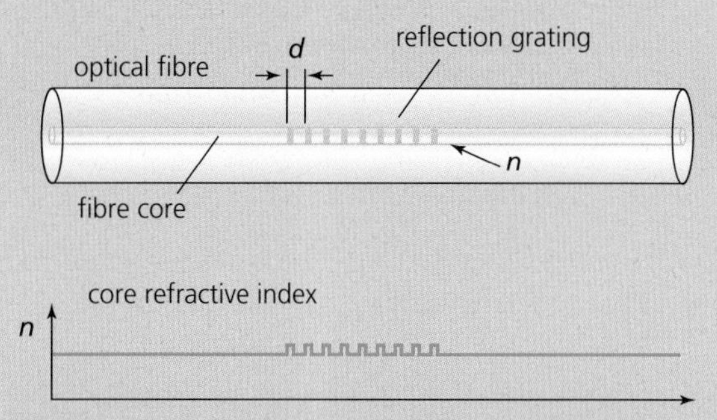


***Figure A1*** *Optical fibre with integrated reflection diffraction grating*

Similar techniques are used to monitor temperature and pressure. Optical fibres are resistant to high temperatures and pressures, and so can be used in many situations where conventional electronic sensors could not function. Since they are not metallic, optical fibres can be used in areas where high voltages are present. Optical fibres are also immune to interference from strong magnetic fields, which would cause electrical interference in conventional sensors. Optical fibres carry no electrical signals, so there is no danger of a spark, which is an important advantage in areas where aviation fuel is present.

## Secure communications systems

Optical fibres in communications systems carry information in a more secure and efficient way than copper wires, so security of data in the air transport business is improved. Electrical wires can be 'tapped' in a number of ways, without breaking the wire – for example, by placing an induction coil around the cable. The signal inside an optical fibre can only be detected by breaking the fibre at some point.

## Medical applications

Modern X-ray, nuclear magnetic resonance and body-scanning techniques have drastically reduced the need for exploratory surgery. However, in some instances nothing can replace the chance to take a good look inside an organ. Doctors can now take advantage of endoscopes – sometimes without anaesthetic – instead of more intrusive exploration. An **endoscope** is a thin, flexible tubular structure containing a cluster of optical fibres. Some of the fibres carry light into the patient, while others carry reflected light back out of the patient to a TV camera. This allows the surgeon to see inside the body through a tiny hole made to insert the endoscope.

The most recent developments allow surgeons not only to look but also to carry out operations through this tiny hole, which gives the technique its name – keyhole surgery. Tools are mounted on the endoscope to let the surgeon cut and seal wounds deep inside the body. Recovery from the procedure takes days rather than the weeks needed for traditional surgery.

Endoscopes are commonly used to examine the upper digestive tract (gastroscopy) or the rectum and colon (colonoscopy). Light is carried to the site of the examination through an **incoherent bundle** of glass fibres (Figure A2). As many as 30 000 individual fibres make up the bundle. An incoherent bundle cannot be used to form an image because the ends of the individual fibres are arranged randomly.

In a **coherent bundle**, the fibres have the same spatial position at each end of the bundle (Figure A2). The light emitted from the end of the bundle is an exact copy of the incident light and an image can be reproduced. Coherent bundles are used in an endoscope to form an image of the area inside the body under examination. As these are expensive to manufacture, incoherent bundles are used for illumination.

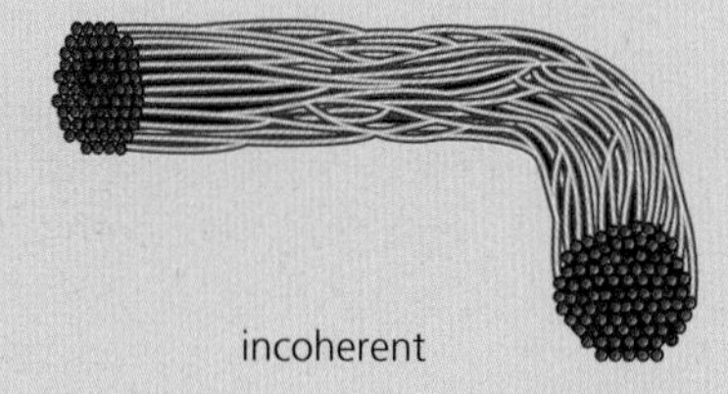


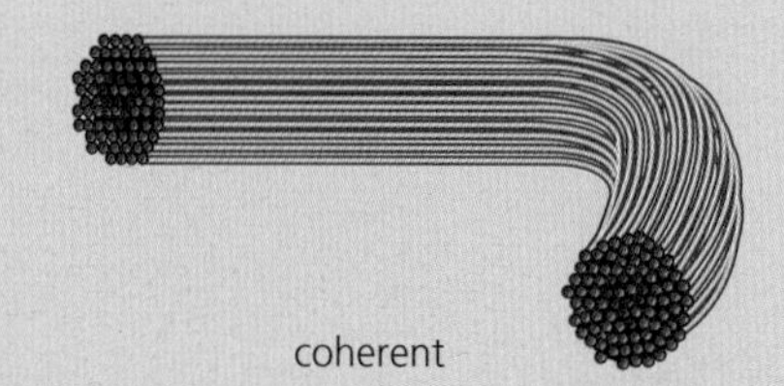


***Figure A2*** *Coherent and incoherent bundles*

## Questions

**A1** List the advantages of using fibre-optic sensors and cables compared with using traditional electronic sensors. Are there any disadvantages?

**A2** One optical fibre is able to carry several different signals at the same time. These could be in different wavelength ranges, or simply pulsed signals interspersed with one another in a process known as multiplexing. Why might this be useful?

**A3** Diffraction gratings can be integrated into an optical fibre at certain points, or throughout its entire length. Why is this useful?

**A4** The diameter of each fibre in an incoherent bundle is about 50 μm. Fibres in a coherent bundle are much thinner, with a diameter of only 5 μm. Why do you think this is?

**A5** Fibre-optic microphones and cables are used in MRI (magnetic resonance imaging) suites in hospitals, so that the medical staff can communicate with the patient in the scanner. Why can conventional equipment not be used?

### Stretch and challenge

**A6** Plan, research and produce an article describing the use of optical fibres in a particular application, such as monitoring the stresses and strains on an aircraft in flight, a bridge during use, or processes in a chemical plant.

Your article should not exceed 350 words (diagrams do not count in the word limit).

You should make sure that your article explains the physics involved, especially the advantages of using optical fibres as compared to traditional electronic sensors.

## PRACTICE QUESTIONS

1. Figure Q1 shows three transparent glass blocks A, B and C joined together. Each glass block has a different refractive index.

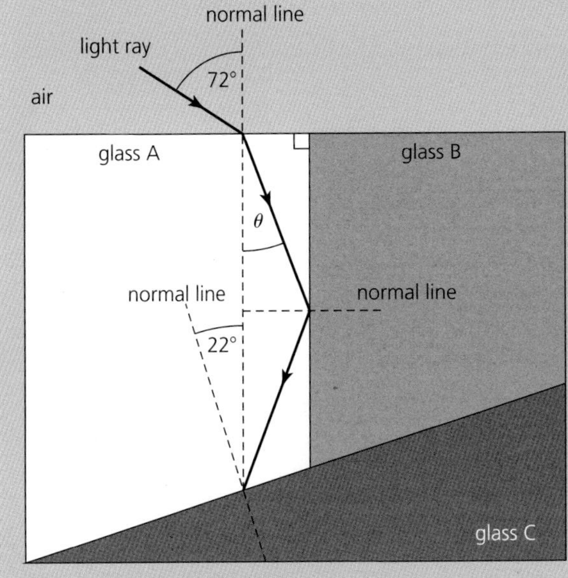


***Figure Q1*** *(Not to scale)*

a. State the **two** conditions necessary for a light ray to undergo total internal reflection at the boundary between two transparent media.

b. Calculate the speed of light in glass A. (Refractive index of glass A = 1.80.)

c. Show that angle $\theta$ is about 30°.

d. The refractive index of glass C is 1.40. Calculate the critical angle between glass A and glass C.

e. i. State and explain what happens to the light ray when it reaches the boundary between glass A and glass C.

ii. Sketch Figure Q1 and continue the path of the light ray after it strikes the boundary between glass A and glass C.

*AQA Unit 2 June 2013 Q5*

2. Figure Q2 shows a section of a typical glass step-index optical fibre used for communications.

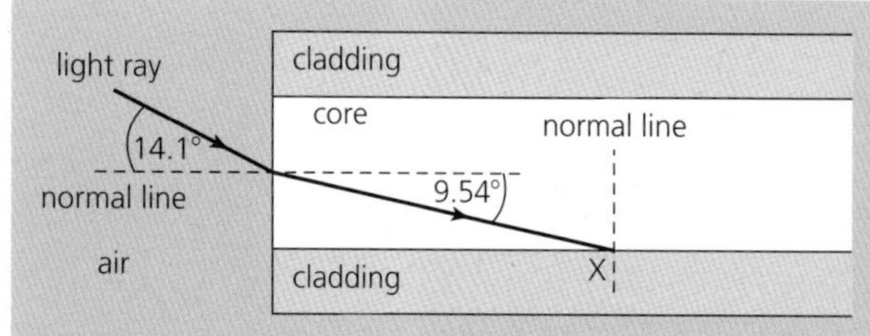


*Figure Q2*

a. Show that the refractive index of the core is 1.47.

b. The refracted ray meets the core–cladding boundary at an angle exactly equal to the critical angle.

   i. Copy and complete Figure Q2 to show what happens to the ray after it strikes the boundary at X.

   ii. Calculate the critical angle.

   iii. Calculate the refractive index of the cladding.

c. Give **two** reasons why optical fibres used for communications have a cladding.

*AQA Unit 2 Jan 2013 Q5*

3. Figure Q3 shows a cross-section through an optical fibre used for communications.

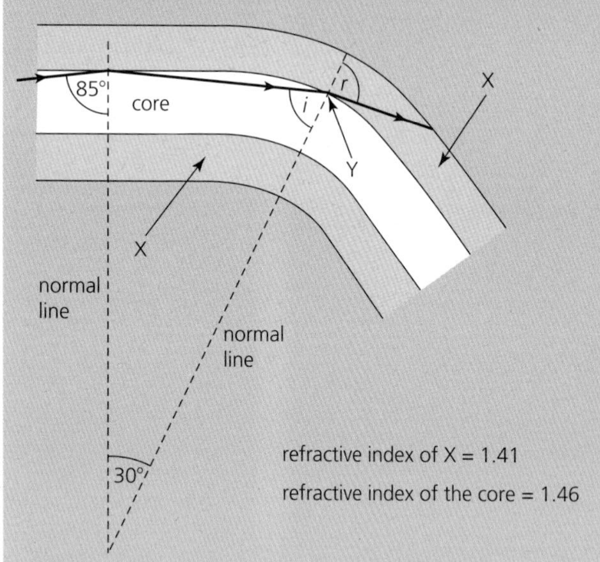


*Figure Q3*

a. i. Name the part of the fibre labelled X in Figure Q3.

   ii. Calculate the critical angle for the boundary between the core and X.

b. i. The ray leaves the core at Y. At this point the fibre has been bent through an angle of 30° as shown in Figure Q3. Calculate the value of the angle *i*.

   ii. Calculate the angle *r*.

The core of another fibre is made with a smaller diameter than the first, as shown in Figure Q4. The curvature is the same and the path of a ray of light is shown.

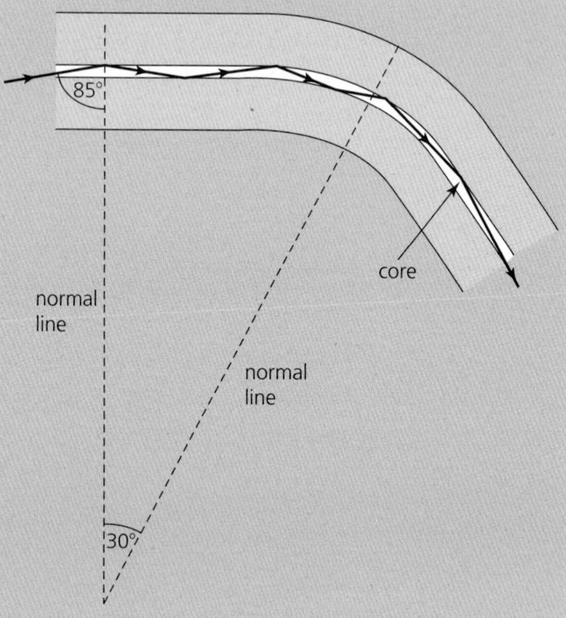


*Figure Q4*

c. State and explain **one** advantage associated with a smaller-diameter core.

*AQA Unit 2 June 2012 Q4*

4. Which of these statements about optical fibres is **false**?

   **A** Multimode fibre has several separate cores inside the cladding.

   **B** Step-index fibre has a core with refractive index that is slightly higher than that of the cladding.

   **C** Monomode fibre has an exceptionally thin core so that only one mode can propagate down the fibre.

   **D** Graded-index fibre has a refractive index that is a maximum at the centre of the core and gets gradually smaller towards the cladding.

5. Optical fibres are used in keyhole surgery to image inside a patient. Two types of optical cable are used. Which of the rows in Table Q1 gives accurate statements?

| | Coherent bundle | Incoherent bundle |
|---|---|---|
| A | The fibres have a larger diameter than those in an incoherent bundle. | Up to 30 000 fibres are used to carry the light. |
| B | Coherent bundles are cheaper to manufacture than incoherent bundles. | There is no relationship between the fibre's position at each end of the bundle. |
| C | The fibres have a smaller diameter than those in an incoherent bundle. | The spatial position of each fibre is the same at each end of the bundle. |
| D | The spatial position of each fibre is the same at each end of the bundle. | Up to 30 000 fibres are used to carry the light. |

*Table Q1*

6. A laser beam travels from air through a glass-sided fish tank filled with water. The laser beam is incident on the first glass surface at an angle of 20°. Which of the diagrams in Figure Q5 shows the correct path of the laser beam through the tank?

7. On a visit to the countryside, Claire notices that it is possible to hear sounds from further away at night than in the daytime. Which row in Table Q2 gives a plausible explanation for this?

| | Sound waves travel ... | The air temperature is ... | The sound waves ... |
|---|---|---|---|
| A | faster in cooler air | lower near the ground | refract away from the ground |
| B | faster in warmer air | lower near the ground | refract towards the ground |
| C | faster in cooler air | higher near the ground | refract away from the ground |
| D | faster in warmer air | higher near the ground | refract towards the ground |

*Table Q2*

8. Diamond has an exceptionally high refractive index, $n = 2.42$. Which of the following statements about a cut diamond is **false**?
   - **A** The diamond twinkles because incident light rays are refracted into the diamond and undergo numerous internal reflections before leaving.
   - **B** Light travels much more slowly inside the diamond.
   - **C** The critical angle for diamond is unusually large.
   - **D** The diamond twinkles because most light rays are reflected from its surface.

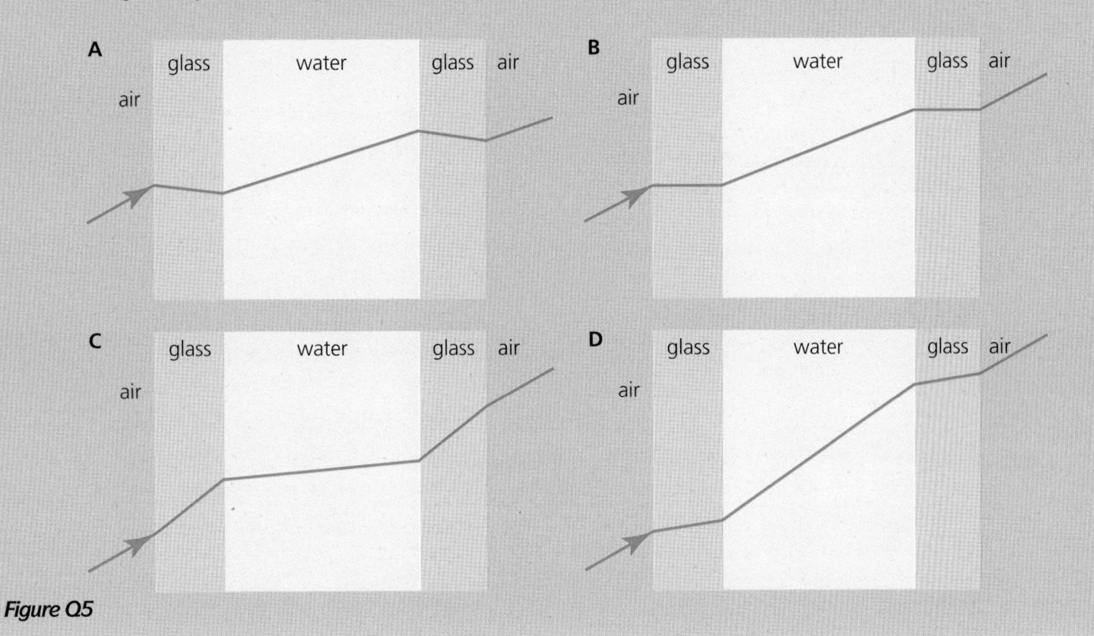


*Figure Q5*

**9.** A ray of light strikes a glass prism (refractive index = 1.45) at an angle of 30° to the normal (see Figure Q6).

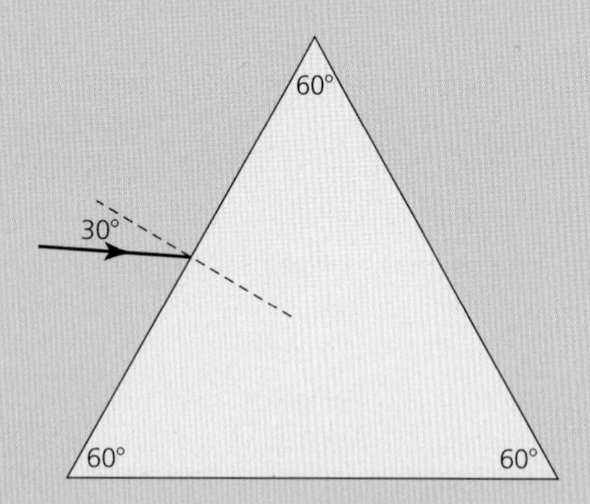


*Figure Q6*

Which row in Table Q3 correctly describes the path of light through the prism?

| | Angle of refraction / degree | Angle of incidence at glass–air boundary / degree | Angle of refraction in air / degree |
|---|---|---|---|
| **A** | 20.2 | 50.2 | 32 |
| **B** | 20.2 | 50.2 | total internal reflection |
| **C** | 46.5 | 76.5 | total internal reflection |
| **D** | 46.5 | 76.5 | 32 |

*Table Q3*

**10.** Read this information about optical fibres and the way that they carry data.

Optical fibres have enormous bandwidth. In 2016, the highest data transmission rate along a single strand of fibre was just over 1 THz, fast enough to download the entire *Game of Thrones* series in High Definition (HD) video in less than 1 second.

**Time division multiplexing** (TDM) uses a switching device that allocates a time slot, known as a channel, to each signal (Figure Q7). These digital signals are then transmitted along the fibre. The receiving end has a similar switch, which directs the correct signal to the correct device. The data rate carried by the fibre has to be significantly greater than that required by the individual signals it carries. High-speed systems run at up to 50 GB per second.

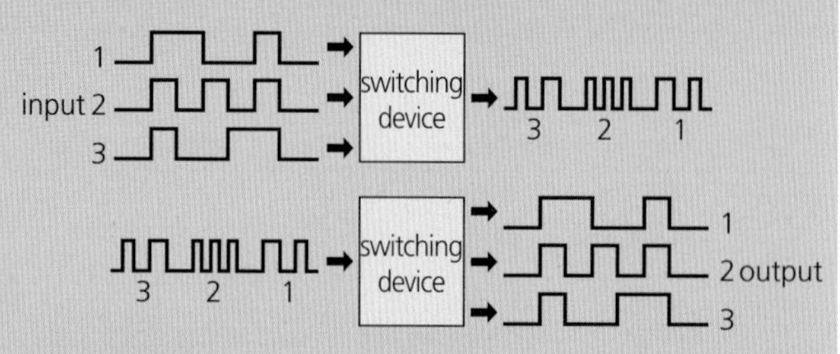


*Figure Q7*

**Wavelength division multiplexing** (WDM) uses several different wavelengths of light to carry signals (Figure Q8). Each wavelength carries one channel of data. Up to 80 channels are carried by the same fibre, with a wavelength difference of about 0.4 nm between each channel. This method tends to be four or five times more expensive than the TDM solution.

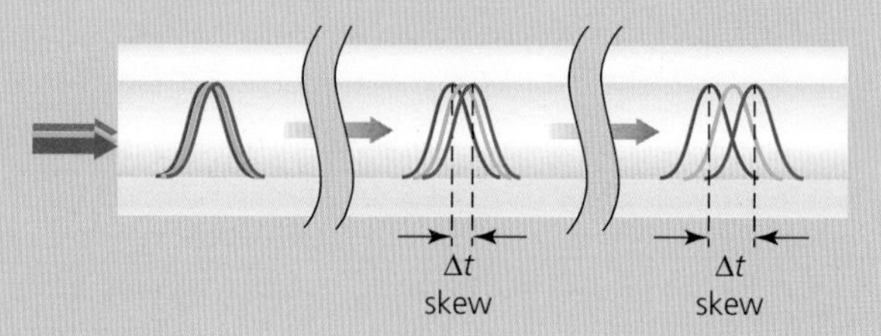


*Figure Q8*

Data transmission rates in an optical fibre are usually limited by the devices that change electrical signals to optical signals and back again. However, there are also limitations set by modal dispersion in the fibre.

**a. i.** Explain what is meant by *modal dispersion* and explain why it limits the maximum data transmission rate in the fibre.

**ii.** Why might this be a particular problem for time division multiplexing?

**b.** Wavelength division multiplexing suffers from 'skew' (see Figure Q8). What causes skew?

c. i. What does *attenuation* mean in the context of optical signals travelling along fibres?

ii. Explain the causes of attenuation.

iii. How might it affect TDM and WDM signals?

### Stretch and challenge

d. An optical fibre is to be used to carry high-definition (HD) video. Time division multiplexing is to be used with a maximum of eight channels. The optical fibre used has a core of refractive index 1.60 and cladding of a refractive index 1.50.

i. HD video has 1920 × 1080 pixels in each frame and 60 frames per second. Assume that each pixel requires 24 bits to define its colour. Calculate the maximum data rate required to stream eight channels.

ii. Use the refractive index values above to calculate the modal dispersion of the optical fibre, and hence determine whether the fibre will be able to handle the necessary data rate (as calculated in part d i).

iii. In practice the required rate will be less. Explain why.

# 8 SPECTRA, PHOTONS AND WAVE–PARTICLE DUALITY

## PRIOR KNOWLEDGE

You will be familiar with the electromagnetic spectrum and the terms used to describe a wave: wavelength, frequency, amplitude and wave velocity.

## LEARNING OBJECTIVES

In this chapter you will learn about spectra, the photoelectric effect and electron diffraction, and what these tell us about the dual nature of light and particles.

**(Specification 3.2.2.1, 3.2.2.2, 3.2.2.3, 3.2.2.4)**

Astronomy has a strong claim to be the oldest science. Systematic observations of the sky were being made in China around 6000 years ago. For almost all the time since then, astronomy has depended on visible light and the human eye to make observations. But only a small fraction of the total energy that reaches us from space is in the form of visible light. We are bathed in radiation from the entire electromagnetic spectrum. It is only in the past 70 years or so that we have developed other ways of 'seeing'. In addition to optical telescopes, we now have huge radio telescopes and satellites that can see the Universe at X-ray, ultraviolet and infrared wavelengths.

ALMA (Figure 1, background) the Atacama Large Millimetre Array, is a large radio telescope array designed to study the Universe at millimetre wavelengths, that is, in the microwave and infrared regions of the spectrum. It is being used by astrochemists to analyse the chemicals in interstellar space. Many organic compounds, such as sugars and alcohols, have been detected by the radiation they emit. Every compound emits its own specific set of wavelengths, dependent on its temperature and the molecules it contains. ALMA has helped scientists to show that these chemicals are present in the Orion Nebula, a region of space where new stars are forming (Figure 2). Any planets that orbit these stars could already have some of the organic molecules necessary for life.

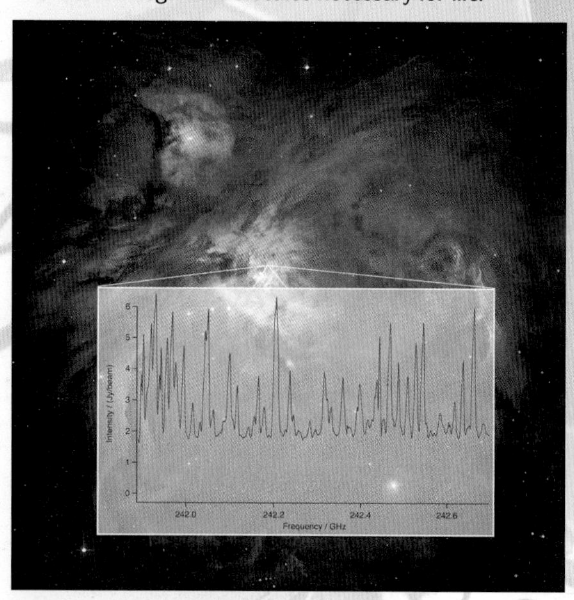


*Figure 2* *The Orion Nebula, with the spectrum produced by ALMA*

Measurement of the energy emitted across the whole range of wavelengths gives detailed information about stars and galaxies: their temperature, their speed relative to us, and their chemical composition.

*Figure 1* *(background) ALMA (the Atacama Large Millimetre Array) was built in the Atacama Desert in northern Chile, 5 km above sea level. This is one of the driest places on Earth, ideal for observations.*

# 8.1 EMISSION SPECTRA

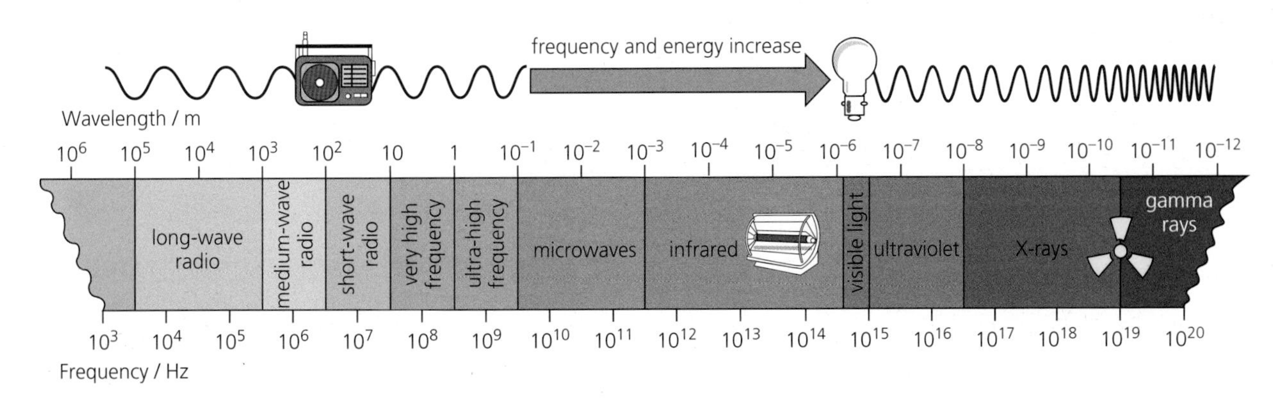


***Figure 3*** *The electromagnetic spectrum*

Visible light is part of the **electromagnetic (EM) spectrum** (Figure 3). The EM spectrum covers an enormous range of wavelengths, from radio waves, which can be hundreds of kilometres long, to gamma radiation with wavelengths of $10^{-12}$ m . Our eyes can detect only the very small range of wavelengths between 400 nm (violet visible light) and 700 nm (red visible light). It is no coincidence that this is the part of the electromagnetic spectrum that penetrates the Earth's atmosphere most effectively.

***Figure 4*** *People are not hot enough to glow visibly in the dark. A surface temperature of around 27 °C (300 K) means that we emit radiation in the infrared part of the spectrum. This false-colour infrared image shows the cool parts of the body as blue and the hotter parts as red. Such images can be used to study the effects of exercise on athletes.*

All objects emit electromagnetic radiation as a result of the thermal motion of their molecules. This is thermal radiation. The radiation is emitted across a range of wavelengths, known as an **emission spectrum**. The peak wavelength (the wavelength at which most radiation is emitted), and the overall intensity of the emitted radiation depend on the object's temperature – see Figure 16 in section 8.3. When an object is heated, more radiation is emitted, particularly at shorter wavelengths. At room temperature, objects emit most of their radiation in the infrared region of the spectrum (Figure 4). If the object gets hot enough, it will emit visible light (Figure 5). A star like our Sun, with a surface temperature of more than 6000 °C, emits radiation across the entire EM spectrum, but most of the radiation is in the visible part of the spectrum, with a peak in the yellow region. The link between wavelength (colour) and temperature only applies to emitted light, not light that is reflected from other sources.

***Figure 5*** *The hottest part of the metal is radiating strongly across the visible part of the spectrum, and appears yellow. Cooler parts emit most of their light in the red parts of the spectrum.*

## QUESTIONS

1. a. The constellation of Orion (Figure 6) has two very bright stars in it, Rigel (bottom right) and Betelgeuse (top left). Rigel appears slightly brighter than Betelgeuse, but which has the higher surface temperature?

*Figure 6* The Orion constellation

   b. Suggest which other factors determine how bright a star looks to us.

2. Why do very hot objects appear to be white?
3. Place the following in order of temperature, with the highest/hottest first: heating element of a toaster; surface of the Sun; central heating radiator; flash of lightning; filament of a car headlight bulb.
4. On a clear night the surface of the full Moon looks white. Does that mean the Moon's surface is hot? Explain your answer.

## Line spectra

The electromagnetic radiation from an object can tell us about the temperature of its source, and it can also tell us about its composition. Light can be analysed by using a prism or a diffraction grating (see section 6.2 of Chapter 6) to disperse it into its separate wavelengths. In some cases this produces an emission spectrum, which appears as a series of sharp bright lines against a darker background, called a **line spectrum** (Figure 7).

Line spectra are produced by hot, low-pressure monatomic gases. Experiments on such gases have shown that each element produces a unique set of lines. The emission lines from hydrogen include a bright red line, as well as some fainter blue and purple lines. The combination, seen without a diffraction grating, appears red to the human eye (Figure 8).

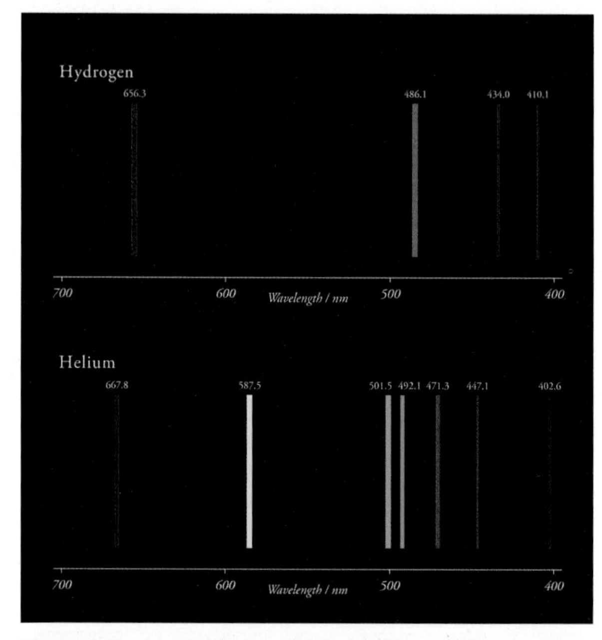


*Figure 7* The emission line spectra of hydrogen and helium

*Figure 8* The red colour of the flare seen during a solar eclipse is due to emission from hydrogen atoms.

The spectra from atoms that are closer together, such as in high-pressure gases, are more complex and appear as bands rather than lines. Spectra from hot solids (Figure 9) and liquids are continuous spectra, like the rainbow, which is the spectrum of visible light from the Sun.

***Figure 9*** *Observing the spectrum emitted by a molten alloy gives us detailed information about its composition.*

The line spectrum of hydrogen was first observed in 1853. Those of other elements followed. Classical physics, the laws of physics as they were understood at the end of the 19th century, could not explain why each element emitted only certain specific colours. At the beginning of the 20th century Niels Bohr (Figure 10) and Max Planck made advances that were critical to our understanding of spectra and atoms.

## 8.2 BOHR'S HYDROGEN ATOM

Niels Bohr built on Rutherford's theory of the nuclear atom (section 2.2 of Chapter 2). The hydrogen atom was thought to contain one electron orbiting around a positively charged nucleus. But there was a problem. According to classical physics, all charged particles emit radiation when they accelerate. An orbiting electron is constantly changing its velocity – in fact, it is accelerating towards the centre of its orbit. So an orbiting electron should be radiating, thereby losing energy. This is similar to a satellite orbiting Earth. If the satellite loses energy as it travels through the upper atmosphere, it will lose height and eventually crash to the ground. Rutherford's hydrogen atom would be unstable in the same way – as the electron radiated energy, it would spiral down into the nucleus.

In 1913 Bohr made the bold suggestion that classical physics did not apply to electrons in atoms. He argued that an electron in an atom could only move in certain 'allowed' orbits. In these allowed orbits the electron could exist without losing energy. Bohr called these orbits 'stationary states'. Each stationary state corresponds to a specific electron energy. These are represented in an **energy level diagram** (Figure 11).

***Figure 10*** *Niels Bohr*

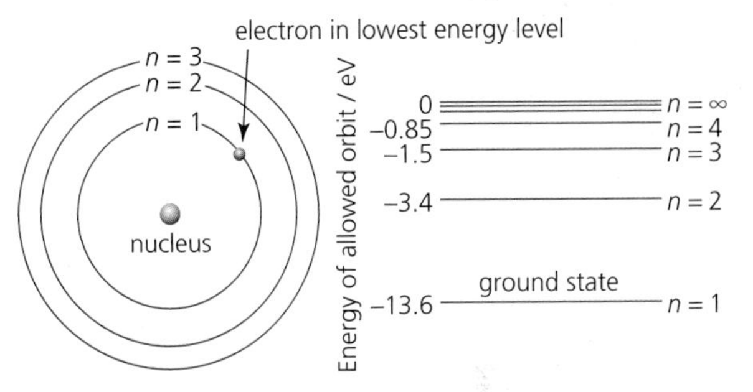


The ground state, $n = 1$, is the lowest energy level. An electron in this level needs an energy transfer of 13.6 eV to free it from the atom.

***Figure 11*** *Allowed orbits (left) and energy levels (right) in Bohr's hydrogen atom*

Bohr suggested that the electron in a hydrogen atom can only gain or lose energy when it jumps from one energy level to another. The electron is not allowed to be between levels. If the electron is to make a jump between orbits it has to lose, or gain, *exactly* the right amount of energy. If the electron moves to a lower energy level it has to lose energy. It does so by emitting electromagnetic radiation, light for example. In the **Bohr model**, the line spectrum of an element is caused by electrons making transitions from higher allowed energy levels to lower ones, emitting light in the process.

An electron in an atomic energy level is considered to have negative energy. This is because the electron is in a 'bound' state, held in the atom by electromagnetic attraction. The electron needs to gain energy to become

## ASSIGNMENT 1: OBSERVING SPECTRA FROM DIFFERENT SOURCES

**(PS 1.2)**

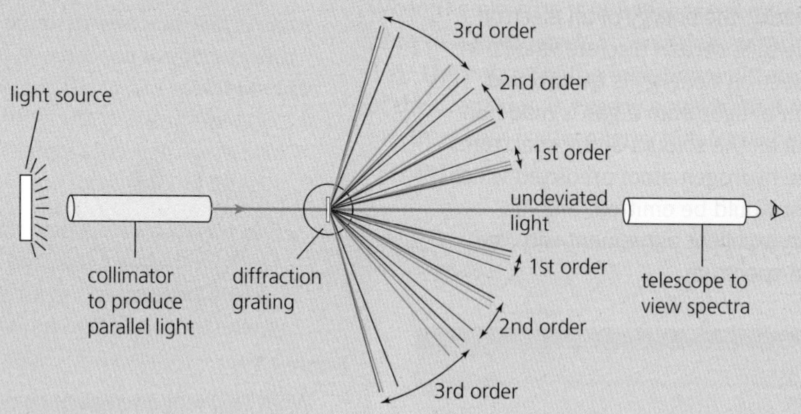


*Figure A1 Producing observable spectra*

A light source, such as an electric bulb, emits light across a range of wavelengths. If the light is passed through a prism, or a diffraction grating (see section 6.2 of Chapter 6), it will be dispersed into its component wavelengths, forming a spectrum. The type of spectrum depends on the light source. A rainbow is a solar spectrum, formed by the Sun's light passing through raindrops, which act as prisms and disperse the white light (see section 7.5 of Chapter 7).

The following light sources will all produce different spectra when viewed with the apparatus shown in Figure A1:

- candle
- Bunsen flame
- fluorescent light
- white LED
- mercury vapour lamp
- sodium vapour lamp
- old-fashioned incandescent light bulb.

### Questions

**A1** Figure A2 shows that some spectra are a set of sharp coloured lines. Others are continuous, with no gaps between the colours.

**a.** Which of the sources listed would give rise to a line spectrum?

**b.** Which would have a continuous spectrum?

**c.** Explain the reason for the difference in the spectra from these sources.

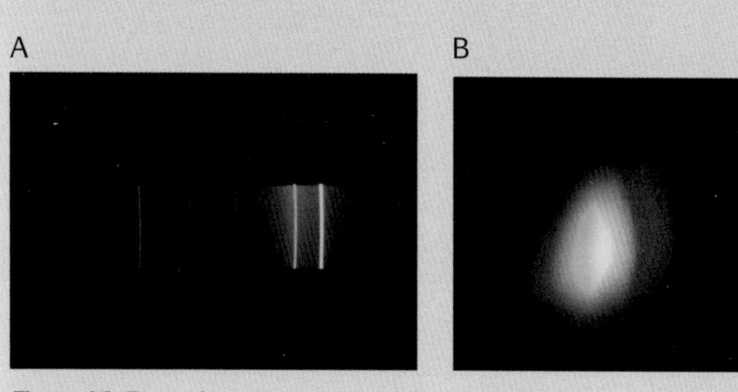


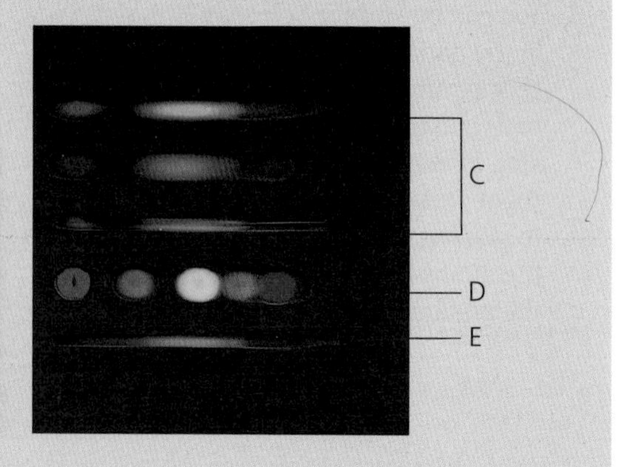


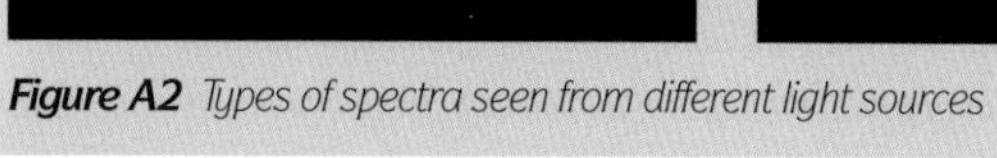

*Figure A2 Types of spectra seen from different light sources*

***Figure 9*** *Observing the spectrum emitted by a molten alloy gives us detailed information about its composition.*

The line spectrum of hydrogen was first observed in 1853. Those of other elements followed. Classical physics, the laws of physics as they were understood at the end of the 19th century, could not explain why each element emitted only certain specific colours. At the beginning of the 20th century Niels Bohr (Figure 10) and Max Planck made advances that were critical to our understanding of spectra and atoms.

## 8.2 BOHR'S HYDROGEN ATOM

Niels Bohr built on Rutherford's theory of the nuclear atom (section 2.2 of Chapter 2). The hydrogen atom was thought to contain one electron orbiting around a positively charged nucleus. But there was a problem. According to classical physics, all charged particles emit radiation when they accelerate. An orbiting electron is constantly changing its velocity – in fact, it is accelerating towards the centre of its orbit. So an orbiting electron should be radiating, thereby losing energy. This is similar to a satellite orbiting Earth. If the satellite loses energy as it travels through the upper atmosphere, it will lose height and eventually crash to the ground. Rutherford's hydrogen atom would be unstable in the same way – as the electron radiated energy, it would spiral down into the nucleus.

In 1913 Bohr made the bold suggestion that classical physics did not apply to electrons in atoms. He argued that an electron in an atom could only move in certain 'allowed' orbits. In these allowed orbits the electron could exist without losing energy. Bohr called these orbits 'stationary states'. Each stationary state corresponds to a specific electron energy. These are represented in an **energy level diagram** (Figure 11).

***Figure 10*** *Niels Bohr*

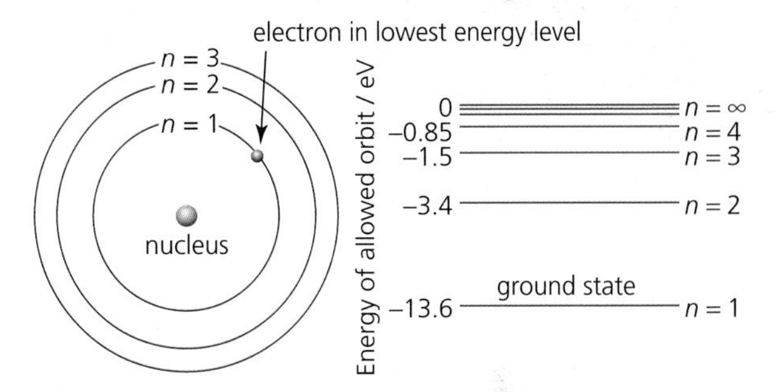


The ground state, $n$ = 1, is the lowest energy level. An electron in this level needs an energy transfer of 13.6 eV to free it from the atom.

***Figure 11*** *Allowed orbits (left) and energy levels (right) in Bohr's hydrogen atom*

Bohr suggested that the electron in a hydrogen atom can only gain or lose energy when it jumps from one energy level to another. The electron is not allowed to be between levels. If the electron is to make a jump between orbits it has to lose, or gain, *exactly* the right amount of energy. If the electron moves to a lower energy level it has to lose energy. It does so by emitting electromagnetic radiation, light for example. In the **Bohr model**, the line spectrum of an element is caused by electrons making transitions from higher allowed energy levels to lower ones, emitting light in the process.

An electron in an atomic energy level is considered to have negative energy. This is because the electron is in a 'bound' state, held in the atom by electromagnetic attraction. The electron needs to gain energy to become free. An electron that is *just* free of the atom has zero energy. On an energy level diagram, like that in Figure 11, the energy levels are labelled with negative energies, which indicate how much energy the electron would need to escape from the atom. An electron with positive energy is free of the atom and has some kinetic energy. The energy values are given in the unit of **electronvolt** (eV). One electronvolt (1 eV) is the energy gained by an electron when it is accelerated through a potential difference of one volt (1 V) (see section 3.1 of Chapter 3).

A hydrogen atom has only one electron. When the electron is in the lowest possible energy level, the hydrogen atom is said to be in the **ground state**. In this orbit the electron cannot lose any more energy. This is the most stable state.

The electron can move to a higher energy level in a process called **excitation**. The energy could come from either of the following:

- collisions – energy can be transferred to an atomic electron when there is a collision with another particle, such as a free electron;
- radiation – energy can be transferred to the atomic electron if it absorbs exactly the right amount of energy from EM radiation, such as light.

When the electron is in a higher energy level, $E_2$, it can drop to a lower level, $E_1$, emitting radiation (Figure 12). The energy of the radiation emitted as a result of this transition is exactly equal to that lost by the electron, that is, the energy difference between the two levels $E_2 - E_1$.

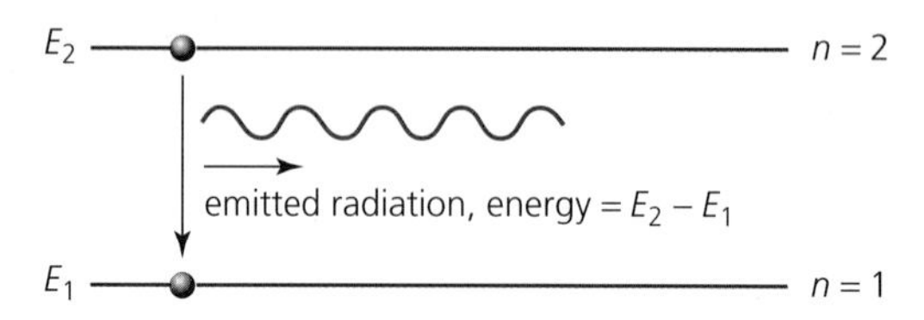


***Figure 12*** *An electron transition*

Sometimes the energy transferred to an atomic electron during a collision is so great that it is completely knocked out of the atom. This process is called **ionisation** (see Figure 13).

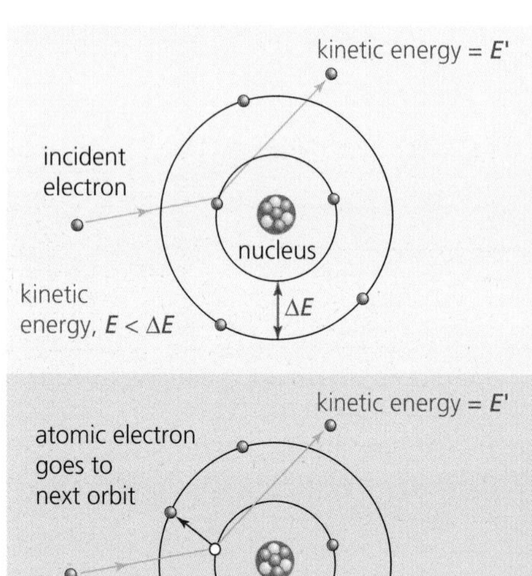


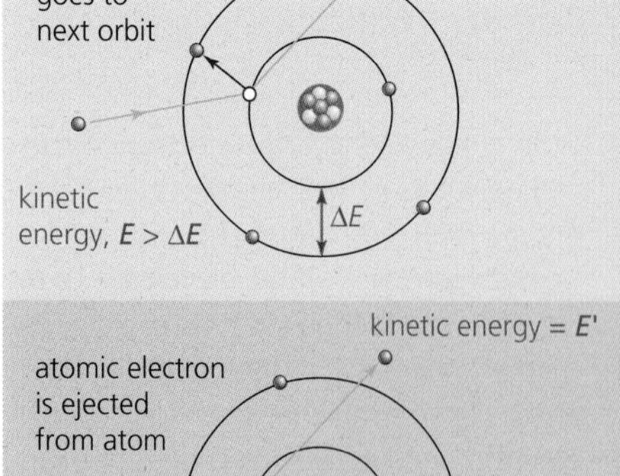


***Figure 13*** *Excitation and ionisation caused by an electron colliding with an atom*

Bohr's theory, which allows electrons in atoms to gain or lose energy only in steps of an exact size, was a revolutionary idea. In classical physics, energy is a continuous quantity – it can take any value – but in Bohr's quantum theory, the energy of an electron in an atom can only take certain discrete values. In other words, the electron's energy is **quantised**. This is why the spectrum of light from a gas is made up of lines, rather than all the colours of the spectrum. Bohr's model of the hydrogen atom predicted which wavelengths of light would be emitted, and his calculations were in excellent agreement with the observed hydrogen spectrum.

## QUESTIONS

5. Old-fashioned TV sets (Figure 14) were particle accelerators. Electrons were accelerated through a tube and made to collide with atoms in the fluorescent coating on the inside of the screen. The screen emitted light. Use the ideas of excitation and energy levels to explain this.

***Figure 14*** *A TV before flat screens*

6. You can buy plastic star shapes that glow in the dark to decorate a child's bedroom ceiling. After dark, the stars glow brightly at first, and then grow dimmer and eventually stop glowing altogether. If you shine a light on them, from a torch for example, they will begin to glow again. Suggest what is happening in the atoms in these plastic stars. What does the gradual dimming suggest to you?

7. In Figure 15 an incoming electron collides with an atomic electron in the lowest energy level.

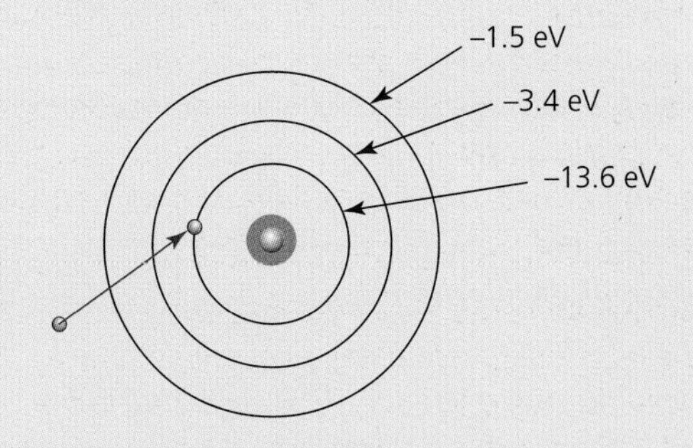


***Figure 15***

What will happen in each case after the collision if the incoming electron has the following energy?

a. 3 eV

b. 11 eV

c. 15 eV

## KEY IDEAS

- All objects emit electromagnetic waves due to the thermal motion of their atoms. The amount of radiation, and the wavelengths emitted, depend on temperature.
- Low-pressure gases or vapours emit defined wavelengths, producing an emission line spectrum.
- Bohr's theory of the atom explained line spectra by suggesting that, in any atom, electrons can only occupy discrete energy levels or orbits.
- Electrons can be excited to a higher energy orbit by absorbing a specific amount of energy.
- When electrons move to a lower orbit they emit a specific amount of energy as radiation.
- Electrons must gain, or lose, an exact amount of energy to move between allowed orbits.

## ASSIGNMENT 1: OBSERVING SPECTRA FROM DIFFERENT SOURCES

**(PS 1.2)**

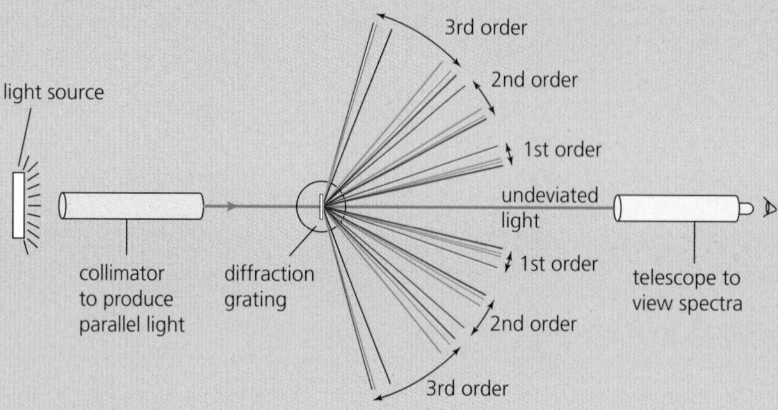


***Figure A1*** *Producing observable spectra*

A light source, such as an electric bulb, emits light across a range of wavelengths. If the light is passed through a prism, or a diffraction grating (see section 6.2 of Chapter 6), it will be dispersed into its component wavelengths, forming a spectrum. The type of spectrum depends on the light source. A rainbow is a solar spectrum, formed by the Sun's light passing through raindrops, which act as prisms and disperse the white light (see section 7.5 of Chapter 7).

The following light sources will all produce different spectra when viewed with the apparatus shown in Figure A1:

- candle
- Bunsen flame
- fluorescent light
- white LED
- mercury vapour lamp
- sodium vapour lamp
- old-fashioned incandescent light bulb.

### Questions

**A1** Figure A2 shows that some spectra are a set of sharp coloured lines. Others are continuous, with no gaps between the colours.

**a.** Which of the sources listed would give rise to a line spectrum?

**b.** Which would have a continuous spectrum?

**c.** Explain the reason for the difference in the spectra from these sources.

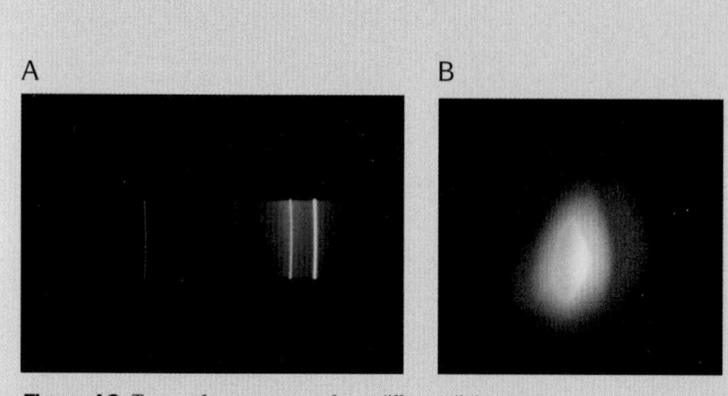


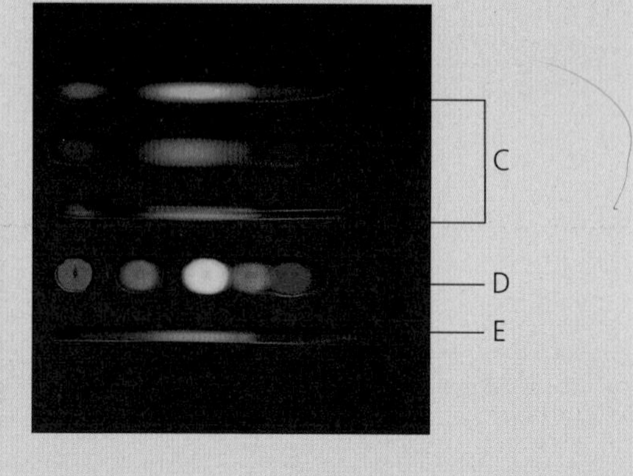


***Figure A2*** *Types of spectra seen from different light sources*

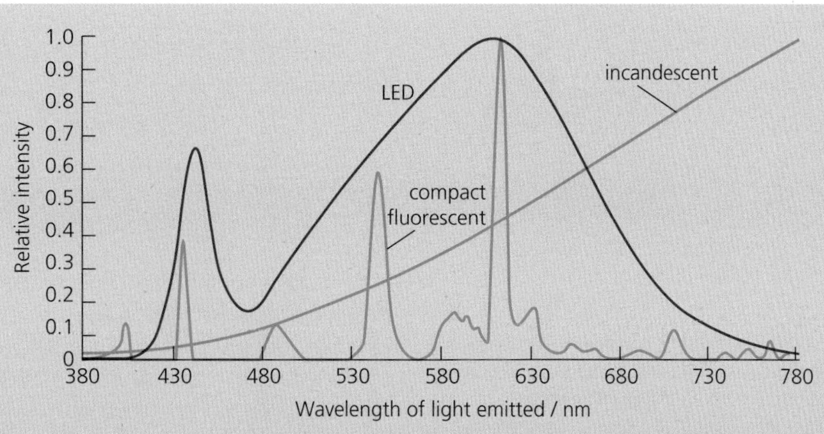


***Figure A3*** *Spectral intensity curves*

**A2** Refer to Figure A3, which shows spectral intensity curves of sources used to light a house (LED, incandescent and fluorescent).

**a.** What are the differences between them?

**b.** Which one would be best for a bedroom?

**c.** Which one would be best for a study?

**d.** Which might you choose to light a kitchen?

**A3** Try to identify the sources of the spectra in Figure A2.

## 8.3 THE PHOTON

In July 1900, new and accurate measurements of the thermal radiation from hot objects were published. These confirmed that there was a large discrepancy between theory and experiment. Physics had no satisfactory explanation for the relationship between intensity and wavelength (Figure 16).

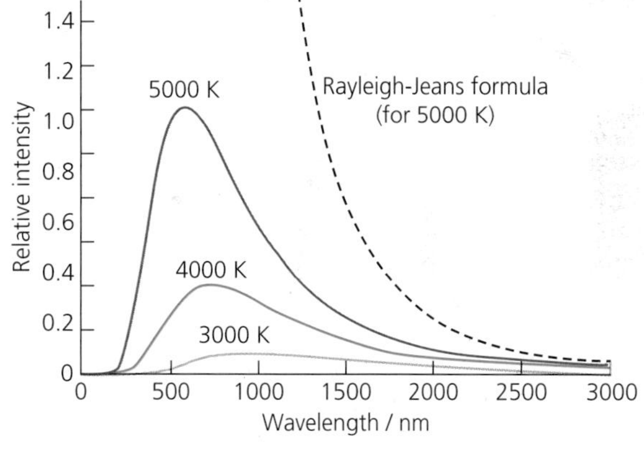


***Figure 16*** *Experimental curves for the continuous spectra from hot objects at different temperatures. Classical physics could not explain the shape of these curves. A theory known as the Rayleigh–Jeans formula produced the dashed curve, which did not agree at all well with experimental results. Worse, it predicted that infinite energy would be emitted at short wavelengths. This became known as the 'ultraviolet catastrophe'.*

Max Planck managed to solve the problem by suggesting that electromagnetic radiation could only be absorbed or emitted in 'packets' of a certain energy. Classical physics treated radiation energy as continuous, and any quantity of energy could be transferred. Planck's revolutionary idea was to assume that radiation energy was transferred only in discrete 'lumps'. These 'lumps', or **energy quanta**, became known as **photons** (Figure 17). Planck suggested that it was only possible to emit energy as a whole number of photons. His idea was the beginning of the quantum theory of radiation. By assuming this, Planck could generate a theoretical graph of intensity against wavelength. His curve matched the experimental results well. His assumption was justified: electromagnetic radiation energy can only be emitted in discrete amounts.

According to Planck's theory, the amount of energy transferred by a photon depends on the frequency of the radiation. A photon of ultraviolet light transfers more energy than a photon of infrared light. Planck stated that the energy of a photon, $E$, is related to the frequency, $f$, of the electromagnetic radiation (see section 3.3 of Chapter 3) by the equation

$$E = hf$$

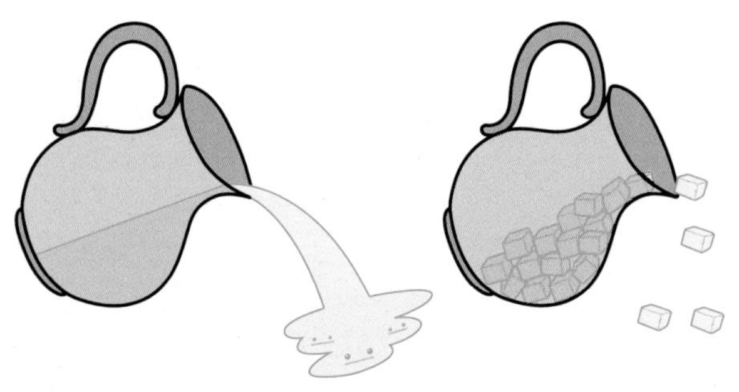

***Figure 17*** *Classical physics treated EM radiation energy as continuous. Any amount of energy could be transferred, just as any amount of water can be poured from a jug. Planck's theory suggested that radiation energy can only be transferred in discrete amounts, more like ice cubes. In this new 'quantum' theory, you could not have half a photon.*

where $h$ is a constant known as the **Planck constant**, $h = 6.626 \times 10^{-34}\,\text{J s}$. As the frequency of a wave is related to its wavelength, $\lambda$, by the formula $f = c/\lambda$ where $c$ is the speed of electromagnetic radiation in a vacuum, Planck's formula for the energy of a photon can also be written as

$$E = \frac{hc}{\lambda}$$

## QUESTIONS

**8.** The Planck constant determines the 'size' of the packets of energy that a particular frequency of radiation transfers. The shortest wavelength of electromagnetic radiation that is visible to humans has a wavelength of around 400 nm.

- **a.** What is the smallest amount of energy that this violet light could transfer, in electronvolts?
- **b.** What is the next lowest amount of energy that could be transferred?
- **c.** And the next?

**9.** **a.** "If radiation energy is 'lumpy', then the lumps are very small." What is meant by this?

- **b.** What difference would it make if the Planck constant were much larger?

## Explaining emission line spectra

Planck's photons and Bohr's energy levels (section 8.2) together give us a way of explaining line spectra. Each time an electron in an atom falls to a lower energy level, a photon is emitted. The energy of the emitted photon, $hf$, must be equal to the energy lost by the electron, $\Delta E$:

$$\Delta E = hf$$

Since $c = f\lambda$, we can write this as

$$\Delta E = \frac{hc}{\lambda}$$

In a hydrogen atom, small energy changes lead to the emission of photons in the infrared region (the Paschen series of spectral lines, Figure 18). Larger energy changes emit visible light (the Balmer series). The largest energy changes, where an electron falls back to the ground state, lead to the emission of ultraviolet radiation (the Lyman series).

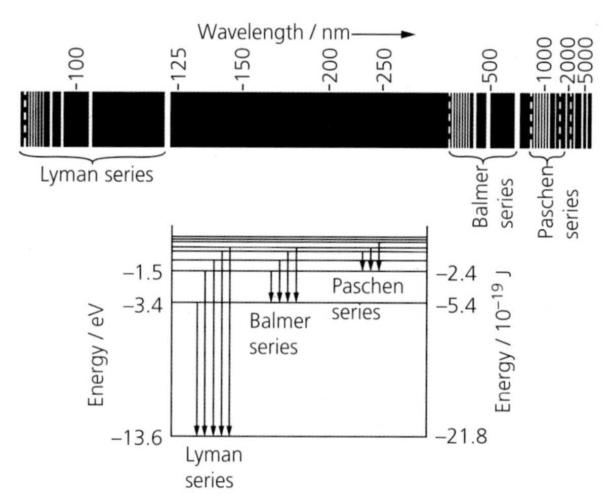


***Figure 18*** *Line series in the hydrogen spectrum*

### Worked example

An electron in the ground state of the hydrogen atom has an energy of −13.6 eV. The electron gains energy from a collision with another electron and is raised to the next energy level, −3.4 eV. It has gained $-3.4 - (-13.6) = 10.2$ eV. Later the electron will fall back to the ground state, emitting a photon of energy 10.2 eV. Calculate the frequency and the wavelength of the emitted radiation.

Using Planck's equation:

$$\Delta E = hf$$

$$f = \frac{\Delta E}{h}$$

But $\Delta E$ has the unit eV, whereas $h$ is in J s, so we must convert $\Delta E$ to joules.

$$\Delta E = 10.2\text{ eV} = 10.2 \times 1.6 \times 10^{-19}\text{ J}$$
$$= 1.63 \times 10^{-18}\text{ J}$$

$$f = \frac{1.63 \times 10^{-18}\text{ J}}{6.63 \times 10^{-34}\text{ Js}}$$
$$= 2.46 \times 10^{15}\text{ Hz}$$

We can calculate the wavelength using the wave equation, $c = f\lambda$, where $c$ is the speed of light ($3.00 \times 10^{8}$ m s$^{-1}$), $f$ is frequency (Hz) and $\lambda$ is wavelength (m). So

$$\lambda = \frac{c}{f} = \frac{3.00 \times 10^{8}\text{ ms}^{-1}}{2.46 \times 10^{15}\text{ s}^{-1}}$$
$$= 1.2 \times 10^{-7}\text{ m} \quad (\text{to 2 s.f.})$$

which is in the ultraviolet part of the spectrum.

## QUESTIONS

**10. a.** How many different frequencies of light could be emitted by an electron moving between the lowest three energy levels in hydrogen (see Figure 11)?

**b.** Calculate those frequencies.

**11.** What is the shortest wavelength of light that can be emitted by an electron transition in a hydrogen atom?

Hydrogen is the simplest atom – it has only one electron. The energy level diagrams for other atoms are more complicated, but the same principles apply. The atoms of each element have their own set of allowed energy levels, and so every element has a unique line spectrum that identifies it. Niels Bohr said that a line spectrum was "like a stained glass window looking into the heart of the atom".

## Fluorescent lights and electron energy levels

Fluorescent lights have long been used as strip lights in kitchens, classrooms and offices. Nowadays small, light-bulb-sized versions are common as 'low-energy bulbs'. These are compact fluorescent lights (Figure 19), which are much more efficient than filament lights. They do not rely on getting hot to emit light.

Fluorescent lamps have an electrode at both ends of a tube containing a low-pressure unreactive gas, such as argon or neon, and some mercury vapour.

***Figure 19*** *A compact fluorescent lamp (CFL)*

Electrons accelerate down the tube, colliding with mercury atoms as they do so (Figure 20). The collision can knock an electron out of a mercury atom, ionising it. If the missing electron came from a lower energy level, another electron will fall from a higher level to fill the gap. This will cause the emission of a photon. Elements such as mercury, with a high atomic number, have large energy gaps between allowed levels. So high-energy photons are emitted, usually in the ultraviolet region.

Fluorescent lamps are very efficient at producing ultraviolet radiation, but that is not really what we need from a light bulb. A second stage of photon emission happens when the ultraviolet photons are absorbed by the phosphor that is coated on the inside of the lamp (see Figure 20). As the photons are absorbed, electrons in the atoms of the phosphors are raised to a higher energy level. These electrons fall back to lower energies, but not necessarily in one transition. So rather than re-emit ultraviolet radiation, two or more photons of visible light could be emitted (Figure 21).

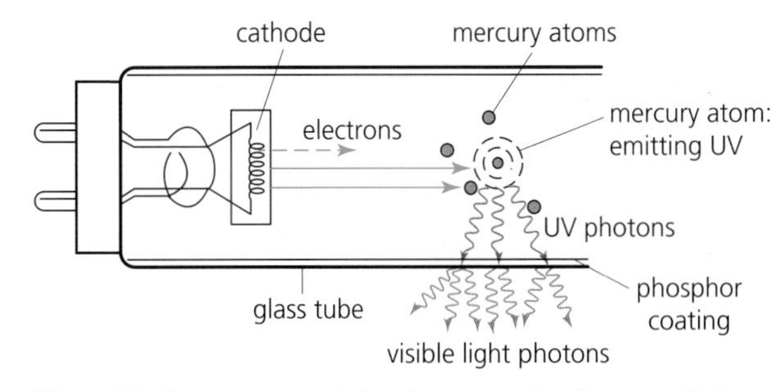


***Figure 20*** *Fluorescence and phosphorescence in a fluorescent tube*

By choosing a phosphor with the right spacing of energy levels, the fluorescent tube can be made to emit just about any colour we choose. These photons of visible light pass straight through the glass tube, but any ultraviolet radiation that is not absorbed by the phosphor is likely to be absorbed by the glass.

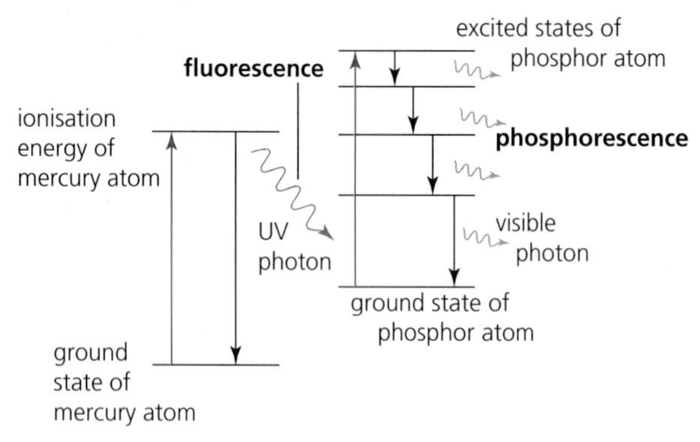


***Figure 21*** *Energy level diagram showing fluorescence and phosphorescence*

## QUESTIONS

**12.** **a.** You cannot get a suntan by sitting under a fluorescent tube. Why not?

**b.** What changes would you make to the tube for a tanning bed?

**13.** Most fluorescent lamps give out white light. How is that done?

**14.** Glass is transparent to light, but opaque to ultraviolet. What makes a material transparent to some wavelengths and not others?

**15.** Niels Bohr said that a line spectrum was "like a stained glass window looking into the heart of the atom". What did he mean by this?

## ASSIGNMENT 2: COMPARING LIGHT BULBS

There are at least four different types of light bulbs in current use in British homes:

- fluorescent tubes or compact fluorescent lights (CFLs)
- incandescent or filament lights
- halogen lights (often used as spotlights)
- light-emitting diodes (LEDs).

Read the section of this chapter on fluorescent lights, and the following text on the three other types of light bulb mentioned, and then carry out task A1 at the end.

The **incandescent light bulb** was first commercially manufactured in 1880 (Figure A1). An electric current is passed through a conducting filament, originally made from carbon but later made from tungsten. The filament becomes hot, quickly reaching its operating temperature of around 2800 K (approximately 2570 °C). At this temperature the filament glows

***Figure A1*** *The first practical light bulb was made by Joseph Swan, an English chemist, who patented his invention a year before Thomas Edison in the USA. His electric lamp had a carbonised thread filament in low-pressure air inside a glass bulb. Swan's own home became the first to be fitted with electric lights in 1879. The basic design has changed little since then.*

white. The efficiency of a 100 W incandescent light is only around 3%. A large proportion of the energy is radiated as infrared, not very useful for reading by but quite good at heating the room. These light bulbs were banned from sale in Europe in September 2012.

**Halogen lights** also use tungsten filaments (Figure A2). The gas inside the bulb is a halogen, which slows down the evaporation of tungsten from the filament and prolongs the life of the bulb. It also means that the bulb can be run at a higher temperature, with the result that more energy is radiated at short wavelengths (in the visible) and relatively less at longer wavelengths. The halogen spotlight gets very hot, though. They are much smaller than ordinary incandescent bulbs, and so the envelope is quartz instead of glass, which would melt. Because the tungsten filament is protected by the halogen gas, the lifetime of the bulb is almost twice the 500 hours or so of an ordinary incandescent bulb. It is also more efficient, though manufacturers' claims are often misleading, partly due to the fact that halogen lights are often used in spotlights, which produce a directed beam of light, whereas incandescent ones tend to radiate their light almost isotropically (the same in all directions).

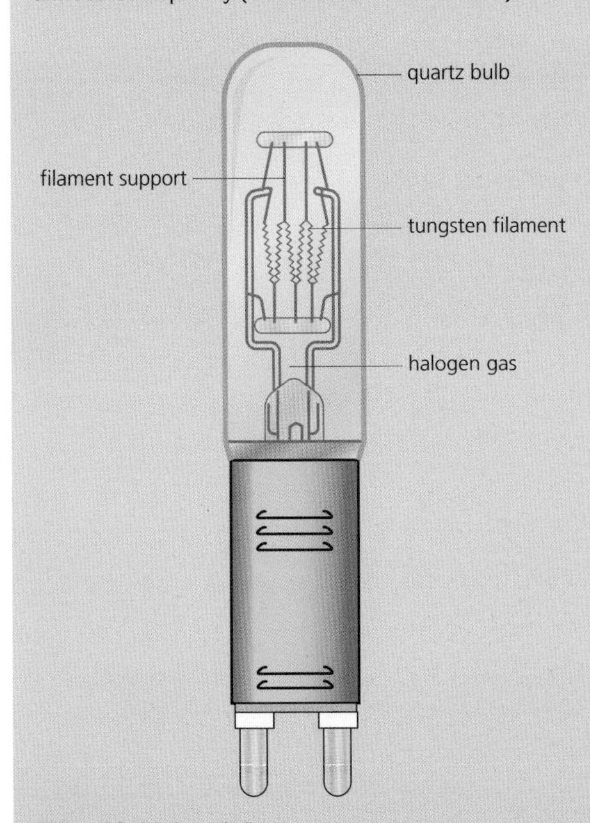


***Figure A2*** *A halogen bulb*

A **light-emitting diode** (**LED**) is made from two pieces of semiconducting material stuck together. A **semiconductor** is a material that is an electrical insulator under normal conditions but will conduct if enough energy is transferred to it, perhaps by heating or by applying a large potential difference. In an LED, the allowed electron energy levels in one semiconductor are higher than in the second material. As the electrons cross the junction between the two types of material, they drop from a higher energy level to a lower energy level, emitting photons of light in the process. LED bulbs (Figure A3) do not need high temperatures to work and are therefore extremely efficient, about six times as efficient as an incandescent light bulb and almost twice as efficient as a fluorescent lamp.

***Figure A3*** *An LED bulb*

The typical spectra of some of these types of light bulb are shown in Assignment 1.

## Questions

**A1** Your task is to produce a leaflet for the public to inform them of the benefits and problems of using each type of light bulb. Start by reading the section on fluorescent lights and the information on incandescent, halogen and LED bulbs. Decide what extra information you need and carry out some research. Useful search terms could be illuminance, spectrum, efficiency, lifetime. Combine these terms with the type of bulb (incandescent, LED, fluorescent, halogen). Your leaflet should aim to give a broad understanding of the way that each bulb works, as well as explaining their comparative advantages and disadvantages.

## Absorption spectra

### Stretch and challenge

The spectrum from the Sun is continuous, but a higher-resolution image shows a series of dark lines superimposed on it (Figure 22). The set of dark lines are the Sun's **absorption spectrum**. The lines represent the wavelengths at which photons have been absorbed by cooler gases in the outer regions of the Sun. Only radiation that has exactly the right amount of energy to lift an electron in the cooler gases' atoms from one allowed energy level to another will be absorbed (Figure 23).

***Figure 22*** *The solar absorption spectrum*

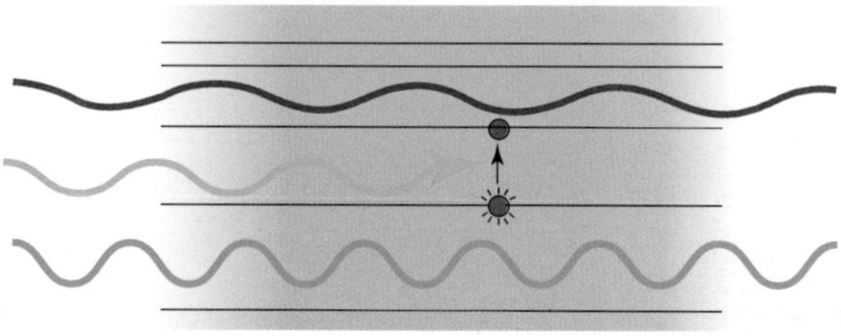

energy levels of atoms in vapour

Only a photon with precisely the right energy will be absorbed. Other wavelengths pass through unaffected.

***Figure 23*** *The absorption of photons*

When the atoms in the cooler gases return to their ground state, they may do so via intermediate energy levels, emitting radiation of other frequencies. The atoms also emit radiation in all directions. These effects reduce the intensity of the absorbed radiation travelling towards Earth, so producing dark lines at specific frequencies in the spectra.

Absorption lines can also be observed in the laboratory by shining a strong white light through a vapour. As early as 1859, Gustav Kirchhoff related the dark lines to particular elements. He observed the Sun's spectrum through a flame into which he sprinkled some kitchen salt (sodium chloride). He found that some of the lines in the Sun's spectrum darkened. Kirchhoff concluded that there must be sodium in the Sun. Just as in emission, each element has its own unique absorption spectrum. Helium was discovered in the Sun, by its absorption spectra, before it was found on Earth.

#### QUESTIONS

**Stretch and challenge**

**16.** Explain why absorption lines arise from the cooler outer atmospheres of the stars and not the hotter centre.

**17.** When Kirchhoff observed the Sun's light through a sodium flame, he noted, "If the sunlight is sufficiently damped, then two luminous lines appear at the position of the two dark lines." Explain why this happens.

**KEY IDEAS**

- Electromagnetic radiation is emitted and absorbed in quanta (discrete amounts) of energy known as photons.
- The energy $E$ of a photon is related to its frequency $f$, and wavelength $\lambda$, by

$$E = hf = \frac{hc}{\lambda}$$

  where $h$ is the Planck constant.
- A photon is emitted from an atom when an electron moves to a lower energy level.
- The larger the gap, $\Delta E$, between energy levels, the higher the frequency of electromagnetic radiation emitted. The energy of the photon is given by

$$hf = \Delta E = E_1 - E_2$$

  where $E_1$ and $E_2$ are higher and lower allowed energy levels in the atom.
- The lines in an emission spectrum show the frequencies at which radiation is emitted. These are characteristic of the element that produced them.

## 8.4 THE PHOTOELECTRIC EFFECT

The idea that radiation is emitted in discrete amounts of energy, as photons, casts the first doubts about the wave theory of light. Another discovery, the **photoelectric effect**, was also making physicists re-examine their ideas about the nature of light. Heinrich Hertz observed the photoelectric effect in 1887 when he noticed that his spark transmitter (Figure 24) worked better when illuminated with ultraviolet light.

The photoelectric effect occurs when light knocks electrons out of the surface of a metal. The effect depends on the type of metal and, crucially, on the wavelength of light used.

A gold leaf electroscope (see section 2.4 of Chapter 2) can be used to demonstrate the photoelectric effect (Figure 25). If you touch the metal disc at the top with a charged object, the gold leaf will rise. If you then touch the disc again, and transfer more charge, the leaf will rise further. If you touch the disc with a wire connected to earth (or touch it with your finger), the charge can escape and the leaf will fall.

*Figure 24* Hertz's spark transmitter

Suppose the electroscope is charged with negative charges, so that the disc has excess electrons. Some of these will travel down to the gold leaf, which will rise. If we could knock electrons out of the surface of the metal, we should be able to remove the extra electrons, and so the gold leaf would fall.

The type of metal used for the disc of the electroscope is important. A clean piece of zinc is often used to demonstrate the effect in a school laboratory. The zinc is placed on the cap of the electroscope, which is then charged from the negative terminal of an HT supply, so that the gold leaf rises. Light from several different sources is then shone onto the zinc. Some typical results are shown in Table 1.

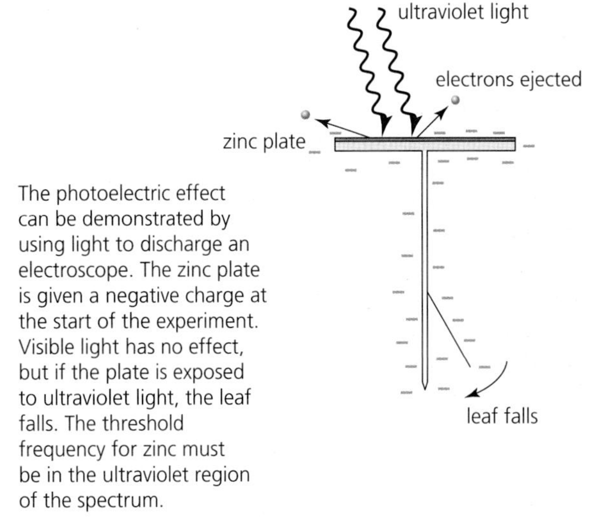


*Figure 25* Demonstrating the photoelectric effect with a gold leaf electroscope

| Light source | Effect on the electroscope |
|---|---|
| Dim red light | No effect |
| Bright red light | No effect |
| Visible light laser | No effect |
| Ultraviolet lamp | The gold leaf falls immediately |

*Table 1*

Not all wavelengths of light can cause the photoelectric effect. For electrons to be emitted from a metal surface, the frequency has to be above a certain minimum value, called the **threshold frequency**. If the frequency of the incident radiation is below the threshold, the electrons remain bound to the positive ions in the metal surface. Each metal has a different threshold frequency. For sodium, it is $5.5 \times 10^{14}$ Hz, which is in the yellow part of the spectrum. Blue or violet light can eject electrons from sodium, whereas red light cannot. An incident photon has to provide enough energy to free an electron from the metal's surface. This energy is known as the **work function**, $\phi$, of the metal.

The work function of a metal is the minimum photon energy required to free one electron from the surface of the metal:

$$\phi = hf_T$$

where $f_T$ is the threshold frequency.

It is difficult to explain the photoelectric effect using a wave theory of light. There are several problems for wave theory:

- The energy carried by a wave depends on its amplitude. Think of a water wave. A wave of larger amplitude carries more energy. In the case of a light wave, a larger amplitude means a brighter light. If the light has insufficient energy to knock electrons out of the metal, wave theory says that a brighter light would work. However, experiments show that light below the threshold frequency cannot dislodge electrons, no matter how bright it is. Several intense red lasers will not prise a single electron from zinc, whereas a less intense ultraviolet glow does so easily.
- The problem of prompt emission. Provided that the light is above the threshold frequency, electrons are emitted immediately. If light was a wave, we might expect the electrons to gradually pick up energy from the wave, so that a weak light would take some time to dislodge an electron.
- When electrons are emitted from the surface of the metal, their kinetic energy depends on the frequency of the light, but not on the amplitude. Wave theory would suggest that the electrons would be emitted with more energy if the light was brighter.

## QUESTIONS

18. Why is it important that the electroscope in Figure 25 is negatively, rather than positively, charged?
19. Why does the gold leaf fall when the zinc is exposed to ultraviolet light?
20. Why is it that red light, even bright red light, has no effect?
21. Explain how the results in Table 1 show the dependence of the photoelectric effect on the frequency and the intensity of light.

Experiments also show that, provided the incident light has a frequency above the threshold frequency, the number of electrons emitted per second (the photoelectric current) is proportional to the intensity of the light. The brighter the light, the larger the photoelectric current. This result can be explained using either the wave model or the photon model of light.

The photon model of electromagnetic radiation is successful at describing the photoelectric effect. Instead of waves of light hitting the metal, we need to think of 'particles' of light, each one a small packet of energy that depends on the frequency of the light The photon theory proposes:

- Below the threshold frequency, a photon does not have enough energy to free an electron. So illuminating a piece of negatively charged zinc with red light will have no effect. Making the red light brighter just means more low-energy photons, none of which have sufficient energy to dislodge an electron.
- At the threshold frequency, one photon has just enough energy to free one electron. There would be no excess energy left to give the electron any kinetic energy.
- When a photon is absorbed, all its energy is immediately transferred to one electron, which is ejected from the surface with no delay.
- Above the threshold frequency, all of the photon's energy is absorbed. Some of this energy is used to liberate the electron from the surface, and any energy left over is transferred as kinetic energy of the electron. Energy is conserved.
- A more intense light can be thought of as more photons per second, so above the threshold frequency a brighter light means that more electrons will be ejected per second.

One analogy for the photoelectric effect is a coconut shy. The coconuts are the electrons, held in their stands like the electrons are held in the surface of the metal. Someone throws ping-pong (table tennis) balls at the coconuts, which has no effect, however many balls are thrown. Someone else is shooting bullets at them, each one knocking off a coconut. The ping-pong balls are photons of light below the threshold frequency, while the bullets are high-energy photons.

### QUESTIONS

**22.** Explain how the coconut-shy analogy fits the experimental observations of the photoelectric effect.

In 1905, Einstein extended Planck's theory of photons and derived an equation that describes the photoelectric effect. Einstein realised that electromagnetic radiation (such as light) is not only *emitted* in discrete amounts, or quanta, of energy, but it is also *absorbed* in discrete amounts, too. When a photon strikes a metal surface, either all or none of its energy is absorbed by an electron. It is not possible to absorb part of a photon.

Einstein expressed this in terms of energy conservation, as follows.

photon energy = minimum energy needed to remove an electron from the surface of a metal (work function)

+ maximum kinetic energy of emitted electron

This is written as

$$hf = \phi + E_{k,max}$$

The observed kinetic energy of an emitted electron, $E_k$, may be less than $E_{k,max}$ if the electron has come from below the surface of the metal.

### QUESTIONS

**23.** A particular metal surface has a threshold frequency in the green part of the spectrum. Suppose you could measure the number of electrons emitted. What would you notice when the metal is exposed to:

**a.** faint red light

**b.** bright red light

**c.** faint blue light

**d.** bright blue light?

**e.** How would the maximum kinetic energy of the electrons alter in each case?

## ASSIGNMENT 3: ANALYSING MILLIKAN'S EXPERIMENT

**(MS 0.1, MS 0.2, MS 3.1, MS 3.3, MS 3.4, PS 1.2, PS 3.1, PS 3.2)**

### Stretch and challenge

When electrons are emitted from a metal by the photoelectric effect, they have a range of kinetic energies up to a maximum value. This value depends on the frequency of light.

Robert Millikan devised an ingenious way of finding the maximum kinetic energy of the electrons $E_{k,max}$ (Figure A1). If the collecting electrode of a photocell (a device in which the photoelectric effect is made to produce a current in an external circuit) is made slightly negative with respect to the photocathode, it will repel the emitted electrons. Only the most energetic will reach the collecting electrode. The potential difference between the cathode and the collecting electrode can be gradually increased until all the electrons are stopped and the current drops to zero. The value of the potential difference when this happens is called the stopping potential.

Since

work done = charge × potential difference

we can say that

maximum kinetic energy of electron = electron charge × stopping potential

That is

$$E_{k,max} = eV_{stop}$$

But from above we also have

$$hf = \phi + E_{k,max}$$

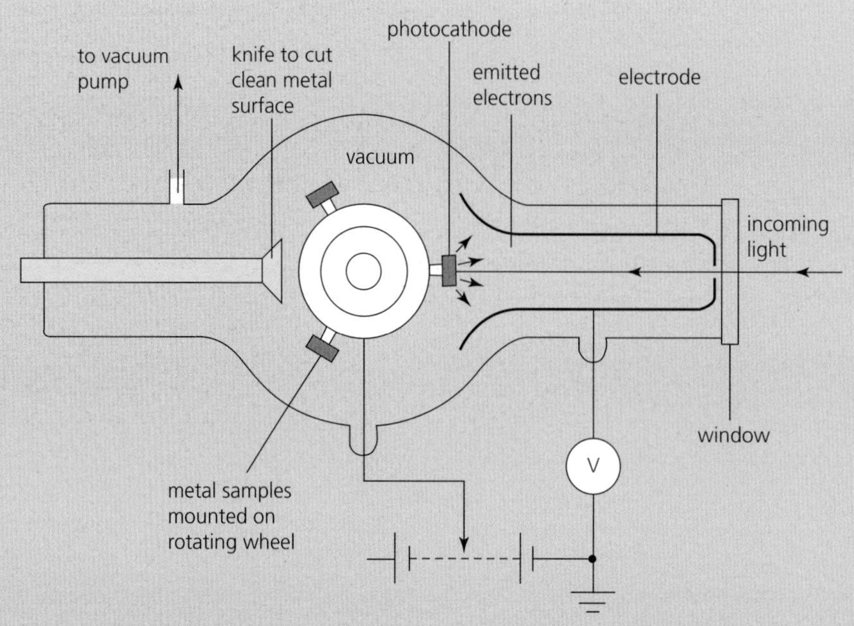


***Figure A1*** *Millikan's photoelectric experiment*

and so substituting for $E_{k,max}$:

$$eV_{stop} = hf - \phi$$

$$V_{stop} = \frac{h}{e}f - \frac{\phi}{e}$$

A graph of $V_{stop}$ versus $f$ is a straight line of gradient $h/e$ with a $y$-axis intercept of $-\phi/e$ (Figure A2).

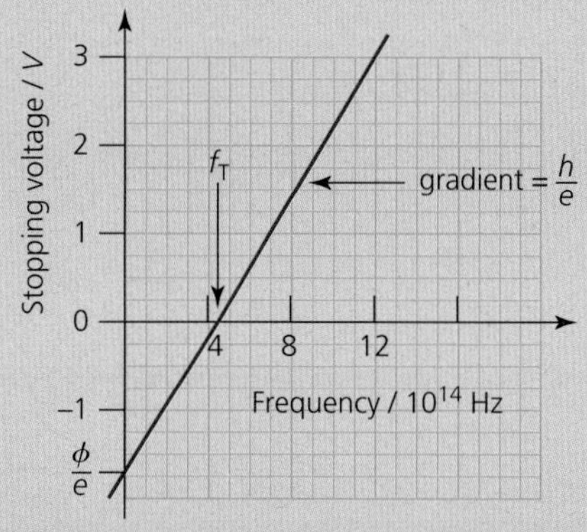


***Figure A2*** *Graph of stopping potential against frequency of incident radiation*

## Questions

**A1** Use the graph in Figure A2 to estimate the Planck constant.

**A2** Use the graph in Figure A2 to estimate the work function of the metal.

**A3** How would the graph look if the metal were changed to one with a higher work function?

## KEY IDEAS

- There is a minimum frequency of electromagnetic radiation that will produce photoelectrons from the surface of a metal. This is called the threshold frequency, $f_T$.
- The minimum energy needed to free an electron from a metal surface is called the work function, $\phi$.
- The absorption of one photon of electromagnetic radiation releases one electron.

$$\phi = hf_T$$

- If the photon energy $hf > \phi$, the excess energy is transferred as kinetic energy of the emitted electron, $E_k$.
- Einstein's photoelectric equation is

$$hf = \phi + E_{k,max}$$

where $E_{k,max}$ is the maximum kinetic energy of an electron.

## ASSIGNMENT 4: HOW DO NEW IDEAS BECOME ACCEPTED?

Read through the following and then answer the questions A1 to A7 at the end.

The following is part of a song, to the tune of 'Men of Harlech', written by the physicist Sir Gilbert Stead in 1920, at a time when physicists were wrestling with the new quantum hypothesis of radiation. (Here $hv$, pronounced 'aitch-nu', is equivalent to $hf$, the energy of a photon.)

*All black body radiations,*

*All the spectrum variations,*

*All atomic oscillations*

*Vary as 'hv'.*

*Ultraviolet vibrations,*

*X- and gamma-ray pulsations,*

*Ordinary light sensations*

*All obey 'hv'.*

*There would be a mighty clearance,*

*We should all be Planck's adherents*

*Were it not that interference*

*Still defies 'hv'.*

Source: *Physics Today*, American Institute of Physics, July 2005

By the start of the 20th century, the wave theory of light had been established for around 100 years, ever since Thomas Young had demonstrated interference effects in 1802. Before that, Newton's corpuscular (particle) theory of light had been accepted by most scientists. Newton's theory was able to describe the phenomena of reflection and refraction, though it predicted that light would travel faster in denser materials. In 1851 Fizeau showed that, in fact, light travelled *slower* in water than in air, but the wave theory was already well established by then. Newton's corpuscular theory had no explanation for the phenomena of interference, diffraction or polarisation of light, and so fell out of favour.

Then, at the turn of the 20th century, several new discoveries were beginning to cast doubt on the wave theory. It struggled to explain the thermal radiation spectrum (section 8.3), emission line spectra (section 8.1) and later the inelastic scattering of X-rays by electrons, called the Compton effect. The photoelectric effect was also an issue, though it was not apparent to most people at the time.

The photoelectric effect had been discovered in 1887 by Heinrich Hertz, during his experiments on radio transmission. Hertz was working with a spark transmitter (Figure 24). A large potential difference applied across a gap between two electrodes caused a spark, which in turn caused an electromagnetic wave to be transmitted (Figure A1). The receiving aerial was simply a loop of wire with a gap.

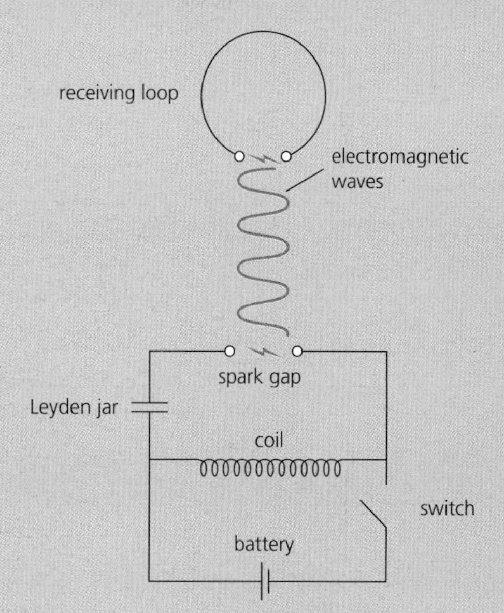


**Figure A1** *The circuit for Hertz's spark transmitter*

As the electromagnetic wave passed through the loop, a potential difference was induced across the gap, which led to a second spark. The gap in the receiving aerial had to be quite small, and the spark was difficult to see. Hertz tried to make the spark more visible by shielding it from other light. He noticed when he did this that he had to reduce the gap between the electrodes still further, so that the second spark was smaller.

"I occasionally enclosed the spark B in a dark case so as to more easily make the observations; and in so doing I observed that the maximum spark-length became decidedly smaller in the case than it was before ... a phenomenon so remarkable called for closer investigation."

Hertz observed that the light emitted by the first spark somehow made it easier to get a spark in the receiver. It was not until 10 years later, in 1899, that J. J. Thomson showed that ultraviolet light, part of the spark's light, could eject electrons from a metal.

At the time, the photoelectric effect seemed to be readily explainable by the wave theory of light. The oscillating electric field of the light wave was able to shake electrons free from the metal surface. Wave theory suggests that a more intense light would provide more energy, free more electrons and give them extra kinetic energy. If a low-intensity light source was used, then it may take more time for the electrons to acquire enough energy to be emitted. It was Philip Lenard, a colleague of Hertz, who conducted a series of experiments on the photoelectric effect in 1902, which eventually proved to be strong support for the quantum nature of light. But Lenard, and many other scientists, did not see it that way for many years.

Lenard first conducted experiments on the photoelectric effect in 1892. His apparatus (Figure A2) enabled him to illuminate a plate in a vacuum tube and collect any charges, though he was not sure of the nature of the charges. He considered several possibilities:

1 The charges were due to the break-up of molecules in the metal surface.

2 The charges came from molecules of gas in the tube.

3 The charges came from within the metal.

Lenard designed an experiment to test the first suggestion. He used a sodium compound as the emitter plate. He ran the experiment for some time, and then removed the collector plate (anode), which he then heated strongly so as to produce a spectrum. He saw no trace of the yellow sodium lines in the spectrum, so he abandoned the molecule theory. Lenard's tube had a much higher vacuum than other experimenters had used, and he felt it was most unlikely that gas molecules could carry charge from the plate, which ruled out the second possible explanation.

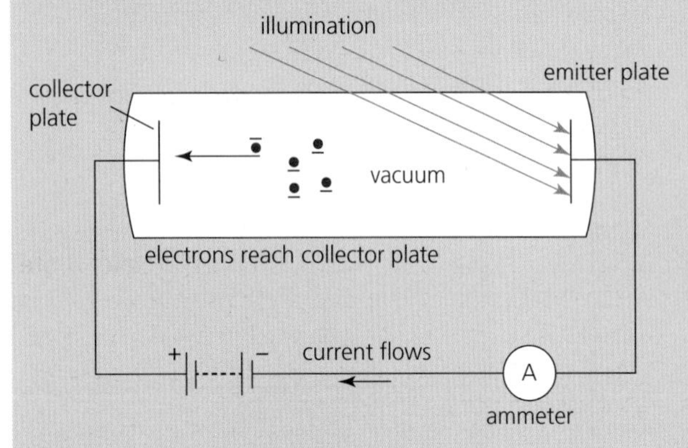


***Figure A2*** *Lenard's photoelectric experiment*

That only left the third hypothesis. The light must be releasing 'something' from the metal, possibly the 'cathode rays' that J. J. Thomson would later identify as electrons. One difficulty was that most German physicists, including Lenard and Hertz, believed that cathode rays were waves, similar to X-rays, which could not carry charge. Lenard did not publish the results of his early experiments. However, when Thomson identified cathode rays as particles (electrons) with a fixed charge-to-mass ratio, Lenard returned to his experiments with the new understanding that ultraviolet light was releasing electrons from the metal. If the vacuum was good enough, these electrons could travel across the tube, causing a current to flow. He found that, even if the collecting plate was slightly negative, a current could be measured. The electrons must be ejected from the metal with some kinetic energy.

Lenard used a very intense carbon arc lamp to illuminate the emitter plate. He found that increasing the intensity of the light increased the number of electrons emitted, but had no effect on their kinetic energy. He also used a prism to separate the colours, and found that the short-wavelength blue light caused electrons to be emitted with more energy. These results were not compatible with wave theory, but Lenard was reluctant to admit that. Instead, he suggested that the electron's kinetic energy arose from its place in the metal. The light wave simply 'triggered' the release of an electron, which then sprang, as if from a catapult, across the tube.

Lenard's experimental work on cathode rays earned him the Nobel Prize in Physics in 1905. He later joined the German National Socialist Party, becoming Hitler's 'Chief of Aryan Physics'.

In 1905 Einstein provided his more radical explanation. Light had to be regarded as a stream of particles (photons), each of energy $hf$. The electrons, held in the metal surface, needed a certain amount of energy, known as the work function, to release them. Low-energy photons (that is, low-frequency radiation) could not eject an electron. These photons were scattered by the metal with no loss of energy. Radiation with a frequency above the threshold was carried by photons with sufficient energy to free an electron. Einstein argued that the photon is completely absorbed – any excess energy would appear as kinetic energy of the electron. Lenard's results were in excellent agreement with Einstein's theory. But physicists were still reluctant to adopt the photon model of light.

The American physicist Robert Millikan designed his photoelectric experiment (see Assignment 3) in the hope that it would disprove Einstein's interpretation of the photoelectric effect. Millikan thought that the idea of photons was a 'reckless hypothesis'. In fact, his painstaking experimental work not only supported Einstein's equation, but enabled the Planck constant to be measured more accurately than before. Even then, Millikan was not convinced. In 1916 he wrote, "Einstein's photoelectric equation ... cannot in my judgement be looked upon at present as resting upon any sort of a satisfactory theoretical foundation [*even though*] it actually represents very accurately the behaviour [*of the photoelectric effect*]."

Eventually, in 1921, Einstein was rewarded with the Nobel Prize in Physics for his work on the photoelectric equation. It is interesting that Einstein's paper on special relativity, also written in 1905, did not win the award. Einstein himself felt that his work on quantum physics had a much more radical effect on our view of the world than did relativity.

### Questions

**A1** Lenard had access to a very bright carbon arc lamp and an effective vacuum pump. Why were these important in his investigations?

**A2** Why might Lenard have been reluctant to publish his early experimental results? Why is it important that all results are published?

**A3** Explain why some features of the photoelectric effect are problematic for wave theory.

**A4** Lenard's 'triggering' theory was rather 'ad hoc'. Find out what the phrase 'ad hoc' means, and explain why it is not a satisfactory way of building scientific theories.

**A5** Einstein's Nobel citation originally said "... for his work on the theory of the photoelectric effect" but, by the time of the ceremony, 'theory' had been exchanged for 'equation'. Why do you think that this was done?

**A6** Planck was dismayed when he realised the implications of his own theory and spent years trying to reconcile his results with classical physics. Why do you think that many physicists were so reluctant to accept the idea of the photon? The last line of the song at the start of this assignment may give you a clue.

### Stretch and challenge

**A7** When the Compton effect was observed in 1923, serious opposition to the photon model of radiation finally disappeared. Find out what the Compton effect is, and explain how it supports the photon model of radiation.

# 8.5 SEEING WITH PARTICLES

## Wave–particle duality

The photoelectric effect suggests that light behaves as particles – photons. Other phenomena, such as diffraction (section 6.2 of Chapter 6), show that light can also behave like a wave. It is possible to carry out an experiment in which light behaves as a wave in one part of the apparatus and as a particle in another (Figure 26). This seemingly contradictory behaviour is an example of what is called **wave–particle duality**. This 'double nature' is not just restricted to light. Electrons and protons, which we have traditionally pictured as particles, turn out to have wave-like characteristics as well.

## De Broglie waves

In 1924, a young French duke, Louis de Broglie, wrote in his PhD thesis "all material particles have a wave nature". He predicted that a particle of momentum $mv$ (mass × velocity) would have a wavelength $\lambda$ given by where $h$ is the Planck constant.

$$\lambda = \frac{h}{mv}$$

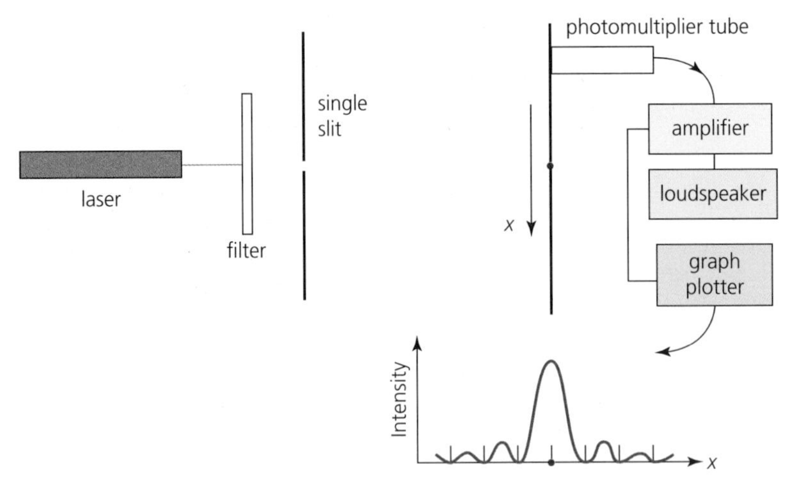


*Figure 26* *The laser light passing through the slit creates a diffraction pattern, which can only be produced by waves. The photomultiplier tube is a light detector that works by the photoelectric effect, which relies on the particle nature of light. With suitably chosen apparatus, the loudspeaker allows you to hear the photons arriving one by one.*

This wavelength $\lambda$ is called the **de Broglie wavelength**. Note that this 'wave nature' of particles does not mean that they are electromagnetic waves.

De Broglie suggested that it should be possible to diffract electrons through crystalline matter, if their velocity was such that their de Broglie wavelength was comparable with the atomic spacing (about $10^{-10}$ m). Clinton Davisson and Lester Germer demonstrated this four years later when they successfully diffracted electrons through a crystal. Diffraction is a wave characteristic, so this confirmed the theory that electrons travel as waves. Today, this principle is used in electron microscopes (see Figure 27).

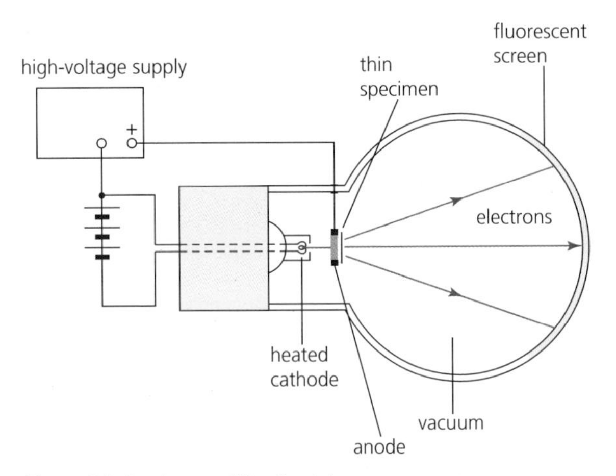


*Figure 28* *An electron diffraction tube*

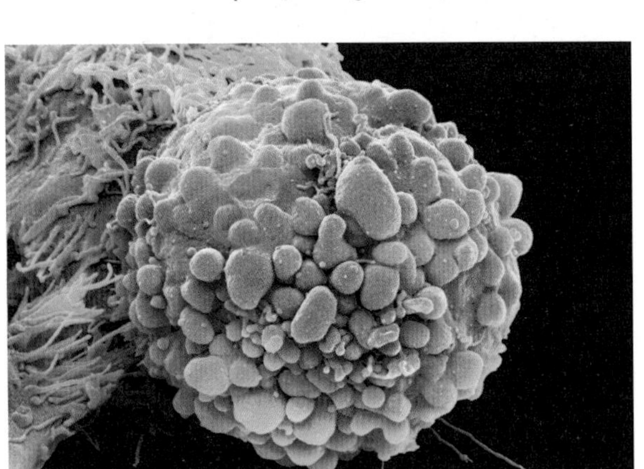

*Figure 27* *A human cancer cell, imaged by an electron microscope*

An electron diffraction tube (Figure 28) is a primitive form of electron microscope. It produces a diffraction pattern on its fluorescent screen (Figure 29). If the potential difference across the tube is known, it is possible to work out the de Broglie wavelength of the electrons.

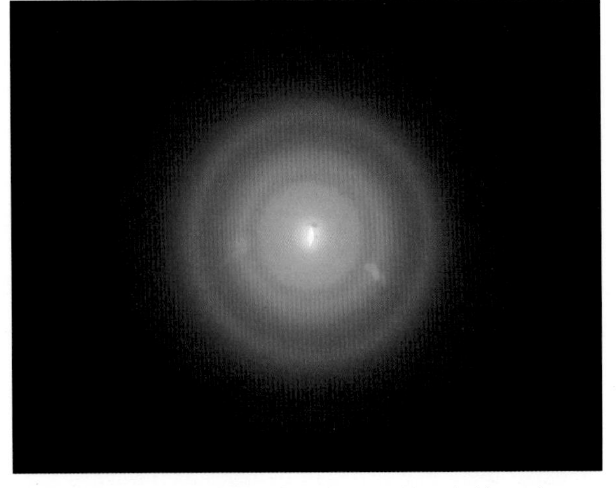

*Figure 29* *The electron diffraction pattern produced with an electron diffraction tube*

### Worked example

The potential difference across an electron diffraction tube is 5.0 kV. What is the de Broglie wavelength of the electrons?

The energy gained by an electron as it crosses the tube is 5000 eV. In joules, that is

$$E = q_e V$$
$$= 1.6 \times 10^{-19}\,\mathrm{C} \times 5000\,\mathrm{V}$$
$$= 8.0 \times 10^{-16}\,\mathrm{J}$$

This is the electron's kinetic energy, so its velocity is given approximately by

$$\tfrac{1}{2} m_e v^2 = 8.0 \times 10^{-16}\,\mathrm{J}$$
$$v^2 = \frac{2 \times 8.0 \times 10^{-16}}{9.1 \times 10^{-31}}$$
$$v = 4.2 \times 10^{7}\,\mathrm{m\,s^{-1}}$$

This is an extremely high speed, but at about one-tenth the speed of light it is low enough to be able to ignore relativistic effects. The momentum of the electron is

$$m_e v = 9.1 \times 10^{-31} \times 4.2 \times 10^{7}$$
$$= 3.8 \times 10^{-23}\,\mathrm{kg\,m\,s^{-1}}$$

So the de Broglie wavelength of the electron is

$$\lambda = \frac{h}{m_e v}$$
$$= \frac{6.6 \times 10^{-34}}{3.8 \times 10^{-23}}$$
$$= 1.7 \times 10^{-11}\,\mathrm{m}$$

The de Broglie wavelength of a particle decreases as its momentum increases. An electron that is accelerated through a potential difference of 60 kV gains a momentum of about $1.3 \times 10^{-22}\,\mathrm{kg\,m\,s^{-1}}$. This gives it a wavelength of about 0.005 nm, compared with an average value for visible light of about 500 nm. The smaller the wavelength used to form an image, the smaller the detail that we can see – the greater the resolution of the optical instrument. Imaging with electrons, rather than with light, increases the theoretical resolution by 100 000 times.

### QUESTIONS

**24.** All matter can show wave-like properties. Duality is not confined to just electrons and protons. However, everyday objects tend to have rather short de Broglie wavelengths. Estimate your de Broglie wavelength when you are walking and hence explain why you do not get diffracted when you pass through a doorway!

**25.** Electron beams fired at crystals can produce a diffraction pattern. X-rays can also be used. Explain why visible light cannot be used.

**26.** What happens to the angle of diffraction in an electron diffraction pattern (between the central maximum and the first minimum) as the momentum of the electrons is increased?

### KEY IDEAS

- Diffraction suggests that light is a wave, and the photoelectric effect suggests that light also behaves as a particle. This is called wave–particle duality.
- Beams of particles such as electrons and protons can be diffracted as if they were waves. All particles have a wavelength associated with them. The de Broglie wavelength is given by

$$\lambda = \frac{h}{mv}$$

where $mv$ is the particle's momentum and $h$ is the Planck constant.

## ASSIGNMENT 5: OBSERVING ELECTRON DIFFRACTION

**(PS 1.2, PS 2.1, PS 3.3)**

The apparatus shown in Figure A1 is used to demonstrate electron diffraction. The vacuum tube has a heated cathode at one end, which emits electrons. The electrons are accelerated towards the anode and they strike a target of a thin slice of graphite on the way. There is a phosphorescent screen at the end of the tube.

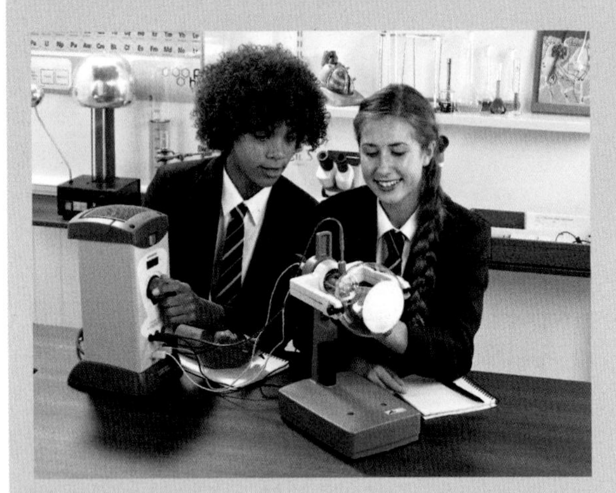

***Figure A1*** *An electron diffraction tube and its connections*

The experiment uses a high potential difference, around 4 kV to 5 kV. Appropriate cables and insulated connectors have to be used. The vacuum tube is expensive and quite fragile. The heater current is kept low to prevent the filament burning out, and the beam is switched off when not in use, to protect the graphite target.

The accelerating voltage is turned down to start with and gradually increased as observations are made. As the potential is increased from 1 kV to 5 kV, a series of rings appear on the screen, as in Figure 29.

### Questions

**A1** Explain the safety precautions you would need to take when carrying out this experiment.

**A2** Diffraction patterns for monochromatic light passing through an ordinary diffraction grating, a grid and a multilayered grid are shown in Figure A2. Compare these with the electron diffraction pattern in Figure 29. What do the rings tell us about the nature of the graphite target?

**A3** Measurements of ring diameter and tube voltage are needed in order to calculate the crystal spacing. Explain how you would make those readings accurately and precisely.

**A4** What does this experiment tell us about the behaviour of electrons?

**A5** Explain how changing the tube voltage changes the appearance of the diffraction pattern.

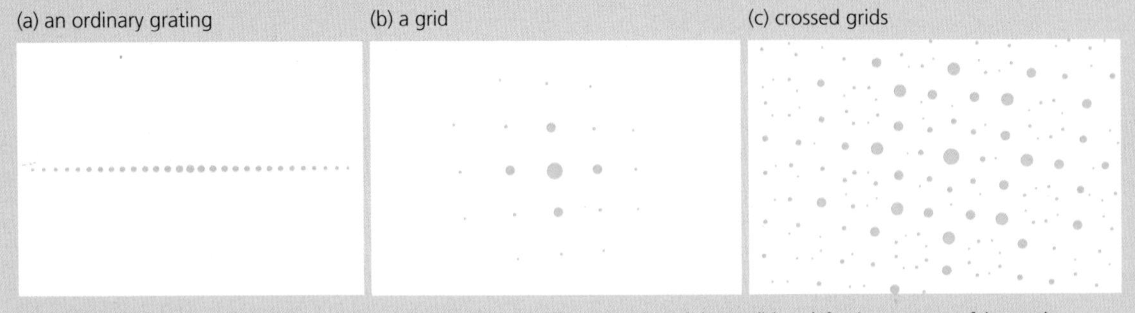


***Figure A2*** *Diffraction patterns of monochromatic light through different gratings. It is possible to infer the structure of the grating by looking at its diffraction pattern.*

## PRACTICE QUESTIONS

**1. a.** The photoelectric effect suggests that electromagnetic waves can exhibit particle-like behaviour. Explain what is meant by threshold frequency and why the existence of a threshold frequency supports the particle nature of electromagnetic waves. The quality of your written communication will be assessed in this question.

**b. i.** An alpha particle of mass $6.6 \times 10^{-27}$ kg has a kinetic energy of $9.6 \times 10^{-13}$ J. Show that the speed of the alpha particle is $1.7 \times 10^{7}$ m s$^{-1}$.

**ii.** Calculate the momentum of the alpha particle, stating an appropriate unit.

**iii.** Calculate the de Broglie wavelength of the alpha particle.

*AQA Unit 1 Jan 2011 Q4*

**2.** Figure Q1 shows the lowest three energy levels of a hydrogen atom.

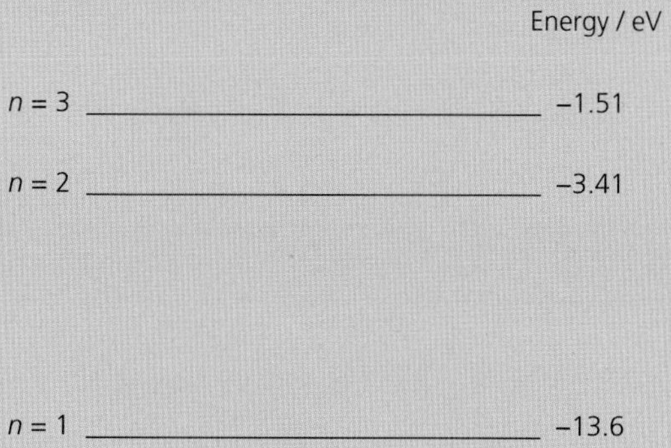

| | Energy / eV |
|---|---|
| $n = 3$ | −1.51 |
| $n = 2$ | −3.41 |
| $n = 1$ | −13.6 |

*Figure Q1*

**a.** An electron is incident on a hydrogen atom. As a result, an electron in the ground state of the hydrogen atom is excited to the $n = 2$ energy level. The atom then emits a photon of a characteristic frequency.

**i.** Explain why the electron in the ground state becomes excited to the $n = 2$ energy level.

**ii.** Calculate the frequency of the photon.

**iii.** The initial kinetic energy of the incident electron is $1.70 \times 10^{-18}$ J. Calculate its kinetic energy after the collision.

**iv.** Show that the incident electron cannot excite the electron in the ground state to the $n = 3$ energy level.

**b.** When electrons in the ground state of hydrogen atoms are excited to the $n = 3$ energy level, photons of more than one frequency are subsequently released.

**i.** Explain why different frequencies are possible.

**ii.** State and explain how many possible frequencies could be produced.

*AQA Unit 1 Jan 2013 Q4*

**3. a.** When monochromatic light is shone on a clean cadmium surface, electrons with a range of kinetic energies up to a maximum of $3.51 \times 10^{-20}$ J are released. The work function of cadmium is 4.07 eV.

**i.** State what is meant by *work function*.

**ii.** Explain why the emitted electrons have a range of kinetic energies up to a maximum value.

**iii.** Calculate the frequency of the light. Give your answer to an appropriate number of significant figures.

**b.** In order to explain the photoelectric effect the wave model of electromagnetic radiation was replaced by the photon model. Explain what must happen in order for an existing scientific theory to be modified or replaced with a new theory.

*AQA Unit 1 June 2012 Q4*

**4.** Niels Bohr famously said that the emission spectrum from a low-pressure vapour was "like a stained glass window looking into the heart of the atom". Which of the following statements correctly explains what he meant?

**A** The colours from the spectrum were caused by emission of radiation as electrons continuously accelerated in orbits around the nucleus of an atom.

**B** The colours emitted showed how quickly the atoms were moving in the vapour.

**C** The colours emitted revealed the spacing of allowed electron orbits in the atom.

**D** The colours emitted revealed the spacing between atoms.

**5.** The wave theory of light cannot explain the photoelectric effect. Which one of the following statements is **not** an explanation for this?

**A** Electron emission happens instantaneously as light strikes a metal surface; wave theory suggests that delayed emission would be possible.

**B** Photoemission does not happen below a threshold frequency; wave theory suggests that a threshold intensity (brightness) would exist.

**C** Above the threshold frequency, the energy of emitted electrons depends on the frequency of the incident light; wave theory suggests that it should depend on the intensity.

**D** Above the threshold frequency, the brightness of the light determines the number of electrons emitted; wave theory suggests that photocurrent should not depend on brightness.

**6.** The Rutherford model of the nuclear atom with orbiting electrons would not be stable. Which is the correct reason for this?

**A** The orbiting electrons would repel one another so that the orbits would not be stable.

**B** The orbiting electrons would be accelerating and therefore radiating energy.

**C** The orbiting electrons are too far away from the nucleus to be held in place by the electrostatic force.

**D** The orbiting electrons would be decelerating and would therefore crash into the nucleus.

**7.** Figure Q2 shows a photograph of the spectrum of a star very similar to our Sun.

*Figure Q2*

Which one of the following statements is **incorrect**?

**A** The black lines are due to light that is absorbed by atoms in the cooler outer parts of the star.

**B** The black lines are due to the fact that some atoms are not present in the core of the star.

**C** The spacing of the black lines is characteristic of particular elements. This allows us to deduce the composition of the star.

**D** The black lines in the spectrum of the Sun led to the discovery of helium.

**8.** Light can be modelled as a wave or as a stream of particles. Which of the rows in Table Q1 gives correct evidence for each theory?

| | Evidence for wave theory only | Evidence for photon theory only |
|---|---|---|
| A | Light can cause the emission of electrons from metal | Light can be reflected |
| B | Light can be refracted | The existence of emission spectra from hot objects |
| C | Light can be reflected | Light can be refracted |
| D | Light can be diffracted | The discrete nature of line spectra from vapours |

*Table Q1*

### Stretch and challenge

**9.** This question is about Millikan's photoelectric experiment.

Figure Q3 shows the apparatus that he used. The central turntable could be rotated so that any one of three metals (sodium, lithium or potassium) could be exposed to the beam of monochromatic light. If the light was above the threshold frequency, electrons would be emitted from the metal surface due to the photoelectric effect. Those with sufficient energy would strike the cathode, causing a small current to flow through the microammeter. A potential divider was used to increase the potential difference between the metal plate and the cathode. Note that this potential difference acted to repel any photoelectrons away from the cathode and, at a certain voltage known as the stopping voltage, $V_{stop}$, no electrons were able to cross and the current dropped to zero. Millikan varied the wavelength of the monochromatic light and found the stopping potential at each wavelength. He repeated this for each metal in turn.

**a.** Explain why it is necessary for the apparatus to be evacuated.

**b.** What is the purpose of the scraper?

**c.** Millikan carried out this experiment in 1914, almost 50 years before the invention of the laser. How might he have produced monochromatic light?

Millikan was not convinced that Einstein's photoelectric equation proved the existence of quanta of light energy (photons). However, his experiment succeeded in finding the most accurate value for the Planck constant so far. Einstein's photoelectric equation is

$$hf = \phi + E_{k,max}$$

where $E_{k,max}$ is the maximum kinetic energy of an emitted photoelectron.

When the stopping potential is just enough to prevent any current flowing, all the kinetic energy of the electron has been transferred as work against the stopping potential $V_{stop}$, so that $E_{k,max} = eV_{stop}$. Substituting into Einstein's equation gives

$$hf = \phi + eV_{stop}$$

**d.** Millikan plotted the stopping voltage against incident light frequency and obtained a straight-line graph. Explain how this enabled him to find values for the Planck constant and for the work function of the metal.

**e.** The graph in Figure Q4 is a replica of Millikan's original. He plotted the first portion of the graph like this, presumably to make it fit. You need to imagine that it is one straight line. Use the graph to find the Planck constant and the work function of the metal.

**f.** How might the graph have looked for the other metals?

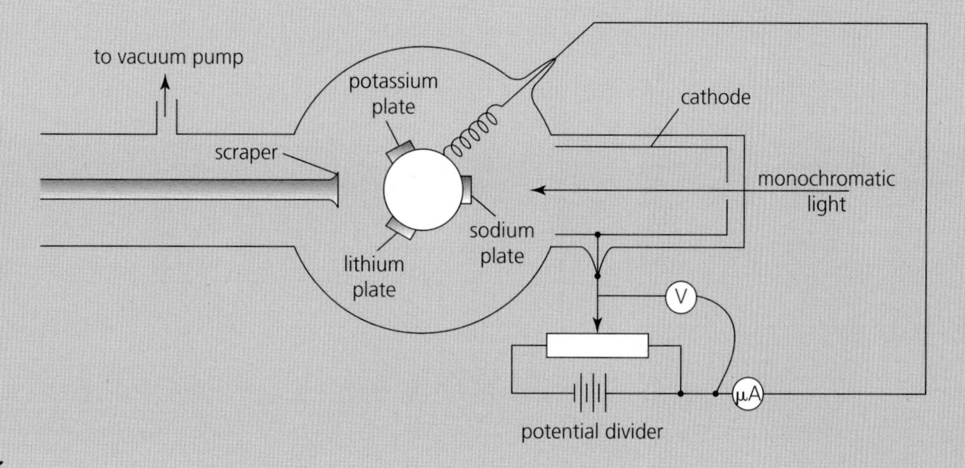


*Figure Q3*

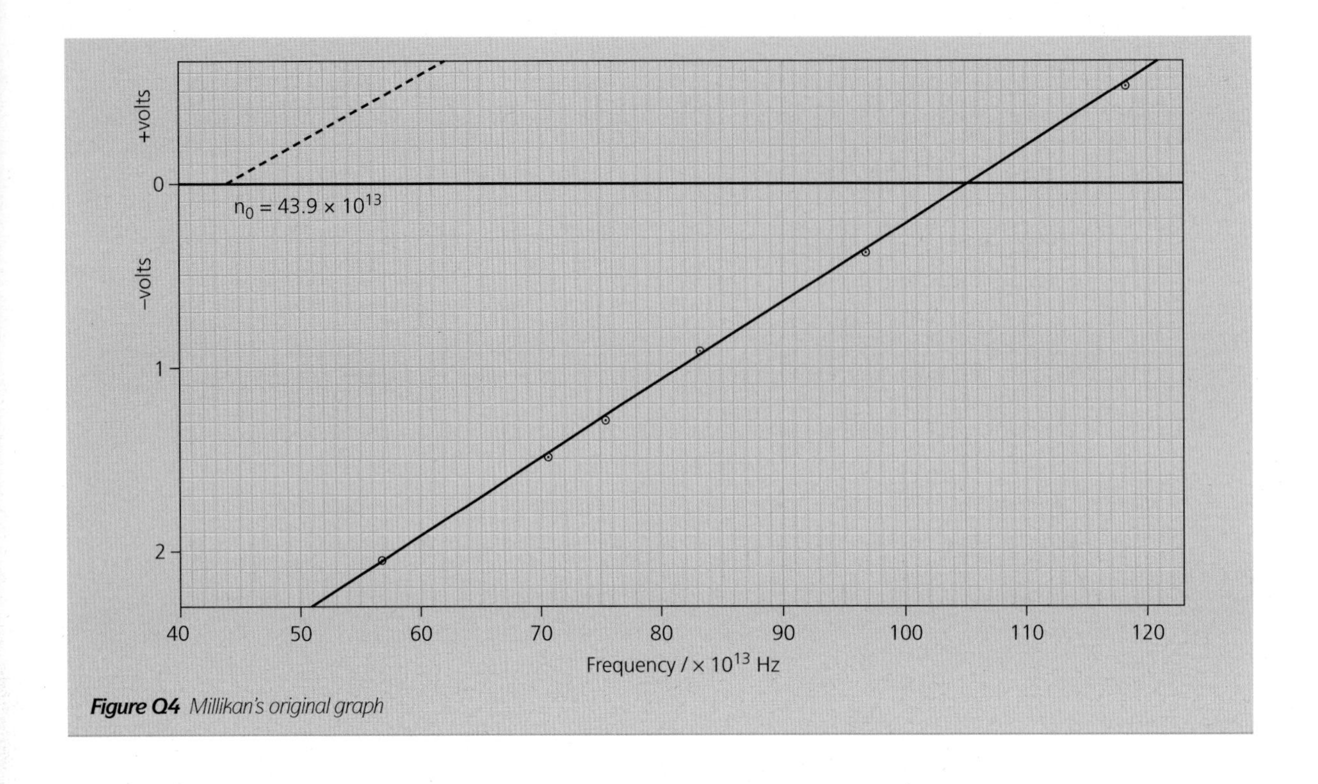


***Figure Q4*** *Millikan's original graph*

# 9 THE EQUATIONS OF MOTION

## PRIOR KNOWLEDGE

You will know how to calculate average speed and average acceleration. You will have met displacement–time graphs and velocity–time graphs.

## LEARNING OBJECTIVES

In this chapter you will learn about scalar and vector quantities and how to use them in the study of motion. You will derive the equations of linear motion and learn how to apply them to the solution of problems.

**(Specification 3.4.1.1 part, 3.4.1.3, 3.4.1.4 part)**

'Citius, Altius, Fortius' (Faster, Higher, Stronger) is the Olympic motto. It is a challenge that athletes have continued to meet since the modern games began in 1896. Most of the current world records have been set in the past 10 years. How is it that modern athletes have been able to keep breaking previous records (Figure 1), and how long can it continue?

A major factor is the increasing impact of science. As modern athletes prepare for competition, they are supported by a team of scientists. Physiologists design training routines, nutritionists plan special diets, and psychologists work on an athlete's state of mind.

***Figure 1*** *World-record times for the men's 100m sprint have continued to improve. Usain Bolt is the fastest man in the world today – possibly the fastest man ever. In 2015 he holds the world record time for the 100 m sprint: 9.58 s.*

Physicists have designed monitors for heart rate, temperature and even brain activity. The latest generation of sports clothing has integrated monitors that link wirelessly to a mobile phone or computer. High-speed cameras linked to computers are used to analyse an athlete's motion, so that performance can be optimised. Physics has also improved the precision and accuracy of the measurements used to judge performance. In track events, for example, electronic timers look for records in thousandths, rather than tenths, of a second. Pressure pads in the starting blocks detect the slightest false start, while ultrasonic meters monitor wind speed. The athlete's equipment, everything from bicycles to swimming costumes, uses the very latest technology in a bid to shave a few milliseconds off the record time.

# 9.1 SCALARS AND VECTORS IN MOTION

## Distance and displacement

In an Olympic 400 m race (Figure 2), the distance covered is 400 m. Distance is a **scalar** quantity. Its size is described by a number and a unit of measurement. The race is run over one complete lap of a running track. At the end of the race, the runners are, approximately, back where they started. We say that their **displacement** is zero. Displacement and distance have the same unit, the metre, but displacement, symbol $s$, is defined as the distance covered *in a certain direction*. This means that displacement is a **vector** quantity. It has a **magnitude** (size) and a direction. A vector quantity therefore needs a number and unit *and* a direction to describe it fully – for example, 10 m north, or 5 m in the direction 20° east of north.

A 100 m sprint is run on a straight part of the track. The distance covered is 100 m and the magnitude of the displacement is also 100 m. In the special case of a journey in a straight line, the distance and the magnitude of the displacement are the same, but in general they will be different (Figure 3).

*Figure 2* In one lap of the 400 m track, a runner's displacement is zero.

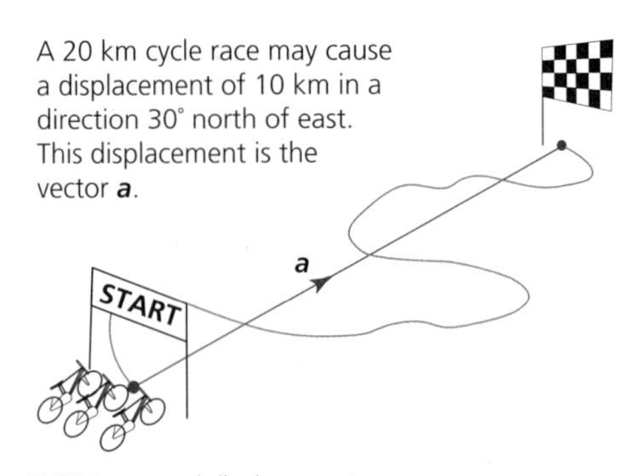


*Figure 3* Distance and displacement

All physical quantities are either scalar or vector quantities. A few examples are shown in Table 1.

| Scalar | Vector |
|---|---|
| Energy | Velocity |
| Distance | Displacement |
| Speed | Acceleration |
| Mass | Force |

*Table 1* Scalar and vector quantities

### QUESTIONS

1. Which of the following physical quantities are vectors? (Ask yourself, "Does it make sense to say that this quantity is in a certain direction?")

   density, momentum, temperature, electric current, power

## Speed and velocity

**Speed** is a scalar quantity. The direction of travel is not important. It is simply the distance covered per unit time. Since speed is likely to vary over a period of time, its value is actually the average speed in that time period:

$$\text{average speed (m s}^{-1}) = \frac{\text{total distance travelled (m)}}{\text{time taken (s)}}$$

Speed is measured in the unit of metre per second (m s$^{-1}$), although it is often given in kilometres per hour (km h$^{-1}$ or kph) or sometimes miles per hour (mph). In these cases, you may need to convert to the unit of metre per second (Figure 4).

*Figure 4* *Speed restrictions are still given in miles per hour on roads in the UK. 1 mile = 1.6 km, so 40 mph = 40 × 1.6 = 64 km h*$^{-1}$ *and 64 km h*$^{-1}$ *= 64 000/(60 × 60) = 17.8 m s*$^{-1}$

**Velocity** is defined as the change in displacement per unit time, and is also measured in the unit of metre per second (m s$^{-1}$):

$$\text{velocity (m s}^{-1}\text{)} = \frac{\text{change in displacement (m)}}{\text{time taken (s)}}$$

It is a vector quantity. The velocity vector has the same direction as that of the *change* in displacement, $\Delta s$. For the magnitudes of the vectors, we can write the relationship as

$$v = \frac{\Delta s}{\Delta t}$$

where $v$ is the average velocity in the time interval $\Delta t$.

Note that the delta symbol, $\Delta$, represents a change in a quantity. Remember that $s$ is the symbol for displacement – a common mistake is to confuse it with the unit s for seconds. The symbol for time is $t$.

Since velocity is a vector quantity, it needs two quantities to fully define it, a magnitude and a direction, for example, $v = 30\ \text{m s}^{-1}$ in an easterly direction. A negative sign, as in $-v$, means that the vector is in the opposite direction.

## Worked example

In 1985 Marita Koch of East Germany ran 400 m in a record-breaking time of 47.60 s. Find her average speed and her average velocity during the race.

Her average speed was

$$\text{average speed (m s}^{-1}\text{)} = \frac{\text{total distance moved (m)}}{\text{time taken (s)}} = \frac{400\,\text{m}}{47.60\,\text{s}} = 8.40\,\text{m s}^{-1}$$

Her average velocity is given by

$$\text{average velocity (m s}^{-1}\text{)} = \frac{\text{change in displacement (m)}}{\text{time taken (s)}}$$

Because she ended where she started, her change in displacement was zero, and therefore her average velocity was zero.

## QUESTIONS

2. The straight-line distance from Clevedon to Cardiff, as the seagull flies, is 14.4 miles, but the shortest road route is 45 miles (Figure 5), a journey that takes 1 hour and 10 minutes. Calculate the average speed and estimate the average velocity of the road journey.
3. Your average velocity for a journey can be less than your average speed. Is it possible for your average speed to be less than your average velocity?

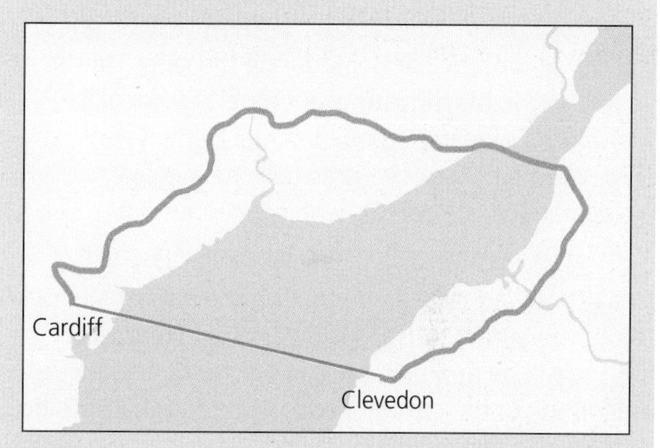


*Figure 5*

4. Athletes in a 1500 m track race could conceivably run a lap at a steady speed, but it is impossible to do so at constant velocity. Explain why this is so.
5. In 1986, Ingrid Kristiansen of Norway won a women's 10 000 m race (25 laps of a 400 m track) in a record time of 30 min 13.74 s.
   a. What was her average speed during the race?
   b. What was her average velocity?

## Average and instantaneous values

The average speed of an athlete does not reveal the full picture. A sprinter's coach will want to know how his/her speed changed during the race, so the time taken to cover intermediate distances is recorded. These are known as split times. Usain Bolt's split times for each 20 m of his 100 m world record sprint are shown in Table 2.

| Distance | Time / s | Cumulative time / s |
|---|---|---|
| Reaction time | 0.146 | |
| 0–20 m | 2.74 | 2.89 |
| 20–40 m | 1.75 | 4.64 |
| 40–60 m | 1.67 | 6.31 |
| 60–80 m | 1.60 | 7.92 |
| 80–100 m | 1.66 | 9.58 |

***Table 2*** *Split times for Usain Bolt's record-breaking 100 m run*

From Table 2 you can see that Bolt was fastest during the 60–80 m interval, when the magnitude of his velocity was

$$v = \frac{\Delta s}{\Delta t} = \frac{20}{1.60} = 12.5\,\mathrm{m\,s^{-1}}$$

The calculation is still an average, although it is over a shorter time. We might want to know Bolt's highest velocity at a particular point during the race. This is known as the **instantaneous velocity** – ideally, the velocity at a single point in time. But any real measurement of velocity is actually an average, a measurement of displacement in a certain time, $v = \frac{\Delta s}{\Delta t}$. If we make that time interval very small, $\Delta t \to 0$, we get close to measuring the instantaneous velocity (Figure 6). Also, since $\Delta t$ is very small, any motion will be approximately in a straight line, so the magnitude of the instantaneous velocity will always be equal to the instantaneous speed.

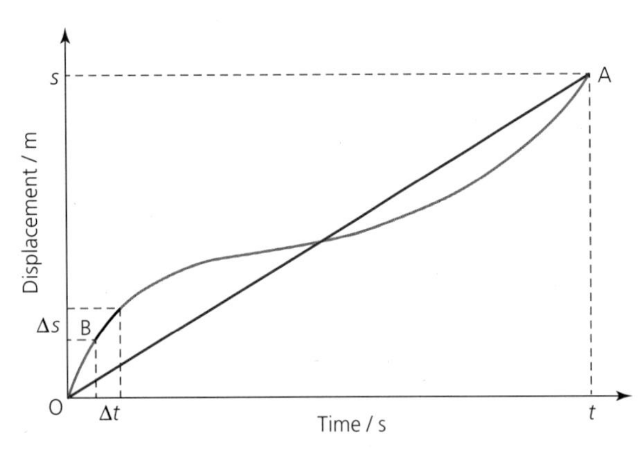


The average velocity between O and A is $s/t$. The average velocity around B is $\Delta s / \Delta t$. If $\Delta t$ is small we can regard this as the instantaneous velocity at B.

***Figure 6*** *Average and instantaneous velocity*

## QUESTIONS

6. Florence Griffith-Joyner was timed at 0.91 s over each 10 m from 60 m to 90 m in the 1988 Olympics 100 m final. Give a quantitative description of her motion over this distance.
7. Who wins the race, the athlete with the highest average speed or the one with the highest instantaneous speed? Explain your answer.
8. Is the average speed or the instantaneous speed more important in a long jumper's run-up? Explain your answer.

## KEY IDEAS

- All physical quantities are either vectors or scalars.
- Vector quantities have a magnitude (size) and a direction.
- Scalar quantities have magnitude only.
- Velocity and displacement are vectors.
- Distance and speed are scalars.
- Average speed ($\mathrm{m\,s^{-1}}$) = $\frac{\text{total distance travelled (m)}}{\text{time taken (s)}}$.
- Velocity ($\mathrm{m\,s^{-1}}$) = $\frac{\text{change in displacement (m)}}{\text{time taken (s)}}$

  or

  $$v = \frac{\Delta s}{\Delta t}$$
- Instantaneous velocity is given by $v = \frac{\Delta s}{\Delta t}$ when the time interval is very small, $\Delta t \to 0$.

# 9.2 VELOCITY AND VECTOR ARITHMETIC

## Relative velocity

Measurements of velocity depend on the motion of the observer as well as the velocity of the object. For example, two athletes running side by side at a velocity of 8 m s$^{-1}$ as measured by a trackside observer will regard each other as stationary, velocity = 0 m s$^{-1}$. Both answers are right. There is no absolute value of velocity. Velocity measurements are always relative to something else. Every time we record a velocity, strictly we should specify what it is with respect to, or relative to – in other words, the **frame of reference** (Figure 7). In practice, we rarely do this because most measurements are made with respect to the surface of the Earth.

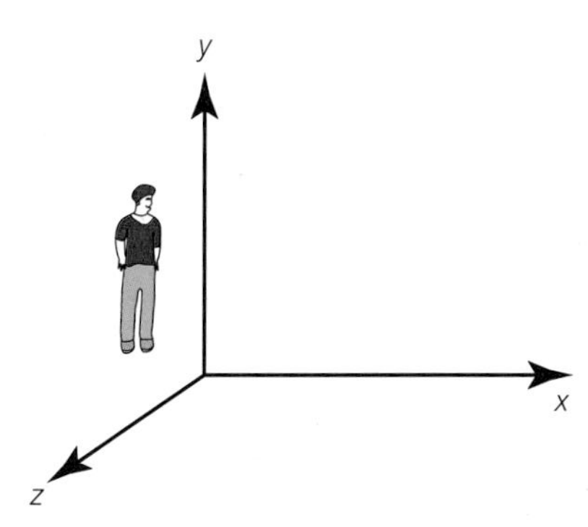


***Figure 7*** *Imagine a set of axes that are at rest with respect to the observer. This is known as a frame of reference.*

Sometimes, however, frames of reference are moving relative to one another, and we need to take account of this. Consider moving walkways, as at an airport (Figure 8). In Figure 9 person A is walking at 1 m s$^{-1}$ relative to the walkway, and the walkway is moving at 2 m s$^{-1}$ as measured by an observer, O, who is not on the walkway. The observer sees person A moving at 2 + 1 = 3 m s$^{-1}$. The velocity vectors are added together. Person B is walking in the opposite direction, the wrong way on the walkway. We need to take account of direction, so her velocity relative to the walkway is −2 m s$^{-1}$. The negative sign indicates the opposite direction. So the observer sees her moving at 2 − 2 = 0 m s$^{-1}$, so she is stationary relative to the observer.

We can show the addition of these velocities as vector diagrams, where the velocities are represented by arrows drawn to scale (Figure 10).

***Figure 8*** *Moving walkways can help us to understand relative motion.*

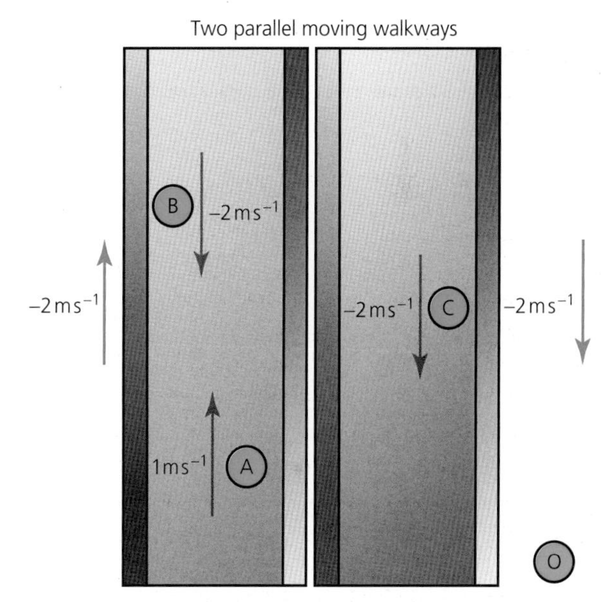


***Figure 9*** *Relative velocity on moving walkways*

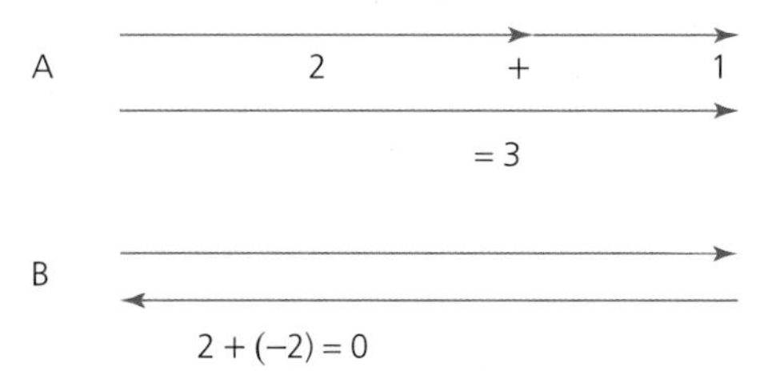


***Figure 10*** *Addition of velocity vectors*

## QUESTIONS

9. In Figure 9, person C is walking at $-2\ \mathrm{m\,s^{-1}}$ on the walkway that has velocity $-2\ \mathrm{m\,s^{-1}}$ relative to the observer, O. What is C's velocity relative to each person?
   a. relative to person O
   b. relative to person B
   c. relative to person A
10. Is there such a thing as a 'stationary' object? Explain your answer.
11. Imagine you are on a train travelling in a straight line at $40\ \mathrm{m\,s^{-1}}$ and you throw a ball at a speed of $10\ \mathrm{m\,s^{-1}}$ down the aisle of a railway carriage towards the front of the train, where it hits the guard.
    a. At what speed does it hit the guard?
    b. Vengefully, he throws it back at you at $20\ \mathrm{m\,s^{-1}}$. At what speed does it reach you?
    c. You hurl it back at $10\ \mathrm{m\,s^{-1}}$, but inaccurately so that the ball sails through the open window, striking a boy on the trackside on the backside. At what speed does it strike him?

### Stretch and challenge

12. The equator moves at a speed of $465\ \mathrm{m\,s^{-1}}$ relative to the Earth's axis, due to the daily rotation of the Earth. Imagine that you are standing on the equator and then you jump vertically up off the ground. You manage to stay in the air as long as an Olympic high jumper, about 1 s. Where will you land?
13. Question 12 may seem a little foolish until you realise that in 1633 a very similar issue almost caused Galileo to be tortured by the Inquisition. Galileo said the Earth moved round the Sun. The Catholic Church insisted that the Earth was self-evidently stationary. In addition to their religious arguments, it was felt that, if the Earth was moving around the Sun at such high speed (about $30\ \mathrm{km\,s^{-1}}$), the birds of the air, maybe even the air itself, would be left behind. What arguments and evidence could Galileo use to defend his case?

## Vector addition

All vectors can be added together in the way shown in Figure 10. The vectors are drawn as arrows, with the length to scale and pointing in the relevant direction. They are added by placing them 'nose to tail'. The answer, or **resultant** vector, is then the combined effect of all the vectors, found by joining the start of the first vector to the end of the final one (see Figure 11).

The resultant, ***a*** + ***b***, of adding two vectors, ***a*** and ***b***, can be found by:

placing them 'nose to tail', so that the arrows follow on. The resultant is the straight line that connects the start of ***a*** to the end of ***b***;

or

by drawing ***a*** and ***b*** from the same point and constructing a parallelogram. The diagonal of the parallelogram is the resultant.

***a*** ***b*** ***a*** + ***b***

***Figure 11*** *Adding vectors to find the resultant*

For example, a displacement of 4.0 m north followed by a displacement of 3.0 m in an easterly direction is equal to a single displacement of 5.0 m in a direction 37° east of north (Figure 12).

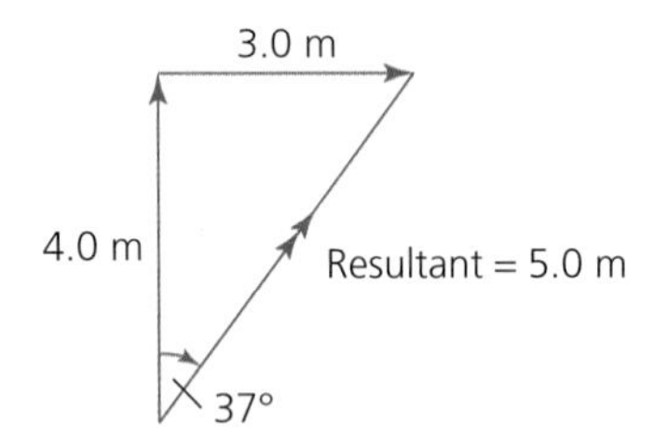


***Figure 12*** *Adding vectors by scale drawing*

Vector addition is more complicated than scalar addition, because we have to take direction into account. Adding together two *distances* of 3.0 m will always produce the answer 6.0 m. Adding together two *displacements* of 3.0 m could give any result from 0 to 6.0 m. You can add any number of vectors by drawing a scale diagram and measuring the length and direction of the resultant.

Vector additions can also be solved using trigonometry (see Figure 13 and Worked example 1).

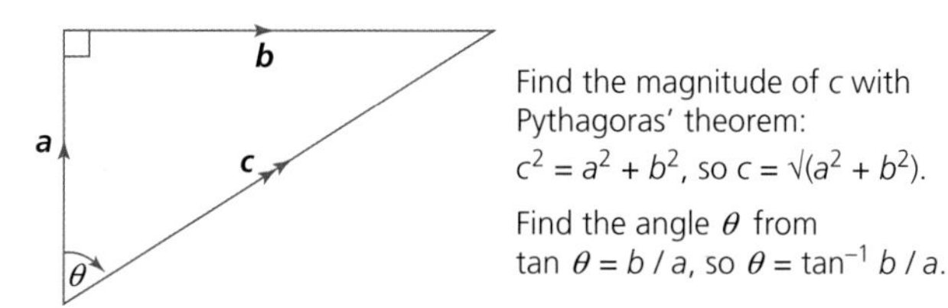


Find the magnitude of c with Pythagoras' theorem:
$c^2 = a^2 + b^2$, so $c = \sqrt{(a^2 + b^2)}$.

Find the angle $\theta$ from
$\tan \theta = b / a$, so $\theta = \tan^{-1} b / a$.

***Figure 13*** *Adding perpendicular vectors using trigonometry*

### Worked example 1

A bird is flying, or at least trying to fly, due south for the winter (Figure 14). It flies at a speed of $20\,\mathrm{m\,s^{-1}}$ relative to the air.

Unfortunately for the bird, the air is moving to the east also at a speed of $20\,\mathrm{m\,s^{-1}}$ relative to the ground. Find the velocity of the bird as seen by a birdwatcher on the ground.

***Figure 14***

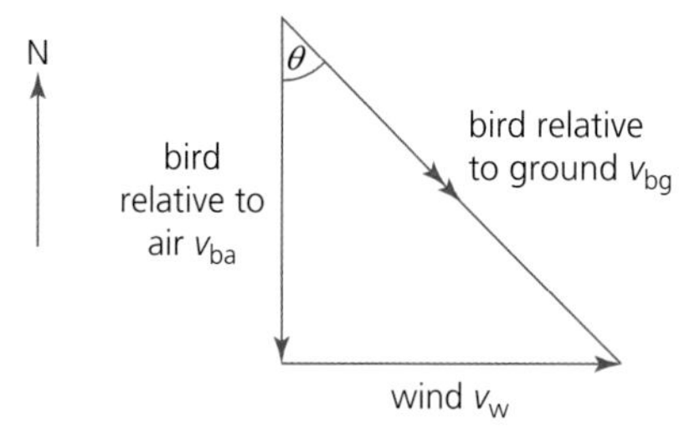


***Figure 15*** *Adding velocities*

First, draw a diagram of the situation described (Figure 15). From the diagram, $v_{bg}^2 = v_w^2 + v_{ba}^2$, so the birdwatcher sees the bird flying at a speed of

$$\sqrt{20^2 + 20^2} = \sqrt{800} = 28\ \mathrm{m\,s^{-1}}\ \text{(to 2 s.f.)}$$

The direction can be found from the tangent of the angle $\theta$:

$$\tan \theta = 20/20 = 1$$

so

$$\theta = \tan^{-1} 1 = 45^\circ$$

So the bird's velocity relative to the observer on the ground is about $28\,\mathrm{m\,s^{-1}}$ in a south-easterly direction.

## QUESTIONS

**14.** "From Skull Rock take five paces due West and then turn and take ten paces South East. The treasure lies eight paces due South." If only pirates had known about vector addition! Use a scale diagram to find the resultant and then simplify the instructions.

**15.** A swimmer is attempting to cross the Severn Estuary. She can swim at $3.0\,\mathrm{m\,s^{-1}}$ relative to the river bank. The tide is running directly downstream, at $2.6\,\mathrm{m\,s^{-1}}$ at right angles to the swimmer's velocity.

**a.** She starts swimming across. Calculate her resultant velocity (the combination of the tide and her own velocity).

**b.** If she is trying to cross at a point where the estuary is 5.0 km wide, how long will the swim take?

**c.** In what direction must she swim if her resultant velocity is to carry her straight across the river, at right angles to the banks?

**d.** How long would this journey take?

## Resolving vectors

The release velocity of a javelin (Figure 16) might be about $27\,\mathrm{m\,s^{-1}}$ but its horizontal velocity might only be about $23\,\mathrm{m\,s^{-1}}$. This is because the javelin is thrown at an angle (of about 30°) to the ground.

***Figure 16*** *Launching the javelin*

The javelin's velocity can be thought of as having two parts, or **components**: a horizontal component and a vertical component (Figure 17). All vectors can be divided up into two (or more) components in this way; the process is called **resolving** the vector. It is possible to choose two components at any angle to each other, which add to form the original vector, but usually a vector is resolved into two components at right angles to each other. For a velocity, $v$, at an angle $\theta$ to the horizontal, the components are $v \cos \theta$ (horizontal) and $v \sin \theta$ (vertical).

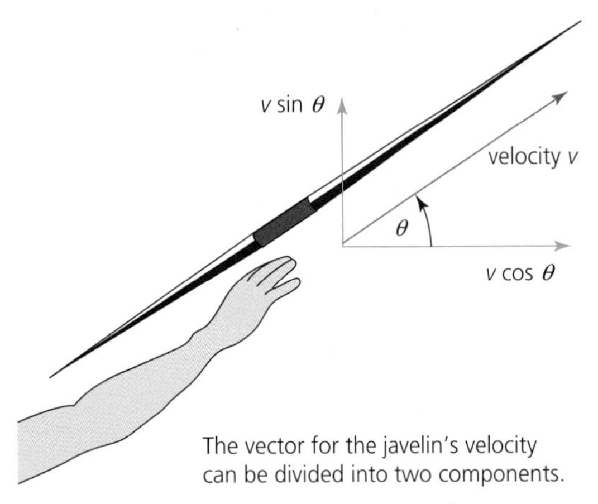


**Figure 17** *Resolving velocity*

### Worked example 2

Records in events such as the javelin are not officially recognised if a wind speed greater than $2.0 \text{ m s}^{-1}$ is blowing in the competitor's favour. Is a wind of $5.0 \text{ m s}^{-1}$, blowing at an angle of 30° to the throwing direction, too strong for a record to be valid?

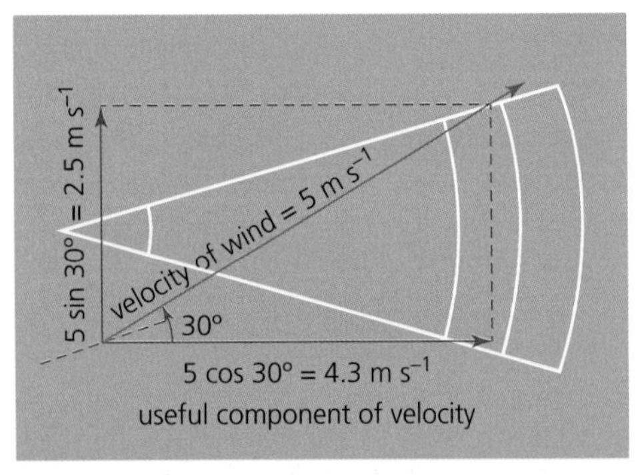


**Figure 18** *Resolving the wind velocity in the direction of throwing*

First draw a diagram showing the situation (Figure 18). From the diagram, the component of the wind in the direction of the throw is $5.0 \times \cos 30° = 4.3 \text{ m s}^{-1}$, which is too strong for a record to be set.

## QUESTIONS

**16.** A shot-putter releases the shot at a velocity of $12.5 \text{ m s}^{-1}$ at an angle of 41° to the horizontal (Figure 19). What is the horizontal velocity of the shot?

*Figure 19*

**17. a.** A cricket ball is thrown at $20 \text{ m s}^{-1}$. Work out the horizontal and vertical components of its velocity if it is thrown at each of the following angles to the horizontal.

**i.** 20° **ii.** 30° **iii.** 40° **iv.** 50° **v.** 60° **vi.** 70° **vii.** 80°

Stretch and challenge

**b.** Suggest which angle you think might give the greatest range. Explain your reasoning.

## KEY IDEAS

- When vectors are added, or subtracted, their direction has to be taken into account.
- Vectors can be added to find their resultant by drawing them to scale, nose to tail.
- The resultant may be found by measurement in the scale drawing, or by using trigonometry.
- A vector may be resolved into two components at right angles to each other.
- For a velocity, $v$, at an angle $\theta$ to the horizontal, the components are $v \cos \theta$ (horizontal) and $v \sin \theta$ (vertical).

## 9.3 ACCELERATION

In a 100 m sprint it is vital to get a good start. A sprinter needs to increase his or her velocity as quickly as possible. The rate at which velocity changes is called **acceleration** – the change in velocity per unit time:

$$\text{acceleration} = \frac{\text{change in velocity}}{\text{time taken for change}}$$

Velocity is measured in metres per second ($m\,s^{-1}$) and time is measured in seconds (s), so acceleration has the unit metre per second per second, $m\,s^{-1}/s$, written as $m\,s^{-2}$. Acceleration takes place in a certain direction and is therefore a vector quantity. The magnitude of the acceleration during a time $\Delta t$ is given by

$$a = \frac{\Delta v}{\Delta t}$$

where $\Delta v$ is the magnitude of the change in the velocity during the time interval $\Delta t$. The acceleration takes place in the direction of the change of velocity (Figure 20).

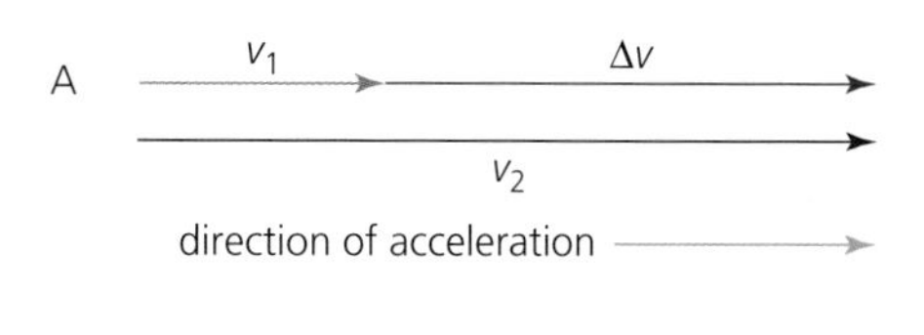


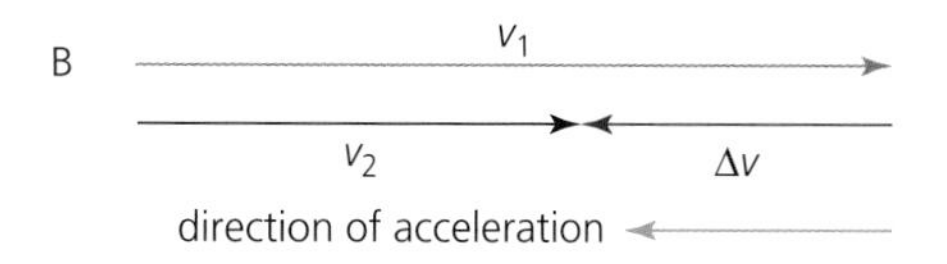


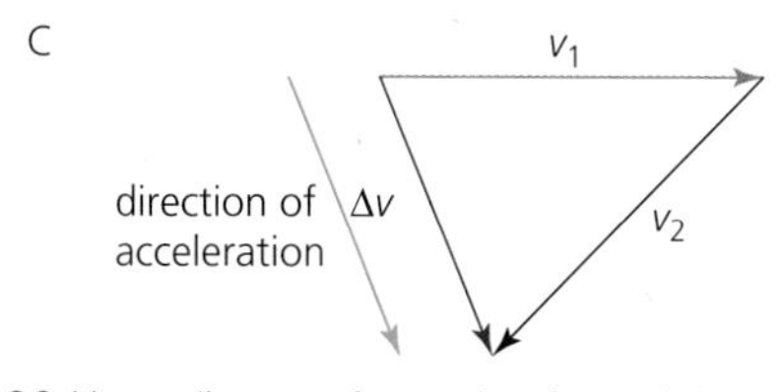


*Figure 20* Vector diagrams for acceleration and change in velocity

The parts of Figure 20 are described here.

A: $v_2 > v_1$ means that speed is increasing, so the acceleration $a$ is positive.

B: $v_2 < v_1$ means that speed is decreasing, so $a$ is negative (an object that was speeding up in the opposite direction would also have a negative acceleration).

C: The magnitudes of $v_1$ and $v_2$ are the same (constant speed) but the direction has changed, so the acceleration is *not* zero.

A change in velocity could be a change in speed *or* a change in direction *or* both. Any change in velocity is an acceleration. Whether you are speeding up, slowing down or just going round a corner at a steady speed, you are accelerating. If the acceleration is in the opposite direction to the velocity, it will reduce the magnitude of the velocity. This is often referred to as deceleration.

### Worked example 1

Two seconds into a 100 m race a top sprinter reaches a speed of 12 $m\,s^{-1}$. Find the magnitude of her average acceleration during the first 2.0 s.

$$\text{acceleration} = \frac{\text{change in velocity}}{\text{time taken}}$$

The motion is along a straight line, so the change in velocity is equal to the change in speed.

$$\text{acceleration} = \frac{\text{final speed} - \text{initial speed}}{\text{time taken}} = \frac{12 - 0}{2.0} = 6.0\,m\,s^{-2}$$

### Worked example 2

A car is travelling at 20 $m\,s^{-1}$ on the M1, going north. The car leaves the motorway on a slip road going north-west, without changing its speed. The manoeuvre takes 2.0 s. Find the acceleration of the car.

First, we draw the situation. This is a vector diagram (Figure 21).

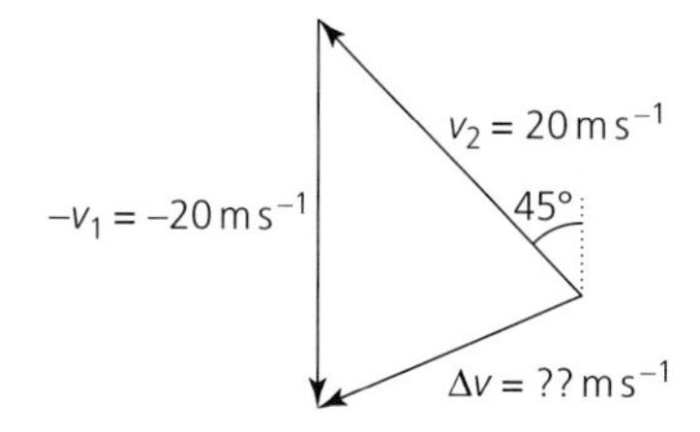


*Figure 21*

The change in the velocity is $v_2 - v_1 = \Delta v$. If the diagram has been drawn to scale, $\Delta v$ can be measured directly. It has a magnitude of 15.3 $m\,s^{-1}$, in a direction of (22.5 ± 1.5) south of west.
So the acceleration is

$$\frac{\Delta v}{\Delta t} = \frac{15.3\,m\,s^{-1}}{2.0\,s} = 7.7\,m\,s^{-2}$$

in a direction of (22.5 ± 1.5) south of west.

The acceleration in each of the Worked examples is actually an average during a time of 2.0 s. But in a real situation the acceleration will vary continuously. For example, a sprint achieves a high acceleration at the instant he or she leaves the starting blocks. The acceleration at a particular moment is defined in a similar way to instantaneous velocity. If we consider the change in velocity, $\Delta v$, in a very small interval of time, $\Delta t$ (Figure 22), then

$$\text{instantaneous acceleration, } a = \frac{\Delta v}{\Delta t}, \text{ where } \Delta t \to 0$$

A velocity–time graph can be used to find the acceleration at a given time. The instantaneous acceleration at any time is given by the slope of the tangent at the point. The tangent at a point is approximated by taking a small change in the velocity, $\Delta v$, over a small time interval, $\Delta t$. The average acceleration is then given by $\Delta v/\Delta t$. As $\Delta t$ gets smaller $\Delta v/\Delta t$ gets closer to the instantaneous acceleration.

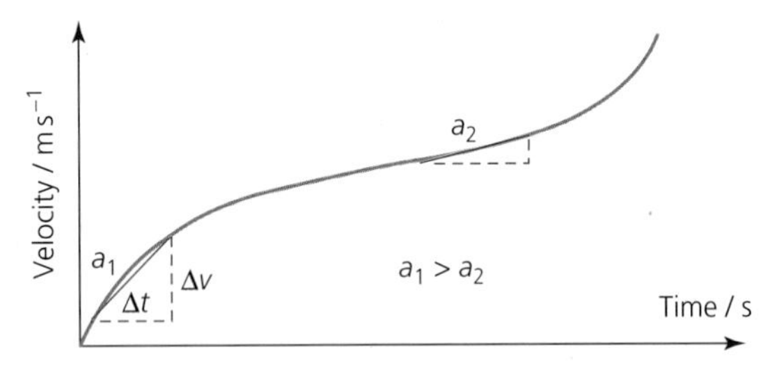


*Figure 22* *Finding the instantaneous acceleration*

**KEY IDEAS**

- Acceleration is a vector quantity.
- Acceleration is the rate of change of velocity and is measured in m $s^{-2}$.
- Average acceleration is given by

$$a = \frac{\Delta v}{\Delta t} = \frac{\text{final velocity} - \text{initial velocity}}{\text{time interval}}$$

- Instantaneous acceleration is the average acceleration as the time interval approaches zero, $\Delta t \to 0$.

**QUESTIONS**

**18.** During a serve, a tennis ball is in contact with the racket for 0.020 s. The ball leaves the racket at 32 m $s^{-1}$. Calculate the ball's average acceleration.

**19.** Estimate the acceleration of a car, pulling away from traffic lights.

**20.** A tennis ball is dropped and bounces up again. Sketch the ball:

**a.** as it falls

**b.** as it bounces up.

Mark in the direction of the velocity and that of the acceleration in each case.

# 9.4 TIME AND MOTION GRAPHS

Graphs showing how the displacement, velocity and acceleration of an object vary with time are used to visualise, and to help analyse, motion.

## Displacement–time graphs

A displacement–time graph is a record of a journey, showing how the distance from a point, in a given direction, changes with time (Figure 23). A negative displacement means a distance from the point in the opposite direction.

We can calculate the average velocity, in the given direction, from the displacement–time graph. Average velocity is given by $v = \frac{\Delta s}{\Delta t}$. This is the gradient of a displacement–time graph, so the greater the gradient, the higher the average velocity. A negative gradient indicates a negative velocity, which means a velocity in the opposite direction.

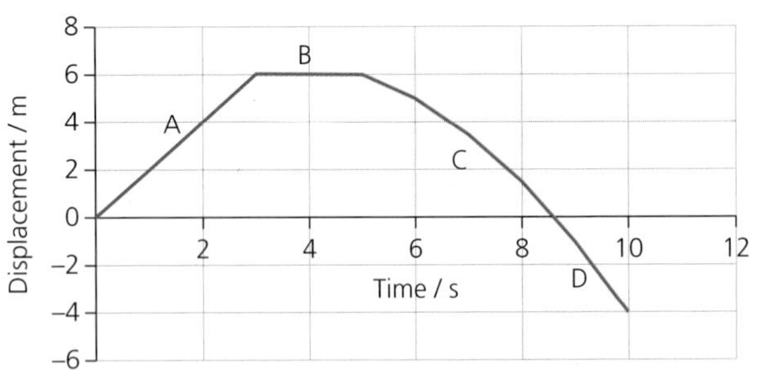


*Figure 23* *A displacement–time graph for a journey*

The motion during each labelled section of Figure 23 is described here.

A: A straight line indicates constant velocity. The gradient gives the magnitude of the velocity. In this case the velocity is

$$v = \frac{\Delta s}{\Delta t} = \frac{6.0 - 0}{3.0 - 0} = 2.0\,\text{m}\,\text{s}^{-1}$$

B: A horizontal line means that displacement is not changing. The velocity is equal to zero, and the object is stationary.

C: The displacement is decreasing. The negative gradient indicates that the velocity is in the opposite direction. The object is returning to the starting position at $s = 0$.

D: A negative displacement means a displacement in the opposite direction.

When the motion is not at constant velocity, the graph is a curve and the average velocity will differ from the instantaneous velocity (see Figure 24). The instantaneous velocity at time $t$ can be found by calculating the gradient of the **tangent** to the curve at that point. The tangent to the curve is the gradient at that point, $\frac{\Delta y}{\Delta x}$, as $\Delta x$ gets very small.

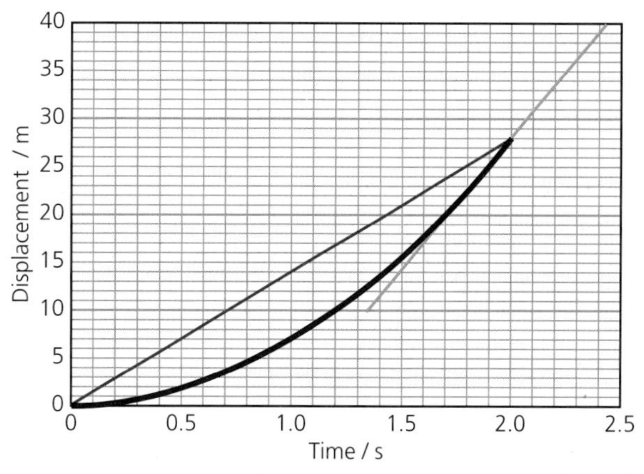


*Figure 24* The black curve shows how the displacement of a Formula One car varies with time at the start of a race.

The car's average velocity over the first 2.0 s is shown by the red line, $v = \frac{\Delta s}{\Delta t} = \frac{28 - 0}{2.0} = 14\ \text{ms}^{-1}$, about 31 mph.

The car's final velocity, that is the instantaneous velocity at $t = 2$ s, is given by the gradient of the curve at that point (shown by the green line), $v = \frac{39 - 11}{2.4 - 1.4}$, very close to $28\ \text{ms}^{-1}$ (63 mph).

## QUESTIONS

21. a. A distance–time graph takes no account of direction, and simply shows the total distance travelled. Can a distance–time graph have a negative section? Can it have a negative gradient? Explain your answers.

    b. Assuming the graph in Figure 23 is for motion in a straight line, how would it look if, instead of displacement, the distance travelled were plotted?

22. Calculate the average velocity between 5 s and 10 s of the graph in Figure 23. Explain how the actual velocity differs from the average velocity.

## ASSIGNMENT 1: ANALYSING THE MOTION OF A SNOOKER BALL

**(MS 3.1, MS 3.6)**

*Figure A1*

Imagine a snooker (or pool) ball travelling across a table (Figure A1).

The diagram in Figure A2 is based on a stroboscopic photograph, and shows a sequence of images of the ball. The images are $\frac{1}{20}$th of a second (0.05 s) apart. This can be used to plot the displacement in a particular direction against time. Before we start to measure any distances, we need to choose a place to measure from, and a direction. We can simply choose the first image of the ball after it was struck. This is marked as O in Figure A2.

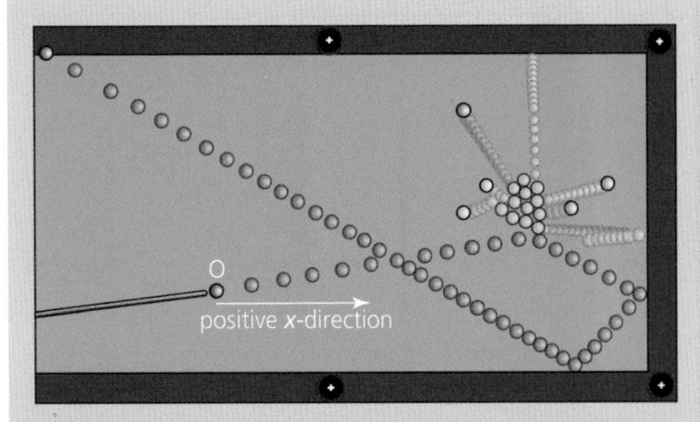


**Figure A2** *Positions of the ball taken by a camera every 0.05 s. The starting point is labelled O and the arrow indicates the positive x-direction.*

Image O has zero displacement ($s = 0$) and it is where we 'start the clock' (so this image was taken at $t = 0$). The direction chosen is the horizontal $x$ direction, marked with an arrow in Figure A2.

To find the displacement, we mark and measure the $x$ position of the ball in each image. The time of each image is the number of images since the start multiplied by 0.05 s. The values of $x$ and $t$ are recorded in a table and used to plot a displacement–time graph for the ball. This is shown in Figure A3.

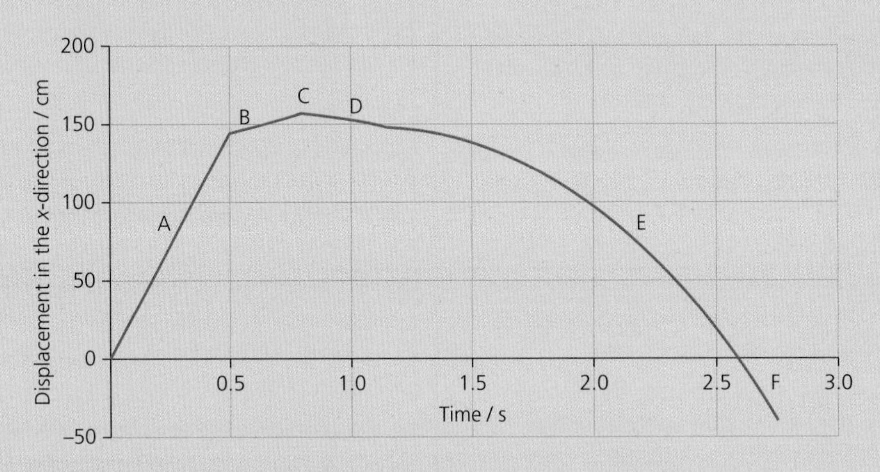


**Figure A3** *Displacement–time graph for the ball in Figure A2*

### Questions

**A1** The displacement–time graph for the ball has several distinct parts, as labelled A–F in Figure A3. Describe the motion of the ball in each section.

**A2** Where is the ball's velocity in the $x$-direction the greatest? Estimate its value.

**A3** Estimate the ball's instantaneous velocity at time = 2.0 s. (You will need to find a tangent to the curve at this point.)

**A4** Something odd is happening to the ball in the last part of its journey. What is it, and why is it odd? Suggest a reason for it.

## Velocity–time graphs

Another way of looking at motion is to plot a graph that shows how velocity changes with time (Figure 25). The gradient of this graph is $\frac{\Delta v}{\Delta t}$, which is equal to acceleration.

A sloping straight line means that the velocity is changing by the same amount over equal time intervals. This is **uniform acceleration**. A horizontal straight line, $\Delta v = 0$, means that the velocity is constant and so the acceleration is zero. A curved graph, whose gradient is changing, indicates a varying acceleration, or **non-uniform acceleration**.

The motion during each labelled section of Figure 25 is described here.

A: A straight line indicates constant acceleration, $a$. In this case, $a = \frac{\Delta v}{\Delta t} = \frac{v_2 - v_1}{t_2 - t_1} = \frac{8 - 0}{4 - 0} = 2\ \text{m s}^{-2}$.

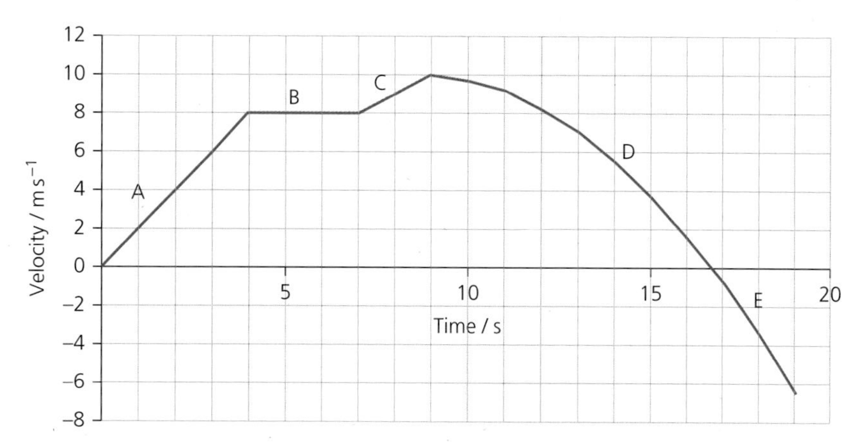


***Figure 25*** *A velocity–time graph for a short bicycle journey, along a straight road*

B: A horizontal line means that $\Delta v = 0$. This represents a constant velocity of 8 $m s^{-1}$. The acceleration is zero.

C: This is straight and the gradient is positive, so this is constant acceleration again, but not as steep as section A. The acceleration is 1 $m s^{-2}$.

D: This has a negative gradient, which is getting steeper, that is becoming more negative. This means it is decelerating at an increasing rate.

E: The object slows down to a stop and then heads back towards its starting point.

Note that, when the velocity is positive, a negative gradient shows the magnitude of the velocity decreasing, in other words the graph is heading for the $v = 0$ ($x$-axis) line, and the object is slowing down. A negative gradient *below* the $v = 0$ ($x$-axis) line shows that the speed is increasing – getting further from zero, but in the opposite direction.

A velocity–time graph can also give information about the displacement:

The displacement after a certain time can be found by calculating the area under the graph up to that time.

A negative velocity indicates motion in the opposite direction, so an area between the time axis and the curve below the axis means a displacement in the opposite direction.

### Worked example

In Figure 12 what is the total displacement between time = 0 and time = 7 s?

To calculate the area under the graph, we divide it into easily calculable sections – in this case by drawing or imagining a vertical line from the graph to the $x$-axis at 4 s and another line at 7 s.

Area from 0 s to 4 s (area of a triangle) =
$\frac{1}{2} \times 4\ s \times 8\ m s^{-1} = 16\ m$

Area from 4 s to 7 s (area of a rectangle) =
$3\ s \times 8\ m s^{-1} = 24\ m$

Total area = total displacement = 16 + 24 = 40 m

## QUESTIONS

**23.** A cyclist starts from rest on a straight road. After 6 s she has reached a speed of 4 $m s^{-1}$. She then travels at constant velocity for 50 s before braking sharply to come to rest in 2 s.

**a.** Draw a velocity–time graph for the journey. What have you assumed?

**b.** Draw a displacement–time graph for the journey.

**24.** Figure 26 shows a velocity–time graph for a sprinter.

**a.** Calculate the acceleration over the first 2 s.

**b.** Calculate the distance covered in the first 8 s.

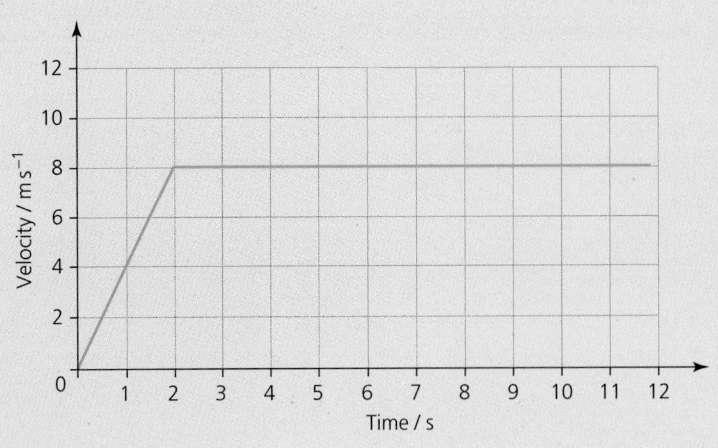


*Figure 26*

**25.** A diver jumps 1 m upwards from a 10 m high board, before diving into the pool below her. Sketch graphs to show how her displacement, velocity and acceleration change with time. Take downwards as negative and use the height of the board as zero displacement. (Assume the motion is all vertical and that air resistance can be ignored.)

## Time and motion graphs for a bouncing ball

We can analyse the motion of a bouncing ball by plotting graphs to show how the acceleration, velocity and displacement vary with time (Figure 27). Imagine a tennis ball, dropped from a height of about 1 m, which bounces three times before being caught again. Because we are dealing with vector quantities, we need to decide which direction is positive, and define the initial conditions. In this case we will take *downwards as negative* and treat ground level as zero displacement.

**Acceleration** (Figure 27a). We can start by considering how the acceleration of the ball changes as it bounces. Whether it is moving up or down, the ball is always accelerating downwards with the acceleration due to gravity ($a = -g = -9.81$ m $s^{-2}$, see section 9.6) except for the very brief time that it is touching the ground, when the acceleration will be upwards. As soon as the ball loses contact with the ground, its acceleration will be $a = -g$ once more. The ball will transfer some of its kinetic energy as internal energy (heating the ball and the ground slightly) each time it hits the ground, so each bounce will be lower than the previous one, and the bounces will get closer together.

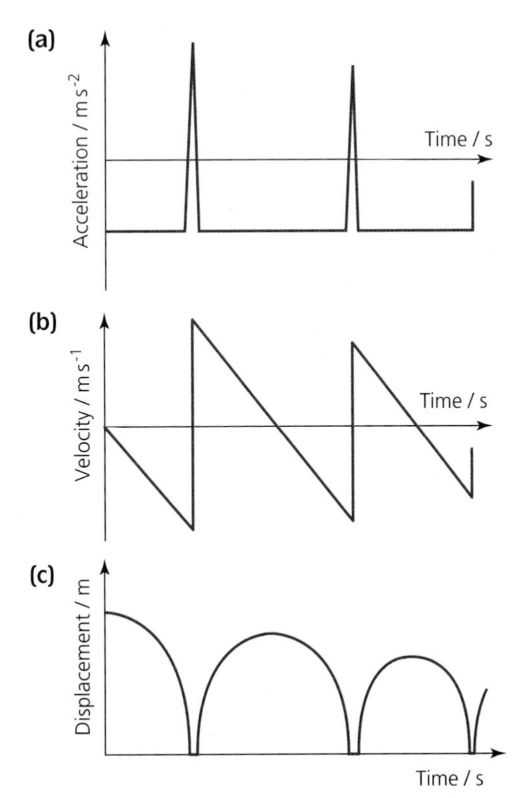


**Figure 27** *Time and motion graphs for a bouncing ball*

**Velocity** (Figure 27b). The ball is released from rest, so its initial velocity is zero. It will then increase its velocity downwards. The gradient of the velocity–time graph is the instantaneous acceleration of the ball. The graph will be straight, with a gradient of $-9.81\ \text{m s}^{-2}$, until the ball hits the ground. Then the ball rapidly slows to a halt and accelerates back upwards. After the ball has left the ground, the velocity will decrease as the ball rises, until it is zero again at the top of the bounce.

**Displacement** (Figure 27c). The gradient of the displacement–time graph is the instantaneous velocity, so the slope starts at zero and becomes increasingly negative until the ball hits the ground. The displacement does not become negative because we have taken ground level to be zero.

## KEY IDEAS

- A displacement–time graph plots the distance from a specified point in a given direction. A negative value indicates a displacement from the specified point in the opposite direction.
- The gradient of the displacement–time graph at a given time gives the velocity at that time. A negative gradient indicates motion in the opposite direction.
- A velocity–time graph plots the speed in a given direction against time. Negative values refer to motion in the opposite direction.
- The gradient of a velocity–time graph at a given time is equal to the acceleration at that time. A positive velocity with a negative gradient indicates deceleration; a negative velocity with a negative gradient indicates acceleration in the negative direction.
- The area between the line of the velocity–time graph and the $x$-axis is equal to the total displacement. Any such area below the $x$-axis indicates displacement in the opposite direction.

## ASSIGNMENT 2: PLOTTING MOTION GRAPHS

**(MS 3.2, MS 3.6, MS 3.7)**

You are going to use acceleration–time, velocity–time and displacement–time graphs to analyse Usain Bolt's 9.58 s sprint. Refer to the data in Table A1 and answer the questions that follow.

| Time / s | Displacement / m |
|---|---|
| 0 | 0 |
| 2.89 | 20 |
| 4.64 | 40 |
| 6.31 | 60 |
| 7.92 | 80 |
| 9.58 | 100 |

***Table A1*** *Usain Bolt's 100 m sprint*

### Questions

**A1** First, copy the table and add a third and a fourth column. Calculate the average velocity for each 20 m section. Enter these velocity figures in the third column of your table, with the correct heading.

**A2** Calculate the average acceleration for each 20 m section. Enter these acceleration figures in the fourth column of your table, with the correct heading.

**A3** Plot a displacement–time graph for Usain Bolt's run.

**A4** Using the same time scales, plot graphs to show how the average velocity, and the average acceleration, vary with time.

**A5** Mark on your graphs the highest velocity and the highest acceleration achieved.

**A6** Describe how Usain Bolt's acceleration changes during the race.

**A7** Estimate his maximum acceleration. Why is this an estimate?

**A8** Describe his motion at the end of the race. Suggest a reason for this.

## 9.5 EQUATIONS OF LINEAR MOTION

There are five important variables that can be used to describe motion (see Table 3).

| Quantity | Symbol |
|---|---|
| Displacement | $s$ |
| Initial velocity | $u$ |
| Final velocity | $v$ |
| Acceleration | $a$ |
| Time | $t$ |

***Table 3*** *Variables that describe motion*

These quantities are linked by four useful equations for uniformly accelerated **linear motion** – that is, motion along a straight line with a constant acceleration.

1. The definition of acceleration gives us the first equation:

$$\text{acceleration} = \frac{\text{change in velocity}}{\text{time taken}}$$

or

$$a = \frac{v - u}{t}$$

$$at = v - u$$

so

$$v = u + at \qquad (1)$$

2. The definition of velocity leads to the second equation:

$$\text{velocity} = \frac{\text{displacement}}{\text{time}}$$

or

$$\text{displacement} = \text{velocity} \times \text{time}$$

When the velocity changes at a constant rate, we take the average velocity as half-way between the initial and final values:

$$\text{displacement} = \text{average velocity} \times t$$

so

$$s = \frac{1}{2}(u + v)t \qquad (2)$$

3. If we substitute equation (1) into equation (2), then equation (2) becomes

$$s = \frac{1}{2}(u + u + at)t$$

$$s = ut + \frac{1}{2}at^2 \qquad (3)$$

Alternatively, you can see how this equation arises by considering displacement as the total area under the velocity–time graph in Figure 28.

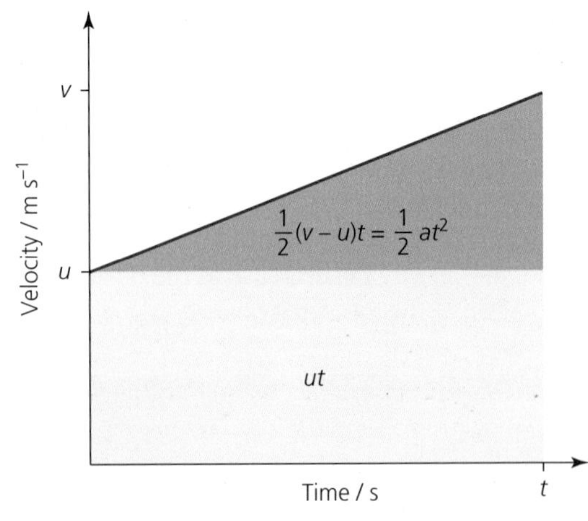


***Figure 28*** *Deriving equation (3)*

4. We can eliminate $t$ from equation (3) by substituting for $t$ from equation (1):

So

$$t = \frac{v - u}{a}$$

$$s = u\left(\frac{v - u}{a}\right) + \frac{1}{2}a\left(\frac{v - u}{a}\right)^2$$

which simplifies to

$$v^2 = u^2 + 2as \qquad (4)$$

Each equation has one of the five variables missing. If you know any three of the five variables, you should be able to calculate the other two. It is important to remember that these equations hold only for uniformly accelerated motion in a straight line.

### Worked example 1

If a 100 m sprinter accelerates from the starting block at 2.0 m s$^{-2}$ and maintains this acceleration over the whole race, we can use the equations to calculate how fast he will cross the finishing line and how long he will take to run the race.

To find the final velocity we could use

$$v^2 = u^2 + 2as$$
$$= 0 + 2 \times 2.0 \times 100.0$$

The final velocity is $v = \sqrt{400} = 20.0\,\text{m s}^{-1}$.

The time taken for the race would be

$$t = \frac{v - u}{a}$$
$$= \frac{20.0 - 0}{2.0}$$
$$= 10.0\,\text{s}$$

In reality, sprinters achieve a higher acceleration than this at the start of the race, though it cannot be maintained throughout the race. We cannot simply apply equations of motion to the whole race, because the acceleration is not constant. But we can apply the equations to each section of the race in turn.

### Worked example 2

Suppose the sprinter from Worked example 1 accelerates at 3.0 m s$^{-2}$ for the first 4.0 s and then runs at constant speed for the rest of the race. What is the time taken to complete the 100 m?

For the first part of the race, 0–4 s, we have:

$$s = ?,\ u = 0\ \text{m s}^{-1},\ v = ?,\ a = 3.0\ \text{m s}^{-2},\ t = 4.0\ \text{s}$$

The velocity after 4 s will be

$$v = u + at$$
$$= 0 + 3.0 \times 4.0 = 12.0\ \text{m s}^{-1}$$

and the distance covered will be given by

$$s = ut + \frac{1}{2}at^2 = 0 + \frac{1}{2} \times 3.0 \times 4.0^2 = 24\ \text{m}$$

For the second part of the race, 24–100 m, we have:

$$s = 76\ \text{m},\ u = v = 12\ \text{m s}^{-1},\ a = 0\ \text{m s}^{-1},\ t = ?$$

We can find $t$ from $s = ut + \frac{1}{2}at^2$, and, because $a = 0$, from $s = ut$, so

$$t = 76/12 = 6.3\ \text{s}$$

So the total time is 4.0 + 6.3 = 10.3 s

## QUESTIONS

**26.** When a golfer hits a ball, the club only touches the ball for about 0.0005 s, but the ball leaves at a speed of 75 m s$^{-1}$. What is the average acceleration of the golf ball while it is in contact with the club?

**27.** A cyclist freewheels from rest down a hill that is 120 m long. By the time he reaches the bottom, he is travelling at 15 m s$^{-1}$.

**a.** Calculate his acceleration.

**b.** At the bottom of the hill, the road levels out and the cyclist brakes gently, coming to a stop in 30 s. Calculate his deceleration and the total distance he has travelled.

### Stretch and challenge

**28.** A girl on a trampoline bounces 3 m above the trampoline. Estimate her time in the air. Explain any assumptions that you have made.

## KEY IDEAS

- The equations of linear motion apply to a body that is moving in a straight line, with uniform (constant) acceleration.
- The equations are

$$v = u + at$$
$$s = \tfrac{1}{2}(u + v)t$$
$$s = ut + \tfrac{1}{2}at^2$$
$$v^2 = u^2 + 2as$$

## ASSIGNMENT 3: USING A MOTION SENSOR

**(MS 1.5, PS 1.2, PS 2.1, PS 2.3, PS 2.4, PS 4.1)**

A motion sensor (Figure A1) uses an ultrasonic beam to measure the distance from an object. Rather like the echo system that bats use, the sensor sends out a pulse of ultrasound and measures the time taken for it to return after it has reflected off an object. The sensor then calculates the distance to the object, using a typical figure for the speed of sound in air. The sensor may be linked to a computer so that real-time graphs of displacement, velocity and acceleration against time may be plotted.

***Figure A1*** *A motion sensor*

### Questions

**A1** Your task is to design an experiment using the motion sensor to verify one of the equations of linear motion, namely

$$s = ut + \frac{1}{2}at^2$$

(*Hint*: You could start from stationary, in which case $ut = 0$, and the equation and the experiment will become more straightforward.)

You will need to think about the following points:

- What will you use as your moving object?
- How will you ensure uniform acceleration?
- What variables might you need to control?
- What readings will you take with the motion sensor?
- What other readings will you take?
- What graph or graphs will you plot?
- Try to identify the potential sources of error in this experiment. Can they be avoided, or can you at least reduce the uncertainty that they cause?
- Can you use your results to find a value for the constant acceleration, $a$?

Your plan should include your proposed method, and your responses to the above points.

If you have the opportunity to do so, carry out the experiment. In the light of your practical experience, suggest how you would improve your plan.

### Stretch and challenge

The speed of sound in air depends on the air temperature. The following equation can be used to calculate the speed $v$ (metre per second, $\text{m s}^{-1}$) at temperature $T$ (degree Celsius, °C):

$$v = 331 + 0.6T$$

### Questions

**A2** Calculate the speed of sound in air at 20 °C.

**A3** Suppose that the motion sensor makes no allowance for a change in temperature. Estimate the uncertainty in distance measurement that this would produce.

# 9.6 FREE FALL AND TERMINAL SPEED

A dense object, like a stone, dropped from a high place, such as a cliff, will accelerate as it falls towards the ground. The air has little effect on the stone and its acceleration will be almost constant. An object is said to be in **free fall** if the only force acting on it is the gravitational force. Experiments show that on Earth the acceleration due to gravity, $g$, is about 9.81 $\text{m s}^{-2}$, though the exact value varies from place to place. For example, the acceleration due to higher is slightly higher at the Earth's poles than at the equator.

Measured at the same place, all objects accelerate under gravity at the same rate. The mass of the falling object makes no difference, as long as air resistance can be neglected. This became obvious when Commander David Scott, of the Apollo 15 mission, stood on the Moon and dropped a hammer and a feather onto the lunar surface. They fell together (Figure 29).

***Figure 29*** *In the absence of air resistance, all objects fall under gravity at the same rate.*

Astronauts are among the very few people to have experienced free fall for any prolonged time. You may have fallen a few metres, perhaps into a trampoline. Sky-divers fall further but they soon have to contend with air resistance, and when the parachute opens they are very glad of that air resistance! An astronaut in orbit around the Earth is often described as 'weightless', but this cannot be true. The Earth's gravity is slightly weaker at that height above the Earth but it manages to keep the Moon in orbit, and it is gravity that keeps a satellite or space station in orbit. An object in orbit is acted on by gravity, towards the centre of the Earth, but it is also moving forward fast enough so that it stays in orbit. Astronauts in the International Space Station are in free fall, continually falling towards the Earth, but (because the acceleration due to gravity does not depend on mass) all the objects in the space station, and the space station itself, are falling, too, all at the same rate. This means that there is no contact force between the astronaut and the space station floor, giving rise to the apparent weightlessness (Figure 30).

***Figure 30*** *Velcro is useful to stop the chess pieces floating off!*

### Worked example 1

If you dropped a coin from the top of The Shard (Figure 31), at a height of 308 m, how long would it take to reach the ground and how fast would it be travelling when it got there? Take $g = 9.81\ \text{m s}^{-2}$.

***Figure 31*** *The Shard, in London, is the highest building in the UK.*

As the initial velocity is zero, you can use the formula $s = \frac{1}{2}gt^2$, giving

$$t^2 = \frac{2s}{g}$$

and so it would take

$$t = \sqrt{\frac{616}{9.81}} = 7.92\ \text{s}$$

Now to find how fast it would be travelling, you can use $v^2 = u^2 + 2as$, giving

$$v^2 = 0 + 2 \times 9.81 \times 308 = 604\ \text{m}^2\text{s}^{-2}$$

and so

$$v = 77.7\ \text{m s}^{-1}\text{, which is 174 mph!}$$

Air resistance would in reality limit this, but the coin would still be travelling fast enough to be very dangerous to someone on the pavement below.

## Worked example 2

Imagine someone is standing on the edge of a 100 m cliff and throws a ball up into the air at 20 m s$^{-1}$. The throw is not quite vertical so when the ball falls again it is out of reach, over the cliff (Figure 32).

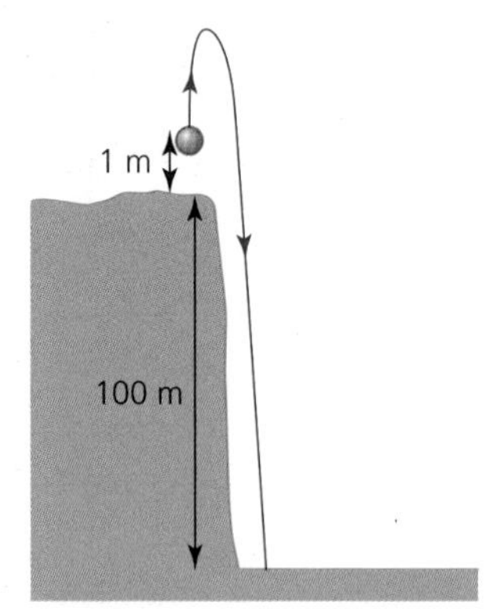


*Figure 32*

**a.** How long does it take for the ball to reach the ground? Take $g = 9.81$ m s$^{-2}$.

**b.** How long does it take the ball to reach a point 4 m above the top of the cliff?

First define the starting position. We will assume the ball starts from 1 m above the top of the cliff, and displacements will be measured from that point (Figure 32). We must also specify which direction is positive. So we will say up is the positive direction.

**a.** The values we know are:

displacement = −101 m

acceleration = $-g = -9.81$ m s$^{-2}$

$u = 20$ m s$^{-1}$

So, using $s = ut + \frac{1}{2}at^2$ gives, after a little rearranging

$$4.905\, t^2 - 20t - 101 = 0$$

This is a quadratic equation and we have to use the formula

$$x = \frac{-b \pm \sqrt{b^2 - 4ac}}{2a}$$

This gives two values for time, $t = 7.0$ or $-3.0$. Negative time means before the ball was thrown, which is impossible. So the answer is 7.0 s.

**b.** As in part **a**, use the same equation but now $s = 3$ m (remembering the 1 m high starting position). This time the equation becomes

$$3 - 20t - 9.81t^2 = 0$$

which leads to $t = 0.16$ s (on the way up) and 4.1 s (on the way down).

Alternatively, these problems can be solved without the use of the quadratic equation formula, by considering first the upward motion and then the downward motion, and adding the times for the two stages of the motion together.

## QUESTIONS

**29. a.** A potholer drops a stone down a mine-shaft and hears it hit the bottom 3.5 s later. How deep is the shaft? Assume that the effect of a finite speed of sound is negligible.

**b.** Estimate the uncertainty in your answer to part **a**.

**c.** Should we have taken the speed of sound, which is 340 m s$^{-1}$, into account?

### Stretch and challenge

**30.** Try this 'thought experiment'. You are in a lift near the top of The Shard (Figure 31) when the lift cable snaps, all the safety systems fail and you plummet downwards.

**a.** Not surprisingly, you let go of your bag. What happens to it?

**b.** More surprisingly, you are standing on bathroom scales. What would they show?

## KEY IDEAS

- All objects, whatever their mass, fall at the acceleration due to gravity when resistance can be neglected. On Earth, neglecting air resistance, this acceleration is approximately 9.81 m s$^{-2}$ downwards.

# REQUIRED PRACTICAL: APPARATUS AND TECHNIQUES

## Determining the acceleration due to gravity, *g*

The aim of this practical is to measure the acceleration due to gravity, *g*, by a free-fall method. The practical gives you the opportunity to show that you can:

- use appropriate analogue apparatus to record a range of measurements (to include length/distance) and to interpolate between scale markings
- use methods to increase the accuracy of measurements
- use a stopwatch or light gates for timing
- use ICT such as a data logger with a sensor to collect data, or software to process data.

There are different methods of measuring *g* in the school laboratory. This method uses a 'trap door' and direct measurements of a falling object (Figure P1).

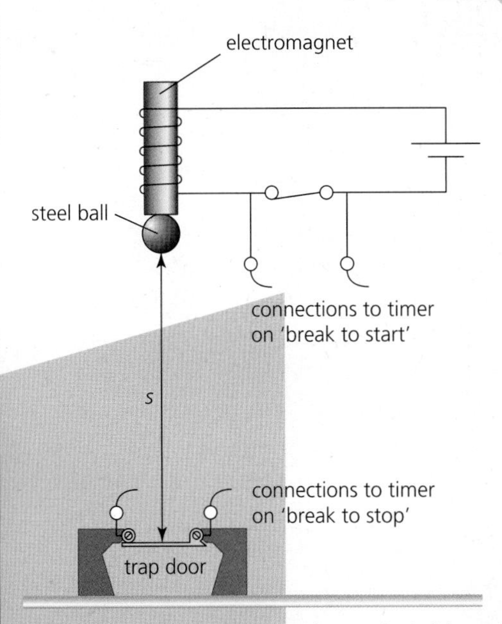


***Figure P1*** *Apparatus for the trap-door method of determining* g

### Apparatus

A steel ball bearing is released from an electromagnet by operating a switch, which also starts an electronic timer (Figure P1). The ball bearing falls through a distance *s*, measured with a metre ruler, until it hits a 'trap door', which opens, breaking a circuit and stopping the timer.

### Technique

The time reading is repeated several times before the distance *s* is changed. Readings are taken for as wide a range of distances as possible.

### Analysis

A sample set of results is shown in Table P1. Using the equation $s = ut + \frac{1}{2}gt^2$, we can plot a graph to find *g*. As initial velocity $= u = 0\ \text{m s}^{-1}$, the equation becomes $s = \frac{1}{2}gt^2$. A graph of *s* (*y*-axis) against $t^2$ (*x*-axis) will give a straight line of gradient $\frac{1}{2}g$.

### QUESTIONS

**P1** Calculate the missing values in Table P1.

**P2** Estimate the uncertainty in the timing readings.

**P3** The timer had a digital readout in milliseconds. Are we justified in using all three decimal places?

**P4** Look at the first column of Table P1. Discuss the use of significant figures in this column. Is it correct?

**P5** Plot a suitable graph, so that you can use the gradient to find *g*.

**P6** Calculate the gradient of the graph, and hence *g*.

**P7** Find the uncertainty in your answer by considering the 'worst-fit' gradient. Express this as a percentage uncertainty.

**P8** Describe sources of error and uncertainty in this experiment and explain how you could improve the accuracy.

| Distance $s$ / m | Time t / s | | | | Mean time squared $t^2$ / $s^2$ |
|---|---|---|---|---|---|
| | Reading 1 | Reading 2 | Reading 3 | Mean | |
| 0.5 | 0.307 | 0.321 | 0.304 | 0.310 | 0.10 |
| 0.6 | 0.342 | 0.345 | 0.357 | 0.348 | 0.12 |
| 0.7 | 0.363 | 0.368 | 0.373 | 0.368 | 0.14 |
| 0.8 | 0.395 | 0.403 | 0.406 | | |
| 0.9 | 0.417 | 0.444 | 0.449 | 0.437 | 0.19 |
| 1.0 | 0.449 | 0.452 | 0.466 | 0.456 | 0.21 |
| 1.1 | 0.457 | 0.455 | 0.464 | 0.459 | 0.21 |
| 1.2 | 0.477 | 0.509 | 0.478 | 0.488 | 0.24 |
| 1.3 | 0.525 | 0.521 | 0.522 | 0.523 | 0.27 |
| 1.4 | 0.543 | 0.537 | 0.544 | | |
| 1.5 | 0.580 | 0.571 | 0.550 | 0.567 | 0.32 |
| 1.6 | 0.594 | 0.583 | 0.597 | 0.591 | 0.35 |
| 1.7 | 0.582 | 0.575 | 0.591 | 0.583 | 0.34 |
| 1.8 | 0.632 | 0.605 | 0.634 | 0.624 | 0.39 |
| 1.9 | 0.636 | 0.644 | 0.625 | 0.635 | 0.40 |
| 2.0 | 0.646 | 0.658 | 0.632 | | |
| 2.1 | 0.673 | 0.641 | 0.672 | 0.662 | 0.44 |

***Table P1*** *Sample results from the trap-door experiment*

## The effect of air resistance on falling objects

***Figure 33*** *The mass of the diver has no effect on her acceleration – heavy or light, she would reach the water in the same time.*

For the diver in Figure 33, air resistance is negligible and she is effectively in free fall. But an object falling on Earth is never *quite* in free fall. The atmosphere always exerts a **drag** force on a moving object. The effect of this drag, due to **air resistance**, is negligible for sprinters, jumpers or even divers, but in some sports, such as cycling or skiing, it is a dominant factor. This is because air resistance increases with speed. The faster you travel, the greater the air resistance. You do not notice air resistance when you walk, but it is more noticeable when cycling at speed. If you travel fast enough, the force due to air resistance will become equal to the force driving you forward and you will stop accelerating. Calculating the size of the drag force on a moving object is a complex matter (Figure 34). It depends on the velocity, area and shape of the moving object, as well as the density and viscosity of the fluid.

***Figure 34*** *Air resistance is a crucial factor in cycling. On a flat road it accounts for up to 90% of the resistance you feel when pedalling. Turbulence leads to low air pressure behind the cyclist. Cycling in the low-pressure zone immediately behind another cyclist can save 30–40% of the required effort. This image was generated using advanced computing techniques at Sheffield Hallam University.*

At low speeds, drag is proportional to velocity.

At higher speeds, drag is proportional to velocity squared.

In the case of a sky-diver (Figure 35), the upward resistive force due to air resistance increases as the diver accelerates, until it is equal to the downward force of gravity. This causes the sky-diver's acceleration to gradually decrease to zero and then the diver falls at a constant speed. We say the sky-diver has reached **terminal speed**. For a sky-diver in a head-down position, the terminal speed is about 83 $m\,s^{-1}$ (186 mph). In the upper atmosphere, where the air is less dense, speeds of up to 280 $m\,s^{-1}$ (about 625 mph) have been reached.

***Figure 35*** *Sky-divers reach their terminal speed when the upward force of air resistance is equal to the downward force of gravity.*

Air resistance also depends on the area and the mass distribution of the object. In general, a more dense object with a smaller surface area will experience lower air resistance, and have a higher terminal speed. Parachutes increase the area, and therefore the air resistance. The terminal speed is therefore reduced.

## QUESTIONS

**31.** High-speed skiing is one of the fastest sports. The record speed is more than 252 km $h^{-1}$ (70 m $s^{-1}$). Why do speed skiers wear streamlined clothes and adopt a crouching position?

### Stretch and challenge

**32.** In *On Being the Right Size*, the biologist J. B. S. Haldane says: "You can drop a mouse down a thousand-yard mine shaft; and, on arriving at the bottom, it gets a slight shock and walks away, provided that the ground is fairly soft. A rat is killed, a man is broken, a horse splashes." Explain the physics behind this.

## ASSIGNMENT 4: INVESTIGATING TERMINAL SPEED

**(PS 1.2, PS 2.1, PS 2.2, PS 2.3, PS 3.1, PS 3.3)**

It can be difficult to measure terminal speed in the laboratory for objects falling through the air, because the speeds reached are quite high unless we use very light objects with a large area, which tend to drift about as they fall. But we can slow things down by using a more viscous medium than air. This experiment measures the terminal speed of a sphere falling through oil (Figure A1).

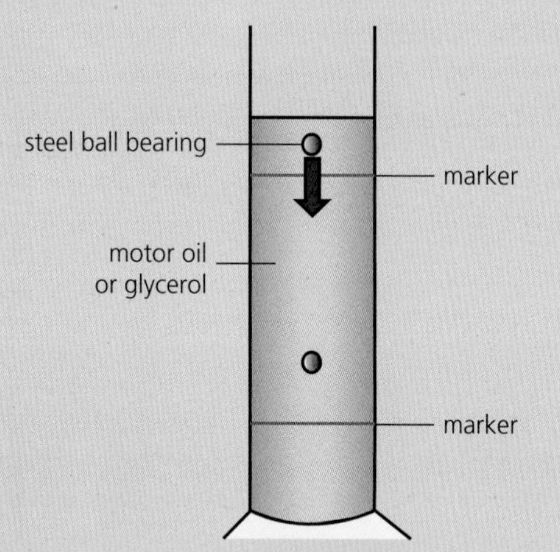


***Figure A1*** *The ball bearing is timed as it passes the two markers. The average speed can then be calculated.*

A very large cylinder is used, filled with motor oil or glycerol. Steel ball bearings are used, because they can be retrieved from the oil without emptying the cylinder or sticking your hand in, by using a strong magnet.

Theory suggests that the terminal speed is proportional to the radius of the sphere:

$$v \propto r$$

It can be difficult to use light gates to start and stop the timer, as the oil is often too opaque, so manual timing is used.

### Questions

**A1** What readings would you need to take to verify the relationship between terminal speed and radius?

**A2** Explain what measuring devices you would use, and estimate the associated uncertainty.

**A3** How would you confirm that the ball bearing had reached terminal speed?

**A4** What experimental errors are likely to arise? Explain which are systematic errors and which are random errors.

## KEY IDEAS

- Air resistance is a drag force that opposes the motion of objects moving through air.
- The size of the air resistance force increases as the object's speed increases.
- The upward force of air resistance acting on an accelerating falling body increases, until it balances the weight (the downward force of gravity). At this point, the object falls at its terminal speed.

## ASSIGNMENT 5: DEVELOPING IDEAS ABOUT FALLING

Until the time of Galileo (1564–1642), the most commonly held idea about the motion of falling objects was mistaken. Almost 2000 years earlier, Aristotle (384 to 322 BCE) had asserted that the speed of fall was proportional to the weight of the object. Although this view had been challenged as soon as 30 years after Aristotle's death, it was still widely accepted when Galileo (Figure A1) questioned it in the 17th century.

***Figure A1*** *Galileo, 'the father of modern science'*

Galileo made the following argument against Aristotle's physics. Suppose there are two objects, one heavy and one light, falling towards Earth. Galileo started by assuming that Aristotle was right, so that the heavy object would be falling faster (Figure A2). Galileo argued that if they were then connected together the lighter one would drag on the faster, heavy one, and slow it down. Similarly, the faster, heavier object would pull on the slower, lighter one, and speed it up. So the combined object would fall at an intermediate speed.

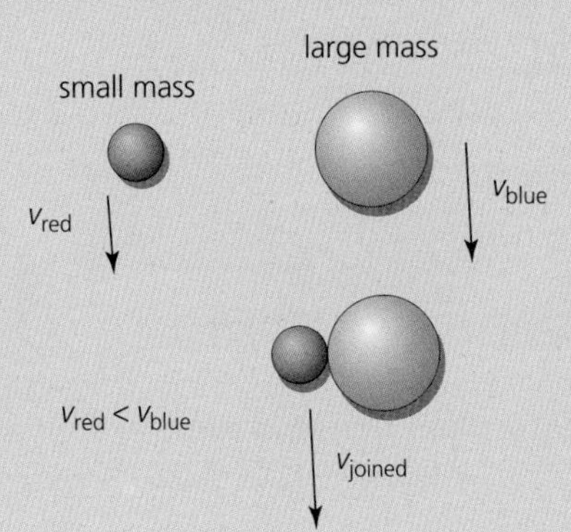


***Figure A2*** *Would $v_{red} < v_{joined} < v_{blue}$? Or would $v_{joined} > v_{blue} > v_{red}$?*

However, Galileo argued that by connecting the two objects together, an even heavier object was formed. Aristotelian physics said that this should be travelling even faster! This is a contradiction: the objects cannot travel faster *and* slower than before they were joined. The initial assumption therefore cannot be true. Galileo correctly inferred that all objects in free fall, no matter what their mass, must fall at the same rate.

Galileo did not just rely on these 'thought experiments'. He saw the power of experimentation and used empirical observations to support his theories. He is often referred to as 'the father of modern science' because of this approach, though he was not the only 'scientist' to be carrying out experimental tests on these ideas. The famous story about Galileo dropping cannon balls off the Leaning Tower of Pisa is probably not true. It seems likely that the experiment was actually done by Simon Stevin in 1586, when he dropped two lead balls from the church tower in Delft, Holland. One ball was 10 times the weight of the other yet the sounds they made on striking the ground were inseparable.

Galileo actually carried out an experiment on 'free fall' by rolling cylinders down inclined planes (slopes). Using this method, Galileo showed that all bodies, irrespective of mass, *accelerate* uniformly when falling under gravity.

Galileo used a water clock to measure time in his experiments. Galileo wrote in *Two New Sciences*:

> *"For the measurement of time, we employed a large vessel of water placed in an elevated position; to the bottom of this vessel was soldered a pipe of small diameter giving a thin jet of water which we collected in a small glass during the time of each descent ... The water thus collected was weighed, after each descent, on a very accurate balance; the differences and ratios of these weights gave us the differences and ratios of the times, and this with such accuracy that although the operation was repeated many, many times, there was no appreciable discrepancy in the results."*

**Questions**

**A1** Why do you think that it took around 2000 years for Aristotle's 'law' of physics to be seriously challenged?

**A2** **a.** Why did Galileo roll cylinders down inclined planes to find a value for $g$, rather than drop weights from the Leaning Tower of Pisa?

**b.** What would affect the precision of Galileo's time measurements?

**c.** Why was it important that the vessel was 'large' and the pipe was of 'small' diameter?

**A3** **a.** Galileo used his experiment to arrive at the equation $s = \frac{1}{2}at^2$, linking the distance, $s$, travelled down the slope to the acceleration, $a$, and the time taken, $t$. Suppose that you were going to repeat the experiment using Galileo's equipment. Explain how you would carry out the experiment and what steps you would take to ensure that your results were as accurate as possible.

**b.** How would you find the acceleration for a particular slope?

**c.** How could you use that value to find $g$, the acceleration of free fall?

**A4** Galileo is often referred to as 'the father of modern science' and he tends to get all the credit for these advances in our knowledge of motion, even though others were also beginning to doubt Aristotle's version of physics. What is it about Galileo's approach that leads us to regard his work as the beginnings of modern science? (In addition to the information here, you may want to use a search engine to look up Galileo and his work. Useful search terms are any of these paired with 'Galileo': law of inertia, empirical approach, Starry Messenger, moons of Jupiter, telescope, Bruno, Pope Urban, Copernican, Simplicus.)

## 9.7 PROJECTILES: MOTION IN TWO DIMENSIONS

In the 1968 Olympic Games in Mexico City, the American Bob Beamon (Figure 36) almost cleared the long-jump pit! His jump of 8.90 m was 70 cm further than anyone had ever jumped at that time.

***Figure 36*** *Beamon's record stood for 23 years until Mike Powell jumped 8.95 m in 1991. No one has jumped further ... yet!*

To understand a long jumper's motion, we need to deal with movement in two dimensions. After take-off, the jumper moves horizontally *and* vertically, and the resultant velocity changes direction throughout the jump. The motion seems complex, but we can simplify it. First, we concentrate on the motion of just one point, the **centre of mass** (see section 10.1 of Chapter 10 for the definition). Once in the air, no matter how the jumper moves his arms or legs, his/her centre of mass will follow a symmetrical curved path, called a parabola (Figure 37).

We can simplify things further by treating the two-dimensional movement as a combination of horizontal and vertical motion. The force of gravity between two objects is not affected by their relative motion, so the acceleration due to gravity is the same whether an object is dropped vertically or thrown horizontally (Figure 38). Vertical motion is totally independent of horizontal motion, and we can treat them separately.

This motion in two dimensions where the vertical motion is subject to gravity is referred to as **projectile motion** (Figure 39).

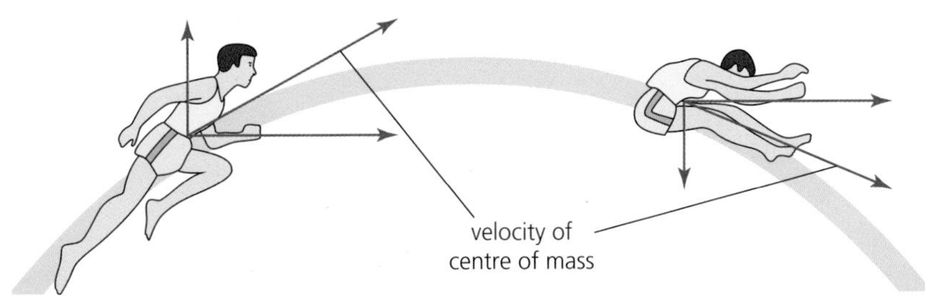


***Figure 37*** *A long-jump trajectory is a combination of horizontal and vertical motion.*

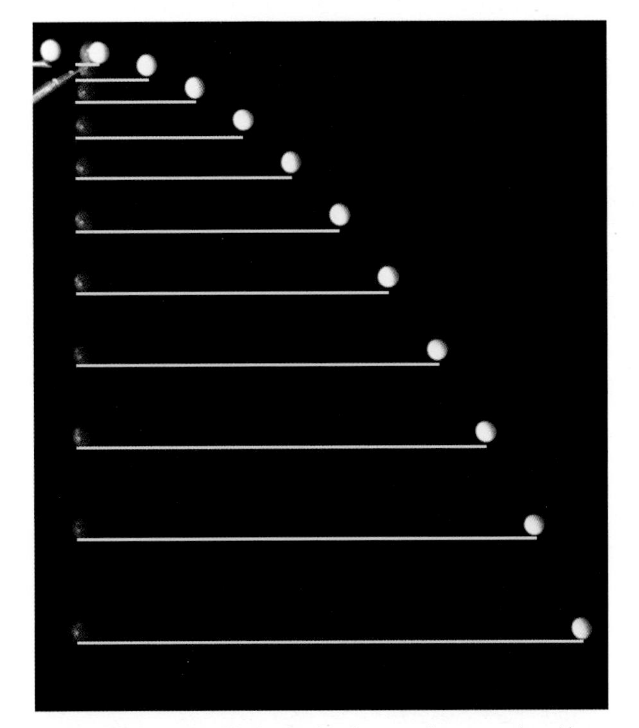

***Figure 38*** *The red ball is dropped at the same instant as the white ball is projected horizontally. Their vertical motions are identical.*

***Figure 39*** *Any object flying through the air is known as a projectile. The centre of mass follows a parabolic path, if we can neglect air resistance (drag).*

The vertical and horizontal components of the motion (see section 9.2) are independent. This means that we can apply the equations of linear motion in each direction, making sure that we keep the two sets of calculations separate. It is a good idea to write down what you know about a problem before you rush into the solution. A table of values such as that in Table 4, or Table 5, may help.

| Variable | Horizontal | Note | Vertical | Note |
|---|---|---|---|---|
| $s$ | The range | | Could be zero for the full trajectory | The greatest height is achieved at half-way through the full trajectory |
| $u$ | Initial velocity | These are equal, if air resistance is neglected | Initial velocity | For a symmetrical full trajectory these are equal in magnitude but opposite in direction, if air resistance is neglected |
| $v$ | Final velocity | | Final velocity | |
| $a$ | Horizontal acceleration | Zero, if air resistance is neglected | $9.81\ \text{m}\,\text{s}^{-2}$ directed downwards | |
| $t$ | Time of flight | This is equal for the horizontal and vertical motion | | |

***Table 4*** *Variables in projectile motion*

| | $s$ / m | $u$ / m s$^{-1}$ | $v$ / m s$^{-1}$ | $a$ / m s$^{-2}$ | $t$ / s |
|---|---|---|---|---|---|
| **Horizontal** | ? | 12 | 12 | 0 | ? |
| **Vertical** | ? | 4.5 | −4.5 | −9.81 | ? |

***Table 5*** *There are three unknown variables (t is the same for horizontal and vertical)*

## Worked example

A record-breaking long jumper has to combine top-class sprinting with a high jumper's spring into the air. A good sprinter achieves a top speed of about 12 m s$^{-1}$; a high jumper leaves the ground with a vertical velocity of about 4.5 m s$^{-1}$. If a long jumper could combine these performances, how far would he be able to jump?

Although the equations of motion only apply to linear movement, we can use them to describe a long jump if we treat the vertical and horizontal velocities separately.

The first steps are to identify what we know about the problem (Table 5) and to specify which is the positive vertical direction. We will choose upwards to be positive, so $a = -9.81$ m s$^{-2}$. Air resistance will be neglected.

The *vertical* take-off velocity will determine the time of flight. We can use $v = u + at$ to find the time spent in the air:

$$t = \frac{v - u}{a} = \frac{(-4.5\,\text{m s}^{-1}) - 4.5\,\text{m s}^{-1}}{-9.81\,\text{m s}^{-2}}$$

$$= \frac{-9.0\,\text{m s}^{-1}}{-9.81\,\text{m s}^{-2}}$$

$$= 0.92\,\text{s}$$

The *horizontal* velocity can be used to find the length of the jump. We can use

$$s = ut + \frac{1}{2}at^2$$

But as $a = 0$, this is just $s = ut$. So length of jump is

$$s = 12\,\text{m s}^{-1} \times 0.92\,\text{s} = 11\,\text{m}$$

The world record for the long jump is about 2 m less than the theoretical result in the Worked example.

There are a number of ways in which our simple model does not match the real event:

- A long jumper has to drop his horizontal speed so that he/she is in contact with the board long enough to gain some vertical speed.
- Air resistance can impose a horizontal deceleration of up to 0.2 m s$^{-2}$ (it has little effect on the vertical motion).
- We have treated the jumper as if he/she was a single point and the distance we have calculated is that travelled by the centre of mass. The actual jump may be longer than this because a good long jumper will have his centre of mass over the board at take-off and behind his/her heels on landing, thereby increasing the time of flight.

This analysis of an athletic event is typical of the way that physics works. We try to understand a complex system by identifying its essential features and creating a theoretical model. Experiments and measurements are used to check how closely our model reflects the real world. The model can then be refined and adapted so that it matches experimental results more closely.

### The effect of air resistance on projectile motion

An important refinement to the simple model of projectile motion is the effect of air resistance (drag). Air resistance always acts against the direction of motion. Consider the effect on the horizontal and vertical components of a projectile's motion:

- The horizontal component of a projectile's velocity cannot be treated as constant. Air resistance tends to reduce it. This reduces the range.

- Air resistance acts downwards as the projectile is on the way up, but upwards as the projectile travels downwards. The overall effect is to reduce the maximum height reached.

When air resistance is taken into account, the path of a projectile is no longer a parabola (Figure 40).

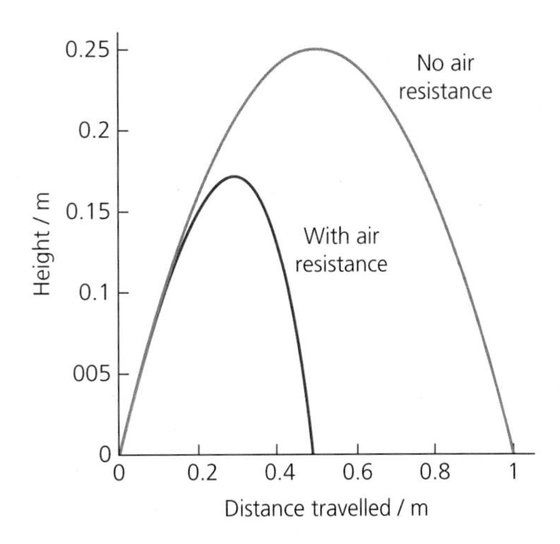


***Figure 40*** *The change in trajectory of a projectile when air resistance is taken into account. Note that the red curve is asymmetric – the projectile falls more steeply than it climbs.*

### QUESTIONS

**33.** A golf ball's flight can be regarded as a projectile.

**a.** Sketch its path after being hit by the club, assuming that air resistance can be ignored.

**b.** Draw arrows to show how air resistance affects the ball's horizontal and vertical motion on the way up, and on the way down.

**c.** Sketch in the trajectory when air resistance is taken into account.

(The spin on a golf ball complicates things, causing the air to provide 'lift' that can increase the range.)

**34.** The best high jumpers achieve a vertical take-off velocity of about $4.8 m s^{-1}$. A typical high jumper's 'centre of mass' is 0.95 m above the ground. Use the equations of motion to estimate the height that this jumper can reach.

**35.** The divers of La Quebrada leap from the cliffs near Acapulco into the sea, 26.7 m below. They need a horizontal range of at least 8.22 m to clear the rocks at the foot of the cliff. Assume that they jump horizontally from the cliff edge.

**a.** Estimate the horizontal velocity needed to clear the rocks.

**b.** At what velocity do they hit the sea?

## Getting the best angle

'Throwing the hammer' (Figure 41) requires great strength. The 'hammer' is a small tungsten ball fastened to the end of a 1.22 m chain. It has a mass of 7.26 kg. The thrower whirls the hammer round three times before releasing it. The velocity of release depends on how quickly the thrower can spin round in the 2 m circle, but the length of the throw also depends on the angle of release. A good thrower will release the hammer at a velocity of about $26 m s^{-1}$. We can resolve the velocity into vertical and horizontal components (see Figure 17, section 9.2) and use the equations of motion to find the best angle.

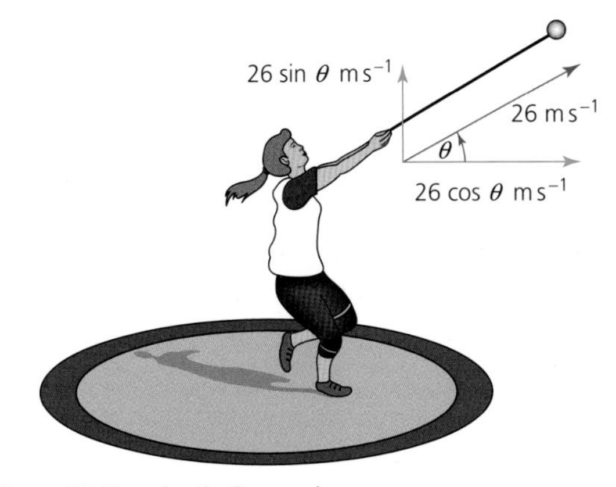


***Figure 41*** *Throwing the 'hammer'*

The distance thrown is determined by the horizontal speed and the time of flight. The time of flight depends on the vertical component of the velocity. The problem can be simplified by considering just the first half of the throw, from the moment of release to when the hammer reaches its highest point.

For the vertical motion, $s$ is unknown, $u = 26\sin\theta\,\text{m s}^{-1}$, $v = 0\,\text{m s}^{-1}$, $a = -9.81\,\text{m s}^{-2}$ and $t$ is unknown. Using $v = u + at$:

$$t = \frac{v-u}{a}$$

$$= \frac{0 - 26\sin\theta\,\text{m s}^{-1}}{-9.81\,\text{m s}^{-2}}$$

$$= 2.65\sin\theta \text{ seconds}$$

The time for the whole flight is:

$$\text{time} = 2 \times 2.65\sin\theta = 5.30\sin\theta$$

The distance thrown is

$$\text{distance} = \text{horizontal velocity} \times \text{time}$$

$$= 26\cos\theta\,\text{m s}^{-1} \times 5.30\sin\theta\,\text{s}$$

$$= 138\sin\theta\cos\theta \text{ metres}$$

This can be shown (see below) to have a maximum value when $\theta = 45°$.

### Stretch and challenge

Using the trigonometric relation $\sin\theta\cos\theta = \frac{1}{2}\sin 2\theta$, we get

$$\text{distance} = 69\sin 2\theta \text{ metres}$$

The greatest range will be achieved when $\sin 2\theta$ has its greatest value. The largest value of $\sin 2\theta$ is 1, so the hammer thrower with a release speed of 26 $\text{m s}^{-1}$ has a maximum range of 69 m.

When $\sin 2\theta = 1$, $2\theta = 90°$ and $\theta = 45°$. Theoretically, the best angle at which to release the hammer is 45°. (In fact, because the hammer is released from above ground level, the best angle of release is slightly less than 45°.)

## QUESTIONS

**36.** **a.** What are the horizontal and vertical components of velocity for a projectile fired at $100\,\text{m s}^{-1}$ at an angle of 45° to the horizontal?

Stretch and challenge

**b.** Theoretically, 45° is the angle that gives the greatest range, for a given projection speed. Can you suggest why this might be?

**c.** Would 45° always be the optimum angle? Suppose the projectile was fired from a cannon that was high on a cliff pointing out to sea? Would this make a difference?

## KEY IDEAS

- In two-dimensional motion under gravity (projectile motion), the horizontal motion and the vertical motion are independent.
- Problems involving projectiles can be solved by resolving the initial velocity into vertical and horizontal components and then applying the equations of linear motion to each component separately.

## ASSIGNMENT 6: ANALYSING PROJECTILE MOTION

**(MS 3.1, MS 3.2, MS 3.12)**

Projectile motion can be studied by examining the path taken by projected water droplets. A stream of water drops is created using the apparatus shown in Figure A1. A constant flow of water is essential. Since the pressure of the domestic water supply can fluctuate, using the water from the tap is not possible. The constant-head device provides a steady pressure. The water stream is broken up into droplets using a ticker-timer device. The result is a stream of projectile water drops.

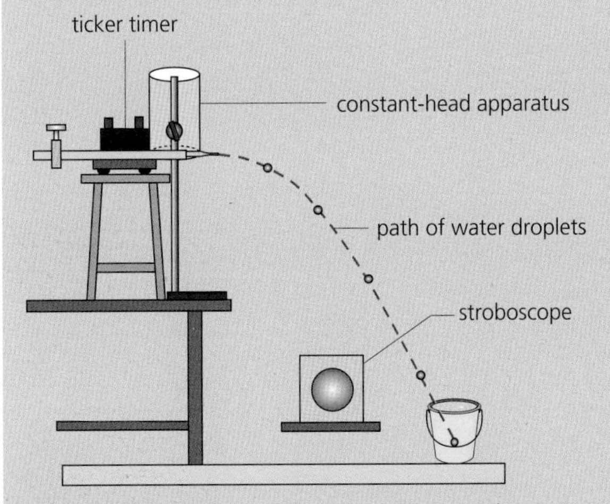


**Figure A1** *Apparatus for producing projectile water drops*

The water stream is illuminated with a flashing stroboscope (that can cause problems for some people, especially anyone who suffers from epilepsy). The stroboscope is set to flash at 50 Hz, to match the rate at that the ticker timer vibrates. This has the effect of 'freezing' the motion of the drops (Figure A2). A transparent screen is placed behind the stream of droplets so that the position of each drop can be marked on the screen. The $x$ and $y$ positions of each drop are recorded (Figure A3).

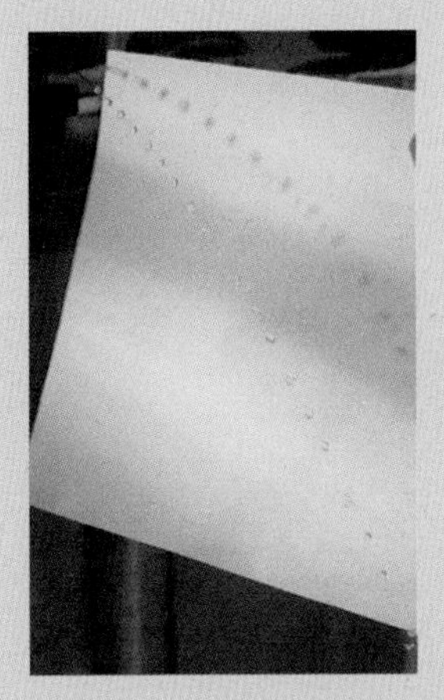

**Figure A2** *The 'frozen' water droplet trajectory*

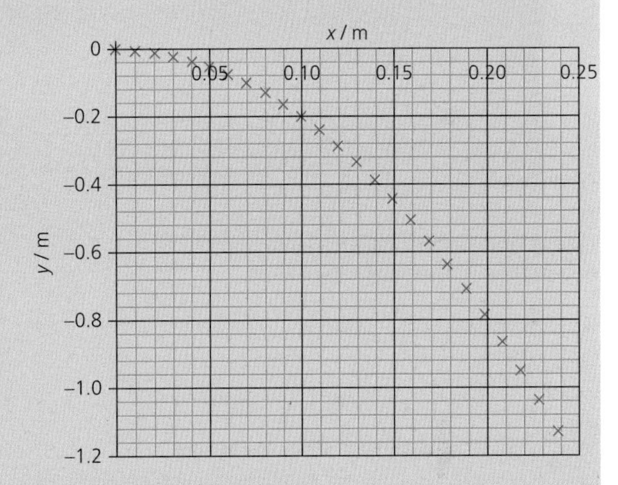


**Figure A3** *A plot of the trajectory*

### Questions

**A1** Use Figure A3 to plot a graph of horizontal displacement against time.

**A2** Is the horizontal velocity constant? Is air resistance a significant factor here?

**A3** If the path is parabolic then the $y$ displacement should be proportional to the $x$ displacement squared, $y \propto x^2$. Plot a suitable graph to test whether the water droplets follow a parabolic path.

## PRACTICE QUESTIONS

1. Figure Q1 shows two different rifles being fired horizontally from a height of 1.5 m above ground level. Assume the air resistance experienced by the bullets is negligible.

*Figure Q1*

a. When rifle A is fired, the bullet has a horizontal velocity of 430 m $s^{-1}$ as it leaves the rifle. Assume the ground is level.

i. Calculate the time that the bullet is in the air before it hits the ground.

ii. Calculate the horizontal distance travelled by the bullet before it hits the ground.

b. Rifle B is fired and the bullet emerges with a smaller horizontal velocity than the bullet from rifle A. Explain why the horizontal distance travelled by bullet B will be less than bullet A.

*AQA June 2013 Unit 2 Q2*

2. A boy throws a ball vertically upwards and lets it fall to the ground. Figure Q2 shows how displacement relative to the ground varies with time for the ball.

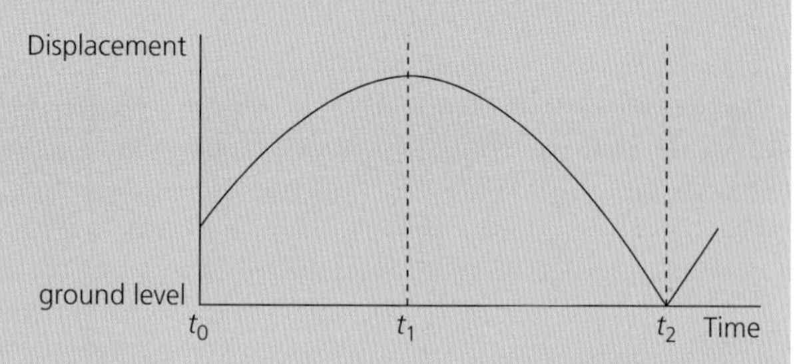


*Figure Q2*

a. State which feature of a displacement–time graph represents the velocity.

b. On a set of axes like those in Figure Q3, draw the shape of the velocity–time graph for the ball from $t_0$ to $t_2$. The starting point is labelled X.

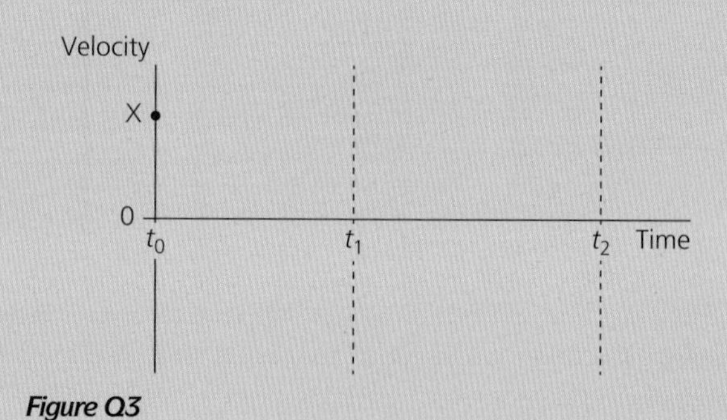


*Figure Q3*

*AQA June 2011 Unit 2 Q1 (part)*

3. A student measures the acceleration due to gravity, $g$, using the apparatus shown in Figure Q4. A plastic card of known length is released from rest at a height of 0.50 m above a light gate. A computer calculates the velocity of the card at this point, using the time for the card to pass through the light gate.

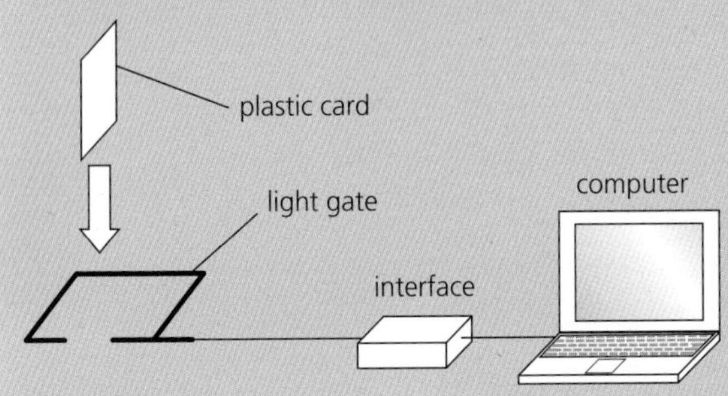


*Figure Q4*

a. The computer calculated a value of 3.10 m s$^{-1}$ for the velocity of the card as it travelled through the light gate. Calculate a value for the acceleration due to gravity, $g$, from these data.

b. The student doubles the mass of the card and finds a value for $g$ that is similar to the original value. Use the relationship between weight, mass and $g$ to explain this result.

c. State and explain **one** reason why the card would give more reliable results than a table tennis ball for this experiment.

*AQA June 2011 Unit 2 Q2*

4. A digital camera was used to obtain a sequence of images of a tennis ball being struck by a tennis racket. The camera was set to take an image every 5.0 ms. The successive positions of the racket and ball are shown in Figure Q5.

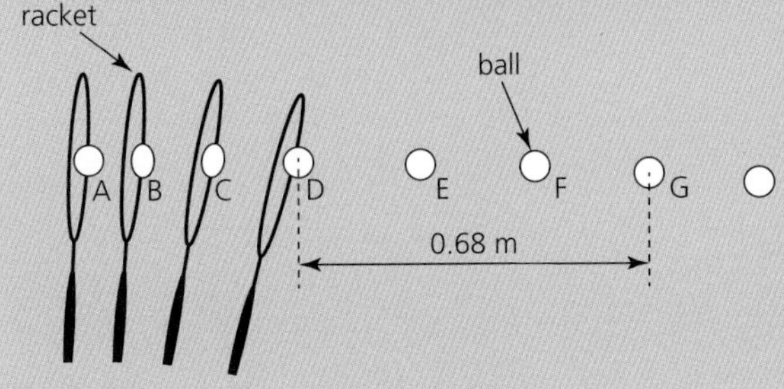


*Figure Q5*

a. The ball has a horizontal velocity of zero at A and reaches a constant horizontal velocity at D as it leaves the racket. The ball travels a horizontal distance of 0.68 m from D to G.

i. Show that the horizontal velocity of the ball between positions D and G in Figure Q5 is about 45 m s$^{-1}$.

ii. Calculate the horizontal acceleration of the ball from A to D.

b. At D, the ball was projected horizontally from a height of 2.3 m above level ground.

i. Show that the ball would fall to the ground in about 0.7 s.

ii. Calculate the horizontal distance that the ball will travel after it leaves the racket before hitting the ground. Assume that only gravity acts on the ball as it falls.

iii. Explain why, in practice, the ball will not travel this far before hitting the ground.

*AQA Jan 2010 Unit 2 Q1*

5. Figure Q6 shows the velocity–time graph for a short bicycle journey.

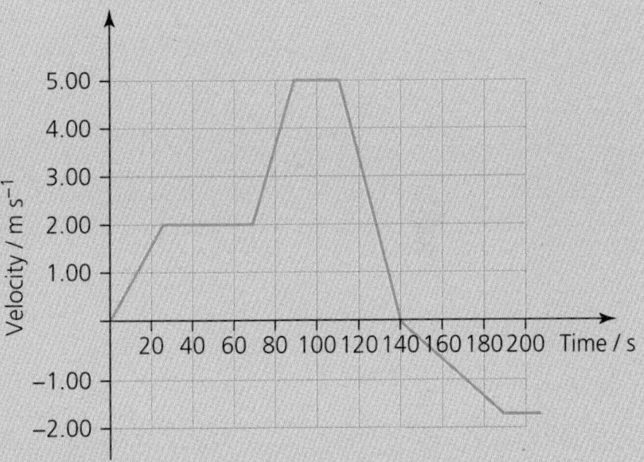


*Figure Q6*

a. Between which times does the bicycle have its greatest acceleration?

b. When does the bicycle come briefly to a stop?

c. What does the negative portion of the graph, after time = 140 s, indicate about the motion of the bicycle?

d. What is the total displacement of the bicycle at 140 s?

6. a. An aircraft is flying at a speed of 300 km $h^{-1}$ due north. A strong wind begins to blow from the west at a speed of 50 km $h^{-1}$. What is the magnitude of the resultant velocity of the aircraft?

   b. If the pilot of the aircraft in part **a** wishes to fly due north, what bearing (angle measured clockwise from north) should he steer on?

7. Which row in Table Q1 is fully correct?

| | Scalar quantities | Vector quantities |
|---|---|---|
| A | Velocity, displacement | Distance, speed |
| B | Acceleration, displacement | Speed, distance |
| C | Speed, distance | Acceleration, displacement |
| D | Velocity, distance | Speed, acceleration |

*Table Q1*

8. Which of the following statements about the Earth's orbit of the Sun is **incorrect**?

   **A** The average velocity of the Earth in one year is zero.

   **B** The displacement of the Earth in one year is zero.

   **C** The Earth is accelerating throughout the year.

   **D** The average speed of the Earth in one year is zero.

9. An aircraft drops two identical crates, X and Y, containing emergency supplies. Crate X has twice the mass of Y. Which of the following statements is true?

   **A** Crate X will have a higher terminal velocity than crate Y.

   **B** Crate Y will have a higher terminal velocity than crate X.

   **C** They will have the same terminal velocity, but crate X will reach it first.

   **D** They will both reach the same terminal velocity at the same time.

10. An archer fires an arrow at a speed of 40 m $s^{-1}$ an angle of 30° to the horizontal. What is the maximum height reached by the arrow?

   **A** 6 m

   **B** 60 m

   **C** 20 m

   **D** 82 m

## Stretch and challenge

11. This question is about analysing motion in sport. Digital video and dedicated computer software are now routinely used to capture data on motion during sporting events, so that coaches can identify ways to improve performance. Table Q2 shows a set of data for five javelin throws, recorded by a student using a motion-analysis system.

| | Throw 1 | Throw 2 | Throw 3 | Throw 4 | Throw 5 |
|---|---|---|---|---|---|
| Launch angle / degree | 46.55 | 40.65 | 45.58 | 43.41 | 43.63 |
| Launch height / m | 1.80 | 2.38 | 1.96 | 2.02 | 1.68 |
| Initial horizontal velocity / m $s^{-1}$ | 14.77 | 21.34 | 13.79 | 16.06 | 15.11 |
| Initial vertical velocity / m $s^{-1}$ | 15.59 | 18.32 | 14.07 | 15.19 | 14.4 |
| Time of flight / s | 2.961 | 3.474 | 2.7 | 2.898 | 2.745 |
| Range / m | 43.731 | 74.133 | 37.233 | 46.539 | 47 |
| Maximum height / m | 12.771 | 17.541 | 10.845 | 12.402 | 11.025 |
| Time to attain maximum height / s | 1.431 | 1.683 | 1.287 | 1.395 | 1.323 |
| Final horizontal velocity / m $s^{-1}$ | 13.293 | 19.206 | 12.411 | 14.454 | 15.022 |
| Final vertical velocity / m $s^{-1}$ | −15.012 | −17.595 | −13.824 | −14.76 | −13.968 |

*Table Q2*

**a.** Which throw achieves the highest initial velocity, and what is its value?

**b.** Consider throw 1 and calculate, using the initial velocity values,

**i.** the maximum height reached

**ii.** the time of flight

**iii.** the range.

**c.** Compare your calculations with the actual values. What effect does air resistance have on the javelin's flight?

**d.** Does air resistance affect all the throws in the same way? Look at throw 5. What do you think happened?

**e.** Comment on the presentation of the data. Are there any mistakes in the table?

**f.** Why are the final vertical velocity results negative?

**g.** Is there a correlation between throwing a javelin a long way and being tall? Use a graph to support your answer. What further research would you do?

# 10 FORCES IN BALANCE

## PRIOR KNOWLEDGE

You will understand what is meant by a force, and be aware that a body can be in equilibrium under the action of several forces. You have already met vector addition and have seen in Chapter 9 how vectors can be resolved into components.

## LEARNING OBJECTIVES

In this chapter you will learn how to use these techniques to analyse the forces that act on an object, including those that cause rotation, and determine whether or not the object is in equilibrium.

**(Specification 3.4.1.1 part, 3.4.1.2)**

In 1829 the Merchant Venturers of Bristol offered a prize of 100 guineas for the best design of a new bridge to span the Avon Gorge, 410 m wide and 75 m above the River Avon. All 22 entries were rejected. One of the judges, the renowned engineer Thomas Telford, believed it was impossible for a suspension bridge to span the river without a central support. Two years later, a second competition was won by an audacious and elegant design from the 24-year-old Isambard Kingdom Brunel, who ignored Telford and submitted plans for a suspension bridge (Figure 1).

The road deck is hung, via metal rods, on chains that pass over the towers at each end of the bridge. The chains then plunge 20 m into the rock on each side of the bridge. The road deck, chains and rods weigh a daunting 1500 tons. On completion the bridge was tested by spreading 500 tons of rocks on the road. This caused the chains to sag by only 180 mm, well within tolerance. The design was, and still is, an undisputed success. Originally intended to carry pedestrians and horse-drawn traffic, it now carries four million vehicles per year.

***Figure 1*** *The Clifton Suspension Bridge bears the inscription 'SUSPENSA VIX VIA FIT' meaning 'A suspended way made with difficulty'.*

In the 150 years since Brunel's bridge was completed, suspension bridges have been built all over the world, many with much longer spans. The world's largest suspension bridge, the Akashi Kaikyō, was completed in 1998, and has a central span of almost 2 km. Since then, the largest bridges in the world have largely used a different design, the cable-stayed bridge (Figure 2).

***Figure 2 (background)*** *The Millau viaduct in the south of France is a cable-stayed bridge. The road deck is hung from cables directly connected to the towers.*

# 10.1 STATICS: IDENTIFYING THE FORCES

The study of forces and motion is referred to as **mechanics**. The branch of mechanics that deals with forces in equilibrium is known as **statics**, since it is often concerned with stationary structures, such as bridges and buildings. You are probably familiar with the idea of a force as a 'push' or 'pull' or 'twist' that can cause an object to accelerate – that is, to speed up, slow down, change direction or rotate. However, in statics, there is often no resultant force and the object does not accelerate – the effect of the forces may be to compress or stretch the object.

The forces involved in a cable-stayed bridge (see Figure 2) are shown in Figure 3. The weight of the deck pulling down on the cables is balanced by the upwards force from the supporting tower. The cable is in **tension** – the forces on it are tending to stretch it. The tower, on the other hand, is under **compression**. The net effect of the cables pulling down and the supporting force of the ground pushing up would be to shorten the tower.

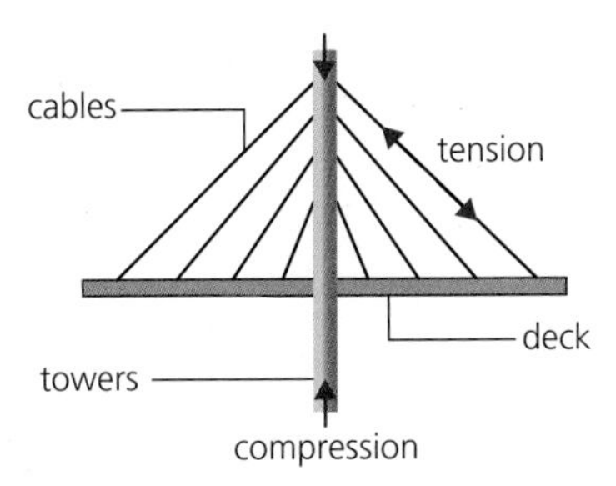


***Figure 3*** *Tension and compression in a cable-stayed bridge*

**QUESTIONS**

1. Figure 4 shows a suspension bridge. State and explain whether the cables and the towers are under compression or in tension.

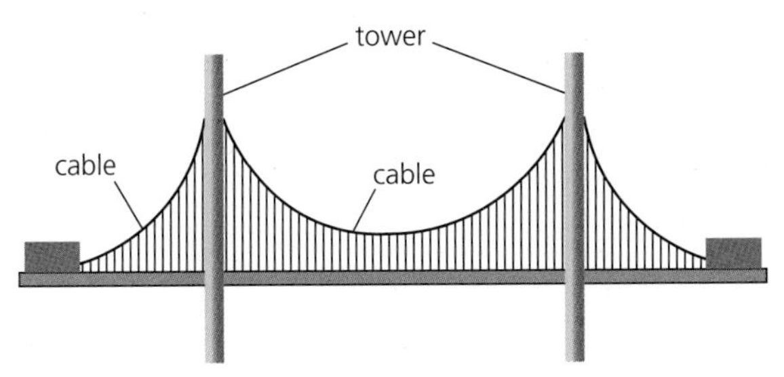


***Figure 4*** *A suspension bridge*

The first step in analysing any statics problem is to identify the relevant forces acting on the object. Forces are vector quantities, with both a magnitude measured in **newtons** (N) and a direction. They can be represented on a diagram by an arrow of suitable length, pointing in the appropriate direction.

Most objects are subject to a range of different forces, such as:

- weight
- contact forces
- surface friction
- tension
- forces due to objects being immersed in or moving through fluids (gases and liquids).

All of these, except weight, are electromagnetic in origin, that is, they are due to the interaction between charged particles. The first four of these forces will be discussed here, while section 10.2 will deal with forces due to fluids (upthrust, lift and drag).

## Weight

Weight, $W$, is the force that acts on a mass due to gravitational attraction, usually that of the Earth. An object's weight depends on its mass, $m$, and the **gravitational field strength**, $g$. This is the gravitational force that acts on each kilogram of mass. The gravitational field strength is measured in newton per kilogram ($N\,kg^{-1}$). The value of $g$ at the Earth's surface is approximately $9.81\,N\,kg^{-1}$. On Earth, a mass of 1 kg has a weight of 9.81 N. The relationship between weight, gravitational field strength and mass is given by

$$\text{weight (N)} = \text{mass (kg)} \times \text{gravitational field strength (N kg}^{-1}\text{)}$$

$$W = mg$$

**QUESTIONS**

2. **a.** An astronaut has a mass of 80.0 kg. What is his weight on Earth?

   **b.** The same astronaut would weigh 128 N on the Moon. Calculate the gravitational field strength on the Moon.

The gravitational attraction of the Earth acts on every particle of mass in an object. Adding up all these forces gives the total weight of the object. This **resultant force** can be thought of as acting at a single point. This point is called the **centre of gravity** of the object (Figure 5). Because weight always acts vertically down, towards the centre of the Earth, we can represent this force by a single vertical arrow from the centre of gravity.

*Figure 5* *The centre of gravity of a regular solid of uniform density, like the ball, is at its geometric centre. For an irregular-shaped object, like an elephant, the centre of gravity is harder to find. In both cases, we represent the entire weight of the object by a single downward arrow acting from the centre of gravity.*

The **centre of mass** of an object is not exactly the same thing as its centre of gravity, though near the Earth's surface, where the gravitational field strength is almost uniform, they can be regarded as the same point. The position of the centre of mass determines what happens to an object when a force is applied to it. If the resultant force acting on an object passes through its centre of mass, it will accelerate without rotating. Imagine trying to push a car parked on extremely slippery ice. If you push through the centre of mass, the car will move forward without spinning. If the line of action of your force does not pass through the centre of mass, the car will rotate, as well as move forward (Figure 6). This turning effect of a force will be considered further in section 10.4.

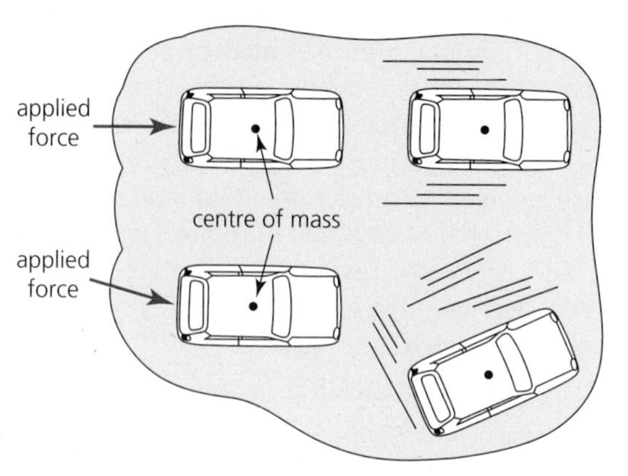


*Figure 6* *The effect of the force depends on where you push.*

**KEY IDEAS**

- The weight of a body (N) = mass (kg) × gravitational field strength (N kg$^{-1}$).
- The centre of gravity is the point at which all the weight can be assumed to act.
- The centre of mass is the point through which an applied force causes no rotation.
- The centre of mass of a regular solid of uniform density is at its geometric centre.

## Contact force

When two solid surfaces are touching, they exert a **contact force** on each other. This force, sometimes called the 'reaction', is electromagnetic in nature and is due to the charges in one surface repelling the charges in the other. It is the contact force between your feet and the ground that prevents the gravitational attraction of the Earth pulling you through the floor.

If the two surfaces are pushed closer together, the contact force increases. As with compressed springs, the atoms exert a greater force as they are pushed together. The resultant force between the surfaces is the sum of all the interatomic repulsions. The component of this force at right angles to the surfaces is called the normal contact force or the perpendicular reaction (see Figure 7).

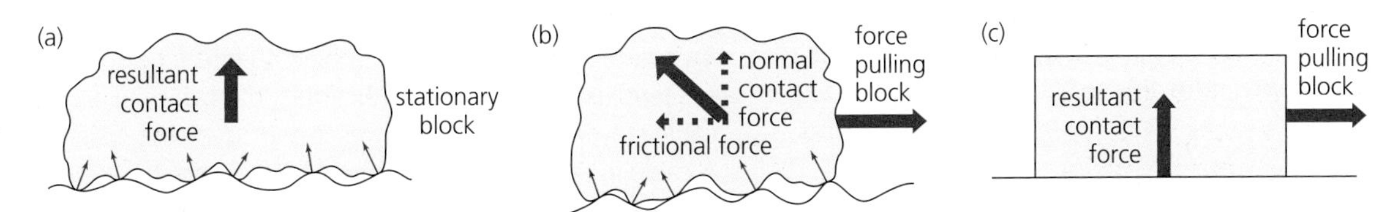


(a) For stationary surfaces with no external forces acting, the resultant contact force is at right angles to the surfaces. At the points where real surfaces touch, the contact force can be at any angle. (b) When surfaces are in relative motion, the frictional force causes the resultant contact force to be non-perpendicular. (c) A perfectly smooth surface is unable to provide any friction at all.

***Figure 7*** *Resultant contact force*

If the surfaces are in relative motion, or if an external horizontal force is acting, the resultant contact force will not be at right angles to the surface (Figure 7b). We can treat the resultant as the combination of two forces. The component at right angles to the surface is the normal contact force. The component parallel to the surface is the **frictional force**, considered below. For perfectly smooth surfaces there would be no friction, and the only force would be the normal contact force (Figure 7c).

### QUESTIONS

3. Make a sketch of the beam bridge with a girl on it, shown in Figure 8. On your sketch, mark in the forces that act *on the bridge*.

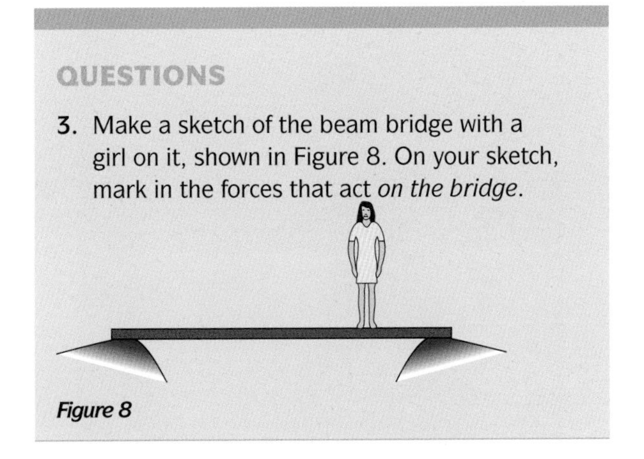

***Figure 8***

## Surface friction

A frictional force acts between two solid surfaces in contact when they are in relative motion, or if a force is trying to slide them across each other. The frictional force is a resistive force that acts to oppose the sliding.

The actual area of contact between two surfaces can be as little as 0.01% of the full surface area (see Figure 9). Since the pressure at a point is the force per unit area, the pressure at these points is extremely high. The high pressure tends to join the two surfaces, actually 'welding' the surfaces together at these small points of contact.

***Figure 9*** *A scanning electron micrograph of crystalline tungsten. Even the smoothest of surfaces look like a mountain landscape at very high magnifications.*

Surfaces will not slide across each other unless an applied force breaks these 'welds'. At the point when the surfaces start to slide, the applied force is just sufficient to overcome the **limiting static friction**. When the surfaces are moving, a frictional force acts against the relative motion. The size of this **dynamic friction** is usually less than the limiting static friction (Table 1). The relative speed of the two surfaces does not have very much effect on the magnitude of the dynamic friction.

The area of the surfaces has no influence on the friction. The frictional force between two surfaces just depends on how hard they are pressed together – that is, the normal contact force between them. The force due to friction, $F$, is proportional to the normal contact force, $N$:

$$F \propto N \quad \text{or} \quad F = \mu N$$

The constant of proportionality, $\mu$, is known as the **coefficient of friction**. Its size depends on the nature of the surfaces (Table 1). Rougher surfaces, such as sandpaper, exert a large frictional force, so the coefficient of friction between sandpaper and wood is high. Although frictional forces tend to prevent surfaces sliding over each other, they do not act to prevent all motion. Imagine trying to cycle on an icy road where friction is low. If your bicycle tyre had no grip on the road at all, you would not be able to speed up, turn or brake.

| Materials | Coefficient of limiting static friction | Coefficient of dynamic friction |
|---|---|---|
| Rubber on concrete (dry) | 1.0 | 0.8 |
| Rubber on concrete (wet) | 0.3 | 0.25 |
| Glass on glass | 0.94 | 0.4 |
| Steel on steel | 0.74 | 0.57 |
| Wood on wood | 0.25–0.5 | 0.2 |
| Ice on ice | 0.1 | 0.03 |
| Human joints | 0.01 | 0.003 |

***Table 1*** *Some values of the coefficients of static and dynamic friction between different surfaces*

## QUESTIONS

**4.** Prof. G: "Without friction, movement would be so much easier, so I have developed school corridors and shoes that have a zero coefficient of friction."

Prof. H: "Nonsense! When you walk, it is friction that pushes you forward."

Who is right? Explain.

**5.** "On icy roads a car is more likely to skid if it is fully loaded." Do you agree with this statement? Explain your answer. Think about driving on a straight, level road and then cornering.

**6.** Estimate the pressure at the points of contact for a 2.0 cm steel cube resting on a steel surface. (Density of steel ≈ 8000 kg m$^{-3}$)

**7.** The coefficient of static friction between snow and modern skis is about 0.02.

**a.** What limiting frictional force acts on a skier of mass 50 kg on horizontal ground?

**b.** Why is the frictional force less than this when the skier is on a steep slope (Figure 10)?

## Tension

An object that is in tension is subject to forces that are tending to stretch it. For example, the cables in a cable-stayed bridge (see Figures 2 and 3) are in tension. A cable in tension exerts an equal force on the object it is attached to, but in the opposite direction (Figure 11).

***Figure 10*** *The frictional force is proportional to the normal contact force.*

***Figure 11*** *The cables are in tension. The red arrows show the forces acting on the cable. The cable exerts an equal but opposite force back on the kite-boarder. The black arrows show the force exerted by the cable.*

When we are analysing the forces in a structure, we need to be clear about which object the forces are acting on.

We can assume that the tension is the same throughout a cable, as long as the weight of the cable is small compared to the tension in it. This is not always the case (Figure 12).

*Figure 12 Nick Wallenda's record-breaking tightrope walk in November 2014 between skyscrapers in Chicago. The tightrope sagged so that the second half of the walk was uphill, with a gradient of almost 20%.*

A solid object, such as a metal rod, can act in tension, pulling on the object it is connected to. A rod may also act in compression, pushing outwards on the objects that are squashing it. A cable or rope cannot act in compression. A fluid normally acts in compression.

The force exerted by a rod is caused by the forces between its molecules. Like the force in a spring, the magnitude of these intermolecular forces varies, depending on whether the rod is being stretched or compressed. When the rod is being stretched, the molecules are slightly further apart and an attractive force between the molecules tries to restore their original separation. When the molecules are pushed closer together, because the rod is being squashed, a repulsive force acts in a direction that would restore the equilibrium separation.

**QUESTIONS**

8. Sketch the radio mast and cables in Figure 13, and mark in the forces acting on the mast.

*Figure 13*

**KEY IDEAS**

- When solid objects are in contact, they exert a force on each other. This is known as the contact force.
- The resultant force between two surfaces is a combination of the contact force, normal to the surfaces, and the friction.
- Friction acts to prevent solid surfaces sliding across each other. The maximum, or limiting, static friction depends on the nature of the surfaces and the magnitude of the contact force.
- Tension is the force that acts in solids to oppose stretching.

## 10.2 UPTHRUST, LIFT AND DRAG

An object that is in a fluid (as we all are!) is subject to fluid forces, although for a dense solid moving at low speed in air these can often be neglected.

### Upthrust

**Upthrust**, or 'buoyancy', is the force that makes boats and hot air balloons float in their respective fluids. The force arises due to the difference in pressure at different heights (or depths) in a fluid.

Imagine an object, a cube say, immersed at a depth $d$ in the fluid, water in this case (see Figure 14).

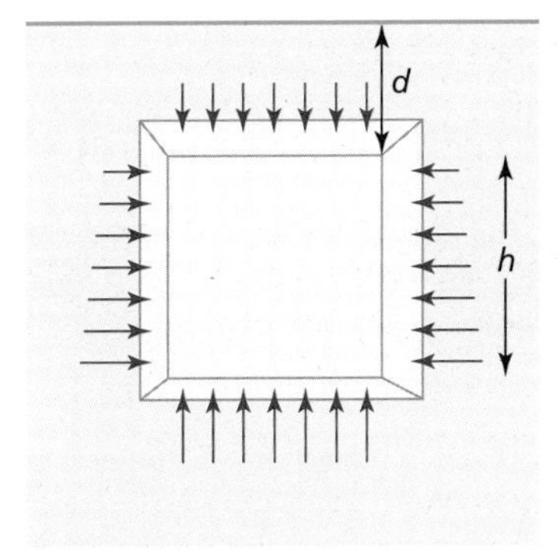


*Figure 14 The pressure in a fluid, such as water, increases with depth.*

At any point in a fluid, the pressure acts isotropically, that is, the pressure is the same in all directions. Pressure is defined as force per normal unit area, $P = F/A$, so the forces on opposite vertical sides of the cube are equal but opposite. But the pressure on the bottom face of the cube is greater than that at the top, and so the resultant force is upwards. This is the upthrust. Before the cube was submerged, that same upthrust was acting on an identical volume of fluid, water in this case. As that 'cube' of water was neither rising nor falling, the upthrust force must be equal to the weight of the cube of water. This is **Archimedes' principle**:

The upthrust on an object in a fluid is equal to the weight of fluid displaced by the object.

If the object weighs more than the weight of fluid that it displaces, then it will sink. If it weighs less than the fluid displaced, it will rise in the fluid until it is only partly submerged and the upthrust exactly matches the weight, when it will float.

QUESTIONS

9. Steel has a density around eight times that of water. Explain, referring to Archimedes' principle, how steel boats are able to float.

## Hydrostatic pressure

### Stretch and challenge

We can go further with the analysis of a submerged object (our cube) to get an expression for the pressure in a liquid. Since

upthrust = difference in pressure × area of cube face

$$\text{upthrust} = (P_{\text{bottom of cube}} - P_{\text{top of cube}}) \times A$$

$$(P_{\text{bottom of cube}} - P_{\text{top of cube}}) \times A = V \times \rho \times g$$

where $V$ is the volume of the cube and $\rho$ is the density of the fluid. So

$$\text{difference in pressure} = \frac{V \times \rho \times g}{A}$$

But $V/A = h$, the length of an edge of the cube. Therefore, the difference in pressure, $P$, due to a depth, $h$, of fluid of density $\rho$ is given by

$$P = h\rho g$$

QUESTIONS

Stretch and challenge

10. The formula for the pressure at a depth $h$ in a fluid works well for liquids, such as water, but rather less well for gases, such as air. Why is this the case?

## Lift

The upward force that keeps aircraft in the air is called **lift**. It results from the aircraft wings travelling at speed through the air. How the main contribution to the lift force arises is complex. In A-level Physics problems, lift can be treated as a force acting upwards, perpendicular to the aircraft's motion, bearing in mind that there are limitations to this simple model.

## Explaining lift

### Stretch and challenge

Until very recently, many textbooks still gave incorrect, or incomplete, explanations of lift. So how does the lift force arise?

We can discount upthrust. Although there is upthrust in air, the force is much smaller than in water. So upthrust can be sufficient to keep boats afloat, but it cannot be responsible for keeping aircraft airborne.

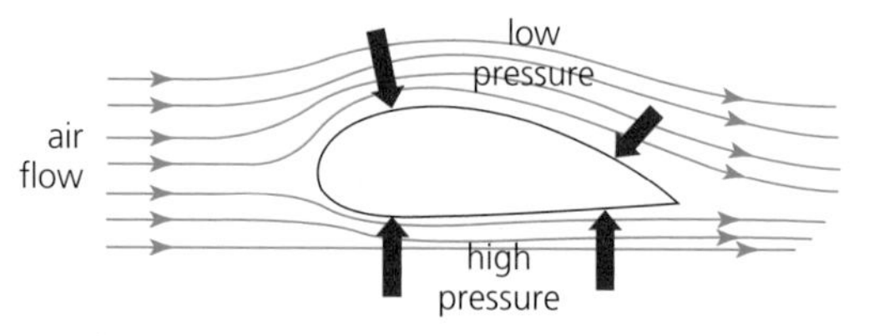


***Figure 15*** *Air flow past an aerofoil*

The traditional explanation of lift focuses on the shape of the cross-section of the wing, the **aerofoil** (seen in section in Figure 15). The argument assumes that air takes the same time to travel over the curved upper surface of the wing as it does to travel along the flatter lower surface. As the distance over the wing is further, the air must travel faster, which causes the pressure to drop. The higher pressure beneath the wing is said to cause lift. Although the air pressure is certainly greater beneath the wing than above, there are several problems with this explanation of lift:

- The force due to this effect is not large enough to lift a large aircraft.
- Experiments in wind tunnels, using high-speed photography, have shown that the air does not take the same time to travel over the upper and lower surfaces of the wing.
- There are some planes that have wings that do not have the usual aerofoil shape but fly perfectly well.
- Planes sometimes deliberately fly upside down (Figure 16).

Another explanation for the cause of the lift force is that an aircraft's wings act as an air pump, forcing air downwards and therefore pushing the wings and the aircraft up, in accordance with Newton's third law of motion (see section 11.1 of Chapter 11).

The full explanation relies on an understanding of motion through a fluid. The lift force depends on a number of factors: the relative velocity of the aircraft and the air; the density, viscosity and compressibility of the air; as well as the size and shape of the aircraft.

***Figure 16*** *According to some theories of flight, this should not be possible.*

## Drag

**Drag** is the force that opposes motion through any fluid, commonly air or water. When the motion is through air, drag is often referred to as **air resistance**. If you are travelling at a low speed through air – walking, for example – drag is barely noticeable. At higher speeds – when cycling, for example – drag becomes significant. This is because air resistance increases with speed (see section 9.6 of Chapter 9).

### QUESTIONS

**11.** What forces are acting on you now?

**12.** What forces act on a hot air balloon as it ascends?

**13.** What forces act on a caravan being towed by a car?

**14.** Draw an aircraft flying at constant speed in a horizontal straight line. Mark in the forces that act *on the plane*.

**15.** An Airbus 310 plane has a maximum speed of 900 km $h^{-1}$ and a ceiling (maximum flying altitude) of 12 500 m.

**a.** Explain why an aircraft has a top speed. What factors would affect this speed?

**b.** Explain why an aircraft has a maximum flying altitude. What factors would affect this altitude?

### KEY IDEAS

- The force of upthrust, or buoyancy, acts upwards on an object in a fluid, and is equal to the weight of fluid displaced.
- Lift is a force that acts upwards on surfaces, such as an aircraft wing, that are moving through a fluid.
- Drag is a force that acts on objects moving through a fluid so as to oppose the motion. Drag increases as the object's speed relative to the fluid increases.

## 10.3 MODELLING THE PROBLEM: FREE-BODY DIAGRAMS

The forces acting on a real object, such as the central span of a bridge, are often complex. We can try to understand the situation by drawing a simplified picture, known as a **free-body diagram** (Figure 17).

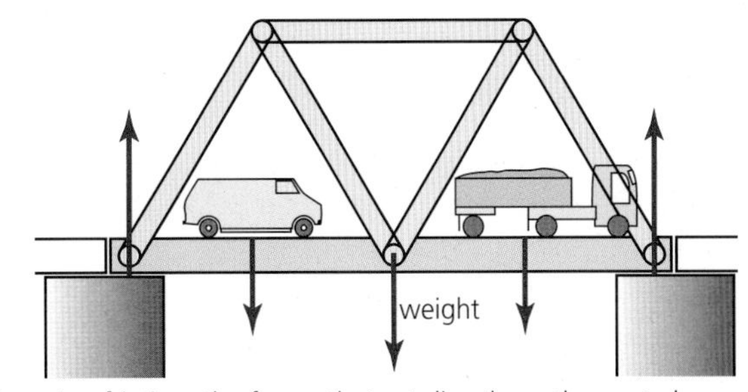


Ignoring friction, the forces that act directly on the central span of a bridge are its own weight, contact forces from vehicles and contact forces from the supports.

***Figure 17*** *A simple free-body diagram*

There are several points to bear in mind when drawing a free-body diagram:

- Forces are vector quantities – they have a size and a direction. We represent a force by an arrow, pointing in the direction of the force and labelled with the magnitude. The base of the arrow should be at the point where the force acts.
- The examples for A-level Physics are restricted to systems of coplanar forces (two-dimensional problems).
- All the forces that do not act directly on an object are ignored. For example, the reaction of the ground on the supports does not act directly on the central span of a bridge.
- We are not concerned with the set of balanced forces that are internal to the system. We can ignore any tension or compression in the span itself.
- We can sometimes combine several forces into one. For example, even though a vehicle on a bridge touches the road at several points, we need only consider a single contact force acting through the centre of mass of the vehicle.
- Take care not to invent forces, such as 'velocity force' or 'centrifugal force'. Everyday forces in mechanics are due to underlying fundamental forces, such as gravity or electromagnetism. Ask yourself, "What is causing the force?"

### Worked example 1

Identify the forces acting on the parasailing person and harness system in Figure 18. Assume that the person is moving horizontally.

*Figure 18*

The person and harness are to be treated together as a single 'system'. We do not need to worry about the tension of the harness pulling the person, for example, because this is an internal force. The forces on the system, shown in Figure 19, are:

- $W$, the weight of the person and harness together, which will act vertically downwards;
- $T_1$, the tension in the cables attached to the boat, which is pulling the person in the direction of the cables;
- $T_2$, the tension in the cables attached to the parasail, which is pulling the person in the direction of the cables;
- $D$, the drag force due to air resistance, which will act in the opposite direction to the motion.

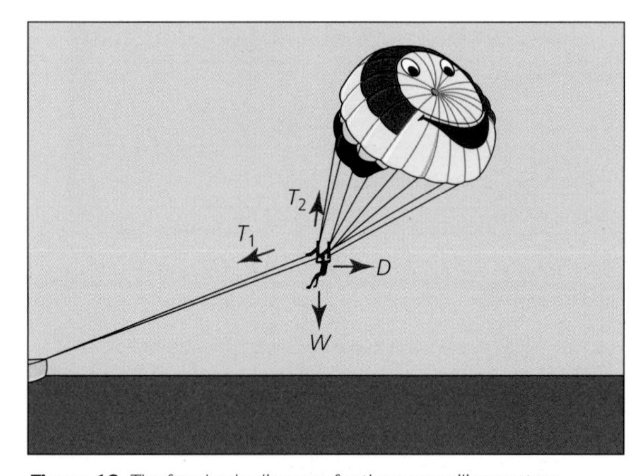


***Figure 19*** *The free-body diagram for the parasailing system*

### QUESTIONS

**16.** Draw a free-body diagram for a child sliding down a playground slide. Label each of the forces.

**17.** Draw a free-body diagram for a cyclist climbing a hill. Treat the cyclist and the bicycle as a single object.

### Forces and equilibrium

Once we have identified all the forces that act on an object, we can simplify the model further by combining them to form a single force. We add the vectors representing the individual forces to give the **resultant force**, as we did with velocity vectors in section 9.2 of Chapter 9.

When there are two forces acting in the same line, if the force vectors are equal and opposite the resultant force is zero.

With three forces, if a closed triangle is formed when the three vectors are placed nose to tail (in any order), then the resultant is zero (Figure 20a). If a closed triangle is not formed (Figure 20b), the resultant is found by joining the start of the first vector to the end of the final one. Similarly, if there are more than three forces, the resultant is zero if the forces form a closed shape (polygon).

In mechanics, we say that the object is in **equilibrium** if it is stationary or if it moves at constant velocity. If an object is in equilibrium, then the resultant force on it must be zero. So, for an object in equilibrium, adding the individual forces together will make a closed **polygon of forces**.

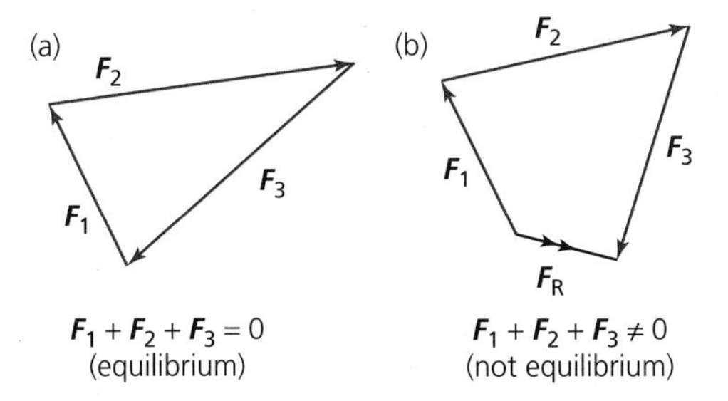


In scale drawing, a number of forces acting on an object can be added together by placing them 'nose to tail'. If the resulting shape is closed, this means that the resultant force is zero and the object is in equilibrium.

***Figure 20*** *The condition for equilibrium*

## QUESTIONS

**18.** The skier shown in Figure 10 in section 10.1 is moving at constant velocity. Draw the polygon of forces.

## Finding unknown forces

To find an unknown force on a system that is in equilibrium, we can either:

- use a scale diagram, or
- use trigonometry (Pythagoras' theorem and trigonometric ratios sin, cos and tan).

Worked examples 2 and 3 show these methods.

### Worked example 2: Using a scale diagram

An oil tanker is using its diesel engine to drive it in a direction due north with a force of 10 kN. It is also being pulled by two tugs, A and B, which exert forces on the tanker as follows:

- force from tug A, 20 kN, 10° east of north
- force from tug B, 25 kN, 20° west of north.

The oil tanker is moving forwards at constant velocity. Use a scale drawing to find the size and direction of the drag force acting on the tanker.

The first step is to identify the forces acting on the tanker and represent them on a free-body diagram (Figure 21a). The force vectors can then be added together by drawing them nose to tail (in any order), making sure that the lengths of the vectors are drawn to scale and that the angles between them are correct. A scale that leads to a large diagram will give a more precise and accurate answer.

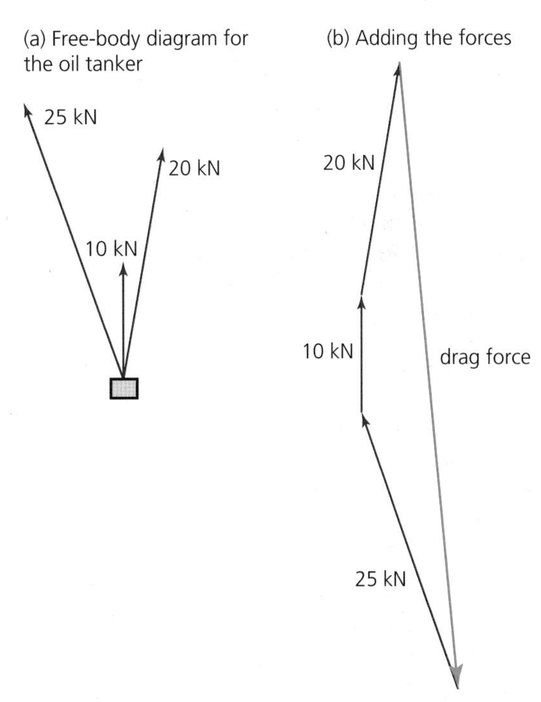


***Figure 21***

The ship is travelling at constant velocity, so it is in equilibrium, and the resultant force must be zero. The vector addition of the forces must make a closed polygon (Figure 21b). The drag force can be measured directly from the diagram. In this case, the drag force = 14.3 kN at an angle of 3.3° east of south.

### Worked example 3: Using trigonometry

Suppose that an engineer suspects that a certain joint in a bridge (Figure 22a) may fail if the weight of traffic increases any further. When the bridge is not loaded with traffic, the joint is in equilibrium under the action of three forces. There is a contact force ($F_s$) from the support, a force of tension ($F_A$) in the horizontal girder A and a force of compression ($F_B$) from the diagonal from girder B. If $F_A$ is 3000 N and $F_B$ is 5000 N, what is the magnitude of $F_S$?

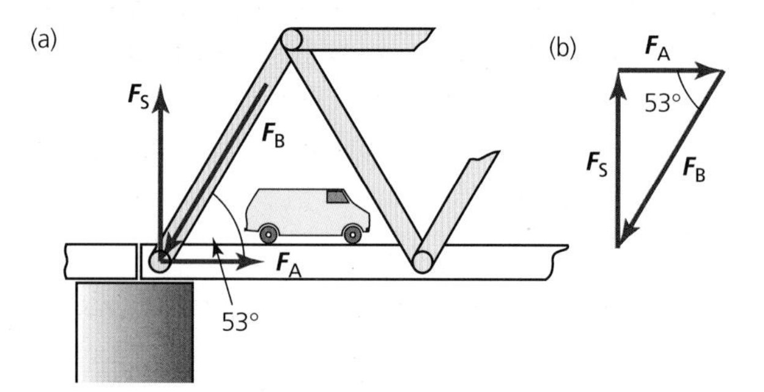


*Figure 22*

Draw the closed vector triangle (Figure 22b). Using Pythagoras's theorem:

$$F_A^2 + F_S^2 = F_B^2$$
$$F_S^2 = F_B^2 - F_A^2$$
$$F_S^2 = 5000^2 - 3000^2$$
$$F_S = 4000\,\text{N}$$

Alternatively,

$$F_S = F_B \sin 53° = 5000 \sin 53° = 3993 \approx 4000\,\text{N}$$

## Using the cosine rule

### Stretch and challenge

In reality, friction will act at the joint in Figure 22, so $F_S$ will not be vertical. Suppose that, with the bridge under load, $F_B$ increases to 8000 N and $F_A$ to 3500 N. In order to find the contact force required to maintain equilibrium, we can use a scale drawing or the cosine rule (Figure 23). (If you are studying A-level Mathematics, you can use the cosine rule, but scale drawing is acceptable in A-level Physics.)

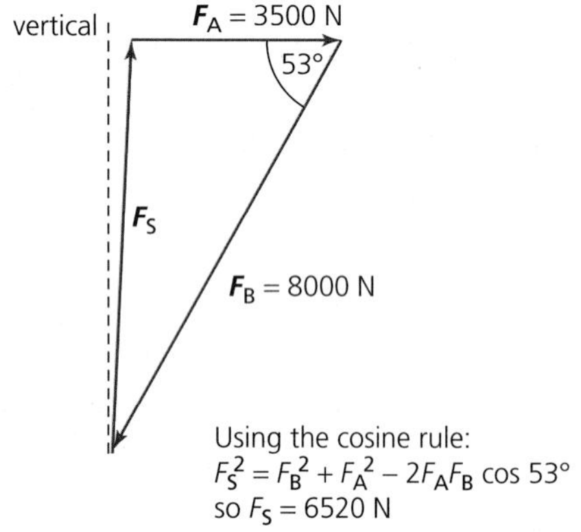


*Figure 23*

More complex problems can be simplified by resolving forces into horizontal and vertical components.

## QUESTIONS

**19.** A baby bouncer (Figure 24) is a type of bungee jump for the very young. A baby (of mass 12 kg) is in a baby bouncer (mass 2 kg) with her right foot pushing down on the floor with a force of 10 N. The angles between the cords of the bouncer and the horizontal bar are 60°.

**a.** Identify the forces on the baby.

**b.** Draw a free-body diagram showing the forces on the baby.

**c.** Identify the forces on the horizontal bar.

**d.** Draw a free-body diagram showing the forces on the horizontal bar.

**e.** If the bar is in equilibrium, calculate the tension in the cords.

*Figure 24*

## Resolving forces

It is often important to know the effect of a force in a certain direction. For example, the tension in the cables of the drawbridge in Figure 25 has two effects:

- horizontally, which puts the road under compression
- vertically, to lift the weight of the bridge section.

We find the size of these effects by splitting the tension into two **components**, at right angles to each other (Figure 26). This is the process of **resolving** a vector, previously discussed in section 9.4 of Chapter 9. Often it is most helpful to resolve the forces into horizontal and vertical components. Resolving all the forces acting in a system into just

***Figure 25*** *A cable-stayed drawbridge*

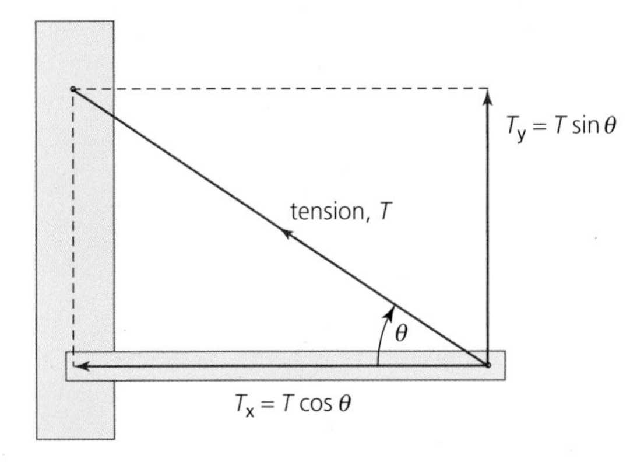


(a) Horizontal component, $T_x$: $\frac{T_x}{T} = \cos\theta$

So $T_x = T\cos\theta$

(b) Vertical component, $T_y$: $\frac{T_y}{T} = \sin\theta$

So $T_y = T\sin\theta$

***Figure 26*** *Resolving a tension force into perpendicular components*

two perpendicular directions can allow us to calculate the forces needed to keep it in equilibrium. If a slope is involved, it is more convenient to resolve the forces along the slope and perpendicular to the slope, as shown in Worked example 4.

## Worked example 4

Suppose a car is unfortunate enough to be on a drawbridge as it lifts. What frictional force is needed to keep it in equilibrium and stop it sliding down?

First, we draw a free-body diagram showing the forces on the car (Figure 27). Then we resolve the forces so that all the components are either parallel or perpendicular to each other. We often do this so that all the force components are horizontal or vertical, but in this case it makes more sense to resolve so that the forces act along the slope or at right angles to it.

For equilibrium, the sum of the forces perpendicular to the slope must be zero,

$$R - W\cos\theta = 0$$

(the minus sign indicates the opposite direction), so

$$R = W\cos\theta$$

and the sum of the forces along the slope must be zero

$$F - W\sin\theta = 0$$

so

$$F = W\sin\theta$$

If the car weighs 12 000 N and the bridge is at an angle of 20° to the horizontal, to keep the car in equilibrium the frictional force needs to be $12\,000 \times \sin 20° = 4100$ N.

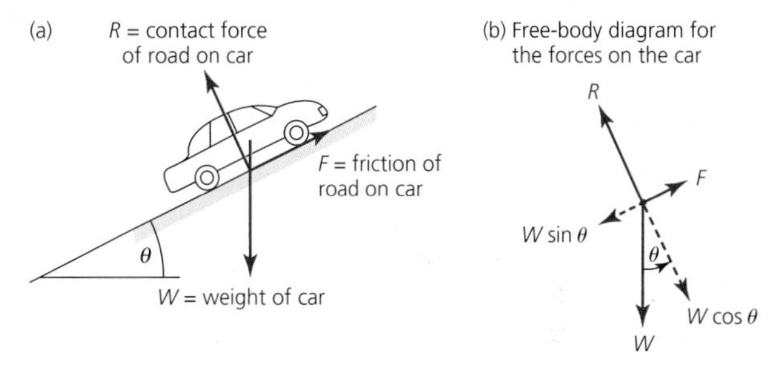


***Figure 27***

In general, for an object in equilibrium, the vector sum of the forces must be zero. An equivalent way of saying this is that, for an object to be in equilibrium:

- the sum of all the force components in one direction must be zero, and
- the sum of all the force components in the perpendicular direction must be zero.

This provides another way of finding the resultant force on a system – or the single force needed to keep a system in equilibrium. We can resolve all the forces acting on an object into perpendicular components, such as horizontal and vertical, or along a slope and at right angles to the slope. We then find the vector sum of the forces in each of the two directions. Then these two perpendicular forces can be added together using trigonometry to find a single force.

### Worked example 5

A boat is travelling in the Severn Estuary. The engines result in a force of 30 000 N in a northerly direction, the wind exerts a horizontal force of 7000 N at an angle of 30° west of north, and the tidal currents exert a force of 2000 N at an angle of 40° east of north. Find the resultant force on the boat.

The vertical forces on the boat must balance one another, or the boat will sink. So it is just the three horizontal forces ($F_{engine}$, $F_{wind}$ and $F_{tide}$) that add together to form the resultant force. We draw a sketch of these forces (Figure 28a).

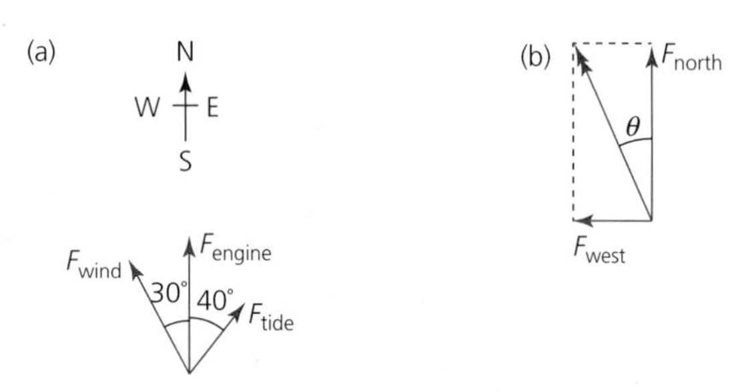


***Figure 28*** *The forces in the horizontal plane*

Resolving the horizontal forces so that the components are parallel or perpendicular to the north–south line:

- $F_{wind}$ resolves into $F_{wind}$ cos 30° (north) and $F_{wind}$ sin 30° (west)
- $F_{tide}$ resolves into $F_{tide}$ cos 40° (north) and $F_{tide}$ sin 40° (east)

Now we add together all the components in the north–south direction:

$$F_{north} = 30\,000 + 7000 \cos 30° + 2000 \cos 40° = 37\,600\text{ N (north)}$$

And in the east–west direction:

$$F_{west} = 7000 \sin 30° - 2000 \sin 40° = 2214\text{ N (west)}$$

To find the resultant force ($F_{res}$), we combine these components together using Pythagoras's theorem:

$$F_{res}^2 = F_{north}^2 + F_{west}^2$$
$$F_{res}^2 = 37600^2 + 2214^2$$
$$F_{res} = 37665\text{N}$$

in a direction given (see Figure 28b) by
$\theta = \tan^{-1}(F_{west}/F_{north}) = \tan^{-1}(2214/37\,600)$
$= 3.3°$ east of north

## QUESTIONS

**20.** In Figure 29, the skier is being dragged up the slope.

**a.** Draw a free-body diagram showing the forces on the skier.

**b.** Resolve the forces into components that are parallel or perpendicular to the slope.

**c.** If the skier is being dragged at a steady speed up the slope, that is, the skier is in equilibrium, write equations connecting the forces.

***Figure 29***

**21.** A car of mass 1000 kg is parked on a ramp inclined at 15° to the horizontal. If the car is in equilibrium, find the normal contact force and the frictional force between the car and the road.

## KEY IDEAS

- The resultant force on an object is found by the vector addition of all the forces acting on it.
- Vector addition can be done using scale drawing.
- Alternatively, forces may be added by resolving them into components in two perpendicular directions. The components are added to find the total force in each of these directions. Pythagoras' theorem is used to find the magnitude of the resultant; sin, cos or tan can be used to find its direction.
- An object that is stationary, or moving at constant velocity, is in equilibrium.
- At equilibrium, there is no resultant force acting on the object; the force vectors form a closed polygon.
- At equilibrium, the sum of all the horizontal force components must be zero, and the sum of all the vertical force components must be zero. (Alternatively, if it is more convenient, components can be taken in two other perpendicular directions.)

## ASSIGNMENT 1: DETERMINING THE COEFFICIENT OF FRICTION

**(MS 0.6, MS 1.2, PS 1.1, PS 1.2, PS 2.1, PS 2.3, PS 3.3)**

A student is given the task of finding the coefficient of friction between a child's shoe and the polished laminate flooring that her parents have just installed.

One way to find the coefficient of friction between two surfaces is to use an inclined plane. If an object is placed on a slope that is gradually made steeper, the angle at which the object slips will give an indication of the static frictional force between the two surfaces.

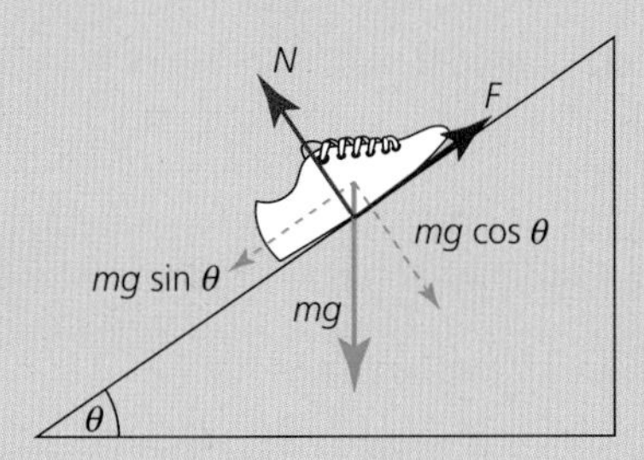


*Figure A1*

Suppose an object is placed on a slope inclined at an angle $\theta$ to the horizontal (Figure A1). The angle could be increased until the object is on the point of slipping. At this point, the friction has reached its limiting value. There are three forces acting on the object: its weight ($mg$) acting vertically downwards, the normal reaction $N$ acting perpendicular to the slope, and friction $F$ acting up the slope so as to prevent the object sliding down the slope.

When the object is on the point of moving, these three forces must add to make zero resultant force. We can therefore construct a vector triangle as in Figure A2.

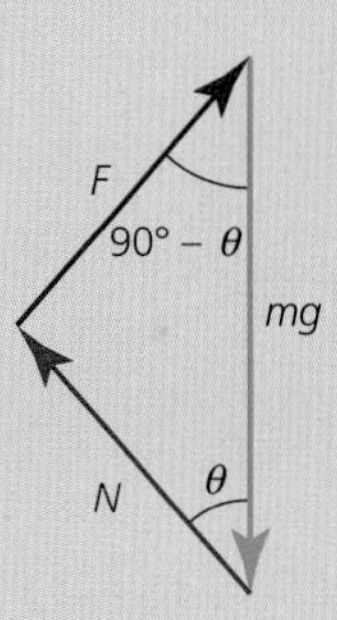


*Figure A2*

Alternatively, we can resolve the weight into two components – one acting down the slope, equal to $mg \sin \theta$, and one acting at right angles to the slope, equal to $mg \cos \theta$. For equilibrium, we must have:

$$N = mg \cos \theta \quad \text{and} \quad F = mg \sin \theta$$

Since the coefficient of friction, $\mu$, is given by $\mu = F/N$, we find that

$$\mu = \tan \theta$$

The student recorded the results shown in Table A1 when carrying out this experiment with the child's shoe. She used a spare piece of the laminate flooring, added masses inside the shoe to give a range of results for the total mass of the shoe, and repeated the experiment three times for each mass.

| Total mass of shoe / kg | Angle $\theta$ at which shoe begins to slide / degree | | |
|---|---|---|---|
| | First trial | Second trial | Third trial |
| 1.00 | 20 | 18 | 24 |
| 2.00 | 24 | 18 | 18 |
| 3.00 | 23 | 21 | 20 |

**Questions**

**A1** Estimate the uncertainty in the results. Discuss whether this is what you would expect.

**A2** How might you improve the precision of the results?

**A3** Use the results to estimate the coefficient of friction between the shoe and the polished floor.

**A4** The student has formed a hypothesis that the weight in the shoe does not affect the angle at which the shoe slips, and so concludes that friction between the shoe and the floor will not be affected by the weight of the child wearing the shoes. Is this correct? Write a short response to the student, discussing the results and conclusions.

## ASSIGNMENT 2: ADDING COPLANAR FORCES ON A FORCE BOARD

**(MS 0.6, MS 1.5, MS 4.1, MS 4.2, MS 4.5, PS 1.1, PS 1.2, PS 2.3)**

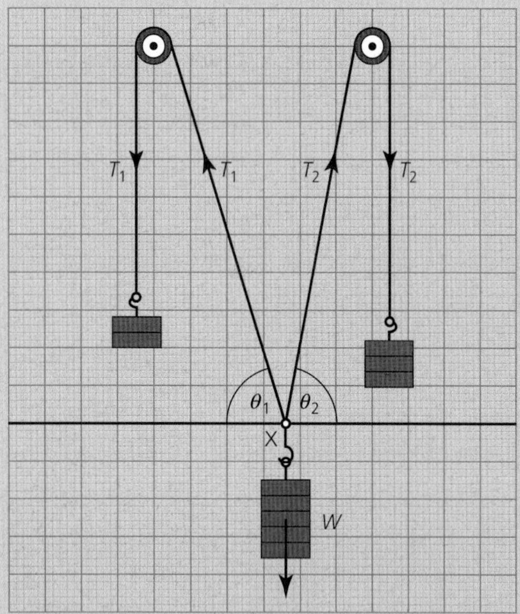


*A force board*

Use of a 'force board' is one way to investigate how forces add together. Three or more strings are attached to a small ring. The other ends of the strings are attached, via a pulley, to a weight to provide a known force. The set-up is shown in the diagram.

The pulleys are fastened securely to the top of the board. The masses are adjusted so that the system hangs in equilibrium. Provided that the pulleys have low friction, the tension on either side of the pulley should be the same. Point X is in equilibrium under the action of three forces, $T_1$, $T_2$ and the mystery weight, $W$. The angles $\theta_1$ and $\theta_2$ can be measured from the force board. There are two possible ways to find the mystery weight:

1 You could construct a vector triangle with sides $T_1$, $T_2$ and $W$. If this is drawn to scale, then $W$ can be measured directly. (Alternatively, the cosine rule could be used to calculate $W$.)

2 You could resolve the forces $T_1$ and $T_2$ into horizontal and vertical components. The sum of their vertical components must equal $W$.

Such an experiment was carried out twice, using the same mystery weight, and the results in the table were obtained.

| $T_1$ / N | $T_2$ / N | $\theta_1$ / degree | $\theta_2$ / degree |
|---|---|---|---|
| 25 | 50 | 44 | 69 |
| 50 | 60 | 28 | 42.5 |

**Questions**

**A1** Using the first set of results, construct a vector diagram of the forces. Find the mystery weight either by scale drawing or by calculation.

**A2** Using the second set of results, find the mystery weight by resolving forces vertically and horizontally.

**A3** What are the potential sources of uncertainty in this experiment? How might you make the experiment more precise?

## ASSIGNMENT 3: FORCES IN WINDSURFING

### Stretch and challenge

Windsurfing can be exhilarating. Even staying upright is a considerable challenge at first. But experienced windsurfers (Figure A1) know how to make use of the wind's force and direction to steer the board and lift it to reduce the drag of the water – speeds of up to 45 km $h^{-1}$ are possible.

***Figure A1*** *You have to control all the forces that are acting on you to get the best motion.*

Some high-performance windsurfing boards cannot displace enough water to provide the upthrust needed to float under the weight of a person. They can only float when the board is moving, relying on a component of the wind's force to stop them from sinking. The flow of air over the sail causes a force at right angles to the sail. When the windsurfer leans back, tilting the sail, the vertical component of this force keeps the board afloat (Figure A2) or even lifts it out of the water.

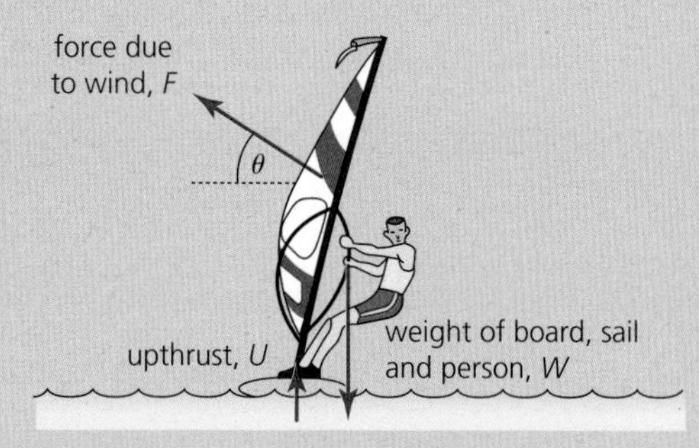


***Figure A2***

### Questions

**A1** Resolve the force of the wind on the sail into horizontal and vertical components.

**A2** Imagine that the board is afloat and moving at constant velocity. Write down two equations that the forces must obey.

**A3** Explain what happens to the vertical component of the wind's force on the sail as the surfer leans back further into the wind, and explain what effect this has on the motion of the board.

## 10.4 THE TURNING EFFECT OF A FORCE

A resultant force on an object will cause it to accelerate. But even if there is no resultant force, the object may still not be in equilibrium. Consider the object in Figure 30. It is acted on by two equal but opposite forces, so there is no resultant force. The object will not have any linear acceleration – but it will rotate if free to do so. Each of the two forces has a turning effect.

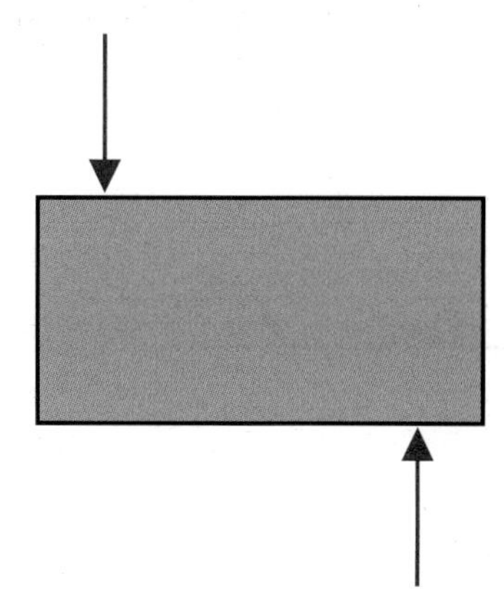

***Figure 30*** *Forces can cause objects to tip or rotate, as well as to accelerate in a straight line.*

## The moment of a force

This turning effect of a force is called the **moment of a force**.

The size of the moment of a force about a point depends on two things (Figure 31):

- the magnitude of the force, $F$
- the perpendicular distance, $x$, between the line of action of the force and the point.

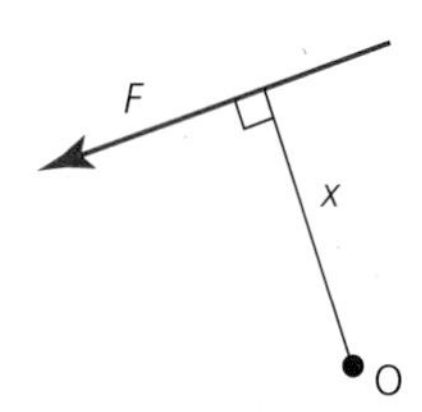


***Figure 31*** *The moment of the force F about the point O is F × x.*

The transverse force of the wind acting on the sail in Figure 32, for example, causes a moment. This acts to tip the sail and mast, and the whole yacht, over.

***Figure 32*** *The force of the wind on the sail has a turning effect on the yacht, which the crew try to balance.*

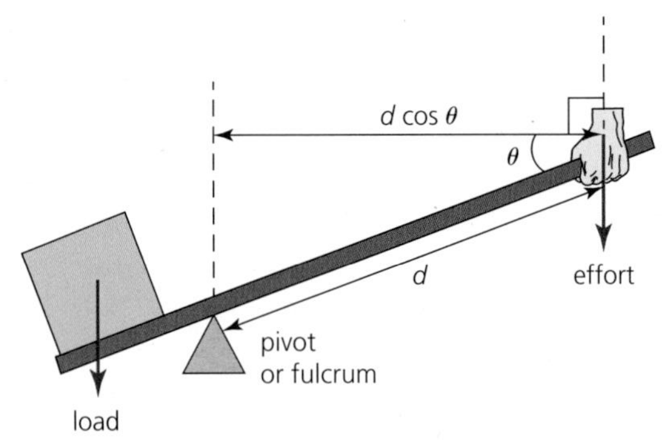


***Figure 33*** *The action of a lever depends on the moment of a force. The further from the pivot that you can apply your force, the greater the turning effect.*

You can calculate the turning effect of a force about *any* point, although in a real situation there is often a **pivot** (sometimes called a fulcrum) or a hinge, about which the object turns (Figure 33).

The **moment of a force** is defined as follows.

The moment of a force about a point is equal to the magnitude of the force multiplied by the perpendicular distance from the line of action of the force to the point.

moment (N m) = force (N) × perpendicular distance of point from line of action of force (m)

The moment of a force has the unit newton metre (N m).

The turning effect of a force can be increased by moving the force further from the pivot. It is the perpendicular distance from the pivot to the line of action of the force that is important; this is marked as $d \cos \theta$ in Figure 33. If the line of action of a force goes through the pivot, it will have no turning effect at all, as shown in Figure 34.

(a)

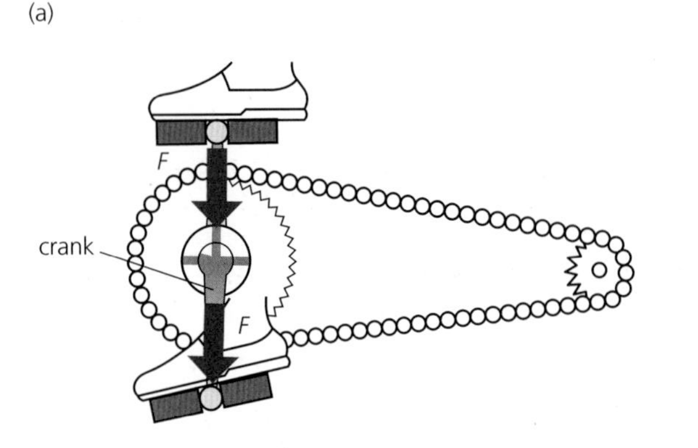


(b)

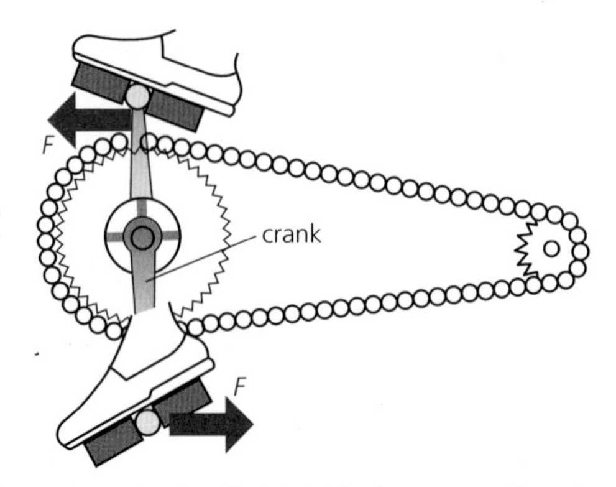


***Figure 34*** *When cycling, it is important to have the correct pedalling style to get the maximum turning effect. In (a) the force passes through the pivot and has zero turning effect. In (b) the force is at right angles to the crank and the turning effect of each force = F × length of crank.*

## QUESTIONS

22. Explain the following observations:
    a. Two people can balance a seesaw, even if one is much heavier than the other.
    b. It is easier to undo a tight screw using a screwdriver with a large-diameter handle.
    c. Old buses and lorries had very large steering wheels.
23. Archimedes is reputed to have said, "Give me somewhere to stand and I will move the Earth". Some sources helpfully add the words "and a long enough lever". The Earth has mass of about $6 \times 10^{24}$ kg. Estimate how long the lever would need to be. (You will need to make some rather sweeping assumptions; state these.)
24. The strongest, possibly heaviest, person in the class tries to push open the door by putting their hand as close to the hinge as possible. The weakest, perhaps lightest, person in the class can keep it closed by pushing on the door, close to the handle, with just one finger. Explain this by showing some simple calculations.
25. Look at Figure 34.
    a. Draw a diagram showing the pedals when the crank is horizontal. Draw in the direction that the force should be applied to each pedal to get the maximum turning effect.
    b. Draw a diagram showing the pedals in an intermediate position, with the crank between vertical and horizontal. Again, draw in the direction that the force should be applied to each pedal to get the maximum turning effect.

### The principle of moments

There are likely to be several forces acting on an object that could have a turning effect on the object. If the object is in equilibrium, that is neither accelerating nor rotating, these turning effects must balance. In other words, the total clockwise turning effect must be balanced by the total anticlockwise turning effect. This is the **principle of moments**, stated as follows.

For an object to be in equilibrium, the sum of the clockwise moments about any point must be equal to the sum of the anticlockwise moments about that point.

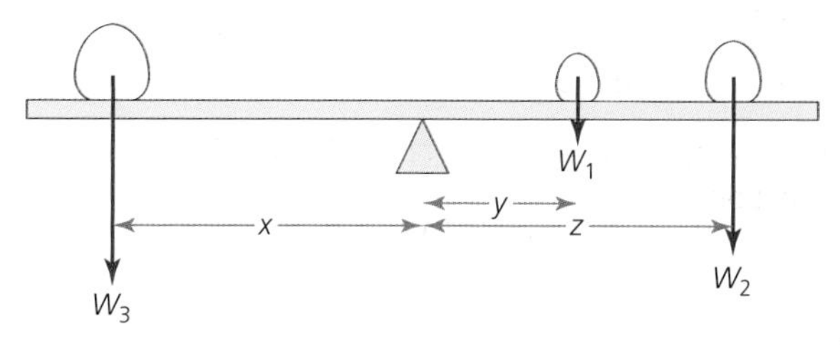


***Figure 35*** *Balancing moments*

Consider the balanced seesaw in Figure 35. If it is to be in equilibrium, then the turning effects must add to zero. First calculate the sum of the clockwise moments about the pivot:

$$\text{clockwise moment} = (W_1 \times y) + (W_2 \times z)$$

There is only one anticlockwise moment:

$$\text{anticlockwise moment} = (W_3 \times x)$$

For equilibrium, the moments must be equal:

$$(W_1 \times y) + (W_2 \times z) = (W_3 \times x)$$

### Worked example 1

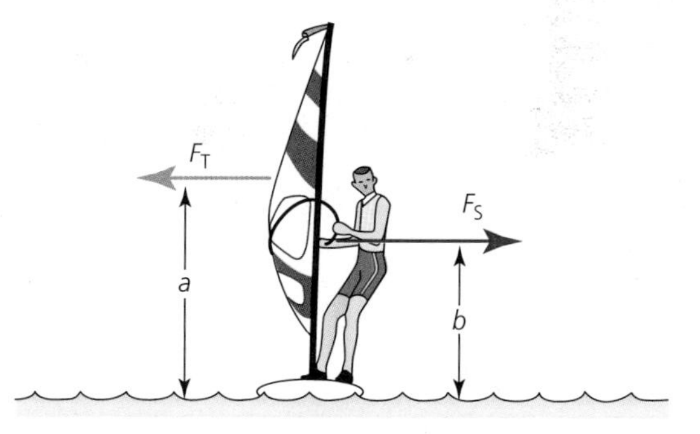


***Figure 36***

The transverse force of the wind, $F_T$, tends to tip the windsurfing board in Figure 36 in an anticlockwise direction. If the board is to stay upright, the windsurfer has to apply a force, $F_S$, that has a clockwise moment. The mast pivots freely where it joins the board. The moments acting about that point are:

$$\text{anticlockwise moment} = F_T \times a$$

$$\text{clockwise moment} = F_S \times b$$

For equilibrium, the moments must balance, so

$$F_T \times a = F_S \times b$$

It is important to remember that the moment depends on the *perpendicular* distance from the force to the pivot. As the wind speed increases, a windsurfer needs to lean at an increasingly large angle into the wind. We can see why this is if we treat the board and the surfer as one object. The important external forces are the combined weight of the board, mast, rig and surfer, $W_S$, and the force of the wind on the sail, $F_T$ (Figure 37).

***Figure 37*** *Leaning to balance moments*

As $F_T$ gets larger, the anticlockwise turning moment increases. To maintain balance, the windsurfer has to increase the clockwise turning moment by increasing the perpendicular distance, $d$, between the line of action of his weight and the pivot, by leaning out as far as possible.

## QUESTIONS

**26.** In Figure 38, what is the maximum weight that the crane can lift before the lorry begins to tip?

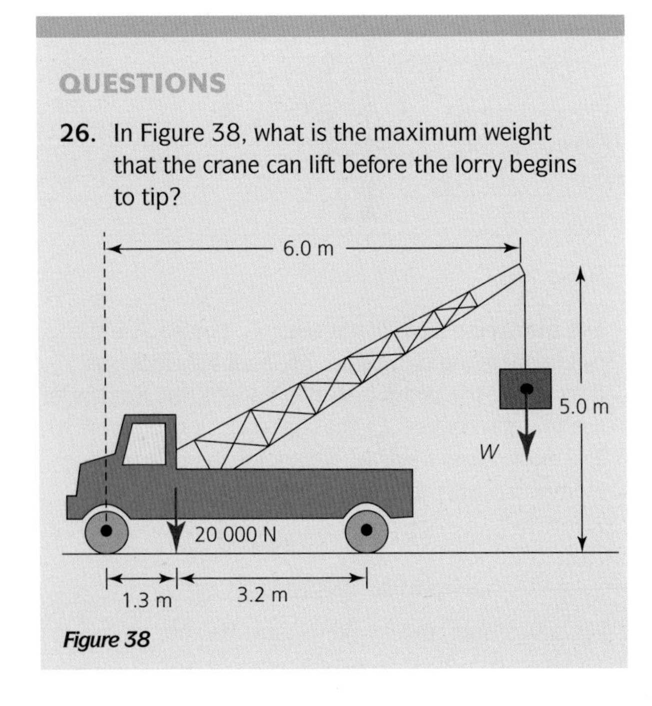


***Figure 38***

**27.** Windsurfing boards use a daggerboard to help them to remain stable (see Figure 39). Use the idea of moments to explain why the daggerboard is useful.

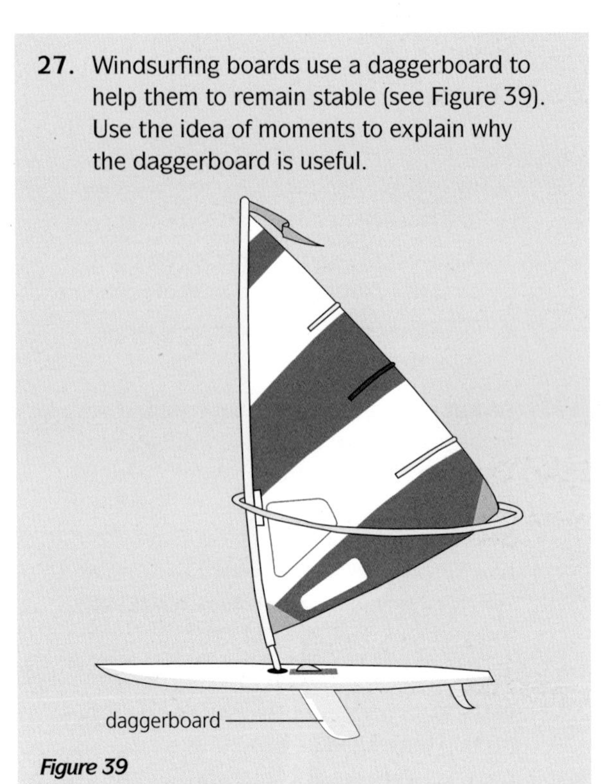


***Figure 39***

### Couples and torques

Pivots and hinges, such as the flexible joint where a mast joins a windsurfing board, are unlikely to be free of friction. The effect of this friction is to oppose rotation. It is known as a frictional **torque**. A torque may also *cause* rotation. When you grip a screwdriver to tighten a screw, you are exerting a torque on the screw – the thicker the handle, the bigger the torque you can exert because the force applied is further from the turning point (the axis). A torque produces or opposes rotation but does not cause any linear acceleration.

Two equal forces that act in opposite directions (Figure 40) cause no linear acceleration. However, if they do not pass through the same point, they both give rise to a turning effect, or moment. The two equal and opposite forces are called a **couple**. Together, they exert a torque, or **moment of a couple**, and cause rotation (Figure 41).

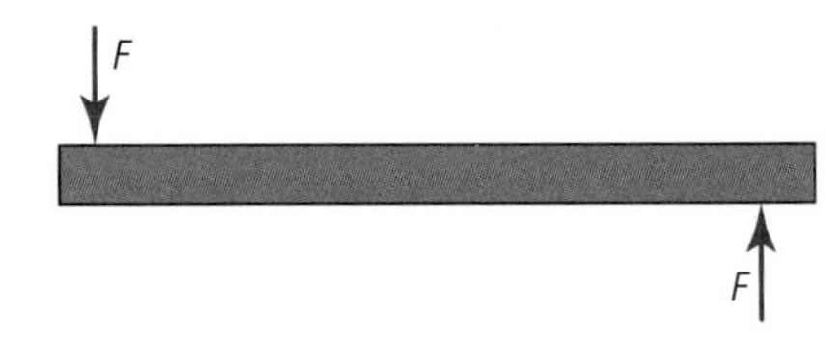


***Figure 40*** *A couple*

***Figure 41*** *Two equal and opposite forces do not necessarily guarantee equilibrium. Unless the forces are aligned, the system will tend to rotate.*

The total moment produced by a couple is equal to the magnitude of the force, $F$, multiplied by the perpendicular distance, $d$, between the two forces (see Figure 42).

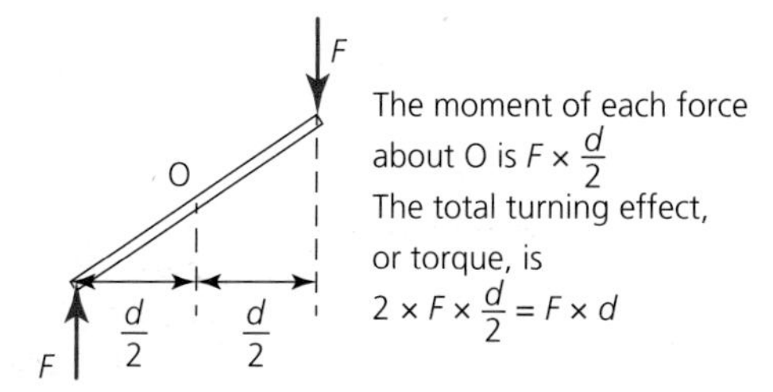


***Figure 42*** *The moment of a couple, or torque*

The forces exerted on the pedals of a bicycle can form a couple (Figure 34b), as can hands on a steering wheel. An important example of a couple is the turning effect produced in an electric motor. Electromagnetic forces acting on each side of a current-carrying coil form a couple, which tends to make the coil spin.

## Worked example 2

Imagine a windsurfer trying to lift the rig and sail out of the water (Figure 43). The principle of moments can be applied. The mass of the rig on the board is 20 kg and its centre of gravity is 2.0 m from the bottom end. The rig is pulled into position using a rope, which is connected to the mast, 1.5 m from the bottom end, and is initially at an angle of 25° to the horizontal. There is a frictional torque of 50 N m in the joint, X, that holds the mast to the board. What force is required to lift the rig from the horizontal?

The first step in all mechanics problems is to draw a diagram (Figure 43) showing all the relevant forces. The next step is to simplify the forces. In this case, the tension, $T$, can be resolved into horizontal and vertical components. (The horizontal component of the tension, $T \cos 25°$, has no moment because it passes through X.) We calculate the moments of the forces about the joint, X. We get

$$\text{sum of anticlockwise moments} = (W \times 2.0)\,\text{N m} + \text{frictional torque}$$

$$= (20 \times 9.81 \times 2.0)\ 392 + 50 = 442\,\text{N m}$$

and

$$\text{sum of clockwise moments} = T \sin 25°\ \text{N} \times 1.5\,\text{m}$$

$$= 0.634T\,\text{N m}$$

These moments balance when

$$0.634T = 442 \text{ or } T = 442/0.634 = 697\,\text{N}$$

The rig can be lifted when

$$T > 700\,\text{N}$$

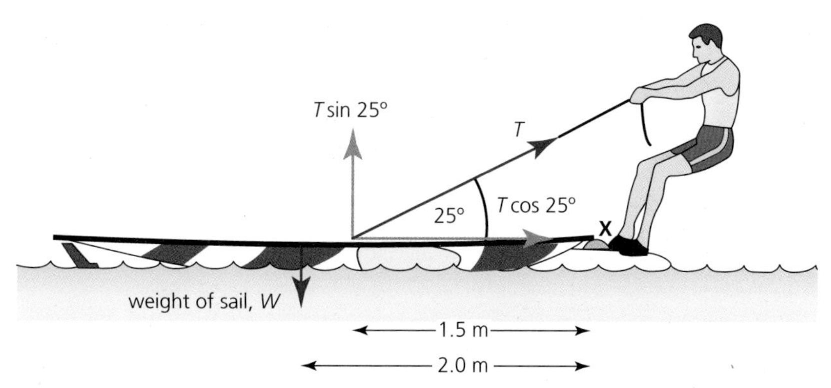


***Figure 43***

### Worked example 3

*Figure 44*

A capstan is a device that was used on old sailing ships or in harbours, to raise the anchor, for example (see Figure 44). Suppose that just two men are operating such a capstan in order to raise the anchor. They exert equal but opposite forces of 200 N and they are 3 m apart (Figure 45).

a. Calculate the moment exerted by the people pushing.

b. Calculate the force exerted on the rope to raise the anchor.

c. Why is the answer to part b likely to be an overestimate?

d. What would be the advantage of making a capstan with longer poles?

e. Does the use of longer poles reduce the energy required to lift the anchor?

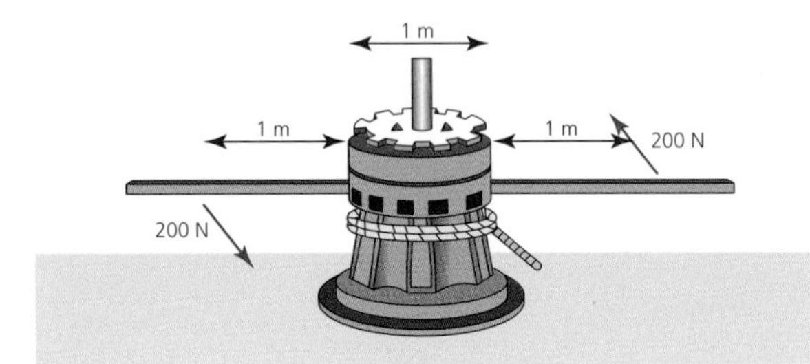


*Figure 45*

a. The forces exerted by the people pushing constitute a couple. The moment of a couple equals force multiplied by the distance between the forces, so

$$\text{moment} = F \times d = 200 \times 3 = 600\,\text{N m}$$

b. The force on the rope equals $F_R$, which exerts a moment $F_R \times 0.5$ N m. This must equal 600 N m if the capstan turns at a constant rate. So the force on the rope is

$$F_R = 1200\,\text{N}$$

c. We have assumed that there is no frictional torque to be overcome.

d. Longer poles would mean a larger torque, and hence a bigger force on the rope.

e. No. Although the people each have to exert a smaller force, they need to move further. Since work done = force × distance, the energy required would at best remain the same. In fact, more will be required since in practice there will be some friction to be overcome.

**KEY IDEAS**

- The moment, or turning effect, of a force about a point is equal to the force multiplied by the perpendicular distance from the point to the line of action of the force.
- The principle of moments states that, for an object to be in rotational equilibrium, the sum of the clockwise moments must equal the sum of the anticlockwise moments.
- For an object to be in equilibrium, that is not accelerating or rotating, it must simultaneously satisfy these two conditions:

  The vector sum of all the forces acting on it equals zero.

  The sum of the moments about any point equals zero.
- A couple is a pair of equal, parallel forces acting in opposite directions, separated by a distance *d*. The moment of a couple, or torque, is equal to the magnitude of one force, *F*, multiplied by the perpendicular distance, *d*, between the forces.

## QUESTIONS

**28.** Calculate the turning moments in situations **a**, **b** and **c** in Figure 46.

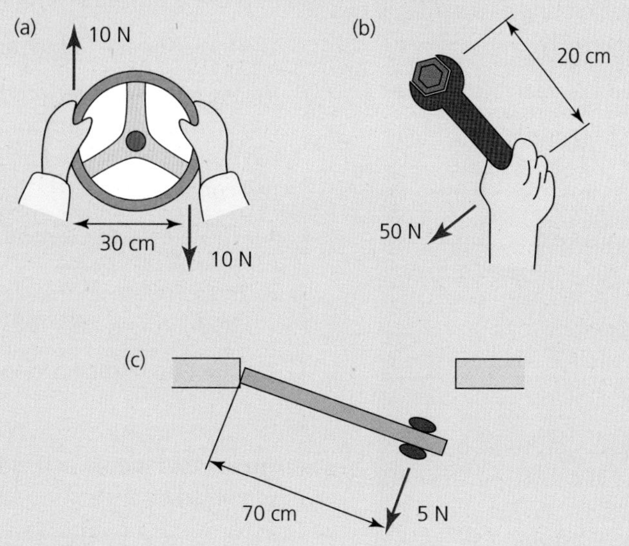


*Figure 46*

**29.** Calculate the unknown forces or distance in situations **a**, **b** and **c** in Figure 47, assuming a state of equilibrium in each case.

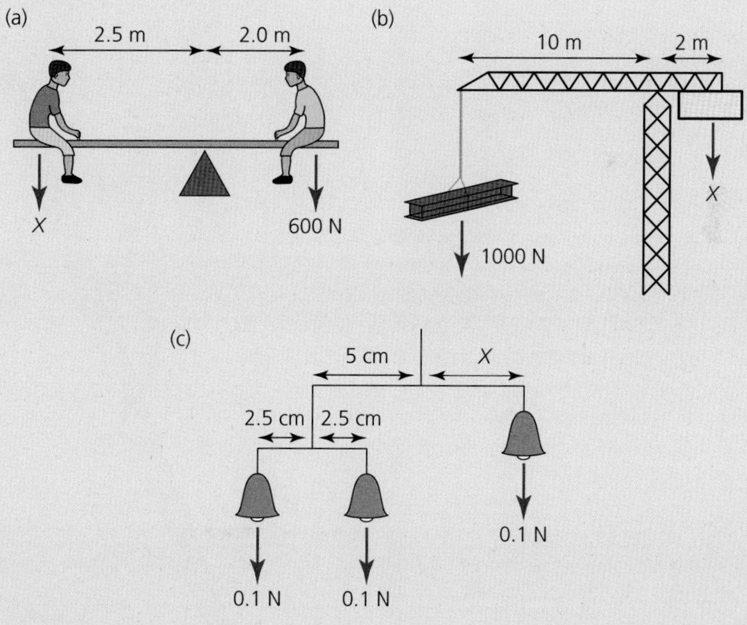


*Figure 47*

**30.** Explain why, in Worked example 2 (see Figure 43), the tension in the board rope decreases as the sail is lifted to a vertical position.

## 10.5 PROBLEMS OF EQUILIBRIUM

From everything that we have considered in sections 10.3 and 10.4, we now know that an object will be in equilibrium, that is, not accelerating or rotating, if two conditions are met simultaneously:

- The vector sum of all the external forces is zero.
- The sum of the clockwise moments about any point equals the sum of the anticlockwise moments acting about that point.

If we apply these conditions to an object in equilibrium, we can often calculate unknown forces or moments.

### Worked example

A window cleaner of mass 55 kg is three-quarters the way up a uniform ladder of length 10 m. The bottom of the ladder is placed 2.0 m from the bottom of the wall. The ladder itself has a mass of 25 kg.

Draw a free-body diagram showing the forces acting on the ladder. The wall is quite smooth, so that friction there can be neglected. Calculate the size of the frictional force on the ground that is just large enough to prevent the ladder from slipping.

The forces on the ladder are shown in Figure 48.

Having identified the forces, we can now apply the conditions for equilibrium. The first condition, that the vector sum of the forces is zero, can be split into two conditions.

1 The sum of the vertical forces is zero:

$$F_W + R_G = W_L + R_M$$

$$F_W + R_G = 25 \times 9.81 + 55 \times 9.81 = 785\,\text{N}$$

2 The sum of the horizontal forces is zero:

$$F_G = R_W$$

We are told that $F_W = 0$, so these equations can be simplified to

$$R_G = 785\,\text{N} \qquad \text{and} \qquad F_G = R_W$$

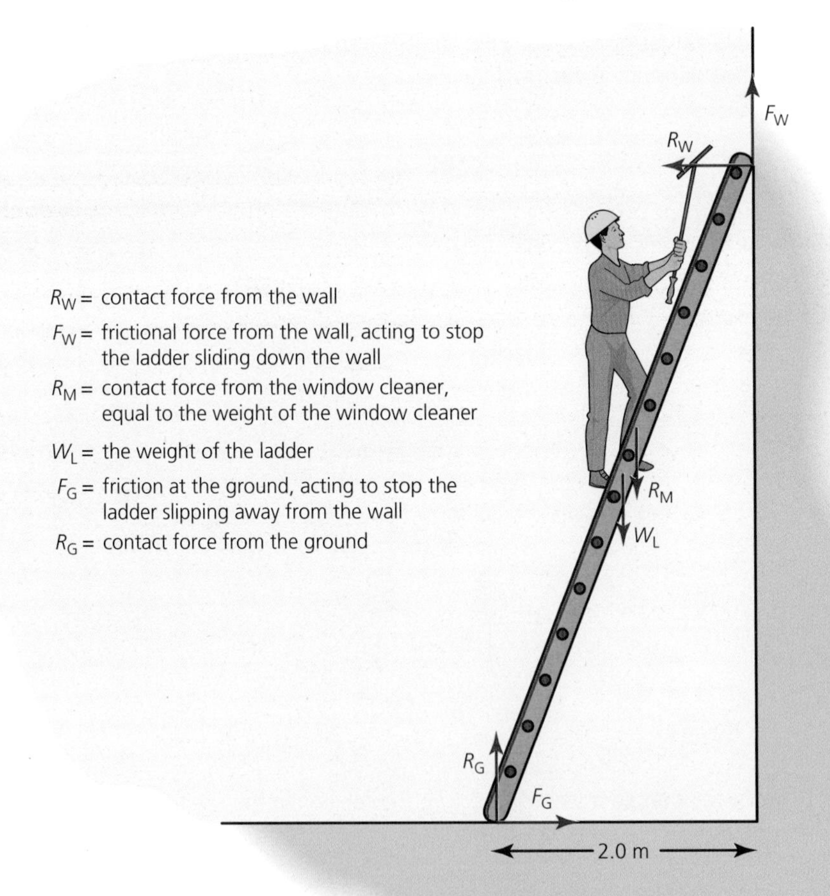


*Figure 48*

In order to find $F_G$ and $R_W$, we need to apply the next condition for equilibrium, namely that the sum of the moments about any point must equal zero. It makes sense to choose a point that eliminates one of the unknown forces. So taking moments about the bottom of the ladder:

$$\text{sum of clockwise moments} = (W_L \times 1.0\,\text{m}) + \left(R_M \times \frac{3}{4} \times 2.0\,\text{m}\right)$$

$$\text{sum of anticlockwise moments} = R_W \times h$$

where $h$ is the height of the top of the ladder. Since the length of the ladder is 10 m and it is 2.0 m out from the foot of the wall:

$$h^2 + 2.0^2 = 10^2$$

so

$$h = \sqrt{96} = 9.8\,\text{m}$$

If the ladder is in equilibrium, these moments must be equal:

$$R_W \times h = W_L \times 1.0 + R_M \times 1.5$$
$$R_W \times 9.8 = 25 \times 9.81 \times 1.0 + 55 \times 9.81 \times 1.5$$
$$R_W \times 9.8 = 245.3 + 809.3$$
$$R_W = \frac{1055}{9.8} = 108\,\text{N}$$

This is also equal to the friction at the ground, so

$$F_G = 108\,\text{N}$$

## ASSIGNMENT 4: BALANCING MOMENTS TO FIND THE DENSITY OF A LIQUID

**(MS 0.1, MS 1.1, MS 1.5, PS 1.1, PS 2.3, PS 3.2, PS 3.3, PS 2.3)**

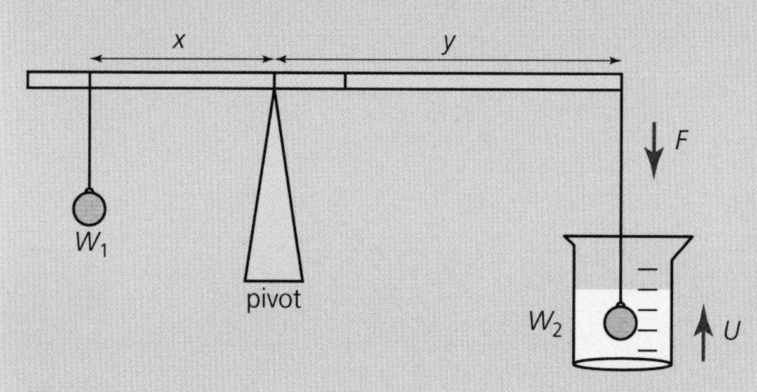


***Figure A1***

The apparatus in Figure A1 can be used to find the density of the liquid in the beaker. Two known weights are suspended from the beam, which is pivoted as shown. The weight $W_2$ is fully submerged in liquid. When the beam is balanced, the anticlockwise moment about the pivot equals the clockwise moment:

$$W_1 \times x = F \times y$$

where $F$ equals $W_2 - U$, and $U$ is the upthrust.

Since all the variables are known except $U$, the upthrust can now be calculated.

Archimedes' principle (section 10.2) states that the upthrust is equal to the weight of fluid displaced, and so, provided that we know the volume of $W_2$, we can calculate the density of the liquid.

### Questions

**A1** Suppose that you are to carry out the experiment described above. Unfortunately, the only known weights available are two aluminium spheres. You measure the diameter of each sphere with vernier callipers and get 10.0 cm for both spheres. Find the volume of the spheres, and estimate the uncertainty in the results.

**A2** Calculate the weight of the aluminium spheres. Density of aluminium = 2700 kg m$^{-3}$.

**A3** The beam is found to balance when $x = 20.0$ cm and $y = 36.7$ cm. Use these values to find the force $F$.

**A4** Find the upthrust $U$, and hence calculate the density of the liquid.

**A5** What is the uncertainty in your final answer?

**A6** Compare your answer with the density values in Table A1. Can you identify the liquid in the beaker?

**A7** Suggest ways in which you might make the experiment more precise.

| Liquid | Density / kg m$^{-3}$ |
|---|---|
| Acetic acid | 1049 |
| Linseed oil | 929 |
| Sea water | 1025 |
| Brine (saturated salt solution) | 1202 |
| Styrene | 903 |
| Turpentine | 868 |
| Pure water | 997 |

***Table A1*** *Density of some liquids at 25 °C*

## PRACTICE QUESTIONS

1. a. State the principle of moments.

   b. Figure Q1 shows a bicycle brake lever that has been pulled with a 35 N force to apply the brake.

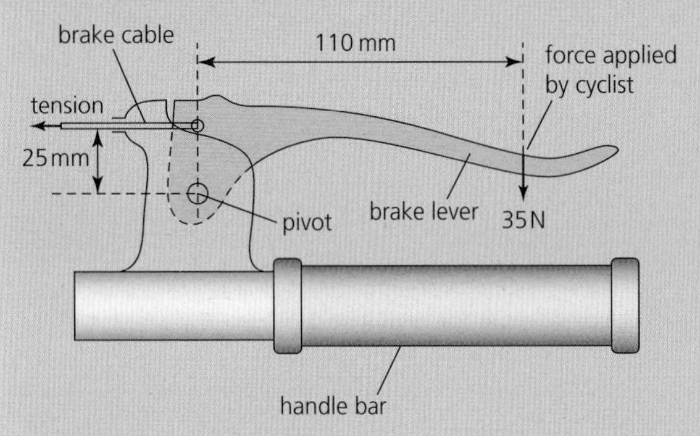


*Figure Q1*

   i. Calculate the moment of the force applied by the cyclist about the pivot. State an appropriate unit.

   ii. Calculate the tension in the brake cable. Assume the weight of the lever is negligible.

*AQA Jan 2013 Unit 2 Q3 (part)*

2. Figure Q2 shows an aircraft designed to take off and land vertically and also to hover without horizontal movement. In order to achieve this, upward lift is produced by directing the jet engine outlet downwards. The engine also drives a vertical lift fan near the front of the aircraft. The weight of the aircraft is 180 kN. The distance between the lift fan and the centre of mass is 4.6 m and the distance between the jet engine outlet and the centre of mass is 2.8 m.

   a. i. Calculate the moment caused by the weight of the aircraft (Figure Q2) about the point X.

   ii. By taking moments about X, calculate the lift fan thrust if the aircraft is to remain horizontal when hovering.

   iii. Calculate the engine thrust.

   b. Having taken off vertically, the jet engine outlet is turned so that the engine thrust acts horizontally. The aircraft accelerates horizontally to a maximum velocity. The forward thrust produced by the jet is 155 kN. The weight of the aircraft is 180 kN.

   i. When the resultant horizontal force is 155 kN, calculate the horizontal acceleration of the aircraft.

   ii. State and explain **one** characteristic of the aircraft that limits its maximum horizontal velocity.

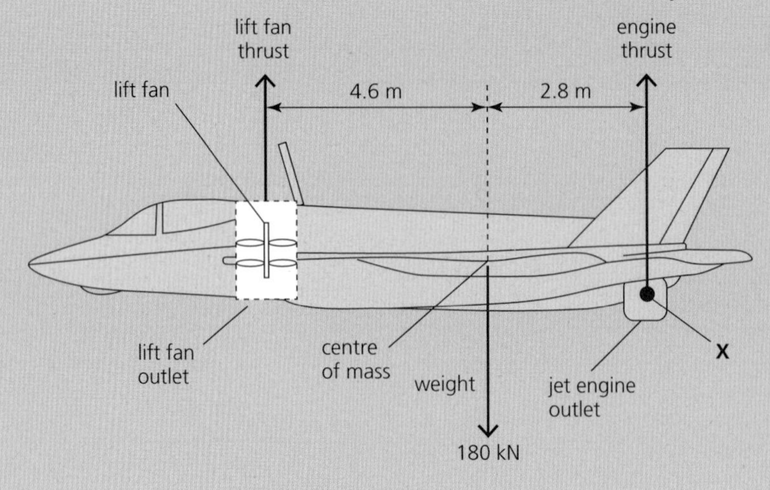


*Figure Q2*

iii. Sketch the velocity–time graph for the horizontal motion of the aircraft as it accelerates from zero to its maximum horizontal velocity.

c. State how a velocity–time graph could be used to find the maximum acceleration.

*AQA Jan 2012 Unit 2 Q3*

3. Ben is pulling a crate along the floor using a rope tied near the top of the crate, as shown in Figure Q3. An equal force of friction is acting at the base of the crate.

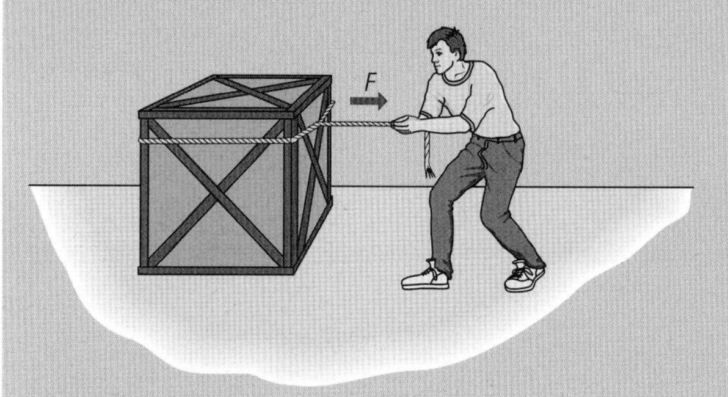


*Figure Q3*

a. Explain whether or not the crate is in equilibrium.

b. How could Ben pull the crate differently to achieve a better result?

4. a. What could happen if the fork-lift truck in Figure Q4 tried to lift a load that was extremely heavy?

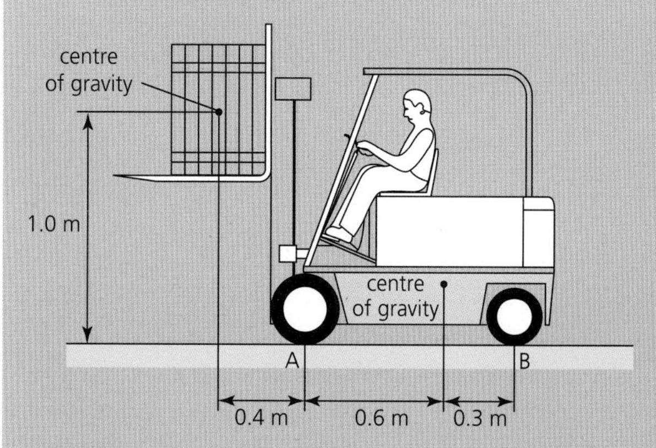


*Figure Q4*

b. What is the maximum load that could be safely lifted by the truck, if the weight of the truck is 2500 N?

5. This question is about the forces on the human body in ballet.

*Figure Q5*

When ballet dancers are '*en point*' (Figure Q5), the contact area between their foot and the floor is very small.

a. What forces are acting on the dancer in the picture?

b. Draw a free-body diagram to illustrate the forces on the dancer and explain what she must do to maintain her balance.

c. Figure Q6 shows the forces acting on the dancer's foot. Assume that the dancer has a mass of 50 kg.

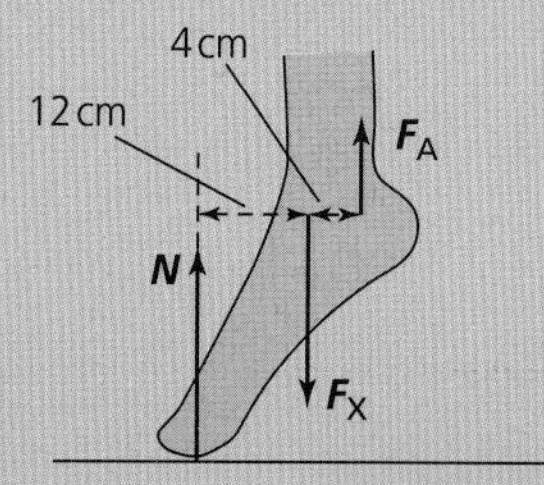


*Figure Q6*

i. Calculate the size of the force $N$.

ii. By taking moments around an appropriate point, find $F_A$, the force in the Achilles tendon.

iii. What is the size of the force $F_X$?

6. A car of mass $m$ is parked on a slope that is inclined at an angle $\theta$ to the horizontal. Which response row in Table Q1 correctly identifies the forces acting on the car?

| | Parallel to the slope | Perpendicular to the slope |
|---|---|---|
| A | $mg$, friction | contact force from the road |
| B | $mg \sin \theta$, friction | $mg \cos \theta$, contact force from the road |
| C | $mg \cos \theta$, friction | $mg \sin \theta$, contact force from the road |
| D | $mg \cos \theta$, contact force from the road | friction |

*Table Q1*

7. Danny Macaskill achieves remarkable balancing positions. Which one of the following statements is false?

   **A** The centre of gravity of Danny and his bike together, must be vertically above the area of contact between the wheel and the ramp.

   **B** All the weight of Danny and his bike can be treated as acting at a single point known as the centre of gravity.

   **C** The balance will be more stable if the tyres are underinflated (slightly flat).

   **D** The balance will be more stable if the combined centre of gravity is as high as possible.

*Figure Q7*

8. Sarah is trying to push a wardrobe across a bedroom floor. She pushes with a gradually increasing force, $F$. Eventually, the wardrobe slides across the floor. The frictional force between the wardrobe and the floor is:

   **A** always greater than $F$

   **B** always less than $F$

   **C** always equal to $F$

   **D** equal to $F$ until the wardrobe starts to move

9. An aircraft is flying horizontally at a constant velocity. Which row in Table Q2 must be true?

| | | |
|---|---|---|
| A | The lift equals the drag | The thrust equals the weight |
| B | The lift equals the weight | The thrust equals the drag |
| C | The drag equals the weight | The thrust equals the lift |
| D | The thrust equals the weight | The drag equals the weight |

*Table Q2*

10. Two 10 N forces are added together. Which of the following is **not** a possible answer?

   **A** 20 N

   **B** 30 N

   **C** 0 N

   **D** 10 N

11. An object is acted upon by two forces, 20 N vertically and 40 N horizontally. Which of the following is the resultant force?

   **A** 45 N at an angle of 26.6° to the horizontal

   **B** 45 N at an angle of 26.6° to the vertical

   **C** 2000 N at an angle of 26.6° to the horizontal

   **D** 60 N at an angle of 26.6° to the horizontal

**12.** A 25 m high television mast is supported by a cable that is connected from the midpoint of the mast to an anchor 10 m from the base of the mast. The tension in the cable is 15 kN. The horizontal component of the tension is equal to:

**A** 11.7 kN

**B** 13.9 kN

**C** 9.4 kN

**D** 5.6 kN

## Stretch and challenge

**13.** The bones of the middle ear (the malleus, incus and stapes) act as a lever connecting the eardrum to the oval window of the inner ear. The relative distances of the eardrum and the oval window from the fulcrum are shown in Figure Q8.

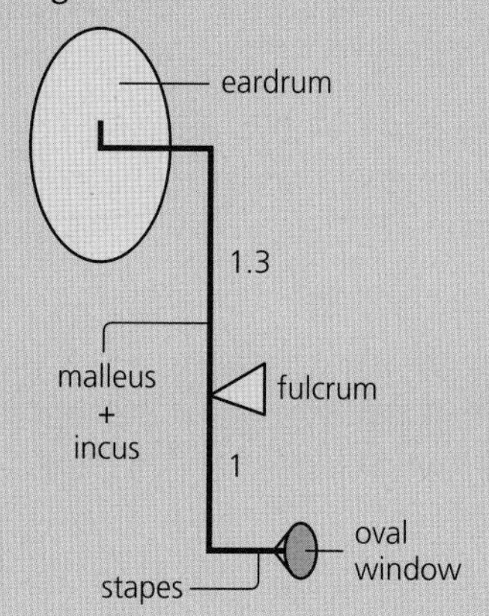


*Figure Q8*

Considering just the action of the bones, how does the size of the force and the amount of movement of the oval window compare to that of the ear drum? Choose the correct row from Table Q3.

| | Force on the oval window | Movement of the oval window |
|---|---|---|
| **A** | Larger | Larger |
| **B** | Smaller | Larger |
| **C** | Larger | Smaller |
| **D** | Smaller | Smaller |

*Table Q3*

# 11 FORCES AND MOTION

## PRIOR KNOWLEDGE

You will be familiar with the definitions of acceleration, velocity and displacement and you will know how to apply the equations of linear motion (Chapter 9) to solve problems. You will be familiar with the formulas for kinetic energy = $\frac{1}{2}mv^2$ and gravitational potential energy = $mgh$.

## LEARNING OBJECTIVES

In this chapter you will learn about Newton's laws of motion, about momentum and impulse, and about the concepts of mechanical work, energy and power.

**(Specification 3.4.1.4 part, 3.4.1.5, 3.4.1.6, 3.4.1.7, 3.4.1.8)**

Britain's roads today are safer than at any time since records began in 1926. That there has been a dramatic reduction in road deaths since 1966, when fatalities reached in the region of 8000 per year, is partly due to general improvements in car design, particularly in the performance of brakes, steering and suspension systems. However, in recent years, increasingly sophisticated safety measures have been introduced with the aim of improving survival in collisions. Crumple zones, safety belts and airbags were all designed to reduce the force experienced by the driver and passengers in a crash.

The emphasis has now turned to avoiding collisions altogether. One simple way is to mark the carriageway with chevrons, designed to keep cars a safe distance apart. But some drivers ignore these. Eliminating driver 'error' is the next challenge. About 70% of accidents are due to the driver speeding, being distracted, driving while tired or intoxicated, or failing to see another vehicle. Automation seems to be the answer. Cars are already being fitted with automatic systems.

- ESC (electronic stability control) detects wheel slide or spin and works with ABS (anti-lock braking system) to get the car back under control.
- V2V is vehicle-to-vehicle communication of position and velocity data, the idea being that cars automatically take evasive action if necessary.
- Computer-controlled headlights can be linked to car speed, the weather, road conditions and the position of the steering wheel.
- Driver monitoring keeps an eye on driving behaviour, concentration and facial expression. It warns you if you are getting too tired to drive.

Soon there may be no need for a driver at all! Prototype driverless cars are already being tested and may soon be on the road (Figure 1).

***Figure 1*** *Nissan plans to have the autonomous Nissan Leaf on sale by 2020.*

# 11.1 NEWTON'S LAWS

## Newton's first law

Head-on collisions are the most common cause of severe and fatal injuries in car accidents. A front-end crash into a stationary object at only 30 km $h^{-1}$ (19 mph) can cause serious injury to the occupants of the car. Modern cars have a range of 'occupant restraint systems' designed to decelerate passengers in a controlled way. Without these, passengers would carry on at their original velocity, until the steering wheel or the windscreen decelerated them in an abrupt and damaging way. It would be wrong to say that the passengers are being hurled forward by the force of the crash. Passengers continue at constant velocity until a force changes their motion. The people in the car are obeying **Newton's first law**:

An object will remain at rest, or continue to move with uniform velocity, unless acted on by an external, resultant force.

At first sight, many objects do not appear to obey Newton's first law. If you remove the driving force from a car, it will not keep moving at constant velocity. Turn off the engine and the car will quickly come to a stop. This is because a combination of air resistance and frictional forces make up the 'external force' that brings the car to rest. On Earth, it is difficult to avoid these forces; but in space, where there is negligible resistance to motion, Newton's first law can be observed more clearly (Figure 2).

***Figure 2*** *Voyager 1 left Earth 36 years ago and is now 18.7 billion kilometres away. It will keep travelling at its current speed of 17 km $s^{-1}$ in a straight line, unless it is acted upon by a force.*

While forces such as friction, drag and gravity are always present on Earth, in some cases they act equally in opposite directions, and so cancel each other out. In these cases, for example a skydiver falling at terminal speed (section 9.6 of Chapter 9), there is no resultant force and the object moves with constant velocity in accordance with Newton's first law.

### QUESTIONS

1. In a campaign to encourage people to wear rear-seat safety belts, the Department of Transport claimed that, in an accident,

   "An adult passenger sitting in the back of a car will be thrown forward with a force of 3 $\frac{1}{2}$ tonnes. That's the weight of an elephant."

   Rewrite this safety warning so that the physics is correct.

2. Car seats have head restraints, which are designed to prevent neck injuries. Why are they especially helpful in rear-impact accidents?

## Momentum

A quarter of all vehicle accidents involve collisions from the side. Many cars now have reinforced bars in the doors to try to maintain a survival space after a collision. The effectiveness of the bars depends on how large the forces are. A greater force is needed to stop a juggernaut moving at 100 km $h^{-1}$ than a bicycle moving at 15 km $h^{-1}$.

Clearly, the size of the forces in any crash will depend on both the mass and the velocity of the vehicles. The relevant property is the product of mass and velocity, which is called **momentum**:

$$\text{momentum} = \text{mass} \times \text{velocity}$$

$$p = mv$$

Momentum is measured in kg m $s^{-1}$. It is a vector quantity.

## Newton's second law

The more momentum an object has, the harder it is to stop. If the object has to lose all of its momentum in a short time, as in a crash, the force needed will be greater. If the object loses its momentum more slowly, as in a normal controlled braking, then less force is needed. The resultant force on an object is related to its rate of change of momentum. This is stated in **Newton's second law**:

> The rate of change of momentum of a body is directly proportional to the external, resultant force acting on it. The change of momentum takes place in the direction of that force.

In symbols, we can write Newton's second law as

$$F \propto \frac{\Delta p}{\Delta t} \quad \text{or} \quad F \propto \frac{\Delta (mv)}{\Delta t}$$

Since the unit of force (newton) is defined as $1\,\text{N} = 1\,\text{kg}\,\text{m}\,\text{s}^{-2}$ (see Figure 3), the proportionality becomes an equality:

$$F = \frac{\Delta p}{\Delta t} \quad \text{or} \quad F = \frac{\Delta (mv)}{\Delta t}$$

If the mass of the object is constant, Newton's second law can be expressed as:

$$\text{force} = \text{mass} \times \text{acceleration}$$

$$F = ma$$

If the mass of the object changes, as in the case of a rocket using up fuel, it is necessary to use the full statement of Newton's second law, involving rate of change of momentum. The mass of a car does not change significantly as it is driven, so the law simplifies to $F = ma$.

We can use $F = ma$ to calculate the average resultant force required by a Lamborghini Diablo (Figure 4) to achieve its acceleration. In SI units:

$$a = \frac{v - u}{t} = \frac{27.8 - 0}{4.09} = 6.80\,\text{m}\,\text{s}^{-2}$$

Therefore, the average resultant force on the car has to be

$$F = ma = 1449\,\text{kg} \times 6.80\,\text{m}\,\text{s}^{-2} = 9850\,\text{N}$$

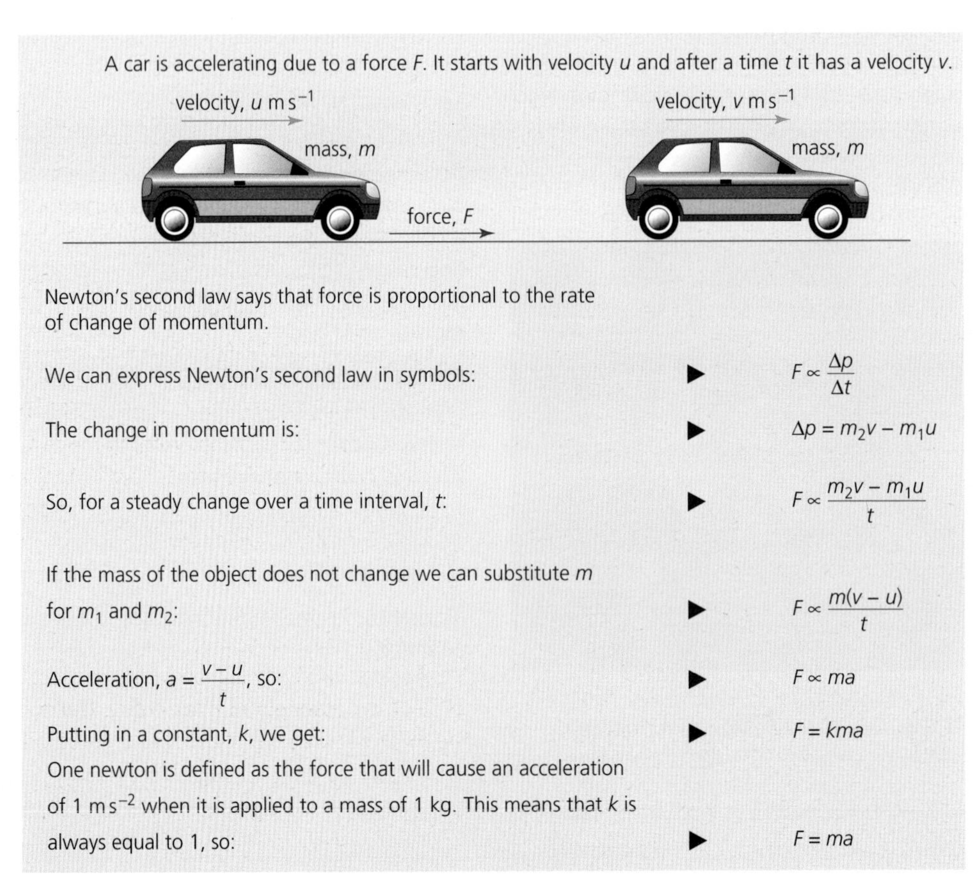


***Figure 3*** *Deriving F = ma*

***Figure 4*** *A Lamborghini Diablo has a mass of 1449 kg and can accelerate from 0 to 100 km h$^{-1}$ (27.8 m s$^{-1}$) in 4.09 s.*

### QUESTIONS

3. a. A cheetah can accelerate at 13 m s$^{-2}$ in very short bursts. It has a top speed of 25.9 m s$^{-1}$. From stationary, how long would it take a cheetah to reach that top speed?
   b. Estimate the force required for this acceleration. Suggest the consequences for the anatomy of the cheetah.
4. Estimate the force on a racing bicycle over the first few metres of a race.

## Resistive forces

In reality, the Lamborghini Diablo must provide a larger driving force than 9810 N because it must also overcome forces holding the car back. When a vehicle is being driven on a flat, straight road, there are two forces opposing its motion (Figure 5):

- Air resistance – this drag force depends on the size and shape of the vehicle, and is roughly proportional to the squared speed of the car (see section 9.6 in Chapter 9).
- Frictional force at the surface – this 'rolling' resistance is roughly proportional to the speed of the car.

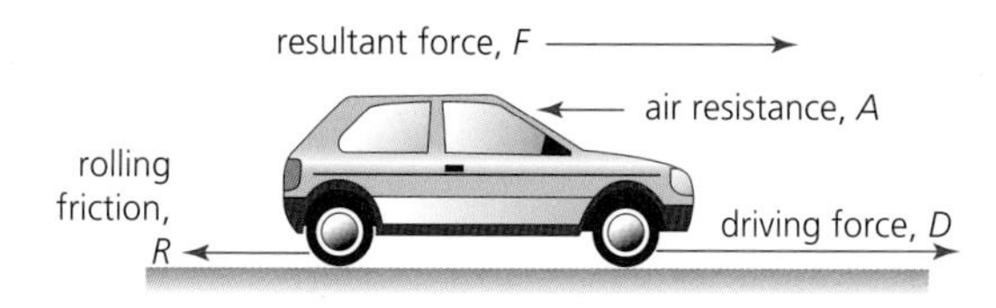


The resultant force on the car, $F = D - A - R$. When a car is travelling at constant velocity, $F$ must be zero, so $D = A + R$.

***Figure 5*** *Horizontal forces acting on a moving vehicle*

## Forces in a crash

The forces that act on a car and its occupants during an accident can be dangerously large. If the Lamborghini (Figure 5) is involved in a crash, its speed could drop from 60 mph (26.8 m s$^{-1}$) to zero in less than one-tenth of a second. This is an acceleration of

$$a = \frac{v - u}{t} = \frac{0 - 26.8}{0.1} = -268 \text{ m s}^{-2}$$

The average force $F$ on the car is given by

$$F = ma = 1449 \times (-268) = -390000 \text{ N} = -3.9 \times 10^5 \text{ N}$$

The negative sign shows that the force acts in the opposite direction to the velocity; it is a decelerating force. The deceleration in this case is about 27 times the acceleration due to the Earth's gravity, $g$. The force on a passenger will be roughly 27 times their weight.

### QUESTIONS

5. Some people believe that a baby can travel safely in a car if it is held by an adult. Estimate the force needed to restrain a baby in the event of a crash.
6. *The Highway Code* claims that it takes a distance of 23 m to come to rest if you make an emergency stop at a speed of 30 mph (48.3 km h$^{-1}$). Of this, 9 m is 'thinking distance', before you apply the brakes. What average braking force is needed if the mass of the car is 1500 kg? (*Hint*: You will need to find the average deceleration of the car.)

## Reducing the force

Newton's second law can be used to calculate the force on a passenger in a car crash. During the crash the force will vary, but the average force will be

$$F = \frac{\text{final momentum} - \text{initial momentum}}{\text{time taken to decelerate}} = \frac{\Delta(mv)}{\Delta t}$$

The initial momentum depends on the velocity of the car; the final momentum is zero. Therefore, all that we can do to reduce the force is to increase the time, $\Delta t$, over which the collision occurs. Design features such as crumple zones (Figure 6), collapsible steering columns and airbags all increase the time taken for a passenger to come to a stop. Some people believe that modern cars are 'weak' and too easily damaged in a crash. In fact, they are designed so that the bonnet and boot deform relatively easily. It is this 'weakness' that saves lives.

***Figure 6*** *Crumple zones reduce the deceleration of the passenger compartment.*

## QUESTIONS

7. Motorway crash barriers are designed to deform as a vehicle hits them (Figure 7). Use Newton's second law to explain why this is safer than a rigid safety barrier.

***Figure 7***

## Newton's third law

When a car hits a stationary fixed object, such as a tree, the front of the car stops almost instantly. The passenger compartment takes longer to stop: its deceleration changes as the front of the car crumples (Figure 8). For a car travelling at 60 km h$^{-1}$ (less than 40 mph), the deceleration may peak at 60$g$ but it is likely to average about 20$g$ over the duration of the crash (around 100 ms).

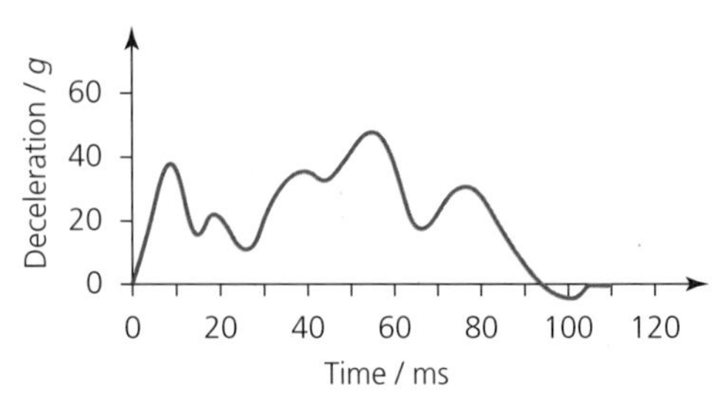


***Figure 8*** *Deceleration of the passenger compartment during a crash*

A seat belt can reduce the average deceleration on the driver to between 10$g$ and 15$g$. The belt achieves this because of its elasticity. It stretches slightly during the collision. The tension in the belt then restrains the driver. As the driver is exerting a force on the seat belt, the seat belt is exerting an equal but opposite force on the driver (Figure 9). The seat belt exerts a force on each of its three mounting points, which in turn pull back on the belt. These pairs of forces are examples of **Newton's third law**:

When body A exerts a force on body B, then body B exerts an equal but opposite force on body A.

This law is often quoted as 'Every action has an equal and opposite reaction'. This can be misleading. It is important to realise that the two forces, the 'action'

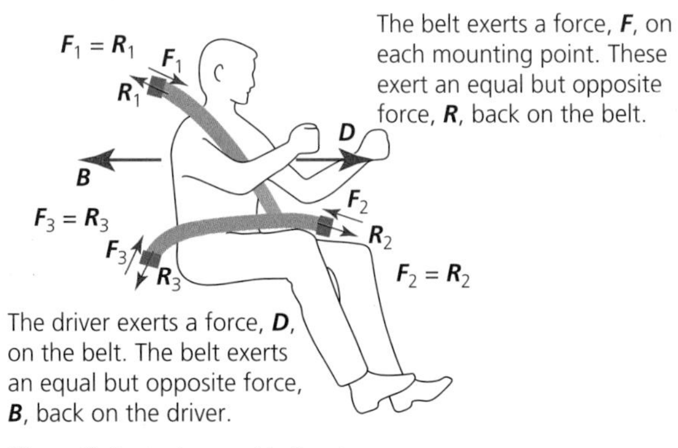


***Figure 9*** *Forces in a seat belt system*

and the 'reaction', act on *different* objects. If the equal and opposite forces acted on the same object, there would never be a resultant force and, by Newton's first law, nothing would ever accelerate.

A contact reaction force was considered in section 10.1 of Chapter 10. But note that Newton's third law is not only relevant when two objects are in contact, it also applies when the interaction is at a distance, for example due to gravitational attraction.

## QUESTIONS

8. a. Sketch a parked car and draw in arrows to show the forces acting on
   i. the car
   ii. the Earth.
   b. Carefully explain which pairs of forces are equal according to
   i. Newton's first law
   ii. Newton's third law.
9. What force pushes a car forward as it accelerates?

## KEY IDEAS

- Newton's first law of motion states:

  An object will remain at rest, or continue to move with uniform velocity, unless acted on by an external resultant force.
- Momentum is defined as mass × velocity: $p = mv$. Its unit is $\text{kg m s}^{-1}$.
- Newton's second law of motion states:

  The rate of change of momentum of a body is directly proportional to the external resultant force acting on it.

  The change of momentum takes place in the direction of the resultant force.

$$F = \frac{\Delta(mv)}{\Delta t}$$

- For constant mass, $F = ma$, where $a$ is the acceleration.

  A force of 1 N will accelerate a mass of 1 kg at a rate of $1\,\text{m s}^{-2}$.
- Newton's third law states:

  When body A exerts a force on body B, then body B exerts an equal but opposite force on body A.

## ASSIGNMENT 1: VERIFYING NEWTON'S LAWS OF MOTION

**(MS 1.5, MS 3.2, MS 3.3, MS 3.4, PS 1.2, PS 2.2, PS 2.3, PS 3.1, PS 3.3, PS 4.1)**

This assignment is about the use of an air track and light gates to verify Newton's first and second laws of motion.

**Newton's first law of motion** says that an object will continue to move at constant velocity, unless acted on by a resultant external force. Getting rid of any external forces is difficult on Earth. Gravity, air resistance and friction are always present. However, we can use an air track to reduce the effect of these forces.

An air track is a horizontal metal rail with holes through which air can be pumped. A glider sits on top of the rail on a cushion of air. This removes contact friction, because the surfaces are not in contact. The glider carries a 'flag', which interrupts the light beam as it passes through the light gate (see Figure A1). The length of the flag is input manually, and the time for which the light is interrupted is recorded by the computer, which can then calculate the speed. Stretched elastic bands are placed at each end of the track and at each end of the glider, so the glider bounces back with very little loss of kinetic energy.

Newton's first law of motion can be verified by simply pushing the glider along the track. The light gates should record a constant velocity.

### Questions

**A1** Identify the sources of uncertainty in calculating the velocity.

**A2** What precautions should be taken in order to get accurate results for the experiment?

**Newton's second law** can be expressed as:

force is proportional to the rate of change of momentum

or, in cases where the mass is constant, as:

$$F = ma$$

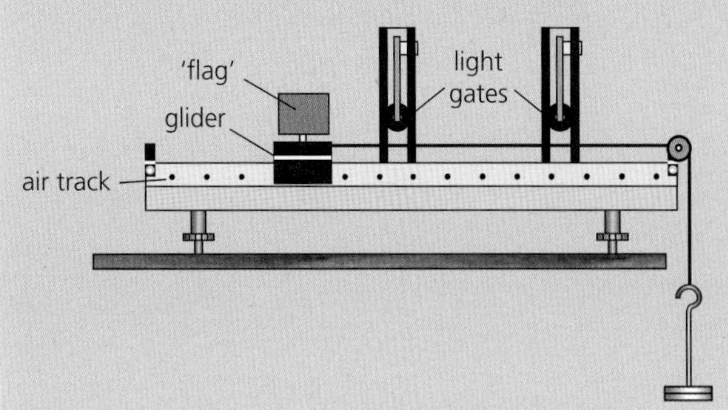


***Figure A1*** *Set-up for verifying Newton's second law of motion*

In this case, a known force can be applied to the glider by hanging masses over the pulley (Figure A2). The light gates measure the velocity of the glider at two different times. The computer also records the time between light gates, and therefore is able to work out the acceleration.

A set of results for the experiment is shown in Table A1. The first column shows the mass hanging over the pulley; the second column converts this to a force in newtons.

### Questions

**A3** Copy Table A1 and complete it by calculating force and acceleration values.

**A4** Plot a graph of force on the $y$-axis against acceleration on the $x$-axis.

**A5** Is this graph consistent with Newton's second law?

**A6** Use the graph to estimate the mass of the glider.

**A7** The graph is not a perfect straight line. Suggest some reasons for this.

| Mass / g | Force / N | Velocity / m s$^{-1}$ | | $t$ / s | $a$ / m s$^{-2}$ |
|---|---|---|---|---|---|
| | | Light gate 1 | Light gate 2 | | |
| 10 | | 0.24 | 0.49 | 1.35 | |
| 20 | | 0.25 | 0.66 | 1.09 | |
| 30 | | 0.25 | 0.78 | 0.96 | |
| 40 | | 0.28 | 0.89 | 0.85 | |
| 50 | | 0.31 | 0.99 | 0.76 | |
| 60 | | 0.34 | 1.08 | 0.70 | |
| 70 | | 0.44 | 1.18 | 0.62 | |
| 80 | | 0.55 | 1.28 | 0.56 | |
| 90 | | 0.66 | 1.39 | 0.52 | |

***Table A1***

## 11.2 CONSERVATION OF LINEAR MOMENTUM IN COLLISIONS

Consider the motion of a car in one dimension; in other words, it is travelling in a straight line. It has linear momentum, $p = mv$. If a moving car A is travelling at high speed and crashes into a stationary car B (Figure 10), Newton's second law can be applied to both vehicles:

- for car A

  change in momentum = force × time

- for car B

  change in momentum = – force × time

Newton's third law states that both vehicles are subject to the same magnitude force. The minus sign shows that they act in opposite directions (momentum is a vector quantity). The time of impact is also the same for both vehicles. Hence, the change in linear momentum has to be the same size for each car. This means that any momentum lost by the first car will be gained by the second car. Therefore, the total momentum before the crash will be the same as the total momentum after the crash (see Figure 10).

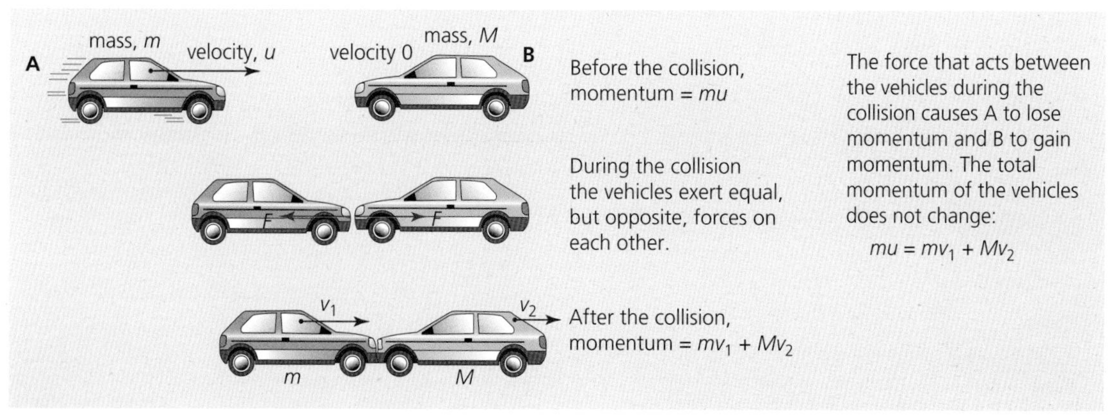


***Figure 10*** *Conservation of momentum in a crash*

In reality, external forces such as friction will slow the cars down. But if there were no external forces, the total momentum in the system would be the same before and after the collision. This fundamental law is known as the **principle of conservation of linear momentum**:

> The total linear momentum of a system does not change, provided that no net external force is acting.

It is not always obvious that this law is being obeyed. When a car's brakes are applied and it comes to rest, where has the momentum of the system gone? The answer depends on what you regard as the system. If you treat the car as the system, then its momentum has not been conserved, but friction between the car and the road is an external force. If you regard the car and the road as the system, there are no significant external forces, since the friction between the two is now an 'internal' force.

The conservation law suggests that any momentum lost by a car slowing down will be gained by the Earth. Hitting the car brakes must speed up the Earth a little! Of course, the reverse is also true; every time a car accelerates away from the traffic lights it pushes the Earth back the other way.

The principle of conservation of momentum applies to all events, from collisions between galaxies to collisions between subatomic particles, and everything in between, including firework explosions (Figure 11) and nuclear decay (Figure 12). In the absence of any external forces, the total momentum before the event is always conserved.

***Figure 11*** *When a firework explodes, the total momentum of all the fragments must equal the momentum before the explosion. If the firework was initially stationary, the total momentum in any direction afterwards would be zero (although gravity acts as an external force, eventually pulling all the pieces back to Earth).*

## Worked example

The principle of conservation of momentum is used in particle physics to calculate the masses and velocities of particles following a decay or a collision. A simple case is the emission of an alpha particle from a radioactive nucleus that is initially stationary (Figure 12).

A nucleus of radium-226 decays to radon-222 by emitting an alpha particle of energy 4.87 MeV. The daughter nucleus, now of radon-222, will recoil as the alpha particle is emitted. We can use the principle of conservation of momentum to calculate the velocity of the recoil.

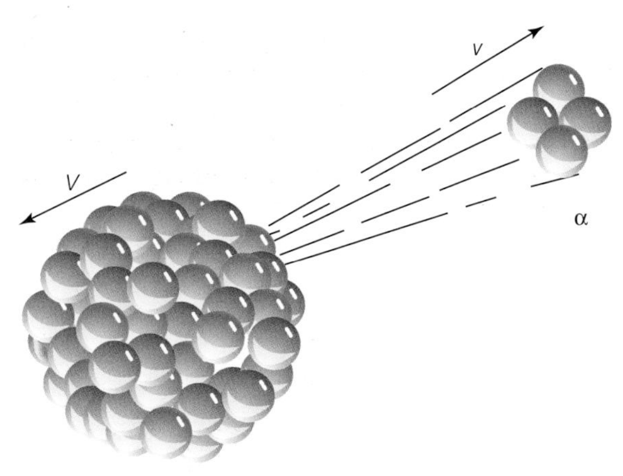


***Figure 12*** *Radium-226 emits an alpha particle as it decays to radon-222.*

The velocity of the alpha particle can be calculated from its energy. The particle is emitted with kinetic energy 4.87 MeV = $4.87 \times 10^6 \times 1.6 \times 10^{-19}$ J = $7.79 \times 10^{-13}$ J. Using the formula for kinetic energy (see section 11.4),

$$E_k = \frac{1}{2}mv^2$$

the velocity is given by

$$v = \sqrt{\frac{2E_k}{m}} = \sqrt{\frac{2 \times 7.79 \times 10^{-13}}{6.64 \times 10^{-27}}} = 1.53 \times 10^7\ \text{m s}^{-1}$$

This is an approximate value, because as speed approaches the speed of light, the kinetic energy equation $E_k = \frac{1}{2}mv^2$ becomes inaccurate. This value is about 5% of the speed of light, so the inaccuracy is relatively small.

Now applying the principle of conservation of momentum:

momentum of the radium nucleus before emission = 0 kg m s$^{-1}$

total momentum after the emission = 0 kg m s$^{-1}$

So the recoil momentum of the radon nucleus must be equal and opposite to the momentum of the alpha particle:

(mass of radon nucleus) × (velocity of radon nucleus) = (mass of the alpha particle) × (velocity of alpha particle)

Given that the relative masses of the alpha particle and radon nucleus are 4 and 222, respectively, we have:

$$\text{recoil velocity of radon nucleus } V = \frac{4 \times 1.53 \times 10^7}{222} = 276\ \text{km s}^{-1}$$

## QUESTIONS

**10. a.** In action films, people who are shot are often shown flying backwards from the impact of the bullet. Is this likely, given that the typical mass of a bullet is 0.04 kg and its speed is around 300 m s$^{-1}$?

**b.** What would you expect to happen to the person firing the gun?

**11.** During a spacewalk, an astronaut, of mass 60 kg, becomes detached from the spacecraft and is floating out into space. She has the presence of mind to throw a spanner, of mass 1 kg, at a velocity of 4 m s$^{-1}$.

**a.** In which direction should she throw the spanner?

**b.** Calculate her velocity after she has thrown the spanner.

**12.** A stationary ice skater pushes his partner away, and finds himself moving in the opposite direction.

**a.** Which of Newton's laws predicts this outcome?

**b.** Which law could be used to predict the acceleration of each skater during the push?

**c.** After he has pushed his partner away, the skater glides at a nearly constant speed across the ice without any effort. Which of Newton's laws predicts this would happen?

**13.** A TV advert showed a car being driven off a tall building and falling to the ground. How does the conservation of momentum apply here?

**14.** A spacecraft of mass 1000 kg is in deep space, and travelling in a straight path at a speed of 10 m s$^{-1}$. The controllers fire one of the rockets at the back of the craft. The burst lasts for 0.2 s, ejecting 100 g of gas at a speed of 2000 m s$^{-1}$. Describe in detail how the motion of the spacecraft will be affected during and after the event. (You may assume that the mass of the spacecraft remains unchanged during the event.)

# 11.3 IMPULSE

Some collisions are designed to create a large change in velocity for one of the objects. For example, in sport, a serve or return in tennis (Figure 13), or a drive in golf or cricket, are often attempts to make the ball travel as fast as possible. Newton's second law is important here. Since force, $F$, is equal to the rate of change of momentum, $\Delta p/\Delta t$, then rearranging gives

$$F\Delta t = \Delta p = \Delta(mv)$$

***Figure 13*** *The impulse changes the momentum of the ball*

The quantity $F\Delta t$ is known as **impulse** of the force. It is the product of the force and the time for which the force acts, and therefore has the unit newton second (N s):

$$\text{impulse} = \text{change in momentum}$$

This refers to *one* object in a collision, since overall the change in momentum is zero. The larger the impulse on an object, the greater its change in momentum.

In the example of a golf drive, the initial momentum of the ball is zero, so the momentum gained by the ball is equal to the impulse. A golfer must try to maximise the force and the time for which it acts. Typically, a golf club exerts a mean force of 5 kN on the ball and the contact lasts for about 0.5 ms. The impulse is therefore

$$F\Delta t = 5\,\text{kN} \times 0.5\,\text{ms} = 2.5\,\text{N s}$$

But we have assumed that the 5 kN is a constant force, acting throughout the 0.5 ms. It is much more likely that the force varies during the impact, as shown in Figure 14. If we consider a small time interval, $\Delta t$, the force, $F$, is approximately constant over that time, so the impulse is $F\Delta t$. The total impulse during the contact is the sum of all such small rectangles, which is the area under the curve.

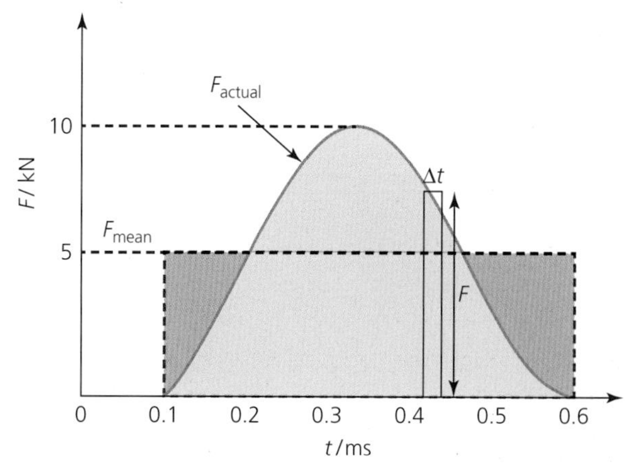


***Figure 14*** *A more realistic model of the force between a golf club and a ball during a drive*

As a first approximation, we could treat the graph in Figure 14 as two triangles of height 10 kN and base 0.25 ms. So the impulse is approximately

$$2 \times \tfrac{1}{2} \times (10\,000 \times 0.25 \times 10^{-3}) = 2.5\,\text{N s}$$

The change of momentum therefore equals 2.5 N s (or 2.5 kg m s$^{-1}$).

Because the ball was initially at rest, the change of momentum, $\Delta p$, is also equal to the final momentum:

$$mv = 2.5\,\text{kg m s}^{-1}$$

As the mass of the golf ball is around 50 g, its final velocity $v$ is

$$v = \frac{\Delta p}{m} = \frac{2.5}{0.050} = 50\,\text{m s}^{-1}$$

## QUESTIONS

**15.** In an annual egg-throwing competition, players have to catch an egg thrown by their partner without breaking it. The winning pair will catch the egg safely over the longest distance. Explain, in terms of momentum and impulse, what the secret of success is!

**16.** A Powerball (a small, very bouncy ball) is placed on top of a basketball (Figure 15). Then the basketball is dropped onto the ground. When the basketball hits the ground, the Powerball shoots up into the air. (This experiment must be done outdoors.) Explain why this happens.

*Figure 15*

**17.** Estimate the force exerted by the racket on the ball when a tennis player returns a serve.

## KEY IDEAS

- The principle of conservation of linear momentum states:

  The total momentum of a system does not change, provided no resultant external force acts.

- Impulse is defined as force × time for which the force acts. It is equal to the change in momentum of the object on which the force acts:

$$F\Delta t = \Delta(mv)$$

- The unit of impulse is newton second (N s) or $\mathrm{kg\,m\,s^{-1}}$.
- The impulse of a varying force is the area below the force–time graph.

## ASSIGNMENT 2: CALCULATING THE IMPULSE WHEN A FOOTBALL IS KICKED

**(MS 0.3, MS 0.4, MS 1.1, MS 1.5, MS 3.1, MS 3.8, PS 2.1, PS 2.3)**

***Figure A1*** *When shooting at goal, the aim is often to transfer as much momentum as possible to the ball, and to score!*

Scientists of the Sports Technology Research Group UK at Loughborough University have studied how footballs behave when they are kicked (Figure A1). To simulate the large impulse during a kick, footballs were fired at a steel plate attached to a force meter. The results are shown in Figure A2. The force between the ball and the steel plate increases after contact as the ball deforms, reaching a maximum value. The ball then returns to its original shape and bounces away from the plate.

The mass of a football is around 500 g. For each impact velocity in Figure A2:

### Questions

**A1** Estimate the total impulse during the collision.

**A2** Calculate the change in momentum.

**A3** Calculate the speed at which the ball leaves the metal plate.

The mathematical model and the experiments are designed to investigate the behaviour of a real football during a match.

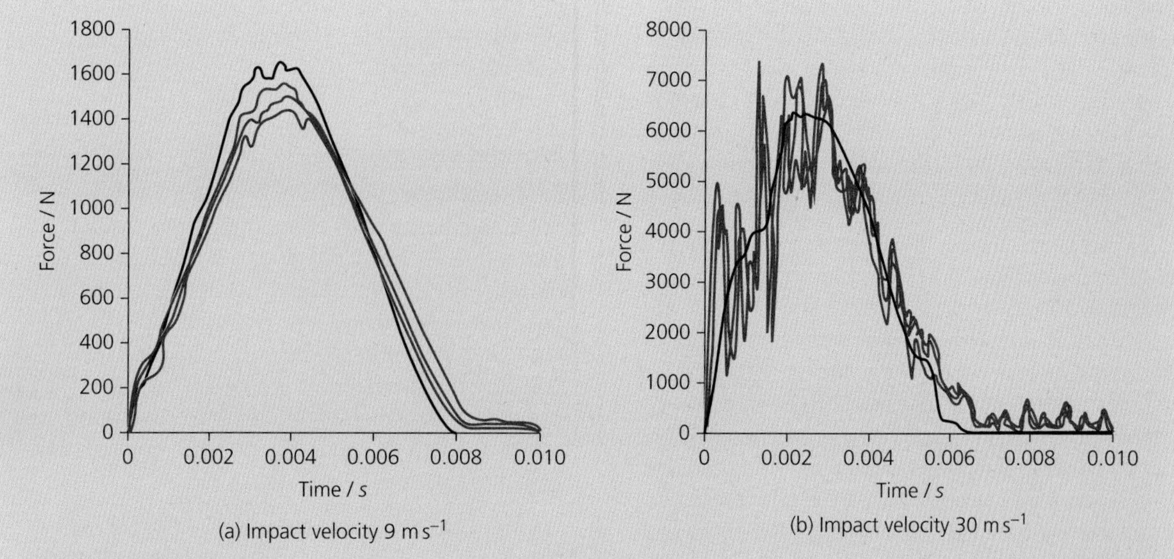


***Figure A2** The black lines show the results of a mathematical model used to investigate the football's performance during an impulse. The red lines are actual experimental results. (Based on information from Loughborough University, Sports Technology Research Group UK, reported in "Journal of Sports Science and Medicine, Supplement 10: 6th World Congress on Science and Football, January 16–20, 2007"; published 1 February 2007.)*

**A4** Comment on the accuracy of the mathematical model.

**A5** What are the advantages and disadvantages of using a mathematical model to investigate a football's behaviour?

**A6** How closely does the experimental set-up mirror the real situation? Discuss any differences that may be relevant.

## 11.4 ENERGY IN COLLISIONS

The study of collisions is very important in physics, quite apart from its relevance to car crashes. On a microscopic level, collisions between constantly moving gas molecules help us to understand the behaviour of gases, and it is collisions between subatomic particles that have illuminated particle physics. At the other end of the scale, astronomers have learned much by studying the collision of comet Shoemaker–Levy with Jupiter (Figure 16).

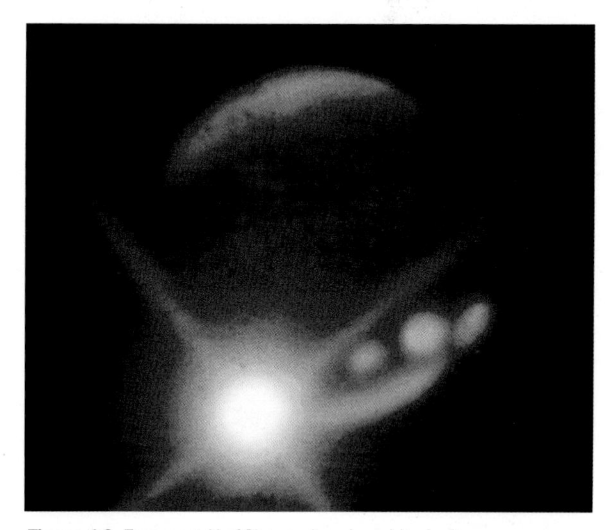

***Figure 16** Fragment K of Shoemaker–Levy hits Jupiter*

In general, in a collision, some of the kinetic energy of the moving objects is transferred to other forms of energy. The **kinetic energy** of a moving object depends on its mass, $m$, and its velocity, $v$. The expression for kinetic energy is

$$E_k = \frac{1}{2}mv^2$$

Collisions are described as either **elastic** or **inelastic**. In an elastic collision, no kinetic energy is lost. Collisions between molecules in a gas can be treated as elastic. Some collisions between subatomic particles are also elastic, such as those that take place between neutrons and carbon atoms in the moderator of a nuclear reactor.

In an inelastic collision, kinetic energy is transferred to other forms (Figure 17). Collisions between everyday objects tend to generate sound and heat and are therefore inelastic. Car crashes certainly fall into this category. Much of the energy is transferred in deformation. Safety features such as crumple zones and airbags (Figure 18) are designed so that the kinetic energy is transferred with as little injury to the occupants as possible. As the crumple zones progressively collapse, the kinetic energy of the car does work against internal forces in the metal and is finally transferred as internal energy in the car body.

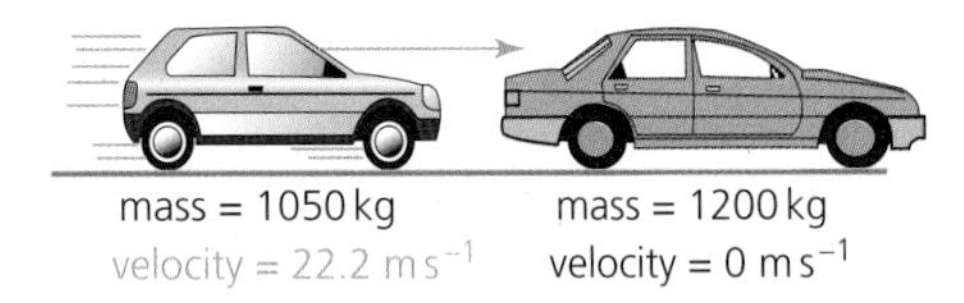


***Figure 17*** *Momentum and energy in inelastic collisions*

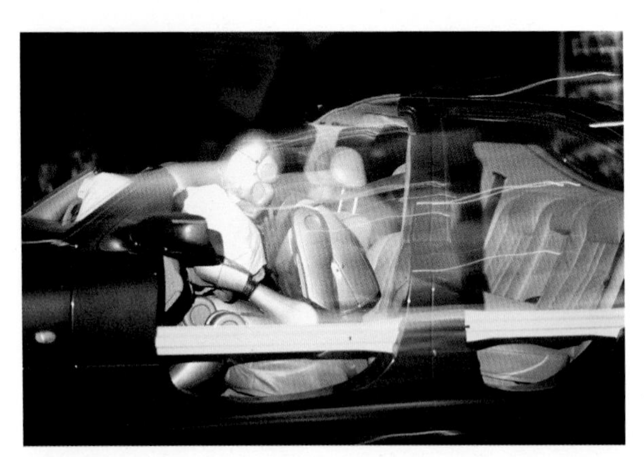

***Figure 18*** *In a collision, your life could be saved by a small, gas-filled balloon. The airbag is triggered by a sudden deceleration, and is then chemically inflated, in less than 50ms. (It takes you 200 ms to blink your eye.) As the driver hits the airbag, the gas escapes and the energy of the driver is transferred to the kinetic energy of the gas molecules.*

It is important to remember that the conservation of momentum applies to both elastic and inelastic collisions.

## Worked example

Suppose that, in poor visibility on a motorway, a car runs into the back of another larger car that is stationary, as shown in Figure 17. The first car is moving at $22.2\,\text{m s}^{-1}$ (about 50 mph) immediately before the impact and has a mass of 1050 kg. The larger car in front has a mass of 1200 kg. The collision, the two vehicles are locked together. We can use the conservation of momentum to calculate the speed at which the two vehicles move immediately after the collision.

The moving car's momentum before the collision = $1050\,\text{kg} \times 22.2\,\text{m s}^{-1} = 2.33 \times 10^4\,\text{kg m s}^{-1}$. The car that is stationary before the collision has no momentum. So the total momentum before the collision is $2.33 \times 10^4\,\text{kg m s}^{-1}$. Conservation of momentum says that the total momentum after the collision must also be $2.33 \times 10^4\,\text{kg m s}^{-1}$.

The vehicles are now locked together so that their combined mass after the collision is $m = 1050\,\text{kg} + 1200\,\text{kg} = 2250\,\text{kg}$. Since

$$\text{velocity} = \frac{\text{momentum}}{\text{mass}}$$

the velocity after the collision is

$$v = \frac{2.33 \times 10^4}{2250} = 10.4\,\text{m s}^{-1}\ (23\text{ mph})$$

The kinetic energy before the collision is entirely due to the moving car:

$$E_k = \tfrac{1}{2}mv^2 = \tfrac{1}{2} \times 1050 \times (22.2)^2 = 2.59 \times 10^5\,\text{J} = 259\,\text{kJ}$$

After the collision:

$$E_k = \tfrac{1}{2} \times 2250 \times (10.4)^2 = 1.22 \times 10^5\,\text{J} = 122\,\text{kJ}$$

So 137 kJ of kinetic energy has been transferred to thermal energy of the surroundings, or **dissipated**, during this inelastic collision.

QUESTIONS

**18.** State whether the following collisions are elastic or inelastic, and explain your answer:

**a.** the collisions between air molecules in a balloon

**b.** a tennis ball bouncing on the ground

**c.** a high-energy electron colliding with a metal target in an X-ray tube.

**19.** In an experiment to measure the speed of a pellet fired from an air rifle, a lead pellet of mass 0.5 g is fired into a Plasticine target mounted on an air track glider. As the pellet embeds itself in the Plasticine, the glider moves off at a speed of 0.2 m s$^{-1}$. The combined mass of the Plasticine and glider is 150 g.

**a.** Calculate the velocity of the pellet.

**b.** Estimate the uncertainty in the result.

**c.** What have you assumed in your calculation?

**d.** If you were carrying out this experiment, what steps would you take to make it both safe and accurate?

KEY IDEAS

- In an elastic collision, kinetic energy is conserved.
- In an inelastic collision, kinetic energy is transferred to other forms.
- The principle of conservation of momentum applies to all collisions, whether elastic or inelastic.

# 11.5 WORK, ENERGY AND POWER

## Work

It is obviously better to stop a car safely, before a crash ever takes place. A reliable braking system is the most important safety mechanism in a car. The frictional forces need to be very large if the car is to stop in a very short distance.

We can calculate how much energy is transferred using the concept of **work**. Physicists say that work is done when a force moves through a distance. If you push a car along a road, you have done some work against resistive forces. If you lift a mass, you have done some work against the force of gravity:

work done (J) = force (N) × distance moved in the direction of the force (m)

$$W = Fs$$

where $F$ is measured in newtons and $s$ is the distance moved in a particular direction, or the **displacement**, in metres. The unit of work is the joule (1 J = 1 N m).

Although the force and the displacement are both vector quantities, work is a **scalar** quantity (see Chapter 9). If the directions of the force and the displacement are at an angle $\theta$, the equation for work becomes (see Figure 19)

$$W = Fs\cos\theta$$

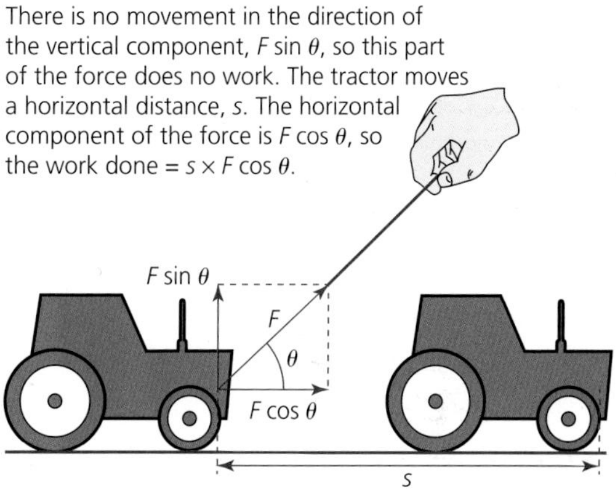

***Figure 19*** *Work done by a force at an angle $\theta$*

### Worked example 1

When the road conditions are good, a car travelling at 70 mph (31 m s$^{-1}$) can usually be brought safely to a halt in 75 m. To calculate the work done, we need to find the force required. The acceleration can be found using one of the equations of linear motion:

$$v^2 = u^2 + 2as$$

so

$$a = \frac{v^2 - u^2}{2s}$$

Since $v = 0$, $u = 31$ m s$^{-1}$ and $s = 75$ m:

$$a = -\frac{31^2}{2 \times 75\text{ m}}$$

$$= -6.4\text{ m s}^{-2}$$

For a car of mass, say, 1000 kg, the average braking force is given by

$$\begin{aligned} F &= ma \\ &= 1000 \times 6.4 \\ &= 6400\,\text{N} \end{aligned}$$

The work done is therefore

$$W = Fs = 6400 \times 75 = 480\,\text{kJ}$$

The method of calculating work in Worked example 1 is adequate for a situation where the force is constant – but forces usually vary. To calculate the work done by a varying force, we need to consider small displacements, $\Delta s$, over which the force is approximately constant. The work done in moving each small displacement, $\Delta s$, is equal to $F\Delta s$, which is the area of the strip shown in Figure 20. The total work done is the sum of the areas of all such strips:

The total work done by a varying force is given by the area under the force–displacement graph.

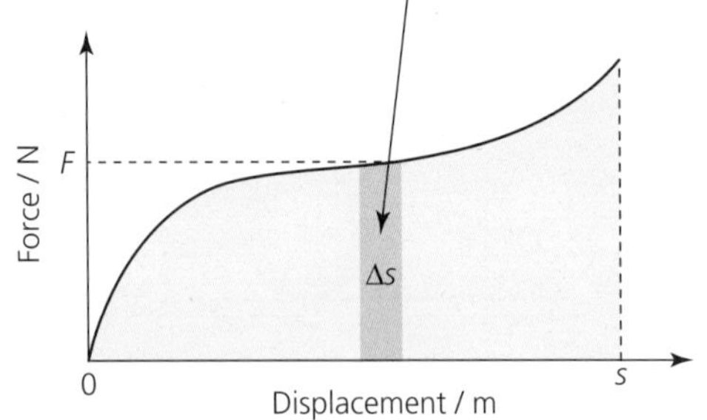


***Figure 20*** *Work done by a varying force*

## Kinetic energy

In physics, an object is said to have energy if it can do work, for example by lifting a weight. A moving car has kinetic energy because of its motion. This energy could be used to lift a weight, perhaps by pulling a rope, passing over a pulley, attached to a weight. Even light, a form of energy, can be used to lift a weight (Figure 21).

***Figure 21*** *Sunlight falling on the solar panels on a Mars rover generates an electric current, which is transferred as kinetic energy in a motor, which operates a robotic arm to lift a weight.*

Whenever work is done, energy is transferred. When a car slows down, work is done by the brakes against friction and the car's kinetic energy is transferred by heating the brakes. The exchange is exact, so the rise in internal energy of the brakes, air and road is equal to the original kinetic energy of the car. This energy balance applies to all situations and is summarised in the **principle of conservation of energy**:

The total amount of energy in any isolated system is constant.

It is important to stress the word 'isolated', because energy can flow into or out of a system. We could not apply the principle just to the car alone, because not all of its kinetic energy transfers to the brakes; some will be transferred to the air and the road as heat.

The kinetic energy, $E_k$, of a moving object depends on its mass, $m$, and its velocity, $v$, *squared*: $E_k = \frac{1}{2}mv^2$. Assignment 3 addresses the derivation of this formula.

How much kinetic energy does a car travelling on the motorway have? We need to know the mass of the car and its velocity. A typical family car has a mass of about 1500 kg. The speed limit on the motorway is 70 mph, which is $31\,\text{m}\,\text{s}^{-1}$ in SI units. The kinetic energy of the car is therefore

$$E_k = \tfrac{1}{2}mv^2 = \tfrac{1}{2} \times 1500 \times 31^2 = 720\,\text{kJ}$$

How many times could we accelerate to this speed, slow to a stop and accelerate to this speed again, using just one litre of fuel? Vehicle fuel is a very concentrated store of energy. This energy is transferred to kinetic energy of the car by combustion of the fuel in the engine. Each time the car slows to a stop, all this kinetic energy is lost, transferred by heating the surroundings. More fuel needs to be burnt in the car's engine to transfer kinetic energy back to the car. If the fuel has an energy density of 50 MJ per litre, one litre of fuel could provide enough energy to bring the car back to 31 m s$^{-1}$ about 70 times. But we have not taken into account two important factors.

1 The work done by the car against resistive forces. This depends on the velocity and the design of the car, particularly the shape and area of the front of the car. For a typical car travelling at 70 mph on the motorway, this work done against resistive forces amounts to about 15 MJ km$^{-1}$.

2 The energy transferred as heat by the engine. It is not possible to turn all the energy released by burning the fuel into kinetic energy of the car. In fact, only about 25% of the energy from the fuel will end up as kinetic energy in the car (see section 11.6).

If we consider just the energy required to accelerate the car to 31 m s$^{-1}$ and then brake to a standstill, we can safely ignore the first factor, because the distance travelled is small. But the second factor means that a litre of fuel will accelerate a car to motorway speed only about 17 times (0.25 × 70). If fuel is £1.20 per litre, every time you come to a stop on the motorway, and then speed up again, it costs about 7p.

## QUESTIONS

**20.** A recent road safety campaign had the slogan "Kill your speed, not a child". Most children hit by a car travelling at 40 mph are killed, while those hit at 20 mph usually escape with only minor injuries. Use the idea of kinetic energy to explain why there is such a major difference.

**21.** **a.** "If you want to drive economically in city traffic, you will need a lighter car. For long-distance motorway driving, this is not as important." Explain why this is true.

**b.** Explain why reducing your speed saves energy in both city and motorway driving.

**22.** In a chip factory, potatoes are being loaded onto a conveyor belt at the rate of 200 kg per minute. The conveyor belt is running at 3 m s$^{-1}$.

**a.** How much kinetic energy is gained by the potatoes in one minute?

**b.** Power is defined as energy transferred per second. What is the minimum power of the conveyor belt?

**c.** Why is this a minimum?

## ASSIGNMENT 3: DERIVING THE EXPRESSION FOR KINETIC ENERGY

**(MS 2.2, MS 3.1, MS 3.2, MS 3.12)**

### Stretch and challenge

The kinetic energy of a moving car depends on its mass and its velocity. We can find the relationship by considering the work done against friction, $F$, as a car brakes to a halt.

Suppose that a constant frictional force, $F$, is used to slow the car down from initial velocity $u$ to a final velocity $v = 0$ (stopped) over a distance, $s$. The work done in terms of $F$ and $s$ is

$$W = Fs$$

In an isolated system, all of this energy must have come from the original kinetic energy, $E_k$, of the car, so

$$E_k = Fs$$

Replace $F$ using $F = ma$:

$$E_k = mas$$

and replace $a$ by using $v^2 = u^2 + 2as$ and the fact that the final velocity $v$ of the car is zero:

$$E_k = m\left(\frac{v^2 - u^2}{2s}\right)s$$
$$= -\tfrac{1}{2}mu^2$$

This is the kinetic energy lost by the car, and so its initial kinetic energy was

$$E_k = \tfrac{1}{2}mu^2$$

For general use, this equation is written with a $v$ for velocity.

### Questions

**A1** What has been assumed in this proof?

**A2** The formula for kinetic energy derived here, $E_k = \tfrac{1}{2}mv^2$, works well for speeds that are significantly lower than the speed of light.

Einstein's theory of special relativity gives the following, which becomes important at speeds approaching the speed of light:

$$E_k = \frac{mc^2}{\sqrt{1-\left(\frac{v}{c}\right)^2}} - mc^2$$

where $m$ here is the rest mass (which means the mass measured in a frame of reference in which it is stationary).

Use a spreadsheet to calculate values of kinetic energy for an object of $m = 1$ kg and velocity values from $v = 0$ to $v = 0.95c$, using both the non-relativistic (low-speed) formula and the relativistic formula. Plot a graph of kinetic energy against velocity showing the relativistic and the non-relativistic formulas.

**A3** From your graph, at what speed does the non-relativistic formula become more than 10% inaccurate?

## Gravitational potential energy

***Figure 22*** *Work done equals gravitational potential energy stored*

When a weight is lifted (Figure 22), work is done against the force of gravity. The energy used to do this work is stored. It could be transferred to another form of energy, for example to kinetic energy, simply by allowing the weight to fall down again. Energy stored in this way is known as **gravitational potential energy**, $E_p$.

The work done against gravity is

$$\begin{aligned} W &= \text{force} \times \text{distance} \\ &= \text{weight} \times \text{height gained} \\ &= mg\Delta h \end{aligned}$$

This is equal to the energy stored, so gravitational potential energy

$$E_p = mg\Delta h$$

This assumes that the weight of the object remains constant as it is lifted through a height $\Delta h$. On Earth, this is only true for objects that are relatively close to the surface. The Earth's gravitational field strength, $g$, decreases with distance from the Earth's surface, so that your weight gets smaller as you go higher. The change is about 1% at a distance of 32 km above the Earth's surface.

If a weight of one newton (1 N) is lifted through a distance of one metre (1 m), then one joule (1 J) of work has been done and 1 J of gravitational potential energy is stored. If the weight is then allowed to fall, the gravitational potential energy is transferred as kinetic energy.

### Worked example 2

A cricket ball is hit upwards so that it reaches a height of 20 m. What speed is it travelling at on its way back down just before it hits the ground?

As the ball will have zero velocity at its highest point, we can assume that the ball has dropped from a height of 20 m. If we ignore energy losses due to air resistance, then the gravitational potential energy has all been transferred to kinetic energy in the ball, so

$$mg\,\Delta h = \tfrac{1}{2}mv^2$$

Cancelling $m$,

$$\begin{aligned} v^2 &= 2g\,\Delta h \\ v &= \sqrt{2g\,\Delta h} \\ v &\approx 20\ \text{m s}^{-1} \end{aligned}$$

### Worked example 3

Gravitational potential energy can be a way of storing energy on a large scale. The upper reservoir in a hydroelectric power station holds water, which can be released when electricity generation is required.

The reservoir behind the Picote Dam in Portugal (Figure 23) holds 205 000 $m^3$ of water. When the water is allowed to flow through the dam, it falls an average distance of 65 m and can be discharged at a rate of 110 $m^3 s^{-1}$, which is 110 000 $kg s^{-1}$, because the density of water is 1000 $kg m^{-3}$. How much electrical energy could be generated every second?

***Figure 23*** *A huge amount of gravitational potential energy can be released by the Picote Dam.*

The mass discharged in one second is $m = 110\,000$ kg, so the gravitational potential energy lost in one second is

$E_p = mg\Delta h = 110\,000 \times 9.81 \times 65 = 70\,000\,000\,\text{J s}^{-1}$

The energy transferred from gravitational potential energy is 70 000 000 J or $7 \times 10^7$ J every second. If this could be transferred entirely to kinetic energy in the water, and then in turbines and generators, $7 \times 10^7$ J of electrical energy could be generated every second.

## QUESTIONS

**23.** An ejector seat in an aircraft is designed to fire the pilot clear of the aircraft in an emergency. The combined mass of the seat and pilot is 150 kg and the height gained on ejection is 20 m. What is the minimum amount of energy needed to achieve this?

**24.** Estimate the minimum energy needed every second to drive an escalator that lifts people between floors in a department store.

**25.** Estimate the impulse exerted by a trampoline on a child landing on it and being bounced back up. (You may want to look back at section 11.3.)

## Power

The 70 000 000 J of electrical energy generated per second at the hydroelectric power station in Worked example 2 is the electrical power output. The **power** of a system is the rate at which energy is transferred, or work is done:

$$\text{power} = \frac{\text{work done}}{\text{time taken}} = \frac{\Delta W}{\Delta t}$$
$$= \text{energy transferred per second}$$

Power is measured in joules per second, or **watts** (W).

### Worked example 4

Most cars can manage a safe deceleration of around $1g$ (10 m s$^{-2}$). For a 1000 kg car travelling at 70 mph (31 m s$^{-1}$), what braking power would be required for this deceleration?

The braking force is

$$F = ma$$
$$= 1000 \times 10$$
$$= 10\,000\text{ N}$$

The distance needed to bring the car to a stop, once the brakes have been applied, is known as the braking distance, which can be found using:

$$v^2 = u^2 + 2as$$
$$s = \frac{v^2 - u^2}{2a}$$

Since $v = 0$, $u = 31$ m s$^{-1}$ and $a = -10$ m s$^{-2}$:

$$s = \frac{0 - 31^2}{-20}$$
$$= 48\text{ m}$$

The time taken to stop can be found from

$$v = u + at$$
$$t = \frac{v - u}{a}$$
$$t = \frac{0 - 31}{-10}$$
$$= 3.1\text{s}$$

The braking power is then given by

$$\text{power} = \frac{\text{work done}}{\text{time taken}}$$
$$= \frac{10\,000 \times 48}{3.1}$$
$$= 155\text{ kW}$$

In reality, the situation is more complex. The safest way of stopping a car is not usually applying the largest possible force to the brakes – the wheels may stop turning but the car might continue in a skid. Modern cars use an anti-lock braking system (ABS), which automatically applies and reapplies the brakes, briefly releasing them if the car skids.

## Maximum speed of a vehicle

The power of a car engine affects the maximum speed of the car and the maximum force that it can exert on the road. Because

$$\text{power} = \frac{\text{work}}{\text{time}}$$

and

$$\text{work} = \text{force} \times \text{distance}$$

then

$$\begin{aligned} \text{power} &= \frac{\text{force} \times \text{distance}}{\text{time}} \\ &= \text{force} \times \frac{\text{distance}}{\text{time}} \\ &= \text{force} \times \text{velocity} \\ P &= Fv \end{aligned}$$

For a given power, an engine can exert a large force at a low speed, or a smaller force at higher speed. A car's gearbox allows control of this balance. A low gear results in a large force at a low speed, useful for climbing a steep hill, whereas a high gear produces a smaller force at a higher road speed, useful for motorway driving, for example. Changing gear cannot increase the power of the engine; in fact, the opposite is true, because some energy will be dissipated as heat in the gearbox.

### QUESTIONS

**26.** A car has a mass of 1445 kg and its engine can deliver a maximum power of 112 kW.

**a.** Estimate the maximum force that it can exert at 20 mph (32 km h$^{-1}$) and at 40 mph (64 km h$^{-1}$).

**b.** What else would you need to know before you could estimate its maximum acceleration at these speeds?

The forces that resist the motion of a vehicle increase as the vehicle travels faster. Air resistance and rolling (frictional) resistance combine to resist the motion of a juggernaut (Figure 24). At steady speed, the force driving the lorry forwards is equal to the resistive forces on the lorry. So if the lorry is to accelerate, the force supplied by the engine – and so its power output – has to increase.

***Figure 24*** *The resistive forces on the juggernaut limit its maximum speed.*

Table 1 shows how the total resistive force increases as the lorry accelerates. Eventually, the tractive force (that is, the driving force of the road on the wheels) and the resistance will be equal again and the lorry will move at a steady speed.

| Road speed / km h$^{-1}$ | Total resistive force / N |
|---|---|
| 80 | 4265 |
| 88 | 4740 |
| 96 | 5250 |
| 105 | 5805 |
| 113 | 6395 |

***Table 1*** *Resistive forces on a 38 tonne juggernaut*

When the engine reaches its maximum power output, $P_{max}$, it will no longer accelerate. The forward thrust on the lorry, $F$, is then equal to the total resistive forces, and the lorry will travel at its maximum velocity, $v_{max}$. The power developed is $P_{max} = Fv_{max}$. The top speed is therefore

$$v_{max} = \frac{P_{max}}{F}$$

where $F$ is the resultant of all the resistive forces.

### Worked example 5

Suppose a 38 tonne juggernaut like the one in Figure 24 is travelling at a steady 80 km h$^{-1}$. From Table 1, the total resistive force at that speed is 4265 N. The driver then pushes the accelerator pedal further down. If the tractive force increases to 5250 N, what would the instantaneous acceleration of the lorry be?

The resultant horizontal force on the lorry is

$$5250\,\text{N} - 4265\,\text{N} = 985\,\text{N}$$

From Newton's second law:

$$a = \frac{F}{m} = \frac{985\,\text{N}}{38 \times 10^3\,\text{kg}} = 0.026\,\text{m}\,\text{s}^{-2}$$

### Worked example 6

For the lorry from Worked example 5, what is the output power of the engine at $80\,\text{km}\,\text{h}^{-1}$ and at $113\,\text{km}\,\text{h}^{-1}$?

The power can be calculated from power = force × velocity. At $80\,\text{km}\,\text{h}^{-1}$, converting to SI units,

$$v = \frac{80\,\text{km}\,\text{h}^{-1} \times 1000\,\text{m}}{3600\,\text{s}} = 22.2\,\text{m}\,\text{s}^{-1}$$

so

$$\text{power} = 4265\,\text{N} \times 22.2\,\text{m}\,\text{s}^{-1} = 94.8\,\text{kW}$$

At $113\,\text{km}\,\text{h}^{-1}$, using the same method, the power is 201 kW, more than twice the output power at $80\,\text{km}\,\text{h}^{-1}$.

### Worked example 7

The lorry from Worked example 5 is travelling at $113\,\text{km}\,\text{h}^{-1}$ when the engine cuts out suddenly. How much work will the lorry do against the resistive forces as it slows to $80\,\text{km}\,\text{h}^{-1}$?

Since the resistive force varies as the vehicle slows down, it is not possible to use the equation work = force × distance directly. An easier approach is to say that the energy to do the work comes from the kinetic energy of the lorry.

The work done against resistive forces is equal to the kinetic energy transferred from the lorry:

$$E_k = \frac{1}{2}m(u^2 - v^2) = \frac{1}{2} \times 38 \times 10^3 \times (31.42^2 - 22.2^2) = 9.4 \times 10^6\,\text{J}$$

## QUESTIONS

**27.** **a.** Explain why a vehicle has a top speed.

**b.** What determines the value of a vehicle's top speed? How could you increase the top speed?

### Stretch and challenge

**28.** A cyclist changes to a lower gear, but maintains her power output. Does the use of a lower gear:

**a.** increase the force exerted on the road by the wheel

**b.** increase the speed of the wheel

**c.** increase the power of the bicycle?

Answer 'yes' or 'no' for each part and explain why.

## KEY IDEAS

- Work is done when a force is moved through a distance in the direction of the force. Work = force × displacement in the direction of the force; $W = Fs\cos\theta$.
- The work done by a changing force is equivalent to the area under a force–displacement graph.
- Energy is the capacity to do work.
- Kinetic energy $E_k = \frac{1}{2}mv^2$.
- Gravitational potential energy $E_p = mg\Delta h$.
- The principle of conservation of energy states that the total amount of energy in any isolated system is constant.
- Power is the rate at which work is done or the rate at which energy is transferred, $P = \frac{\Delta W}{\Delta t}$.
- For a constant force $F$ on an object moving at velocity $v$, the power is given by $P = Fv$.

## 11.6 EFFICIENCY

The principle of conservation of energy tells us that all the energy transferred to a machine, for example a car (Figure 25), must eventually be transferred out again, since energy does not just disappear. However, some of the energy may end up in a less useful form, such as heat in the surroundings. The useful work that can be done by a machine is measured, as a fraction of the total energy supplied to it, by the machine's **efficiency**.

***Figure 25*** *The Volkswagen XL1, powered by a diesel–electric hybrid engine, is claimed to be the world's most efficient car. It can travel 313miles on one gallon of fuel (or 110 km per litre), compared with 40 miles per gallon for a typical family car. It is manufactured from lightweight materials, such as carbon fibre, so its mass is just 795 kg. The streamlined shape, known as a teardrop, the scissor doors and cameras instead of wing mirrors are all designed to reduce drag to a minimum. This not only gives the car a higher top speed for the same engine power but also reduces the demand for fuel, since the power needed is less at all road speeds.*

Efficiency should not be confused with effectiveness, or power. It is the ratio of the useful work done by a machine to the total energy supplied to it:

$$\text{efficiency} = \frac{\text{useful output work (or energy)}}{\text{total input energy}}$$

If we consider the work done and energy supplied per second, we can write:

$$\text{efficiency} = \frac{\text{useful output power}}{\text{total input power}}$$

Efficiency may be expressed as a simple fraction, such as 2/5, or as a percentage, such as 40%, by multiplying the above equations by 100. Because it is a ratio of two values of energy (or power), efficiency has no unit.

For a simple device, such as a solar cell, efficiency is easy to define. The useful output energy of a solar cell is electrical and the input energy is the light energy that falls on the cell. A modern solar cell is approximately 20% efficient: 20% of the light energy falling on the cell is transferred to electrical energy. What happens to the other 80% of the incident light energy? Some may be reflected, and some absorbed as heat, raising the temperature of the solar cell and the surroundings. This energy has not 'disappeared', but it is not what the solar cell is designed to produce, so it is not referred to as 'useful'.

During every energy transfer, some of the input energy is transferred by heat (dissipated) to the environment.

The bicycle is said to be the most efficient means of transport that we have ever invented, with up to 99% of the energy transferred to the pedals being transferred to the wheels. But to compare this efficiency with motor-driven vehicles, we need to include the efficiency of the human 'engine'. Estimates of the efficiency of the human body vary widely, but a typical value for cycling is 25%. A physically fit cyclist can sustain power outputs of around 100 to 200 W for a reasonable length of time. A typical chocolate bar can provide 1100 kJ of energy, which means that one chocolate bar of fuel would keep the cyclist going for $(1100000 \times 0.25)/(150 \times 60) = 30.6$ minutes.

### QUESTIONS

**29.** Estimate how many times you would need to walk up a flight of stairs to 'burn off' the 1100 kJ of energy in a chocolate bar.

**30.** A wind turbine is operating at 60% efficiency and generating 20 kW of electrical power.

  **a.** What form does the input energy take?

  **b.** What is the input power?

  **c.** Why is it impossible for a wind turbine to be 100% efficient?

**31.** The 'recumbent bicycle' is the most efficient bicycle design (Figure 26).

***Figure 26*** *A recumbent bicycle*

a. Identify the input energy and the output energy for a person riding a bike.
b. Where do energy 'losses' occur?
c. Outline how you could compare the efficiency of different bicycle designs.
d. Considering that the recumbent bicycle is the most efficient, suggest why it is not more popular.

The useful output energy for a car is difficult to calculate. It is the work done by the car in moving from A to B, which is equal to the force exerted by the car multiplied by the distance moved. But the force exerted by the car will vary during the journey, depending particularly on the car's speed and acceleration. Figure 27 shows how the force could vary for a short car journey. The car accelerates, quickly at first, and then at a lower rate, until it reaches a steady speed. The force provided by the engine is greater at first, because it has to accelerate the car as well as overcome drag and resistance forces. The force reaches a steady value as the car is driven at constant velocity.

The input energy for a car comes from the fuel, usually petrol or diesel. The total input energy is the volume of fuel used in litres multiplied by the energy density of the fuel in joule per litre.

Suppose that the car journey described in Figure 27 required 0.0275 litres of petrol with energy density of 36 MJ per litre. Then

$$\text{input energy} = \text{volume of fuel (litres)} \times \text{energy density of fuel (joule per litre)}$$

$$= 36\,\text{MJ} \times 0.0275\,\text{litres} = 0.99\,\text{MJ}$$

The output energy, given by the area under the graph, is approximately equal to 300 kJ. So

$$\text{efficiency} = \frac{0.3\ \text{MJ}}{0.99\,\text{MJ}}$$

$$= 0.3 \text{ or } 30\%$$

This is a typical efficiency for an internal combustion engine (petrol or diesel) when it is used to power a car. But a car journey, especially in city traffic, has many stops and starts. The engine is still using fuel when the car is stationary, although the output energy, and therefore the efficiency, has dropped to zero. The overall efficiency of a journey can be as low as 15%.

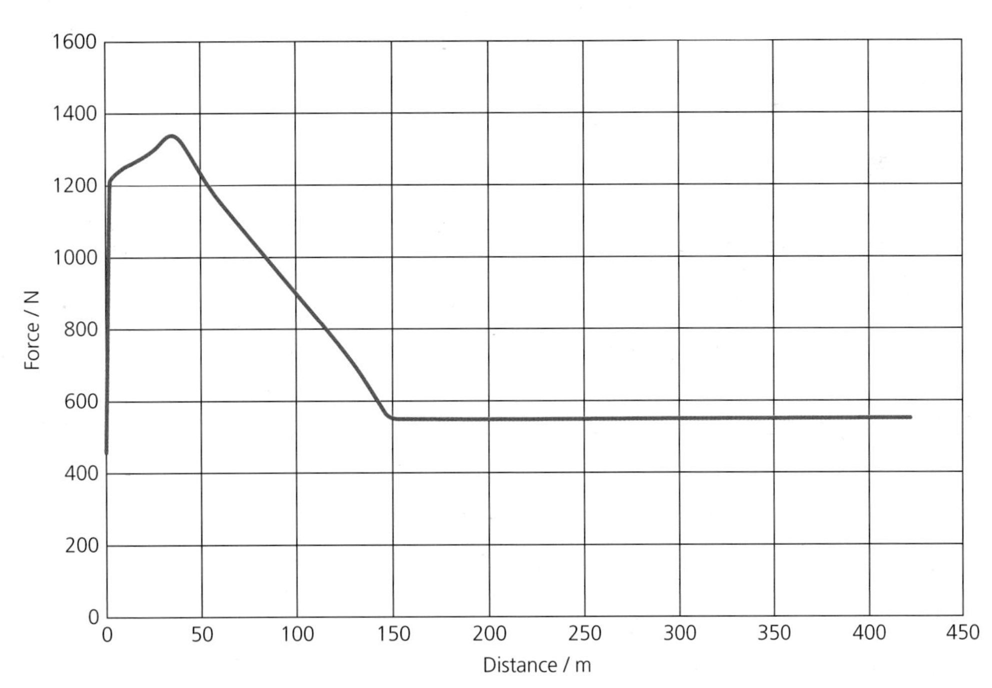


***Figure 27*** *Force from car engine against distance graph for a short journey. The car accelerates over the first 150 m and then drives at a constant speed.*

## QUESTIONS

**32.** A diesel generator has an output electrical power of 30 kW. It burns 3.0 gallons of fuel per hour when it is operating at full power. Calculate its efficiency. (There are 4.55 litres in 1 gallon and the energy density of diesel is around 38.6 MJ per litre.)

**33.** The efficiency of aircraft engines has been steadily improving, but air travel is still one of our biggest energy users. How would you make a fair comparison of the energy use of different modes of transport?

### Improving energy efficiency in the future

Electric motors have several advantages over internal combustion engines when it comes to powering a car. Electric motors do not rely on getting hot in order to work, they do not use energy when the car is stationary, and some of the energy wasted in braking can be used to recharge the car's batteries. Electric cars can achieve efficiencies of around 80%, although this ignores the efficiency of generating the electricity used to charge the car's batteries.

## KEY IDEAS

- Efficiency $= \dfrac{\text{useful output work (or energy)}}{\text{total input energy}}$
- Efficiency $= \dfrac{\text{useful output power}}{\text{total input power}}$

## ASSIGNMENT 4: FINDING THE EFFICIENCY OF AN ELECTRIC MOTOR

**(MS 0.3, MS 1.5, PS 1.1, PS 2.3, PS 3.2, PS 3.3, PS 4.1)**

The efficiency of an electric motor can be measured in the laboratory:

$$\text{efficiency} = \frac{\text{useful output work}}{\text{input electrical energy}}$$

The motor can be used to lift a weight, by winding a string onto a shaft (Figure A1). The work done in lifting the weight, $W$, through a height, $h$, is $W \times h = mg\Delta h$.

The electrical energy supplied to the motor can be measured using a joule meter.

The efficiency of the motor will depend on several factors, including the size of the load – the weight that is being lifted.

Table A1 has a typical set of results for one run of the experiment.

| Joule meter reading / J | 4.5 |
|---|---|
| Height lifted / m | 1.51 |
| Mass lifted / g | 100 |

*Table A1*

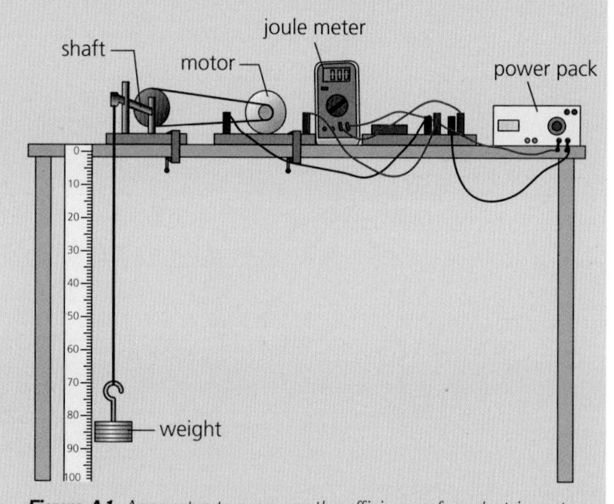


***Figure A1*** *Apparatus to measure the efficiency of an electric motor*

The useful work done by the motor $= 0.1 \times 9.81 \times 2.5$
$= 2.45$

The energy input = 4.5 J

So

efficiency = 2.45/4.5 = 0.54 or 54%

### Questions

**A1** What errors are likely to arise in this experiment? Explain which of these are random errors and which are systematic.

**A2** Estimate the uncertainty in the readings.

**A3** Calculate the uncertainty in the efficiency.

**A4** Suggest what could be done to reduce the uncertainty in the final answer for the efficiency of the motor.

## ASSIGNMENT 5: PHYSICS AND ROAD SAFETY

**(MS 3.1, MS 3.2)**

There are several basic laws of physics that contribute to our understanding of what happens during road accidents. The initial and final velocities, time taken, momentum, acceleration, deceleration, work done (therefore energy transferred), kinetic energy, potential energy and friction all have a part to play in road safety. Designing safer cars requires a great deal of data and a lot of complex calculations. Much of the data is obtained by crash-testing cars in carefully controlled conditions. The crash dummies used today are very sophisticated, with between 50 and 60 sensors called accelerometers built into the body. The principle used in accelerometers is that, when a deforming force is applied to a silicon chip, its electrical resistance changes. This change can be used to produce the data as electrical pulses, which can be analysed later. The same principle applies when the accelerometer is used as a trigger to set off an airbag.

### Questions

**A1** How much kinetic energy does a driver of mass 90 kg transfer if he decelerated from $13\,\mathrm{m\,s^{-1}}$ to rest in a crash?

**A2** Draw a graph of force against distance to show how the kinetic energy calculated in question A1 can be reduced to zero in these two crash scenarios:

**a.** when the driver is using a seat belt, which will move about 50 cm before stopping

**b.** when the driver is not using a seat belt and his head hits the dashboard; the dashboard deforms by 0.5 cm as the kinetic energy of the driver is reduced to zero.

**A3** What conclusions can you come to from the calculations in question A2?

**A4** Explain how crumple zones can help to save the lives of car drivers and passengers.

**A5** In order to stop a car safely, a driver has to consider two factors:

- The 'thinking distance' needed as the driver reacts to the situation. An average reaction time is about 0.7 s. During this time the car can travel quite a distance.
- The 'braking distance' as the car decelerates to a stop. On a dry road with good brakes, a car would decelerate at $8.5\,\mathrm{m\,s^{-2}}$.

Calculate the thinking distance, braking distance and total stopping distance for the following speeds on a dry road:

**a.** $13\,\mathrm{m\,s^{-1}}$

**b.** $22\,\mathrm{m\,s^{-1}}$ (50 mph)

**c.** $30\,\mathrm{m\,s^{-1}}$ (70 mph)

Put your results in a suitable table. Then construct a graph of total stopping distance (m) against speed ($\mathrm{m\,s^{-1}}$) and work out the total stopping distance for a speed of $15\,\mathrm{m\,s^{-1}}$

**A6** Prepare an illustrated presentation to be given at an assembly in your school or college to make students more aware of the dangers of speeding on motorbikes or in cars. Use the information in this chapter and from other sources such as the Internet.

## PRACTICE QUESTIONS

1. In the 17th century, when thinking about forces, Galileo imagined a ball moving in the absence of air resistance on a frictionless track, as shown in Figure Q1.

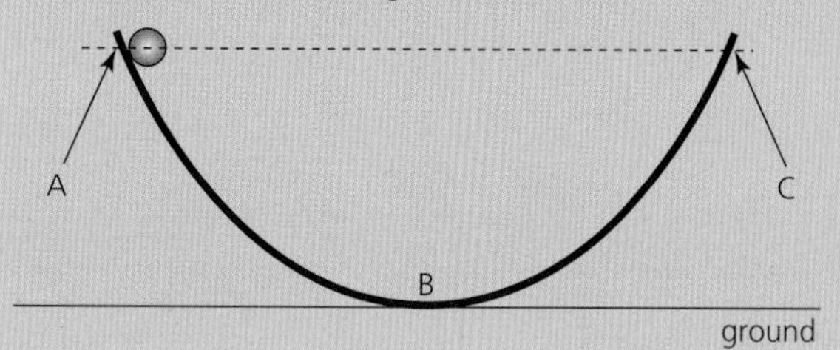


*Figure Q1*

   a. Galileo thought that, under these circumstances, the ball would reach position C if released from rest at position A. Position C is the same height above the ground as A.

   Using ideas about energy, explain why Galileo was correct.

   b. Galileo then imagined that the track was changed, as shown in Figure Q2. The slope beyond B was now horizontal.

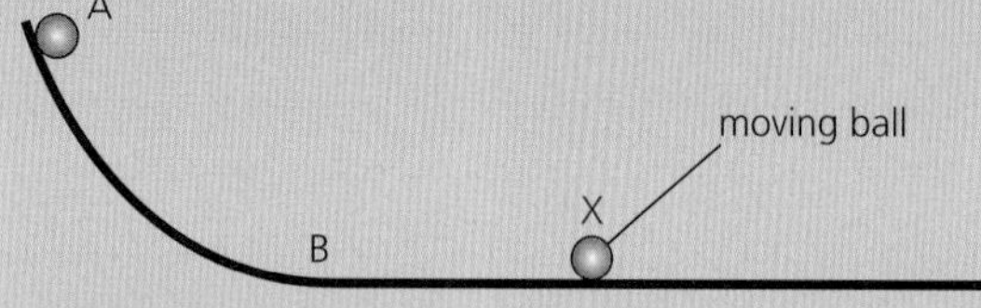


*Figure Q2*

   Sketch a speed–time graph for the ball from its release at A until it reaches the position X shown in Figure Q2. Indicate on your graph the time when the ball is at B.

   c. Newton later published his three laws of motion. Explain how Newton's first law of motion is illustrated by the motion of the ball between B and X.

*AQA Unit 2 June 2012 Q2*

2. Figure Q3 shows a rollercoaster train that is being accelerated when it is pulled horizontally by a cable.

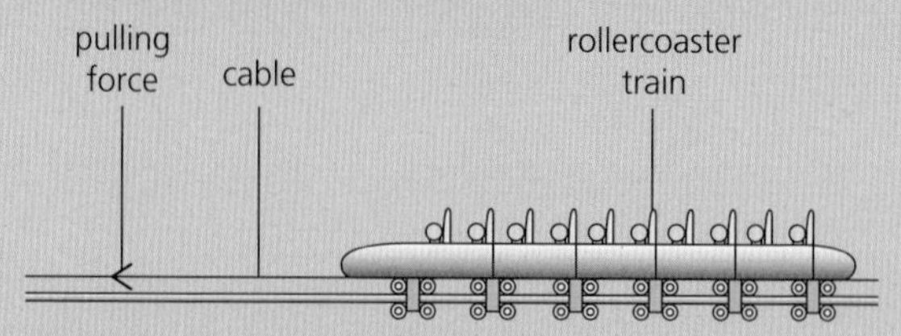


*Figure Q3*

   a. The train accelerates from rest to a speed of $58\,\text{m}\,\text{s}^{-1}$ in 3.5 s. The mass of the fully loaded train is 5800 kg.

      i. Calculate the average acceleration of the train.

      ii. Calculate the average tension in the cable as the train is accelerated, stating an appropriate unit.

      iii. Calculate the distance the train moves while accelerating from rest to $58\,\text{m}\,\text{s}^{-1}$.

      iv. The efficiency of the rollercoaster acceleration system is 20%. Calculate the average power input to this system during the acceleration.

   b. After reaching its top speed the driving force is removed and the rollercoaster train begins to ascend a steep track. By considering energy transfers, calculate the height that the train would reach if there were no energy losses due to friction.

*AQA Unit 2 June 2011 Q3*

3. Figure Q4 shows a gymnast trampolining. In travelling from her lowest position at A to her highest position at B, her centre of mass rises 4.2 m vertically. Her mass is 55 kg.

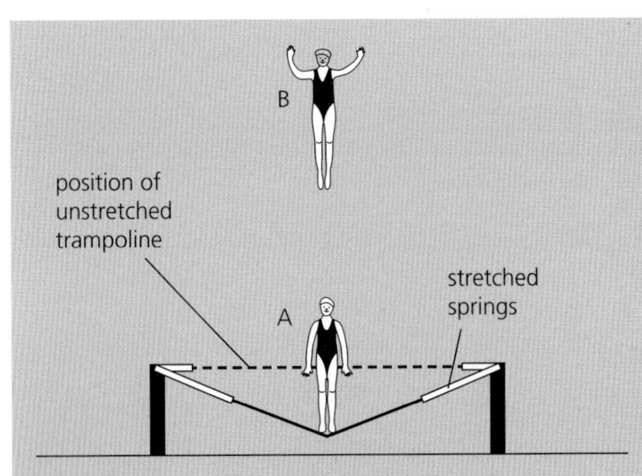


*Figure Q4*

a. Calculate the increase in her gravitational potential energy when she ascends from position A to position B.

b. The gymnast descends from position B and regains contact with the trampoline when it is in its unstretched position. At this position, her centre of mass is 3.2 m below its position at B.

   i. Calculate her kinetic energy at the instant she touches the unstretched trampoline.

   ii. Calculate her vertical speed at the same instant.

c. Sketch the lower part of Figure Q4 and draw an arrow to show the force exerted on the gymnast by the trampoline when she is in position A.

d. As she accelerates upwards again from position A, she is in contact with the trampoline for a further 0.26 s. Calculate the average acceleration she would experience while she is in contact with the trampoline, if she is to reach the same height as before.

e. On her next jump the gymnast decides to reach a height above position B. Describe and explain, in terms of energy and work, the transformations that occur as she ascends from her lowest position A until she reaches her new position above B. The quality of your written communication will be assessed in this question.

*AQA Unit 2 June 2010 Q4*

Questions 4 to 6 refer to Figure Q5, which shows the forces acting on a car travelling in a straight line on a horizontal, level road.

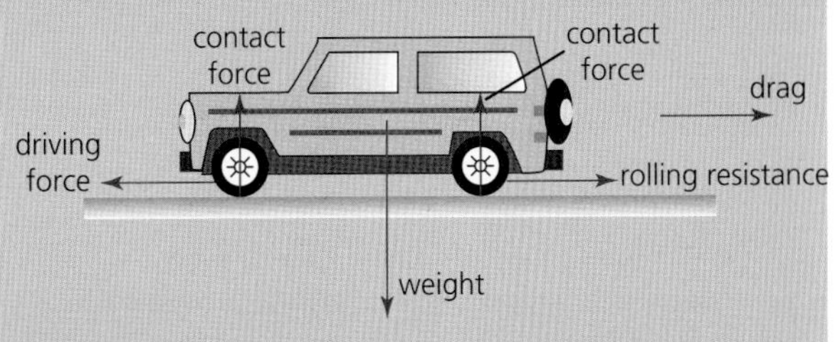


*Figure Q5*

4. If the car in Figure Q5 is moving at constant velocity, which of these statements about the forces on the car is **incorrect**?

   **A** Sum of the contact forces = weight of the car

   **B** To keep the car moving at constant velocity, driving force = drag + rolling resistance

   **C** To keep the car moving at constant velocity, driving force > drag + rolling resistance

   **D** Vector sum of all the forces acting on the car = 0

5. The car in Figure Q5 now accelerates forwards. Which of these statements is **incorrect**?

   **A** Sum of the contact forces = weight of the car

   **B** Driving force = drag + rolling resistance

   **C** Driving force > drag + rolling resistance

   **D** The drag force will increase as the car's velocity increases.

6. Which of the statements A to D, about the forces due to the car in Figure Q5 acting on the Earth, **cannot** be correctly deduced from Newton's third law?

   **A** Sum of the contact forces acting on the Earth = weight of the car

   **B** Sum of the contact forces acting on the car = sum of the contact forces acting on the Earth

**C** Car's gravitational pull on the Earth = weight of the car

**D** Frictional force pushing the car forward (driving force) = frictional force pushing the Earth backwards

**7.** A car of mass 1200 kg drives into the back of a stationary van of mass 1800 kg. After the collision the vehicles are locked together and they move with an initial speed of 10 $ms^{-1}$. The speed of the car just before the collision was:

**A** 25 $ms^{-1}$

**B** 10 $ms^{-1}$

**C** 15 $ms^{-1}$

**D** 16.7 $ms^{-1}$

**8.** Which of these statements is true for an inelastic collision?

**A** Kinetic energy is conserved but momentum is not.

**B** Kinetic energy and momentum are both conserved.

**C** Momentum is conserved but kinetic energy is not.

**D** Neither momentum nor kinetic energy is conserved.

**9.** A 40 kW electric motor is used to raise a service lift in a skyscraper. The lift has a mass of 1500 kg. Assuming that the motor operates at 80% efficiency, how long would it take to raise the lift through 120 m?

**A** 55 s

**B** 44 s

**C** 540 s

**D** 440 s

## Stretch and challenge

**10.** This question is about the use of airbags in cars. Read the following and then answer the questions.

Airbags are designed to protect the occupants of a car in the case of a collision. An airbag is triggered when the car decelerates much more quickly than normal. However, it is important that it does not inflate at the wrong time because this could cause an accident. The electronic module controlling the airbag must be able to distinguish between a potentially dangerous frontal collision and just hitting a pothole. Typically, an airbag should activate if the car collides with a solid barrier at a speed of 8–14 mph. (1 mile = 1.6 km)

The electronic module monitors the car's acceleration every 10 ms. For a small car the airbag is fired if two consecutive measurements are less than approximately $-1.0g$, though the actual value depends on the construction of the car, in particular the strength-to-weight ratio of the car.

There are circumstances where the airbag should activate but fails to. For example, a collision with a lamp post that just deforms part of the car, or one where the impact is more gradual, such as when a car runs into the back of a vehicle moving in the same direction.

**a.** A car travelling at 15 mph runs into a solid wall. The duration of the collision is 120 ms.

**i.** Explain why the car takes 120 ms to stop, given that the wall does not move during the collision.

**ii.** Calculate the average deceleration of the car during the collision.

**iii.** Is the airbag likely to activate in these circumstances? Explain your answer.

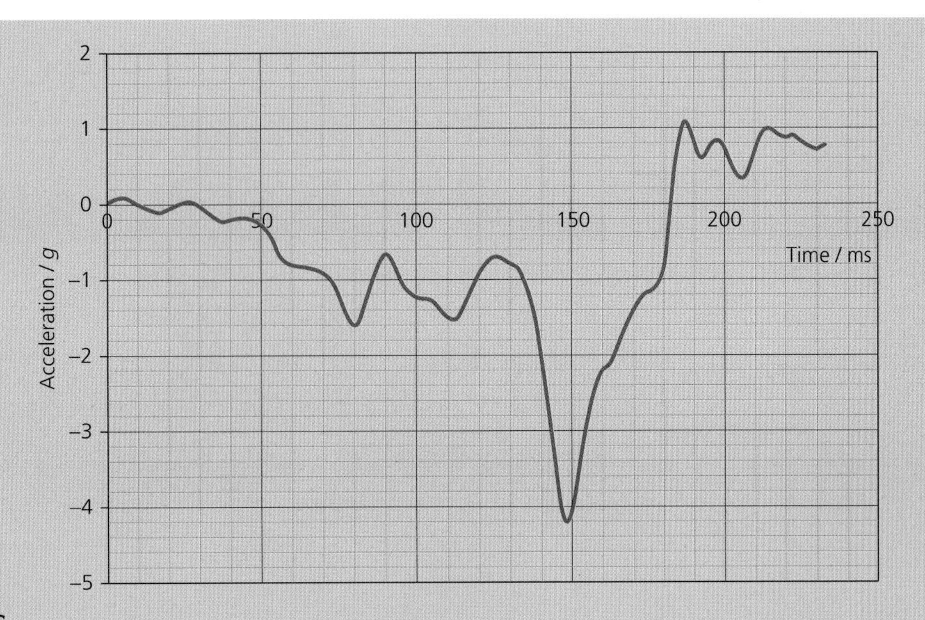


*Figure Q6*

**b.** A particularly dangerous type of collision occurs on a motorway when a car runs into the back of a lorry, sometimes running under the lorry. In some of these cases the airbag does not deploy. Figure Q6 is an acceleration–time graph for a car as it under-rides a stationary lorry.

**i.** Estimate the duration of the collision (the period of sustained deceleration).

**ii.** At what time would the airbag be activated? (Assume that readings are taken every 10 ms starting at $t = 0$.)

**iii.** Estimate the average deceleration during the collision.

**iv.** If the mass of the car and occupants is 1850 kg, what is the average force on the car during the collision?

**v.** Calculate the change in speed of the car.

# 12 THE STRENGTH OF MATERIALS

## PRIOR KNOWLEDGE

You will be familiar with the concepts of force, work and energy, and will have some experience of elastic materials.

## LEARNING OBJECTIVES

In this chapter you will learn how materials behave under stress, compare elastic with plastic behaviour, and link this behaviour with the structure of the material.

**(Specification 3.4.2.1, 3.4.2.2)**

***Figure 1*** *The golden orb spider produces seven different kinds of silk. The longest strands in the web are strong enough to trap small birds.*

***Figure 2*** *In the movie 'Spiderman 2', Spiderman stops a runaway train with eight strands of silk, each about 5 mm thick. Final-year physics students from Leicester University showed that the silk could easily withstand the 300 000 N force needed to stop the train.*

Spider silk is a truly remarkable material (Figure 1). It can be spun into a thread finer than a human hair that is five times stronger than a similar steel thread. Some spider silk threads can stretch to four times their original length without breaking (Figure 2). Spider silk is waterproof and keeps its elastic properties down to temperatures as low as −40 °C. It has hundreds of possible applications, from bullet-proof vests to artificial skin, and from suspension bridge cables to unrippable writing paper. There has been intense competition to discover its exact structure and to develop production techniques that do not require the slave labour of millions of spiders! In any case, spiders cannot be farmed like silkworms because they tend to eat one another!

So far, the silk has proved difficult to synthesise in the laboratory, but genetic engineering may have the answer. Silkworms with spider genes are now starting to produce commercial quantities of this wonder material.

## 12.1 HANGING BY A THREAD: DENSITY, STRESS AND STRAIN

A spider can suspend itself on a thread that is only a few thousandths of a millimetre thick. A human abseiling down a cliff face needs something more substantial (Figure 3). The most important physical property of a climber's rope is its strength, which is a measure of the maximum force that it can support without breaking. Ropes for rock climbing are usually designed to withstand a load 25 times the climber's weight. However, the rope must not be too heavy, or it would be difficult to carry.

***Figure 3*** *Mountain climbers need ropes that are strong and light. This requires a material of low density and high strength.*

An important property of a rope is its **density**, $\rho$. The density of a material is defined as the amount of mass in a given volume:

$$\text{density} = \frac{\text{mass}}{\text{volume}}$$

$$\rho = \frac{m}{V}$$

In SI units, density is measured in kilogram per cubic metre ($\text{kg m}^{-3}$), though gram per cubic centimetre ($\text{g cm}^{-3}$) is sometimes used. For example, at standard temperature and pressure, air has a density of $1.29\,\text{kg m}^{-3}$. Water has a density of approximately $1000\,\text{kg m}^{-3}$ or $1\,\text{g cm}^{-3}$.

### Worked example 1

A typical nylon climbing rope has a density of $1130\,\text{kg m}^{-3}$. If the rope is 50 m long and has a diameter of 10 mm, we can calculate its mass:

$$\begin{aligned}\text{volume of a cylinder} &= \pi r^2 h \\ &= \pi \times (5 \times 10^{-3})^2 \times 50 \\ &= 3.93 \times 10^{-3}\,\text{m}^3\end{aligned}$$

Since mass = density × volume:

$$\begin{aligned}\text{mass of the rope} &= \text{density} \times \text{volume} \\ &= 1130 \times 3.93 \times 10^{-3} \\ &= 4.44\,\text{kg}\end{aligned}$$

### QUESTIONS

1. Show that $1000\,\text{kg m}^{-3}$ is equivalent to $1\,\text{g cm}^{-3}$.
2. The kilogram is defined as the mass of a standard cylinder, which is kept in Sèvres, France. The cylinder is actually made of a platinum–iridium alloy. If it was made of pure platinum, what would its volume be? (Density of platinum = $21\,450\,\text{kg m}^{-3}$.)
3. Calculate the mass of water held in a household hot-water tank. These are usually 0.5 m diameter cylinders, about 1.0 m high.
4. Estimate the average density of a bag of sugar.
5. Estimate the density of butter.

### Under stress

We can assess the strength of a material using the idea of **tensile stress**, $\sigma$, applied to a material. This is the applied tensile (stretching) force per unit cross-sectional area (normal to the force):

$$\text{tensile stress} = \frac{\text{force}}{\text{cross-sectional area}}$$

$$\sigma = \frac{F}{A}$$

Using the force per unit area, rather than the force, enables us to compare ropes made of different materials, even if they have different diameters (Figure 4). Tensile stress has the unit newton per square metre ($N m^{-2}$) or pascal (Pa).

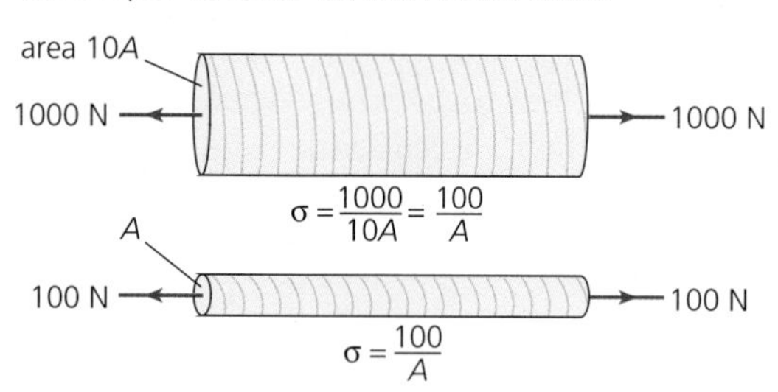


***Figure 4*** *Tensile stress*

The largest tensile stress that can be applied to a material before it breaks is known as its **ultimate tensile stress** (UTS). It is sometimes referred to as the material's **breaking stress**. Materials with a high UTS are referred to as 'strong' materials. The density and UTS values of some materials are shown in Table 1.

| Material | Density / $kg\, m^{-3}$ | UTS / MPa |
|---|---|---|
| Nylon | 1130 | 85 |
| Stainless steel | 7930 | 600 |
| Carbon fibre | 1750 | 1900 |
| Kevlar | 1440 | 3100 |
| Spider silk (typical) | 1250 | 850 |

***Table 1*** *Density and ultimate tensile stress of some materials*

Kevlar (Figure 5) is a strong synthetic **polymer** – a material such as rubber or polythene that is made from long-chain molecules. Cables made from Kevlar are so strong that they are used to secure oil rigs.

***Figure 5*** *Bulletproof vests are currently made using layers of Kevlar. In future it may be possible to make them even stronger using spider silk.*

## Worked example 2

We can use the equation for tensile stress to calculate how thick a Kevlar climbing rope would have to be to support a load of 15 000 N.

From Table 1, the ultimate tensile stress of Kevlar is 3100 MPa. The cross-sectional area A of the rope is found from

$$A = \frac{F}{\sigma} = \frac{15000\,\text{N}}{3100 \times 10^{6}\,\text{Pa}} = 4.84 \times 10^{-6}\,\text{m}^2$$

The cross-sectional area of the rope is $\pi r^2$, so this gives $r = 1.24 \times 10^{-3}$ m. The rope would need to have a diameter of only 2.5 mm. A 50 m length of this rope would have a mass of only 0.35 kg.

## QUESTIONS

**6. a.** What is the maximum weight of a fish that could be lifted out of the water using a 2 mm diameter line made from nylon (UTS = 85 MPa)?

**b.** If the fish struggles, the force on the line may increase to five times its weight. How thick will the nylon line need to be now?

## Taking the strain

The most remarkable property of spider silk is the distance that it can stretch before it breaks. The increase in length of a material, $\Delta l$, caused by a tensile force is called the **extension**. A rope's extension under a given load depends on the original length of the rope. A long rope will stretch further than a short one if they are subjected to the same force, so the extension is usually given as a fraction of the original length. This ratio is known as the **tensile strain**, $\varepsilon$:

$$\text{tensile strain} = \frac{\text{extension}}{\text{original length}}$$

$$\varepsilon = \frac{\Delta l}{l}$$

Strain has no unit, because the extension and the original length are both measured in metres.

Steel wire can undergo a tensile strain of 0.01 before it breaks, and Kevlar can manage 0.04. The silk from a spider has a breaking strain of between 0.15 and 0.30. This enables a web to absorb the kinetic energy of an insect, transferring it to internal energy in the web.

Spider silk is also highly **elastic**, meaning that after a stretching force is removed it will return to its original length. Many materials do not return to their original length after being subjected to a strain. When the tensile force is removed, the material remains deformed. This is called **plastic** behaviour. For example, putty and dough behave in this way.

We tend to classify materials as either elastic or plastic, but many materials show both types of behaviour depending on the stress applied. For small values of stress, polythene behaves elastically, returning to its original dimensions when the stress is removed. Above a critical value of stress, known as the **yield stress**, polythene begins to be plastically deformed. We say that it has passed its **elastic limit**. Materials that have large plastic deformations before breaking are described as **ductile**. Materials that can absorb a lot of energy before they break are referred to as **tough**. Glass fibres can be extremely strong, that is, they have a high ultimate tensile stress, but they show hardly any plastic deformation before they break, or **fracture**. Materials like this are said to be **brittle** (Figure 6).

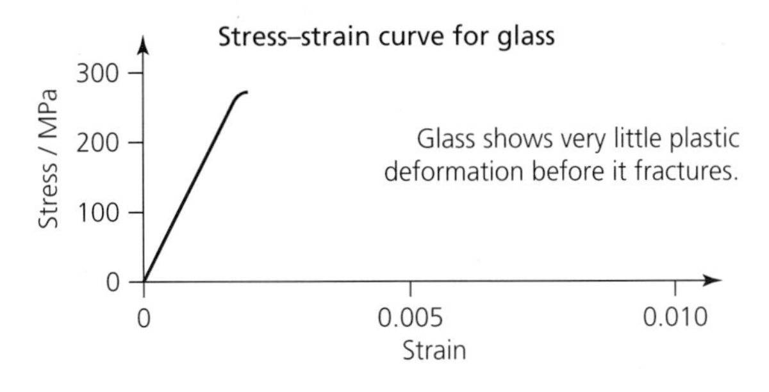


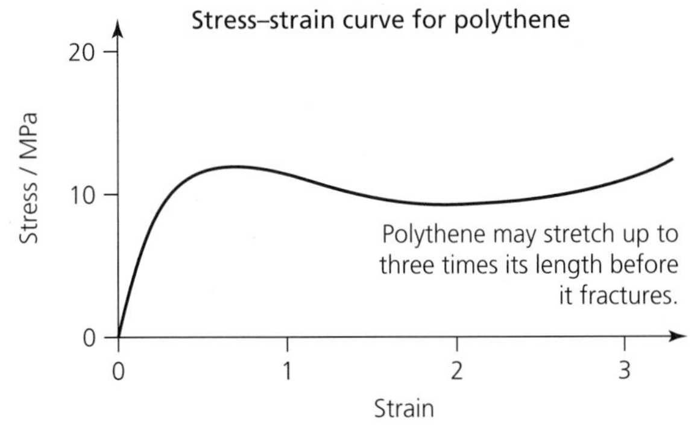


***Figure 6*** *A comparison of glass (brittle) and polythene (ductile) under tensile stress.*

## QUESTIONS

**7.** A 50 m nylon rope will stretch about 7.5 cm when supporting an 80 kg man. What is the size of the tensile strain?

**8.** Give an example of a material that is:

**a.** ductile

**b.** plastic

**c.** elastic

**d.** brittle

**e.** tough.

## KEY IDEAS

- Density is mass per unit volume, $\rho = \frac{m}{V}$.
- Tensile stress is the tensile (stretching) force per unit area acting on a material, $\sigma = \frac{F}{A}$. It is measured in pascal (Pa).
- Tensile strain is the extension of a material on stretching, per unit length, $\varepsilon = \frac{F}{\Delta l}$.
- Elastic materials regain their original shape when an applied stress is removed.
- Plastic materials do not regain their original shape when an applied stress is removed.
- As the tensile stress on a material is increased, plastic deformation begins to occur at the elastic limit.
- Ductile materials show large plastic deformations.
- Brittle materials show little or no plastic deformation before they fracture.

# 12.2 SPRINGS

Strength is not the only important property of a climbing rope. It is important that the rope does not stretch too much when the mountaineer is suspended from it. The amount of strain caused by a given stress depends on the **stiffness** of a material. Solid materials resist being pulled apart, rather like a spring resists being stretched. Stiffness is a measure of that resistance. The stiffness of a tension-coiled spring can be measured in a school laboratory (Figure 7).

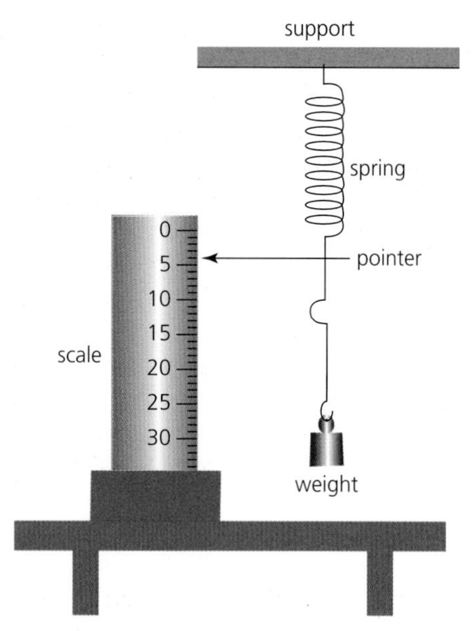


***Figure 7*** *Apparatus used to investigate the extension of a loaded spring*

The force pulling down on the spring, $F$, is increased by adding weights. As the force increases, the spring extends further. The results (Figure 8) show that the extension of the spring, $\Delta l$, is proportional to the force $F$ on the spring, $F \propto \Delta l$, until the force reaches a certain value, known as the **limit of proportionality**. If the force is increased still further, beyond the spring's elastic limit, the spring will be permanently deformed – it will not return to its original length when the external force is removed (see the black dashed line in Figure 8).

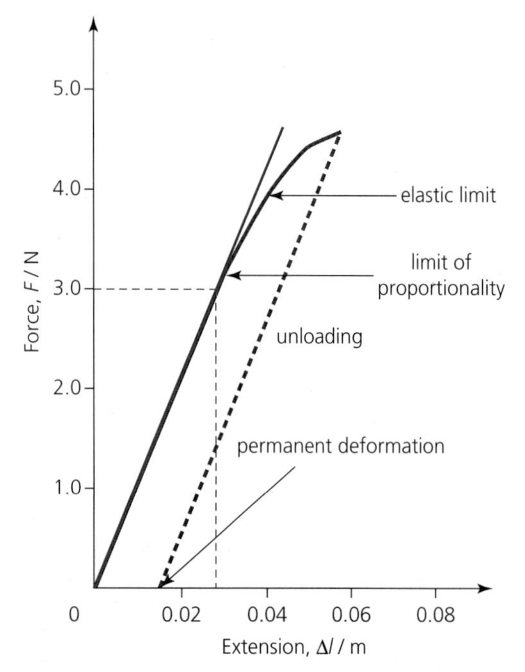


***Figure 8*** *Force–extension graph for a steel spring. The gradient of the straight-line region gives the spring constant.*

Robert Hooke (Figure 9) first proposed in 1678 that the extension of a spring is proportional to the force exerted on it. **Hooke's law** is not a universal law of physics, or even of springs. But many springs, and elastic materials, behave in this way until the force becomes large enough to cause permanent deformation. The spring in Figure 7 follows Hooke's law up to a force of around 3 N. Below this force, $F \propto \Delta l$, and we can write

$$F = k\Delta l$$

where $k$ is a constant known as the **spring constant**. The spring constant quantifies stiffness, and it has a unit of newton per metre ($\mathrm{N\,m^{-1}}$). As $k = \frac{F}{\Delta l}$, it is equal to the gradient of a force versus extension graph. So, to find the spring constant from the graph in Figure 8, we simply calculate the gradient of the straight-line part (in red):

$$k = \frac{3.0}{0.028}$$
$$= 110\ \mathrm{N\,m^{-1}}\ \text{(to 2 s.f.)}$$

***Figure 9*** *A modern impression of Robert Hooke, who achieved much more than the law of springs that bears his name. He devised the vacuum pumps used by Robert Boyle in his work on gases. Hooke made telescopes and observed the rotation of Jupiter. He made microscopes and carried out detailed observations, which led him to support evolution. He did important work on gravitation and light, and played a leading role in rebuilding London after the Great Fire of 1666.*

## QUESTIONS

9. With reference to Figure 8, explain the difference between the limit of proportionality and the elastic limit.

The springs in a car suspension (Figure 10) are compressed, rather than extended. Hooke's law still applies in this case.

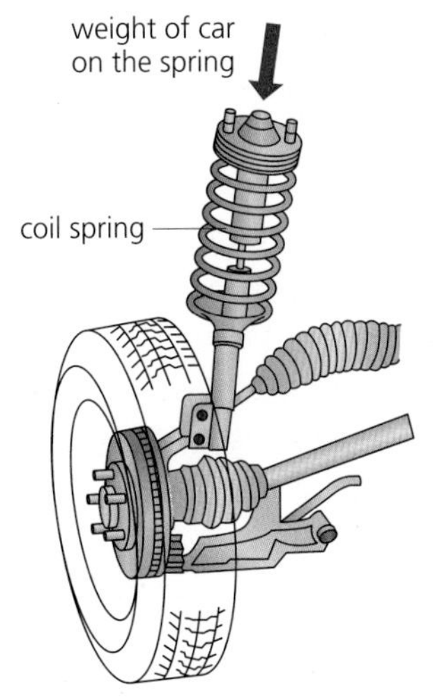


***Figure 10*** *A typical spring constant for a car suspension spring is around 25 kN m⁻¹.*

## Strain energy

A stretched spring stores energy. This **elastic strain energy** can be used to do useful work, perhaps to close a door, or operate a clock mechanism. The amount of energy stored is equal to the work done in stretching the spring. In Chapter 11 you saw that work done, $W$, is defined as force × distance moved in the direction of the force, $W = Fs \cos \theta$. That equation holds true for a constant force. But the force required to stretch a spring varies with the extension.

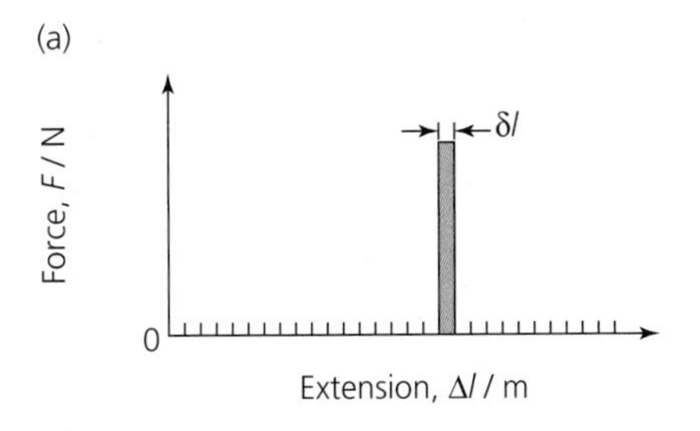


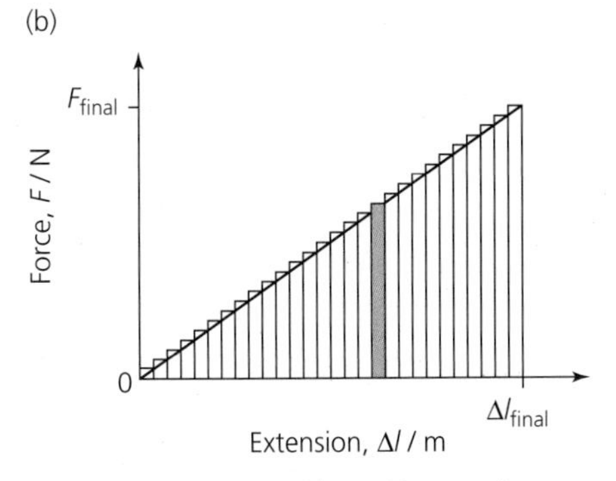


***Figure 11*** *Finding the work done in stretching a spring*

To calculate the work done in this case, we need to consider a small change in extension, $\delta l$, over which the force can be regarded as constant (Figure 11a). The work done in moving that small distance is $\Delta W = F\,\delta l$ (which is the area of that small strip). To find the total work done in stretching the spring, we add together the area of all such strips (Figure 11b). This is equivalent to the area under the force–extension graph. So the total work done in stretching the spring is $\Delta W = \frac{1}{2} \times F_{final} \times \Delta l_{final}$. Conservation of energy means that this is equal to the energy stored in the spring. From Hooke's law, $F_{final} = k\,\Delta l_{final}$, we can combine these equations and calculate the energy stored in a spring, of spring constant $k$ and extension $\Delta l$, using

$$E = \frac{1}{2}k(\Delta l)^2$$

This stored energy could be transferred as kinetic energy (for example when a rubber band is used as a catapult) or as gravitational potential energy (for example in a pogo stick or trampoline).

There are many types of springs (Figure 12). Most obey Hooke's law for part of their extensible range, though some never do. Some are designed to have two linear regions of different stiffness.

***Figure 12*** *Springs come in all shapes and sizes. They can act in tension, compression or rotation (by exerting a restoring torque when twisted through an angle).*

### Worked example 1

A chest 'expander' is made from five identical springs in parallel (Figure 13). When a force of 100 N is applied to the handle, the springs extend by 1 cm. Find the spring constant, and find the energy stored in the springs when extended.

When the force on the handle is 100 N, that force is shared by the parallel springs. Each spring is under a force of 20 N, which extends it by 1 cm, so $k = \frac{20}{1}$ $= 20\,\text{N cm}^{-1}$ or $2000\,\text{N m}^{-1}$.

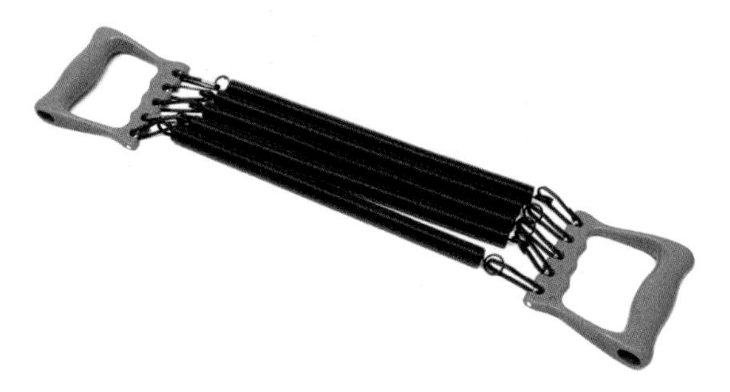

*Figure 13* Chest expander

The energy stored in each spring is $E = \frac{1}{2}F\,\Delta l$ $= \frac{1}{2} \times 20 \times 0.01 = 0.1$ J, so 0.5 J is stored altogether.

### Worked example 2

Chris wants to increase the work he does in stretching the springs of the chest expander in Worked example 1, so he dismantles it and connects the springs in series (end to end) instead. He secures one end to the wall and pulls the other end with the same force as before, 100 N. Will this increase the work done?

The force of 100 N is now applied to all the springs, rather than shared across them. They each have a spring constant of 2000 N m$^{-1}$, so they stretch by $\frac{100}{2000} = 0.05$ m, or 5 cm. The energy stored in each spring is $\frac{1}{2} \times 100 \times 0.05 = 2.5$ J, so that is 12.5 J altogether. The new arrangement means that 25 times more energy is stored by the springs in this configuration than in the parallel configuration, so Chris is certainly doing more work.

### QUESTIONS

**10.** A spring is used to fire a toy rocket vertically up into the air. The mass of the rocket is 200 g and when it is placed on the spring, it compresses it by 15 mm. Assume that the spring obeys Hooke's law.

**a.** Find the spring constant.

**b.** The spring is compressed by another 20 mm. How much energy is stored in the spring?

**c.** Assuming this was all transferred to the rocket, what is the maximum height of the rocket's flight?

**11.** A spring balance used to weigh hand luggage needs to weigh objects of mass between 0 and 10 kg.

**a.** Suggest a suitable value for the spring constant.

**b.** The balance is used to weigh a mass of 5 kg. What would be the uncertainty of the measurement?

**c.** A student says that choosing a suitable spring for a balance is a compromise between the precision of the readings and the maximum mass that can measured. Is she right? Explain your answer.

### KEY IDEAS

- Hooke's law for springs, force ∝ extension, applies to most springs, up to a certain force, known as the limit of proportionality.
- Hooke's law can be written as $F = k\,\Delta l$, where $k$ is the spring constant, which measures stiffness in N m$^{-1}$.
- The work done in stretching a spring is equal to the elastic strain energy stored in the spring:

$$E = \tfrac{1}{2}F\,\Delta l = \tfrac{1}{2}k(\Delta l)^2$$

- The work done in stretching a spring, and so the energy stored in it, is equal to the area under the force–extension graph.

## 12.3 MATERIALS IN TENSION

Whether an engineer is designing a bridge, an aircraft or a humble shopping bag, it is vital to know how the materials used will respond when subjected to a force. Forces can act to stretch, compress, bend, twist or shear an object (Figure 14), and there is likely to be different values of strength and stiffness in each case.

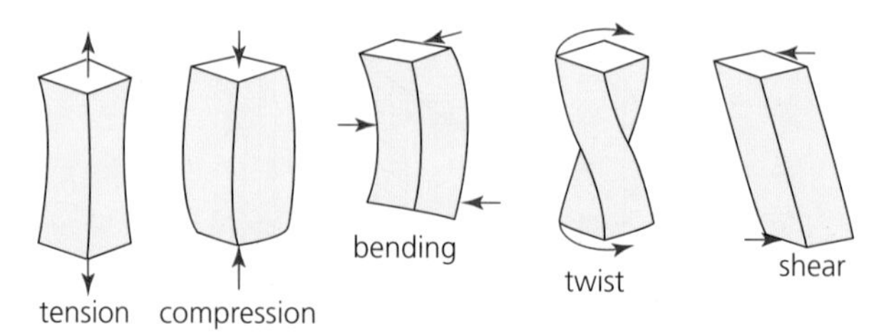


*Figure 14* How forces can act to deform an object

This section concentrates on the behaviour of materials that are stretched by a tensile force. We say that they are under **tension**. Some materials respond in a similar way to springs when they are put under tension. Metals, for example, tend to follow Hooke's law up to a certain applied stress, after which they become permanently stretched.

Samples of metal, in the form of a small bar, can be stretched using a tensile tester (Figure 15). A small bar is gripped between two jaws. A large tensile force (or load) is gradually applied and the extension at each load is automatically measured. For some samples at low loads, a graph of tensile force against extension produces a straight line (Figure 16). The sample follows Hooke's law. The load ($F$) is proportional to extension ($\Delta l$) or

$$F = k\,\Delta l$$

As for springs, the constant, $k$, is measured in $\mathrm{N\,m^{-1}}$, but here $k$ is referred to as the **stiffness**. A large value of $k$ means that the sample is difficult to stretch, and it is said to be stiff. In Figure 16, sample A is stiffer than sample B.

***Figure 15*** *Using a tensile tester*

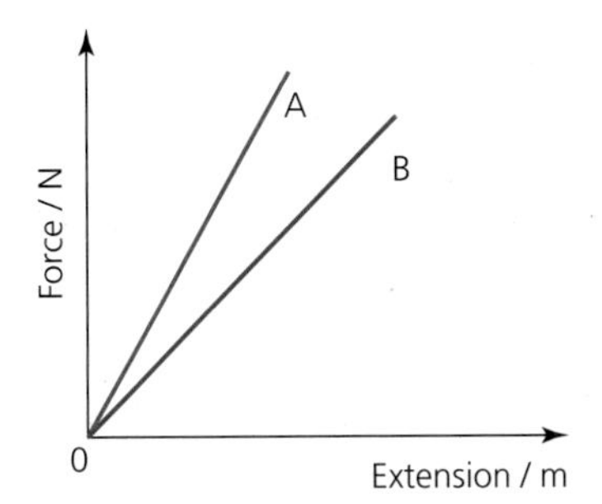


***Figure 16*** *Behaviour of two metals under tensile force*

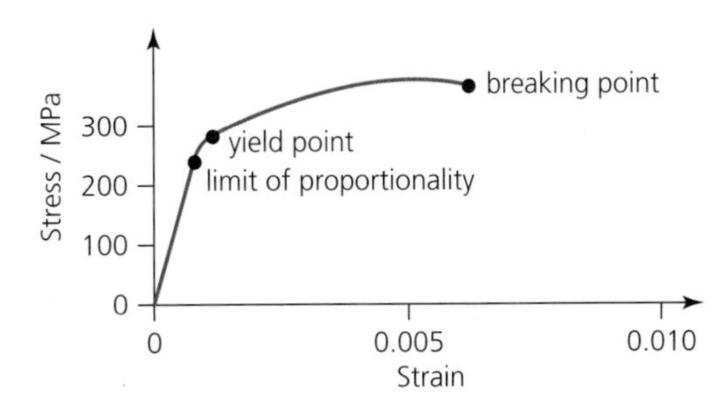


***Figure 17*** *Typical stress–strain graph for a metal wire*

We cannot compare different materials this way unless all the samples have identical dimensions. A sample may stretch less than another one because it is thicker, or shorter. We need to use stress (see section 12.1) rather than force, and strain rather than extension, to compare materials.

A stress–strain graph for a metal wire is shown in Figure 17. The linear part of the graph still indicates how stiff the material is, but note that the gradient here is stress/strain rather than force/extension. This ratio is known as the **Young modulus**.

$$\text{Young modulus} = \frac{\text{tensile stress}}{\text{tensile strain}} = \frac{\text{force} \div \text{cross-sectional area}}{\text{extension} \div \text{length}}$$

Stress is measured in pascal and strain is a ratio of two lengths, so the Young modulus is measured in pascal (Pa), though typically the values will be of the order of gigapascal (GPa). Values of the Young modulus for some materials are shown in Table 2. (The value for some of these materials, rubber for example, may depend on the applied stress.)

| Material | Young modulus / GPa |
|---|---|
| Carbon fibre | 270 |
| Steel | 210 |
| Kevlar | 124 |
| Copper | 117 |
| Bone | 28 |
| Polystyrene | 3.8 |
| Nylon | 3.0 |
| Rubber | 0.02 |

***Table 2*** *Typical values of stiffness of materials, in terms of the Young modulus*

For a metal, the Young modulus does not remain constant at higher values of stress. Beyond the limit of proportionality (see Figure 17), stress is no longer proportional to strain. At a slightly higher stress, the metal begins to be permanently deformed – this is the elastic limit or yield point. Metals obey Hooke's law over most of the elastic region, however, which means that the limit of proportionality and the elastic limit almost coincide. Some polymers, such as rubber or silk, may be elastic right up until they break, but they often do not obey Hooke's law at all.

### Worked example

A 0.75 m long copper wire of diameter 2 mm is used to support a painting. The tension in the wire is 100 N. How much will the wire stretch?

The cross-sectional area of the wire is

$$\pi r^2 = \pi \times (1 \times 10^{-3})^2\,\text{m}^2 = 3.14 \times 10^{-6}\,\text{m}^2$$

The applied stress is therefore

$$\text{stress} = \frac{100}{3.14 \times 10^{-6}} = 31.8\ \text{MPa}$$

The Young modulus of copper is 117 GPa, so

$$\frac{\text{stress}}{\text{strain}} = 117 \times 10^9\ \text{Pa}$$

Therefore the strain is given by

$$\text{strain} = \frac{31.8 \times 10^6}{117 \times 10^9} = 2.72 \times 10^{-4}$$

But because

$$\frac{\text{extension}}{\text{original length}} = \text{strain}$$

we have

$$\text{extension} = \text{strain} \times \text{original length}$$

$$= 2.72 \times 10^{-4} \times 0.75$$

$$= 2.04 \times 10^{-4}\ \text{m}$$

or just over 0.2 mm.

### QUESTIONS

**12. a.** A tough material has to absorb a lot of energy before it breaks. It is usually ductile and strong. Explain why.

**b.** Synthetic rubber is a tough material that only follows Hooke's law for very small deformations. Sketch a stress–strain curve for such a rubber.

**13.** The graphs in Figure 18 show the stress–strain curves for three different materials.

**a.** Which material is the stiffest?

**b.** Which material is the strongest?

**c.** Which material is the most ductile?

**d.** Which material has the lowest yield stress?

**e.** Which material is brittle?

**f.** Which material is tough?

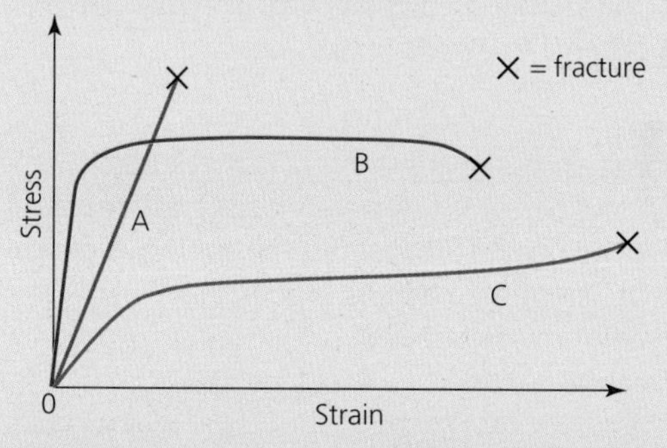


***Figure 18*** *Stress–strain curves*

**14.** A climber's rope needs to be strong, but also light. The ratio UTS/$\rho$, where UTS is the ultimate tensile stress (see section 12.1) and $\rho$ is the density of the material, is a useful measure of strength per unit weight. Similarly, the ratio (Young modulus)/$\rho$ gives a measure of a material's stiffness per unit weight. Use these criteria, and the data in Tables 1 and 2, to suggest the most suitable material for a climber's rope.

### KEY IDEAS

- Under tension, some materials obey Hooke's law, force $\propto$ extension, up to the limit of proportionality.
- The Young modulus is a measure of the stiffness of a material in the region where stress $\propto$ strain:

$$\text{Young modulus} = \frac{\text{stress}}{\text{strain}}$$

It has the unit pascal (Pa).

- The Young modulus is equal to the gradient of the linear section stress–strain graph.

# REQUIRED PRACTICAL: APPARATUS AND TECHNIQUES

## Measuring the Young modulus of a metal wire

The aim of this practical is to arrive at an accurate value for the Young modulus of a metal wire. The Young modulus is defined as

$$\text{Young modulus} = \frac{\text{stress}}{\text{strain}}$$

so

$$\text{stress} = \text{Young modulus} \times \text{strain}$$

If you plot stress ($y$-axis) against strain ($x$-axis), the gradient will give the value of the Young modulus.

This practical gives you the opportunity to show that you can:

- use appropriate analogue apparatus to record a range of measurements (to include length/distance) and to interpolate between scale markings
- use methods to increase the accuracy of measurements, such as use of a fiduciary marker, set square or plumb line
- use callipers and micrometers for small distances, using digital or vernier scales.

### Apparatus

The apparatus is assembled as in Figure P1. Two identical wires, made of the material under test, are suspended from the same support. The wires are held in 'chucks' at each end, which must be tightened carefully. First, identical weights are attached to the bottom of each wire, to pull the wires taut and remove any kinks. Then one wire, the 'test wire', is loaded and its new length is compared with that of the 'control wire' to find the extension.

### Safety considerations

The wires are often under significant tension and do occasionally break, or pull out of the chuck. Eye protection must be worn. The floor beneath the weights should be protected by a padded cardboard box, to prevent damage and to discourage experimenters from standing too close.

### Technique 1: Measuring the stress

Stress is defined as:

$$\text{stress} = \frac{\text{force}}{\text{cross-sectional area of wire}}$$

The wire is put under tension by hanging a mass, $m$, on it. The force is equal to $m \times g$.

The cross-sectional area of the wire $= \pi r^2 = \pi \frac{d^2}{4}$, where $r$ is the radius and $d$ is the diameter of the wire.

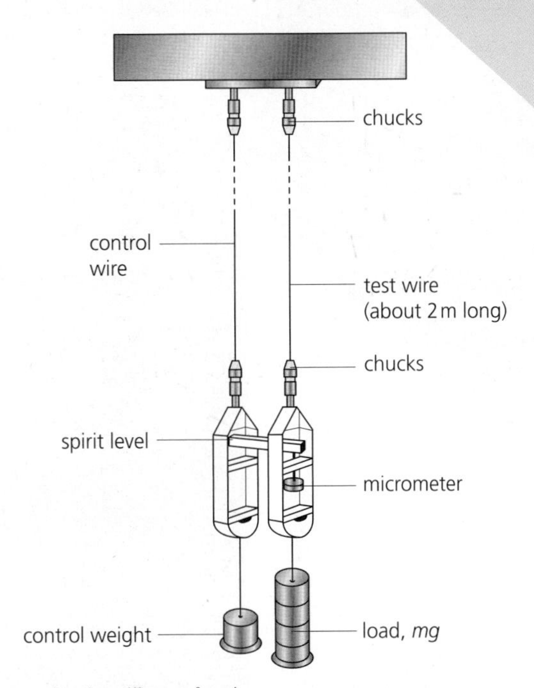


***Figure P1*** *Searle's apparatus for measuring the stiffness of a wire*

The diameter of the wire can be measured using a micrometer screw gauge (Figure 11 in Chapter 1). To allow for fluctuations in the thickness of the wire, the diameter is measured at three different places along the wire. At each of these points, two readings are taken at right angles to each other in case the wire has a non-circular cross-section. All six readings are used to find the mean diameter of the wire.

### Technique 2: Measuring the strain

Strain is defined as:

$$\text{strain} = \frac{\text{extension}}{\text{original length of wire}}$$

The extension of the wire is likely to be small and therefore difficult to measure precisely. The extension of the wire is often so small that the expansion caused by temperature changes is of a similar magnitude. Small movements of the support could also be a problem. There are several steps that can be taken to improve the precision and accuracy of this measurement.

- Use as long a length of wire as possible, so as to increase the extension. Practically, this usually means hanging the wires from a beam in the ceiling of the laboratory. For a particularly stiff material, for example brass, a thinner wire may be needed.
- Two wires are used. One wire is put under tension and extended (the test wire). Its length is compared to a second wire (the control wire), which is hung from the same support (see Figure P1). Any sag of the support, or any expansion caused by temperature changes, will affect both wires equally and will therefore not be measured.
- Use the spirit level and the micrometer to measure the extension. As each weight is added to the test wire, a screw thread is used to bring the spirit level back to the horizontal (Figure P2). A micrometer scale measures the movement of the screw (Figure P3). This allows measurement of the extension with an uncertainty of $\pm 0.01$ mm.

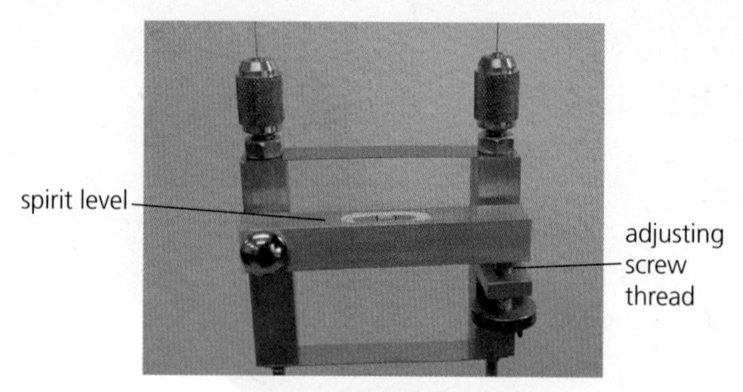


***Figure P2*** *The spirit level arrangement*

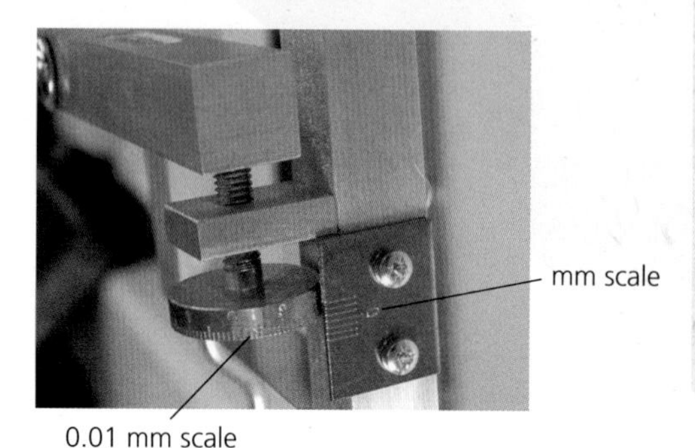


***Figure P3*** *The micrometer scale for measuring the extension of the test wire*

### Technique 3: Taking the readings

Weights are gradually added to the test wire. After each reading, the load is briefly removed and the spirit level is re-aligned to check that the wire has not been permanently extended, or slipped in the chucks. Readings are taken over as wide a range of weights as possible. Practically, the limiting factor may be the elastic limit of the wire, or its breaking point. If neither of these is reached, it is good practice to remove the weights one by one and repeat the readings as each weight is removed.

A set of readings for the experiment is shown in Tables P1 and P2. The original length of the wire was 2.00 m.

| Diameter of the wire / mm | 1.03 | 1.05 | 1.04 | 1.01 | 1.02 | 0.98 |
|---|---|---|---|---|---|---|

***Table P1***

| Mass added / kg | Weight added / N | Micrometer reading / mm | Extension / m | Stress / GPa | Strain |
|---|---|---|---|---|---|
| 0.00 | 0.00 | 0.02 | | | |
| 0.10 | 0.98 | 0.04 | | | |
| 0.20 | 1.96 | 0.06 | | | |
| 0.30 | 2.94 | 0.08 | | | |
| 0.40 | 3.92 | 0.10 | | | |
| 0.50 | 4.91 | 0.12 | | | |
| 0.60 | 5.89 | 0.14 | | | |
| 0.70 | 6.87 | 0.20 | | | |
| 0.80 | 7.85 | 0.22 | | | |
| 0.90 | 8.83 | 0.24 | | | |
| 1.00 | 9.81 | 0.26 | | | |
| 1.10 | 10.79 | 0.28 | | | |
| 1.20 | 11.77 | 0.30 | | | |
| 1.30 | 12.75 | 0.32 | | | |
| 1.40 | 13.73 | 0.36 | | | |
| 1.50 | 14.72 | 0.40 | | | |
| 1.60 | 15.70 | 0.44 | | | |

***Table P2*** *Data for loading the test wire*

## QUESTIONS

**P1** Calculate the mean diameter from the results in Table 1 and hence calculate the mean cross-sectional area of the wire.

**P2** Copy and complete Table P2.

**P3** Plot a graph of stress ($y$-axis) against strain ($x$-axis).

**P4** Write a conclusion to explain the shape of your graph.

**P5** Use your graph to calculate the Young modulus for this material, and identify the material by referring to Table 2 in section 12.3.

## 12.4 THE ENERGY STORED IN STRETCHED MATERIALS

In section 12.2 we saw that elastic strain energy is stored in extended (or compressed) springs. Stretched materials, such as a guitar string or a rubber band, also store elastic strain energy. The energy stored, $E$, is equal to the work done in stretching the material and (provided that the material follows Hooke's law) is given by

$$E = \frac{1}{2} F \Delta l$$

From this it can be shown that

$$E \text{ per unit volume} = \frac{1}{2} \times \text{stress} \times \text{strain}$$
$$= \frac{1}{2} \times \text{Young modulus} \times \text{strain}^2$$

Energy per unit volume is measured in the unit of joule per metre cubed ($J\,m^{-3}$). (For the derivation of this, see Assignment 1, which follows this section.)

This expression is only valid for materials that follow Hooke's law. However, for all materials, the work done in extending the material (which is equal to the energy stored per unit volume of the material) can be found by calculating the area below the stress–strain curve (Figure 19). If the material is tested to destruction, the area below the graph is a measure of the material's **toughness** – that is, how much energy it can absorb before breaking.

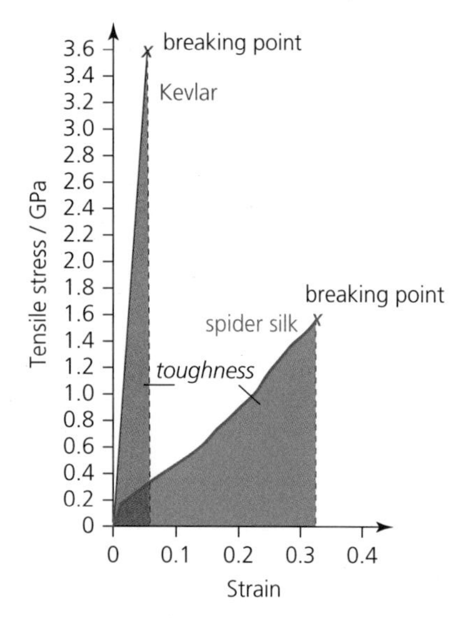


***Figure 19*** *The area below the curve is equal to the energy stored per unit volume, a measure of toughness.*

In most circumstances, we can assume that all the work done in stretching the material is stored as elastic strain energy, and is then available to do work. But in some cases energy is transferred during the stretching as internal energy, raising the temperature of the material. One example of this is rubber. To demonstrate this to yourself, take a rubber band, stretch it and then let it return to its original size. Quickly repeat this a number of times and then hold it to your lips. You should notice how warm it is. More work is done in stretching a rubber cord than is released when the cord is unloaded. This type of behaviour is known as **hysteresis** (see Figure 20). Each time the rubber is stretched and released, some energy is transferred as internal energy in the rubber.

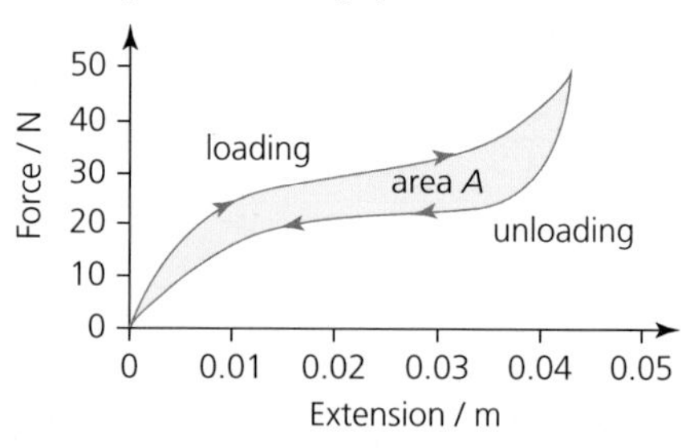


***Figure 20*** *Force–extension graph for loading and unloading a rubber band, showing hysteresis. The same force can cause two different extensions, depending on the history of the sample.*

Most types of rubber obey Hooke's law only over a limited range of extensions. Rubber tends to stretch easily at first and become much stiffer at high extensions, so the expression for strain energy ($\frac{1}{2} \times$ Young modulus $\times$ strain$^2$) will only be approximately correct over part of the cord's extension. However, we can find the work done in stretching the cord by calculating the area below the force–extension graph (Figure 20), plotted as the rubber is loaded. (Alternatively, we can find the work done per unit volume by calculating the area under the stress–strain graph.) The elastic strain energy recovered is the area below the curve as the rubber is unloaded. The area between the two curves is the energy transferred to the rubber as internal energy. The larger the area of this 'hysteresis loop', the greater the rise in internal energy of the rubber, and the hotter the rubber will get. Rubber is said to be **resilient** if it has a hysteresis loop with a small area.

## QUESTIONS

**15.** Spider silk is stiffer than rubber and has a much lower resilience.

**a.** Explain what 'lower resilience' means.

**b.** What might a bungee jump be like if silk cords were used instead of the usual rubber cords?

**16.** Squash balls (made of rubber) are not very bouncy when they are cold, but they get bouncier as they are bashed around the court. Sketch a force versus compression graph for a squash ball when it is cold, and one for when it has warmed up. Why does it warm up in use?

**17.** 'Sandbagging' is a variant on the bungee jump. A jumper holds on to a heavy sandbag, and then drops it at the bottom of the bungee jump. This has been banned in some countries, as it puts spectators and the jumper in danger. Why is it dangerous to the jumper?

## KEY IDEAS

- Elastic strain energy is stored in a stretched material.
- If the material obeys Hooke's law,

  elastic strain energy stored per unit volume $= \frac{1}{2} \times \text{Young modulus} \times \text{strain}^2$
- For all stretched materials, the elastic strain energy is equal to the area under the force–extension graph.
- The energy stored per unit volume is equal to the area under the stress–strain graph.
- For some materials, for example rubber, the work done in stretching them is greater than the energy transferred when they are unloaded.

## ASSIGNMENT 1: UNDERSTANDING ELASTIC STRAIN ENERGY

**(MS 0.1, MS 0.2, MS 0.5, MS 1.1, MS 2.2, MS 2.3, MS 2.4)**

In this assignment you will be looking at the energy transfers that occur in a typical bungee jump (Figures A1 and A2). As a bungee cord stretches, the kinetic energy of the jumper is transferred to elastic strain energy in the cord. Work is done against the tension in the rubber cord, and this is stored as elastic strain energy in the stretched cord.

***Figure A1*** *Modern bungee jumping began on April Fool's Day, 1979, when some members of the Oxford Dangerous Sports Club jumped from the 75 m high Clifton suspension bridge in Bristol.*

***Figure A2*** *"The first half of the jump, when the cable is slack, is horrifying. You free fall for up to 30m, with the ground rushing up at you at an alarming rate. Then the cable begins to stretch and slow your descent. The cords get tighter, until you slow to a stop and begin to accelerate back up. You get quite close to your original height, before you start falling again."*

The strain energy stored in a stretched cord (that obeys Hooke's law) is equal to $\frac{1}{2}F\,\Delta l$. If the cord was originally $l$ (m) long and has a cross-sectional area of $A$ ($m^2$), then its volume, $V$ ($m^3$), equals $A \times l$.

The strain energy stored per unit volume of the cord is given by

$$\frac{W}{V} = \frac{\frac{1}{2}F\,\Delta l}{Al}$$

$$= \frac{1}{2}\frac{F}{A}\frac{\Delta l}{l}$$

$$= \frac{1}{2} \times \text{stress} \times \text{strain}$$

$$= \frac{1}{2} \times \text{Young modulus} \times \text{strain}^2$$

To complete the assignment, you will need to refer to the data and assumptions in the box below, and then answer the questions that follow.

**Data and assumptions**

You may assume that:

- Our poor bungee jumper has only one cord attached.
- The bungee jumper's mass is 70 kg.
- Air resistance can be neglected.
- The mass of the bungee cord and harness can be neglected.
- The Young modulus for the cord is constant throughout the jump.
- The cord diameter is 1.91 cm.
- The cord stretches by 100% of its length under a load of 300 kg.

### Questions

**A1** Describe the energy transfers that take place as a bungee jumper leaps from a bridge, accelerates downwards until the cord becomes taut, and then is slowed to a halt.

**A2** If the bungee cord is 30 m long, how fast will the bungee jumper be travelling when the cord first becomes taut (that is, after falling 30 m)?

**A3** What will be the jumper's kinetic energy at that point?

**A4** What is the jumper's acceleration at that point?

**A5** When does the bungee jumper begin to accelerate back up? Explain why this is not the same point at which the jumper's velocity changes to an upward direction.

**A6** What is the Young modulus for the cord?

**A7** When does the jumper stop falling? What is the overall energy change that has happened?

### Stretch and challenge

**A8** The elastic strain energy stored in the bungee at the lowest point of the jump is

$$\frac{1}{2} \times \text{Young modulus} \times \left(\frac{\Delta l}{l}\right)^2 \times V$$

where $V$ is the volume of the cord (assumed constant) and $\Delta l$ is the extension of the cord. Assume that this is equal to the loss in gravitational potential energy, which is $mg(\Delta l + 30)$, and hence find how far the jumper falls.

**A9** What is the maximum upwards force acting on the jumper, and the acceleration and the velocity of the jumper at that time?

**A10** What effect would air resistance and the weight of the bungee cord have on the answer?

**A11** Real jumpers have several cords attached, often four. How would this change your answers to **A8** and **A9**?

## ASSIGNMENT 2: INVESTIGATING THE STRESS–STRAIN BEHAVIOUR OF DIFFERENT MATERIALS

**(PS 1.2, PS 2.1, PS 2.4, PS 4.1)**

In this assignment you will design a practical investigation. The aim is to subject a range of materials to tensile forces and compare their behaviour by plotting stress–strain graphs.

### Suggested materials

You could use samples cut from various plastic carrier bags, polythene rings that hold drink cans together, rubber cords, rubber bands, copper wire and brass wire.

### Suggested method

Unless the laboratory has a tensile tester, weights should be hung on the samples to put them under tensile stress, in a set-up such as that in Figure A1. It can be difficult to measure the cross-sectional area of strips of polythene, but measuring the thickness of several layers with a micrometer will improve the precision. Samples are often cut in 'I'-shaped pieces to give a wide section at each end to grip, and a narrower section to test.

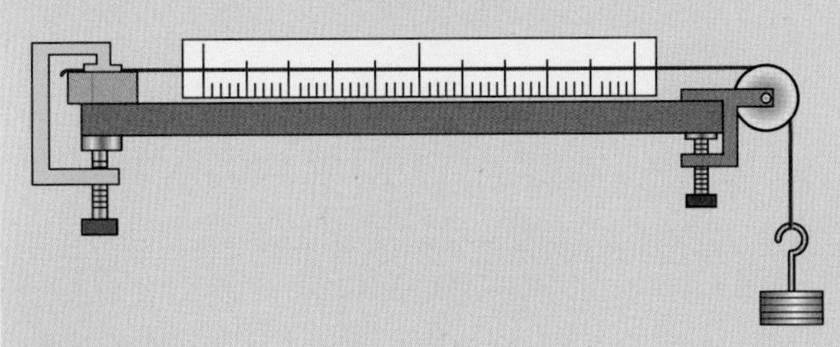

***Figure A1*** *Suggested set-up*

### Questions

**A1** What risks are involved in your experiment? List the safety measures that should be taken.

**A2** What readings will need to be taken? Describe how you would optimise the accuracy and precision of the readings.

**A3** The stress–strain graphs for some materials are shown in Figure A2.

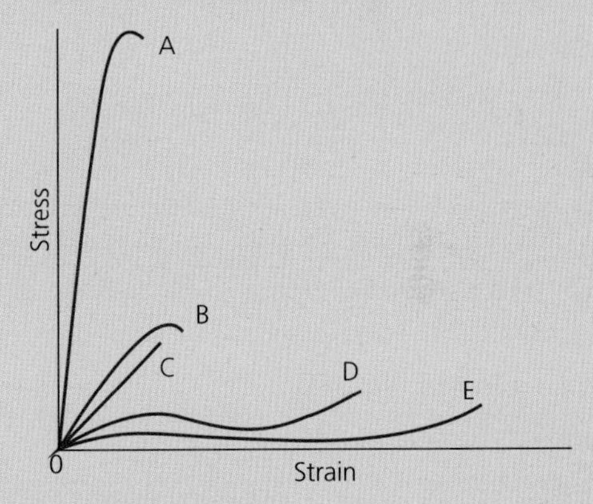


***Figure A2*** *Stress–strain curves*

Match up each graph with the correct material from the list that follows, using the brief descriptors to help. You might also need to do a little research.

- rubber of inflatable boat (ductile, tough)
- mild steel (extremely strong and stiff)
- glass (stiff, brittle)
- nylon fishing line (strong, stiff)
- high-density polythene (very ductile)

## PRACTICE QUESTIONS

1. A type of exercise device is used to provide resistive forces when a person applies compressive forces to its handles. The stiff spring inside the device compresses as shown in Figure Q1.

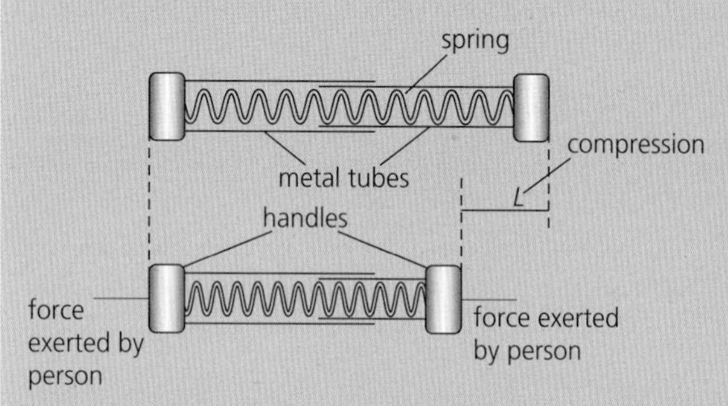


*Figure Q1*

a. The force exerted by the spring over a range of compressions was measured. The results are plotted on the grid in Figure Q2.

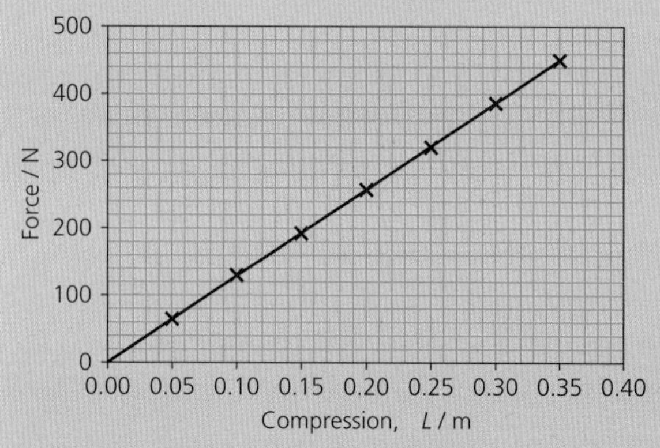


*Figure Q2*

i. State Hooke's law.

ii. State which two features of the graph confirm that the spring obeys Hooke's law over the range of values tested.

iii. Use the graph to calculate the spring constant, stating an appropriate unit.

b. i. The formula for the energy stored by the spring is

$$E = \frac{1}{2} F\Delta L$$

Explain how this formula can be derived from a graph of force against extension.

ii. The person causes a compression of 0.28 m in a time of 1.5 s. Use the graph in part **a** to calculate the average power developed.

*AQA Unit 2 January 2011 Q1*

2. A cable-car system is used to transport people up a hill. Figure Q3 shows a stationary cable car suspended from a steel cable of cross-sectional area $2.5 \times 10^{-3}$ m$^2$.

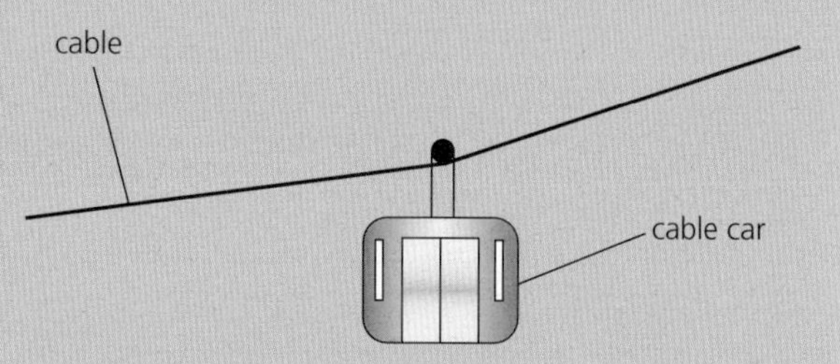


*Figure Q3*

a. The graph in Figure Q4 is for a 10 m length of this steel cable. Use the graph to calculate the initial gradient, *k*, for this sample of the cable.

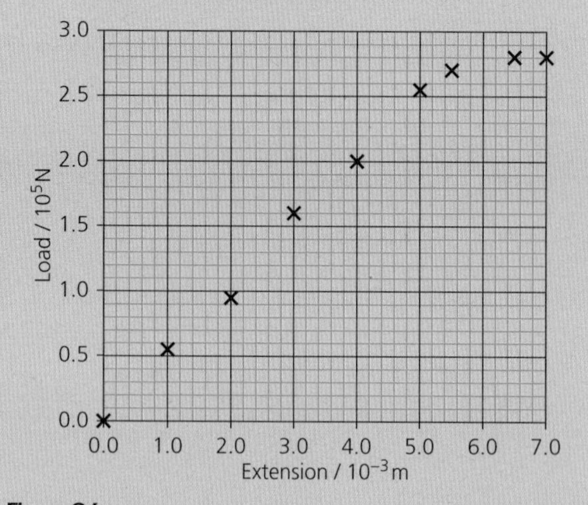


*Figure Q4*

b. The cable breaks when the extension of the sample reaches 7.0 mm. Calculate the breaking stress, stating an appropriate unit.

c. In a cable-car system a 1000 m length of this cable is used. Calculate the extension of this cable when the tension is 150 kN.

*AQA Unit 2 January 2011 Q6 (part)*

**3.** A rubber cord is used to provide mechanical resistance when performing fitness exercises. A scientist decided to test the properties of the cord to find out how effective it was for this purpose. The graph of load against extension is shown in Figure Q5 for a 0.50 m length of the cord.

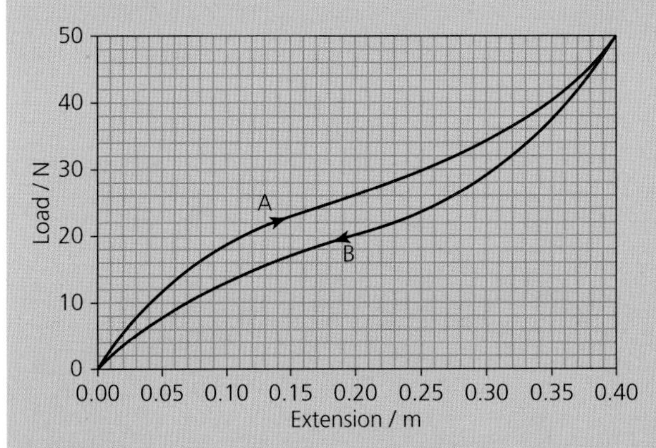


*Figure Q5*

Curve A shows loading and curve B shows unloading of the cord.

a. State which feature of this graph confirms that the rubber cord is elastic.

b. Explaining your method, use the graph (curve A) to estimate the work done in producing an extension of 0.30 m.

c. Assuming that line A is linear up to an extension of 0.040 m, calculate the Young modulus (unit Pa) of the rubber for small strains. The cross-sectional area of the cord = $5.0 \times 10^{-6}$ m$^2$. The unstretched length of the cord = 0.50 m.

*AQA Unit 2 June 2010 Q5 (part)*

**4.** A hollow sphere is made from steel of density 8000 kg m$^{-3}$. The radius of the sphere is 5.00 cm and the thickness of the steel is 0.25 cm.

Which of statements A to D is true?

(Density of North Sea water = 1030 kg m$^{-3}$; density of Dead Sea water = 1240 kg m$^{-3}$.)

**A** The sphere would float in the North Sea but not in the Dead Sea.

**B** The sphere would float in either sea.

**C** The sphere would sink in either sea.

**D** The sphere would float in the Dead Sea but not in the North Sea.

**5.** A spring is used to fire a small toy rocket into the air. The mass of the rocket is 100 g. When the rocket is placed on the spring, it compresses the spring by 0.5 mm. The spring is pushed down by a further 10 mm before firing the rocket.

An estimate of the height that the rocket would rise is:

**A** 1 cm

**B** 10 cm

**C** 1 m

**D** 10 m

**6.** Which of the metals shown in Figure Q6 could be described as *brittle*?

**A** Aluminium alloy

**B** Strong grey cast iron

**C** Magnesium alloy

**D** Pure aluminium

**7.** Which row in the table correctly describes the properties of the metals shown in Figure Q6?

| | Highest yield stress | Stiffest |
|---|---|---|
| A | Aluminium alloy | Pure aluminium |
| B | Magnesium alloy | Pure aluminium |
| C | Aluminium alloy | Aluminium alloy |
| D | Strong grey cast iron | Aluminium alloy |

8. Figure Q6 shows that the Young modulus for aluminium alloy is:

   **A** 68.8 MPa

   **B** 68.8 GPa

   **C** 68.8 kPa

   **D** 688 MPa

Questions 6 to 8 relate to Figure Q6.

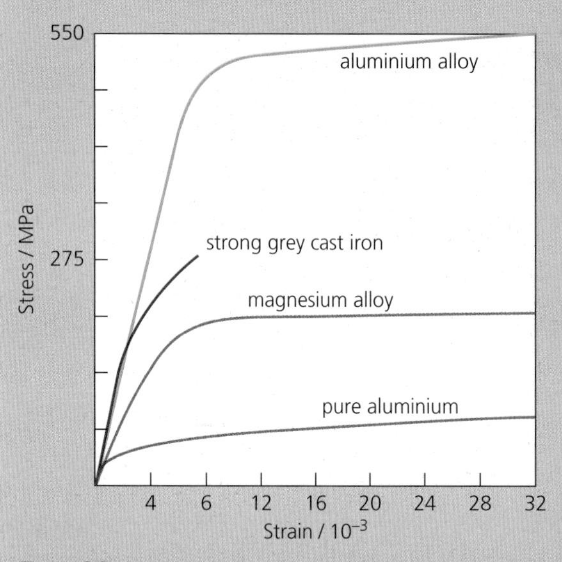


*Figure Q6*

## Stretch and challenge

9. This question is about the mechanical properties of bone. Bone is a composite material with two components – collagen, which is a protein, and a mineral containing calcium phosphate and calcium carbonate. The collagen provides a flexible framework, while the mineral gives bones rigidity and strength.

   A long bone such as the femur (thigh bone) contains two types of bone material – cortical (or compact) bone and cancellous (or spongy) bone. Cancellous bone is 25% to 50% as dense, 10% as stiff, but five times as ductile as cortical bone. Cortical bone is good at resisting torque, while cancellous bone is strong against compression and shear forces.

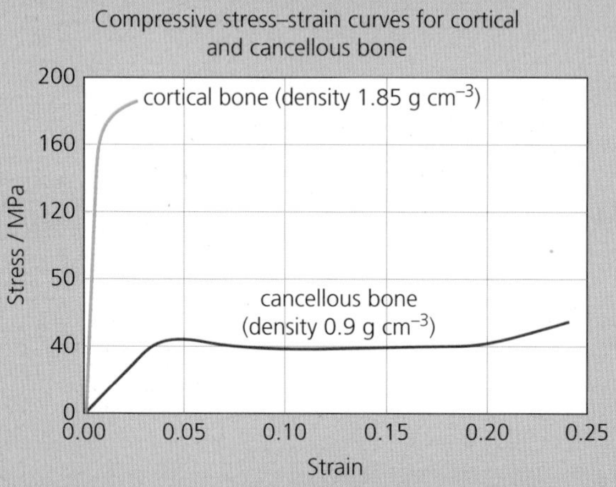

   Compressive stress–strain curves for cortical and cancellous bone

   cortical bone (density 1.85 g $cm^{-3}$)

   cancellous bone (density 0.9 g $cm^{-3}$)

   Stress / MPa: 0, 40, 50, 120, 160, 200

   Strain: 0.00, 0.05, 0.10, 0.15, 0.20, 0.25

   *Figure Q7*

   **a.** Explain what is meant by *ductile* and describe how ductility is shown in Figure Q7.

   **b.** The text above states that cortical bone is 10 times stiffer than cancellous bone. Carry out calculations to check this, using data from Figure Q7.

c. The properties of bone change with age. Figure Q8 shows stress–strain curves for cancellous bone for people of three different ages. The toughness of a material is a measure of its ability to absorb energy without fracturing. Estimate the energy absorbed that is shown by the graphs in Figure Q8 and hence comment on the effect of ageing on bone toughness.

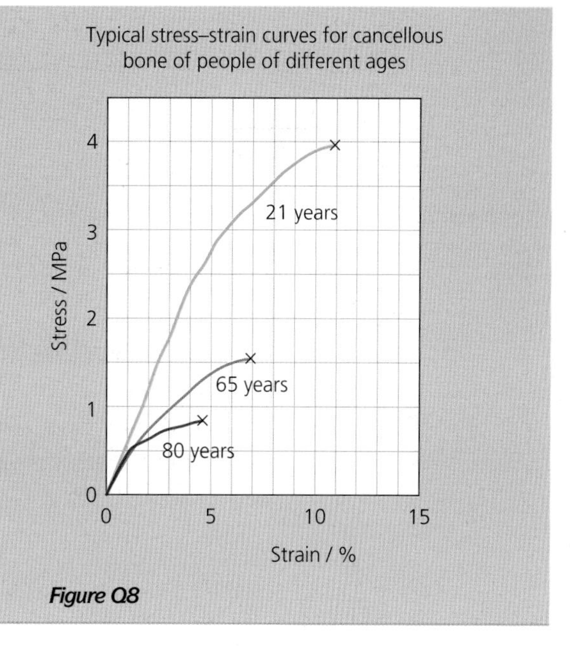


*Figure Q8*

# 13 ELECTRICITY 1

## PRIOR KNOWLEDGE

You will have met electric current, potential difference (voltage) and resistance in series and parallel circuits. You will have constructed simple circuits and used an ammeter and a voltmeter to measure current and potential difference, respectively.

## LEARNING OBJECTIVES

In this chapter you will learn about resistors, and how they are combined in series and parallel circuits. You will learn about current–voltage characteristics for some common components, and about the property called resistivity.

**(Specification 3.5.1.1, 3.5.1.2, 3.5.1.3 part, 3.5.1.4)**

*ELECTRIC LIGHT has its defects … The intense shadows [and] the unsteadiness of the light. [It] has an unfortunate habit of misbehaving itself when it is most wanted … Many have tried the light and abandoned it … but the brilliancy of a well-lighted room is simply enchanting. Its cleanliness is one of its great merits. It emits no smoke. At Glasgow … more work has been got out of the men; the electric light will pay for itself … in this point alone.*

**Figure 1** *Extract from an article by the electrical engineer William Henry Preece, in 'Popular Science Monthly', Volume 19, July 1881*

New technology always has its teething problems. Despite Mr Preece's concerns (Figure 1), the incandescent (hot filament) electric light bulb was an invention that really did change the world. Two men, Thomas Edison in America and Joseph Swan in Britain, claimed the invention as their own. Both men held patents and were ready to use the law to establish precedence. But which of them had the original 'light bulb' moment?

Swan demonstrated his light bulb in Newcastle in February 1879. He passed an electric current through a thick carbon rod inside an evacuated glass bulb (see Figure A1, Assignment 2, Chapter 8). The carbon emitted an impressive bright light, but the bulb had a less than impressive lifetime. Edison demonstrated a very similar lamp later that year in the USA. But Edison had realised that a much thinner filament was the key to success. His bulb lasted longer. Wisely, the two inventors settled for a joint venture, and in 1883 The Edison-Swan Electric Light Company was founded.

Now, more than 130 years later, the incandescent bulb is being retired. Too inefficient and prone to failure, it is being replaced by the light emitting diode (LED). LEDs have a longer lifetime and are much more efficient, though new technology always brings new concerns (Figure 2).

*[LED streetlights have] … a blue-rich colour mix like moonlight or bright daylight, which many people find unpleasant. [They] spread light across the street, creating glare for pedestrians and drivers [and] making the night sky so bright it washes out the stars. Blue light can also suppress melatonin production … affecting sleep and behaviour.*

**Figure 2** *Extract from article by Jeff Hecht, in 'New Scientist', October 2014*

# 13.1 ELECTRICAL POWER AND ENERGY

Lighting accounts for about 20% of the world's electrical energy demand. A change to LED lighting (Figure 3) would dramatically reduce that. (LED stands for light emitting diode – see section 13.5.) LED bulbs use much less electrical energy per second to produce the same amount of light as incandescent light bulbs (Figure 4). The electrical energy transferred per second is the electrical **power** of the bulb:

$$\text{power} = \frac{\text{electrical energy transferred}}{\text{time taken}}$$

The amount of energy transferred is measured in joules (J), and power is measured in joules per second or **watts** (W).

***Figure 3*** *LED light bulbs now come in many shapes and powers, and are getting more and more efficient.*

A typical incandescent light bulb used for room lighting would have a power of 100 W. If the light bulb is on for 8 hours, it would transfer $8 \times 60 \times 60 \times 100 \approx 3$ MJ of electrical energy to light energy and thermal energy. Unfortunately, around 95% of the electrical power supplied to an incandescent light bulb goes directly to heating the surroundings, rather than illuminating them. In contrast, an LED bulb with a power of only 18 W will provide the same output of light, and the latest types of LED lamp may need only 7 W.

| Brightness | 220 | 400 | 700 | 900 | 1500 |
|---|---|---|---|---|---|
| Conventional incandescent | 25 W | 40 W | 60 W | 75 W | 100 W |
| Halogen | 18 W | 28 W | 42 W | 53 W | 70 W |
| Compact fluorescent | 6 W | 9 W | 12 W | 15 W | 20 W |
| LED | 4 W | 6 W | 10 W | 13 W | 18 W |

***Figure 4*** *The electrical power required from different bulbs to give the same brightness of light (values of brightness given in the unit lumen)*

Table 1 shows the power of different types of LED light, used for very different lighting purposes.

| Device | Power |
|---|---|
| LED power indicator on TV | 5 mW |
| LED table light | 10 W |
| LED car headlight | 40 W |
| LED street light | 70 W |
| LED stadium floodlight | 1 kW |

***Table 1*** *Typical input electrical power for different LED light sources*

The joule is rather a small unit when it comes to measuring domestic energy demand. An alternative unit of energy is the **kilowatt-hour** (kWh). This is a power of 1 kW transferring energy for 1 hour:

$$1 \text{ kWh} = 1000 \text{ W} \times 3600 \text{ s} = 3.6 \text{ MJ}$$

## Efficiency

The **efficiency** of an electrical device measures how much of the electrical energy supplied is transferred into a useful form:

$$\text{efficiency} = \frac{\text{useful energy output}}{\text{electrical energy input}} \times 100\%$$

or

$$\text{efficiency} = \frac{\text{useful power output}}{\text{electrical power input}} \times 100\%$$

For example, an incandescent light bulb transfers electrical energy into light and heat. However, a 100 W light bulb transfers only about 5 W as useful light energy – the rest is transferred to the surroundings as heat. The light bulb is therefore 5% efficient. An LED light bulb is typically around 25% efficient.

### QUESTIONS

1. A typical family uses about 1000 kWh of energy every three months. How many joules of energy is that?
2. a. An electric light bulb has a power of 100 W. If it is left on all day (24 hours), how much energy does it transfer? Give your answer in joules and in kilowatt-hours.

b. The same amount of light would be emitted by an 18 W LED lamp. How much energy would that transfer in a day (in J and in kWh)?

3. An electric car has a nickel–metal hydride battery pack, which has a capacity of 26.4 kWh.

   a. How many joules of energy is this?

   b. If it ran for 3 hours, what would its average power be?

4. The latest type of LED tube lamp is twice as efficient as the fluorescent tubes it is designed to replace. Estimate how much money you might save in a year if you replaced one 80 W fluorescent tube in your kitchen.

**KEY IDEAS**

- Power is the rate of energy transfer, in joule per second, or watt (W).
- Electrical energy is often measured in kWh; 1 kWh = $3.6 \times 10^6$ J.
- The efficiency of an electrical device is the percentage of electrical energy (or power) input that is transferred usefully.

## 13.2 ELECTRIC CURRENT

### Charge

Electrical devices work when an electric current passes through them. Electric current is a flow of electric charge. **Charge** is a fundamental property of matter, like mass. We know that different particles have different masses, which are all affected by gravitational force. In a similar way, particles can carry different amounts of electric charge and are affected by electrical force. But there is an important difference. Unlike mass, charge can be positive or negative. Two particles that carry the same sign of charge will repel each other; two particles with opposite charges will attract each other.

These electric forces between charges keep atoms stable and hold atoms together as molecules. Atoms have negatively charged electrons, which are attracted to the positively charged nucleus of the atom. Atoms as a whole are electrically neutral, but they can become **ionised**. If an electron is added to an atom, a negative **ion** is formed. A positive ion is created by taking an electron away from a neutral atom. The size of the force between charged particles such as ions depends on the magnitude of the charges, which we measure in the unit **coulomb** (C). The charge of one electron is $-1.6 \times 10^{-19}$ C. Charge is given the symbol $Q$.

### Conductors and insulators

The flow of electric charges through a material is called **conduction**. A material that does not readily conduct electricity, such as glass, plastic or air (Figure 5), is known as an **insulator**. An insulator has no charges that are free to move through it. The electrons in its atoms are tightly bound, which means that a relatively large amount of energy (on an atomic scale) is needed to free one electron from the atom so that it is mobile and can take part in conduction. In glass, this would take around 3 eV of energy for each electron.

***Figure 5*** *Air is a good electrical insulator. It will conduct only if the forces are great enough to ionise the atoms. The potential difference (see section 13.3) has to be very high – around three million volts per metre.*

It is possible to add extra electrons, or remove them, from an insulator, to leave it with a net charge. This is easily done by rubbing two insulators together. When you brush your hair, or rub a balloon on your jumper, or even stroke the cat, you will remove electrons from one surface (leaving it positively charged) and add them to the other (making it negatively charged). The charges will remain on the surface, and the two surfaces will then attract each other. Objects that carry similar surface charges, either both positive or both negative, will repel each other (Figure 6).

*Figure 6* Hair that has gained a net charge will repel hair that has the same charge.

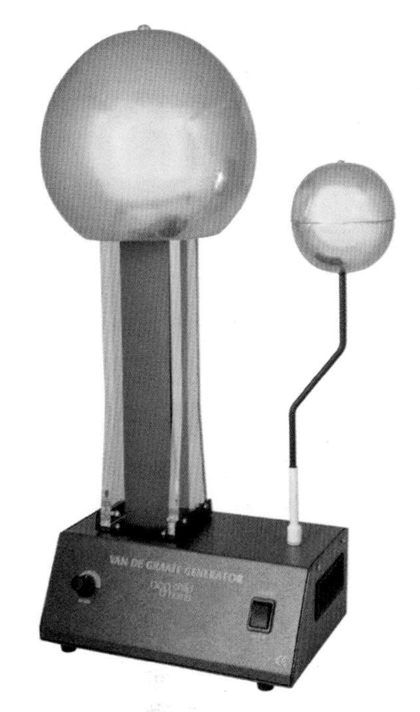



*Figure 7* A Van de Graaff generator is designed to charge up a metal dome. When a large enough charge has accumulated, a spark will jump to a nearby conductor such as the smaller metal sphere here, discharging the Van de Graaff.

### QUESTIONS

5. How many electrons are needed to make a charge of one coulomb (1 C)?
6. Explain why the text describes the removal of electrons by rubbing insulators together as "easily done", although "a relatively large amount of energy" is needed to free electrons in an insulator to form an electric current.

An electrical **conductor** allows charges to pass through it. Metals are good conductors principally because some of their atomic electrons are free to move around between the atoms. It is possible to add charge to a conductor, such as the metal dome on a Van de Graaff generator (Figure 7), and the charges will move and distribute themselves over the surface. If there is a conducting path to earth available, the charge will leak away.

## Current

The effects with balloons, hair brushes, lightning and a Van de Graaff generator, described above, are due to objects gaining a net charge. They are referred to as **electrostatic** phenomena – there is no continuous flow of charge. When charges do flow continuously, like electrons around a closed loop of conducting wire, it is referred to as an electric **current**. Any movement of charged particles is an electric current. It could be a flow of electrons in a metal wire, or positive and negative ions moving through a battery, or a beam of protons in an accelerator. The size of the current, $I$, is given by the amount of charge, $\Delta Q$, passing a point in a given time, $\Delta t$. Current is the rate of flow of charge:

$$I = \frac{\Delta Q}{\Delta t}$$

Electric current is measured in **ampere** or amp (A). In a circuit with a current of one ampere (1 A), a charge of one coulomb (1 C) passes a given point every second.

The direction of a current is conventionally that of positive charge – that is, current goes from a positive terminal to a negative terminal. Electron goes in a conducting circuit of metal wire that is actually in the opposite direction (see Figure 11).

### QUESTIONS

7. A current of 0.25 A flows through a table lamp. If the lamp is on for an hour, how much charge will have flowed through it?
8. The electron beam in a cathode ray oscilloscope carries a current of 0.1 mA. How many electrons strike the screen per second?

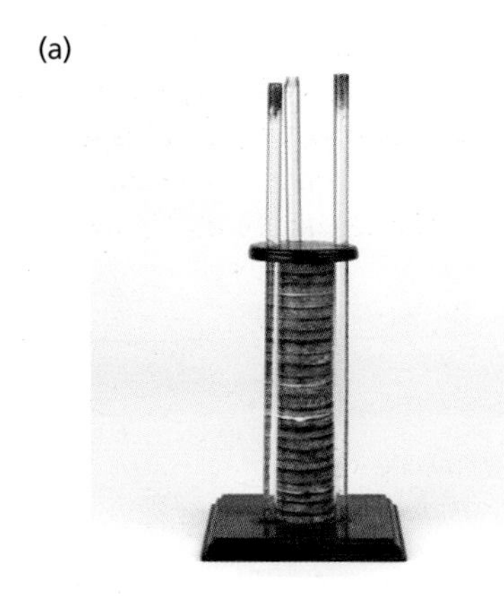


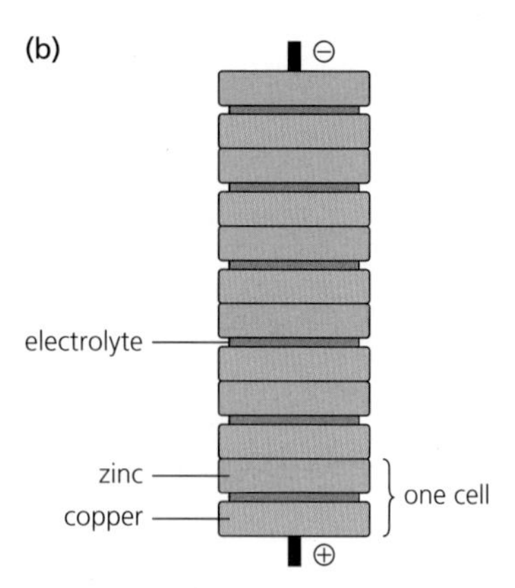


***Figure 8*** *(a) Volta published details of his battery in 1800. This was the first device capable of delivering a continuous electric current to a circuit. (b) It consisted of discs of copper and zinc separated by felt soaked in salt water. Each pair of discs constituted an electrolytic cell. Volta's battery led to the discovery of the electrolysis of water, and of a number of chemical elements, such as calcium and magnesium. Volta's battery and its descendants remained the only usable source of electrical power until the first practical dynamos appeared in the 1870s.*

***Figure 9*** *Different types of battery for different portable devices, and a car battery*

## Electric circuits

An electric current will be set up in a conductor if there is

- an energy source (a cell or a Van de Graaff generator, for example) pushing charge through it *and*
- a continuous conducting path (a circuit) around which the charges can flow.

A **battery** can be used as the energy source to maintain a current. A battery (as Volta's original battery in Figure 8) is formed from a number of connected **cells**. A chemical reaction in the cell produces electrical energy, which results in a current flowing in an external conducting circuit. The 'batteries' that we buy for our portable gadgets (Figure 9a) are properly called 'cells'. A car battery (Figure 9b) is accurately named, because it is usually six cells connected together.

A conducting circuit can be formed by connecting various conducting components, such as lamps and resistors, with metal wires. Figure 10 shows the circuit symbols used for the components needed at AS level.

A simple electric circuit can be formed with a cell, a bulb and some wire. An ammeter can be added to this circuit to measure the current. The simplest arrangement is a single loop (see Figure 11). These components are said to be in **series**; the components are arranged one after the other, with no alternative current paths.

## Conservation of charge in a circuit

In a series circuit, the magnitude of the current must be the same all around the circuit (Figure 12a); current does not get 'used up'. The charge that flows into any point in a circuit in one second must equal the charge per second that flows out of the point. In a series

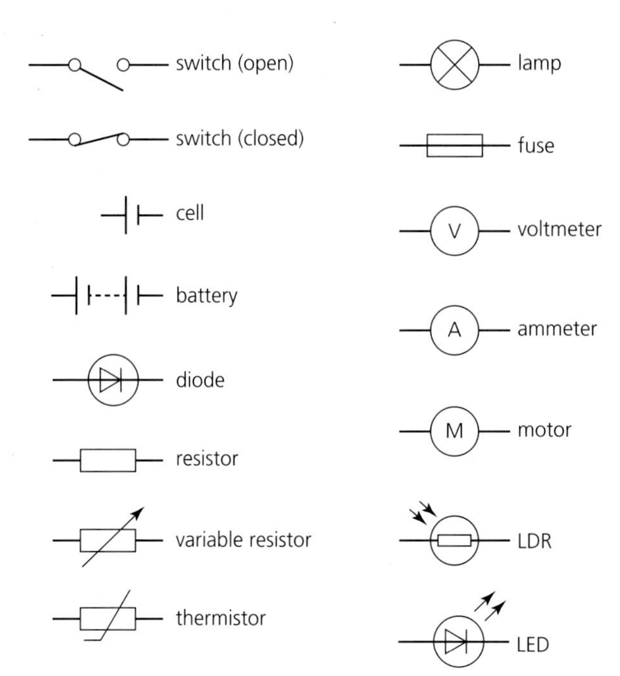


*Figure 10* *Circuit symbols*

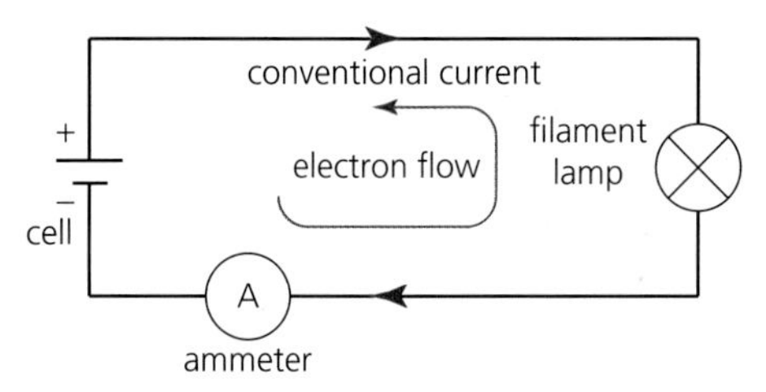


*Figure 11* *Current in a simple series circuit*

circuit there are no alternative paths for the current to take. An ammeter, used to measure current, will record the same reading wherever it is placed in the circuit.

A circuit, or part of a circuit, where there are alternative paths for the current to take is known as a **parallel** arrangement (Figure 12b). The current will divide between the two (or more) paths, but the sum of the currents in the parallel sections must equal the total current flowing in the circuit.

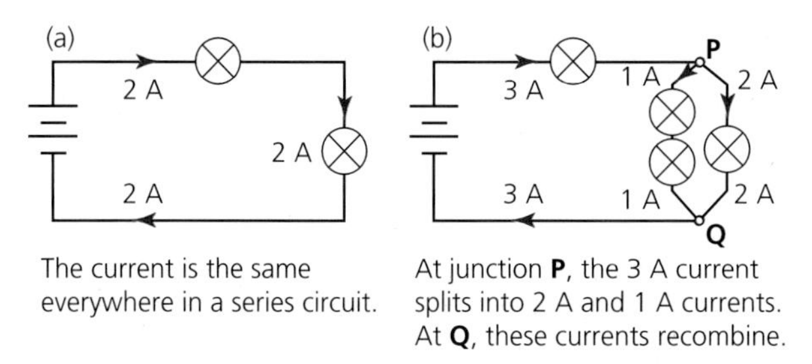


*Figure 12* *(a) A series and (b) a parallel arrangement*

These ideas about the continuity of current, or the 'conservation of charge', are summarised by **Kirchhoff's first law**, which states that the total current entering any point in a circuit must equal the total current leaving that point (Figure 13).

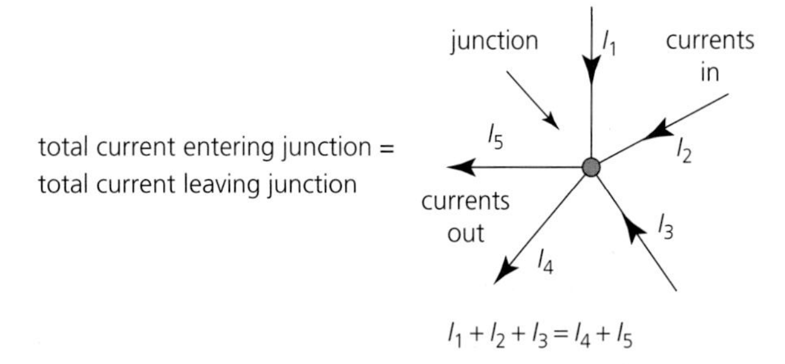


*Figure 13* *Conservation of charge in a circuit (Kirchhoff's first law)*

## QUESTIONS

**9. a.** Look at the circuit in Figure 14. Write down one equation that shows how the currents in each branch, $I_1$ to $I_4$, are related.

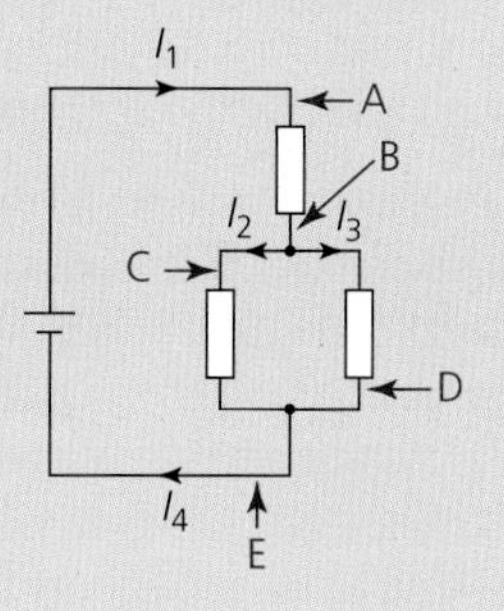


*Figure 14*

**b.** Suppose ammeters were connected in the circuit in Figure 14 at points A, B, C, D and E. Say whether or not each of these statements is true, and explain your answer.

**i.** The current at A is greater than at B.

**ii.** The current through D is equal to the current through C.

**iii.** Ammeter E will read almost zero.

**iv.** Ammeter E will show the same current as ammeter A.

**v.** The current through ammeters C and D will add together to the same as the ammeter reading A.

**KEY IDEAS**

- Charge, $Q$, is a property of matter, carried by charged particles, such as the electron and proton. Charge is measured in coulomb (C). The charge carried by an electron is $-1.6 \times 10^{-19}$ C.
- Current, $I$, is the rate of flow of charge, $I = \frac{\Delta Q}{\Delta t}$, measured in ampere or amp (A).
- The direction of conventional current is that in which positive charges flow – from positive (+) to negative (–).
- Charge is conserved in a circuit – the total current entering any point in a circuit equals the total current leaving that point (Kirchhoff's first law).

## 13.3 POTENTIAL DIFFERENCE

The energy needed to push current round a circuit can be provided by a battery. The battery transfers energy to each charge through a chemical reaction, pushing negative charges 'uphill' towards the negative terminal. This energy is then transferred in the circuit as the charges flow round and back to the positive terminal.

The amount of energy that is transferred when a charge moves between two points in a circuit is known as the **potential difference**, or **pd**, although often the term 'voltage' is used. Electrical potential difference is analogous to gravitational potential difference. When a mass drops from a height, its gravitational potential energy is transferred to kinetic energy as it falls. The further the height through which the object falls, the greater its final velocity and the higher the energy with which it hits the ground. In the same way, a larger electrical potential difference increases the energy that is transferred as a charge moves through it:

The potential difference (pd) between two points is defined as the work done (or energy transferred) per unit charge passing between the points.

The unit of potential difference is the volt (V). When a charge of one coulomb (1 C) passes through a pd of one volt (1 V), it does one joule (1 J) of work:

$$\text{potential difference (V)} = \frac{\text{work done (J)}}{\text{charge (C)}}$$

$$V = \frac{W}{Q}$$

**Worked example**

A potential difference of 2.0 V is applied across an LED, causing a current of 20 mA to flow through it. After 10 minutes, how much charge has flowed through the LED? How much energy is transferred as the charge flows through the LED?

Charge that has flowed is
$\Delta Q = I \times \Delta t = 20 \times 10^{-3} \times 10 \times 60 = 12$ C

Energy transferred = work done = $QV$
$= 12 \times 2.0 = 24$ J

Potential difference is measured using a voltmeter connected in parallel with the component. A voltmeter measures a difference between two points, so it needs to be in parallel, connected *across* a component (Figure 15), whereas an ammeter is placed in the circuit, in series. Ammeters and voltmeters can be assumed to have no effect on the circuit – they have effectively zero and infinite resistance, respectively (see Assignment 1).

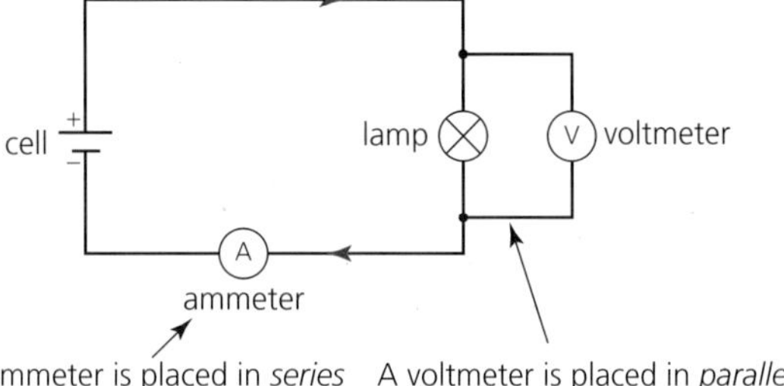


***Figure 15*** *Measurement of current and potential difference*

**QUESTIONS**

**10.** If 30 C of charge per hour flows through an LED with a pd of 2 V across it, how much energy is transferred?

**11.** A 12 V battery is used to power a series circuit containing a light bulb and a motor. A voltmeter placed across the bulb reads 4.0 V. An ammeter in the circuit reads 3.0 A.

**a.** What is the potential difference across the motor?

**b.** How much energy is transferred in the motor in 20 minutes?

**c.** What is the power of the motor?

### Conservation of energy in a circuit

Consider a charge moving round the series circuit in Figure 16a. As it moves round the circuit, it gains energy from the battery and transfers energy as light, movement and heat as it passes through the bulb, motor and resistor. The principle of conservation of energy tells us that the total energy transferred by the charge must be equal to the energy it gained in the battery. So the sum of the energy transferred by a charge as it passes through each component must be equal to the energy transferred to the charge in the battery. (In practice, there would be a small energy transfer across the connections and leads, but this can usually be neglected.) The energy transferred to each charge *in the battery or power supply* is known as the **electromotive force** or **emf**. This will be considered further in Chapter 14. So, because energy is conserved, we can write:

> The sum of the pds around a circuit is equal to the sum of the emfs of the batteries (or other power supply).

This is known as **Kirchhoff's second law**. It is the principle of conservation of energy applied to electrical circuits – it says, in simple terms, that the energy gained by a charge as it passes through any batteries or other energy sources is equal to the energy transferred by the charge as it flows around the circuit.

Now consider the parallel circuit, as in Figure 16b, where there are two possible alternative paths. The potential differences across the two paths are equal. Because the start and end points of the two paths are the same, the potential difference must be the same, whichever route is taken by the charge.

## QUESTIONS

**12. a.** What would the potential difference be across each of the resistors in Figure 17?

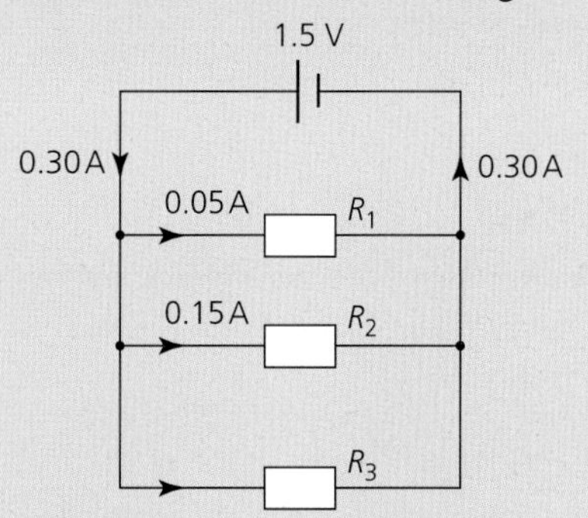


*Figure 17*

**b.** What current flows through resistor $R_3$?

**c.** How much energy (in joule) is transferred by one coulomb of charge as it flows through
  - **i.** $R_1$
  - **ii.** $R_2$
  - **iii.** $R_3$?

**d.** How much energy (in joule) is transferred each second in
  - **i.** $R_1$
  - **ii.** $R_2$
  - **iii.** $R_3$
  - **iv.** all three resistors together?

**e.** How much energy (in joule) is transferred each second by the cell?

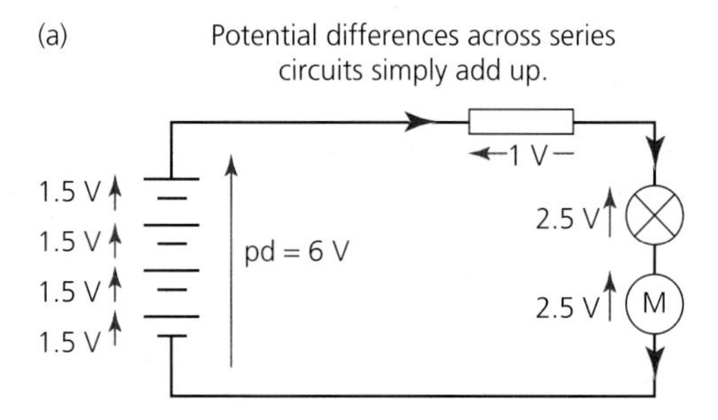


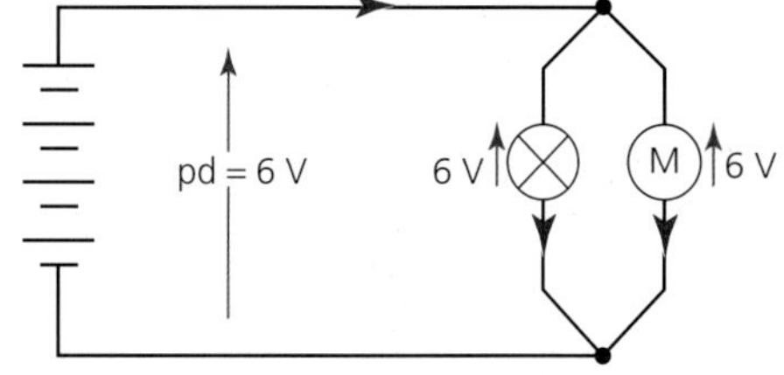


***Figure 16*** *Potential difference in (a) series and (b) parallel circuits*

## Power, potential difference and current

The power delivered to a component depends on the potential difference across the component and the current through it. The definitions of current and potential difference can be used to derive the equation for electrical power.

From $P = \frac{W}{t}$ $V = \frac{W}{Q}$ $I = \frac{Q}{t}$

we get

$$P = \frac{QV}{t} = \frac{ItV}{t}$$

so

$$P = IV$$

power = current × potential difference

The energy transferred by a component is power × time, so

energy transferred $E = IVt$

This formula $P = IV$ can be applied to any component in a circuit, or to the whole circuit. For example, the formula is important when selecting the thickness of cable or the appropriate **fuse**. A fuse is a device containing a wire that breaks when too much current flows in the circuit. An electrical plug used in the UK for mains electricity contains a cartridge fuse. This must be of a type, or rating, suitable for the appliance in use. The standard ratings for cartridge fuses in a three-pin plug are 3 A, 5 A and 13 A. These are the current values that will melt the fuse.

### Worked example

A table lamp has an LED light bulb rated 18 W. What size of fuse should be chosen for the plug?

The domestic mains voltage in the UK is 240 V, so the current passing through the light bulb is

$$I = \frac{P}{V} = \frac{18\,\text{W}}{240\,\text{V}} = 0.075\,\text{A}$$

The smallest possible fuse that will carry the required current should be chosen, so a 3 A fuse should be used.

### QUESTIONS

**13.** Work out the current drawn by a 1400 W vacuum cleaner connected to the 240 V mains. What fuse would you use for the vacuum cleaner?

**14.** Electric showers are high-power appliances, typically 7 kW. Work out the current supplied to an electric shower. Why are electric showers not connected via a standard three-pin plug?

**15.** Estimate the amount of energy used by an electric toaster that draws a current of 5 A to make two slices of toast.

### KEY IDEAS

- The potential difference (pd), $V$, between two points in a circuit is defined as the work done (or energy transferred) per unit charge passing between the points, $V = \frac{W}{Q}$.
- Potential difference is measured using a voltmeter connected in parallel with the component.
- Around any closed loop of a circuit, the sum of the emfs provided by power sources is equal to the sum of the pds across the components (Kirchhoff's second law).
- Electrical power, $P = IV$.
- The energy transferred, $E$, in time $t$ is $E = IVt$.

### ASSIGNMENT 1: READING THE METER

**(MS 1.1, PS 2.3, PS 3.3, PS 4.1)**

#### Ammeters and voltmeters

Ammeters measure the current that flows *through* a circuit, so the circuit needs to be broken to put an ammeter in. The meter should not change the current it is trying to measure, so an ideal ammeter has zero resistance. Putting an ammeter across a battery is effectively a short circuit; a large current will flow through the ammeter and will probably blow a fuse.

Voltmeters, on the other hand, measure the potential difference *across* a component. Voltmeters should not allow any current to bypass the component, so, ideally, a voltmeter will have infinite resistance. Putting a voltmeter in series with the circuit would effectively switch the circuit off.

There are two types of ammeter and voltmeter:

- Analogue meters use a moving coil attached to a pointer that moves in front of a scale.
- Digital meters are electronic devices that have a digital display.

## Analogue meters

Analogue displays may seem old-fashioned but they still have their place. It can be easier to get a quick picture of the information from a set of dials than from an array of numbers. When the reading is fluctuating, for example when taking a measurement of the potential difference generated by a wind turbine, you may need to estimate its average value. This can be easier to do with an analogue meter (Figure A1).

***Figure A1*** *An analogue meter*

The uncertainty in reading the scale on an analogue meter may be taken as the smallest scale division. It is important to view the scale from the correct position to avoid parallax errors (see Chapter 1). Some meters have a mirror behind the scale. The correct viewing position is where the pointer is in line with its image. The meter may have a zero error; this can often be corrected by turning the adjusting screw with a small screwdriver.

A meter can cover different ranges, for example with an ammeter by using different connections or by adding resistors in parallel, called shunts. A meter has to be connected so that the black terminal is closest to the negative side of the circuit.

## Digital meters

Digital meters (Figure A2) have replaced moving-coil meters in most scientific work because they are easier to read accurately. They usually, but not always, give a more accurate and precise reading.

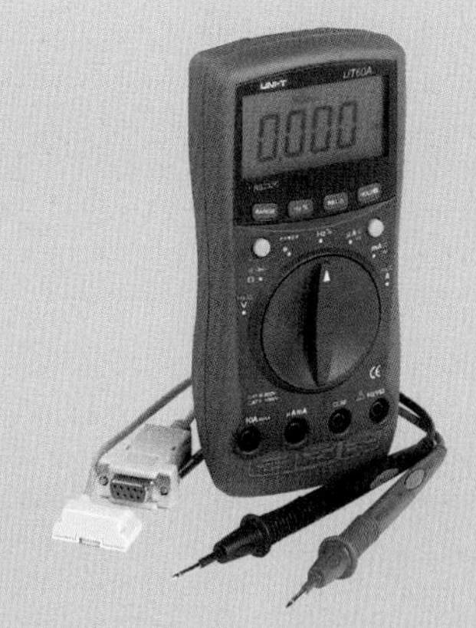

***Figure A2*** *A digital multimeter can be used as an ammeter, a voltmeter or a resistance meter. The dial in the centre is used to select the correct function and range.*

The uncertainty in a digital display is often taken as ±1 in the last displayed digit, though this is an overestimate of their accuracy. The manufacturer's data will often give the expected accuracy and the resolution. For example, from the data in Table A1, a reading of 32.4 Ω would have an uncertainty of 0.4%, which is 0.13 Ω. The meter has a resolution of 0.1 Ω on its most sensitive resistance setting. So when we add these, the total uncertainty is 0.23, giving a reading of 32.4 (±0.2) Ω.

| | Voltage (dc) | Current (dc) | Resistance |
|---|---|---|---|
| **Maximum reading** | 1000 V | 10 A (or 20 A for 30 s max.) | 50 MΩ |
| **Resolution** | 100 μV | 0.01 mA | 0.1 Ω |
| **Accuracy** | 0.1% | 0.4% | 0.4% |

***Table A1*** *Multimeter manufacturer's data. The resolution is the smallest possible change that the meter can show. That is ±1 in the final digit, on the most sensitive setting. The accuracy tells you how close you can expect your reading to be to the true value.*

It is important to select the correct range on a digital meter. You should always choose the most sensitive range possible, making sure that your maximum expected reading is less than the maximum available on the range. Typical ranges on the ammeter setting might be marked 2 mA, 20 mA, 200 mA, 2 A and 20 A. These represent the maximum values of current that can be read on each range. If you expect your maximum reading to be 38 mA, you should choose the 200 mA range. If you do not know the likely maximum reading, it is good practice to start on a high current range (20 A) and turn the dial down as appropriate – you will blow fewer fuses that way!

**Questions**

**A1** If you were using the multimeter described above, what would the uncertainty be in readings of

**a.** 0.5 mA

**b.** 100 V (assuming the resolution on this range = 0.1 V)?

**A2** The range switch is marked 600 mV, 6 V, 60 V, 600 V, 1000 V. Explain what these values indicate.

**A3** The headlights on your car have stopped working. How could you use a multimeter to find the fault?

**A4** A small potential difference arises when there is a temperature difference between two junctions of two different metals. This is known as a thermocouple. For a specific thermocouple, you are told to expect a potential difference of 0.05 mV for every degree (°C), and at 20 °C the pd is 1.109 V.

**a.** What would be the uncertainty in this reading (assume you have the voltmeter described above)?

**b.** The temperature rises to 30 °C. What reading would you expect to get on the meter? What uncertainty would the reading have?

**c.** What change in voltage reading do you get when the temperature goes from 20 °C to 30 °C? What is the percentage uncertainty in your answer?

**d.** Comment on your answer to **c**.

## 13.4 RESISTANCE

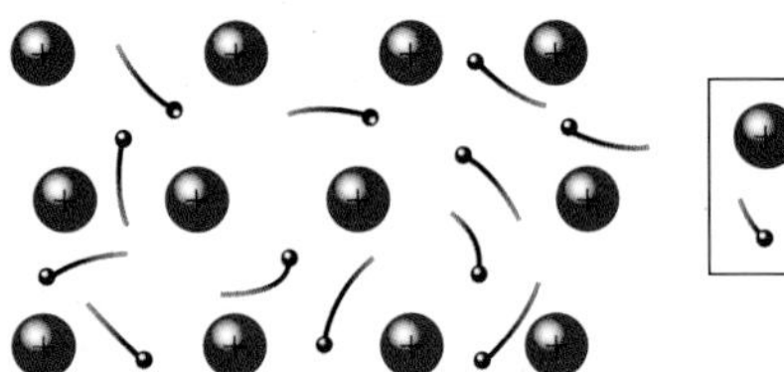


Electrons break loose from their parent atoms, leaving a 'gas' of free electrons in between positive metal ions.

**Figure 18** *Free electrons in a metal*

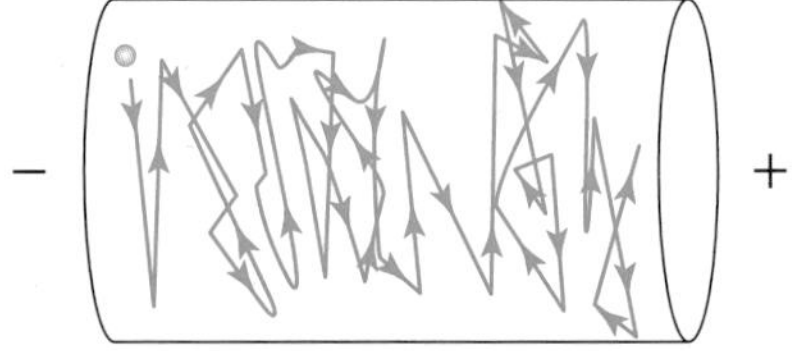


electron path through a metal wire

**Figure 19** *When an electric current flows through a wire, the motion of a free electron is rather like a pinball making its tortuous way through a pinball machine, undergoing many collisions on its way. The instantaneous velocity of an electron is hundreds of kilometres per second, but the average velocity of the electrons towards the positive terminal, the 'drift velocity', is only a few millimetres per second.*

Metals are the most common conductors used in electrical circuits. They have free electrons that can move between the fixed, positively charged metal ions. The free electrons move randomly at high speed (Figure 18), rather like the molecules in a gas. But when a potential difference is applied to the metal, the negatively charged electrons accelerate towards the positive connection. If they were in a vacuum, the electrons would continue to accelerate unimpeded, but in the metal they collide with the positive ions (Figure 19). During each collision, some of the electron's kinetic energy is transferred to the lattice of positive ions, increasing its vibrational energy. This raises the temperature of the metal. If the energy transfer is large enough, the conductor may get hot enough to glow red, or even white. This is the basis of the incandescent light bulb, as well as other electrical heating appliances, such as kettles and toasters (Figure 20).

**Figure 20** *Electrons flow through the nichrome ribbon that forms the heating element in a toaster. Electrons transfer energy to the atoms in the nichrome, which becomes very hot.*

QUESTIONS

16. "If the average velocity of electrons in an electric current is only a few millimetres per second, why do the lights come on instantly when I press the switch?"
17. "If the electrons in this metal pen are whizzing about at hundreds of kilometres per second, why is there no electric current flowing?"
18. Explain why increasing the current through a wire makes the wire get hotter.

## Resistance and Ohm's law

The collisions that occur between electrons and the lattice of positive ions in a metal wire slow the electrons and so reduce the total charge that flows past any point in one second. The wire is said to offer **resistance** to the current.

A potential difference, $V$, is needed across a conductor to maintain a current, $I$, through it. The ratio of potential difference to current is the resistance of a conductor:

$$\text{resistance} = \frac{\text{potential difference}}{\text{current}}$$

$$R = \frac{V}{I} \qquad I = \frac{V}{R} \qquad V = IR$$

Resistance is measured in the unit **ohm** (symbol Ω; a Greek capital omega), named after Georg Simon Ohm, who showed (in 1867) that the current through a metal conductor was proportional to the potential difference across it. This simple rule holds for metals and a few other substances, referred to as **ohmic conductors**, but only under limited conditions. The rule breaks down if the temperature changes. Other physical conditions, such as light intensity, pressure, strain and magnetic fields, can also affect the resistance of some materials. There are some conductors (water, electrolytes, semiconductor diodes, light bulbs) for which the proportionality does not work at all. These are called non-ohmic conductors (see Assignment 2).

**Ohm's law** must be stated very cautiously:

> Provided that temperature and other physical conditions remain constant, the current through a conductor is proportional to the potential difference across the conductor.

The relationship is shown in Figure 21.

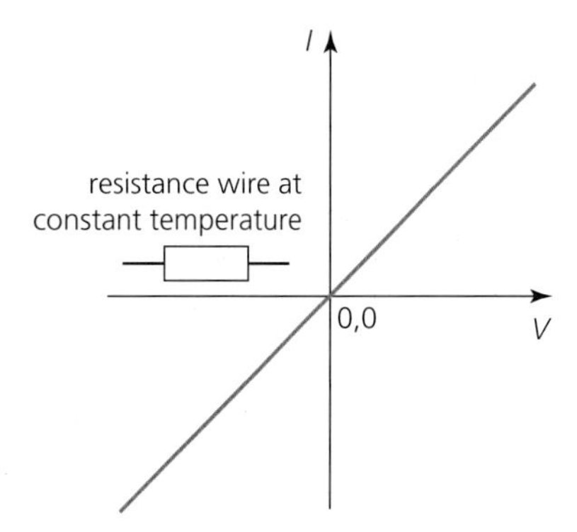


*Figure 21 An ohmic conductor gives an I–V graph that is a straight line through the origin.*

Resistance is always defined as the ratio of potential difference to current, $R = \frac{V}{I}$, even when the ratio does not remain constant with changing potential difference or current.

A resistance of 1 Ω needs a potential difference of 1 V to maintain a current of 1 A through it. Table 2 gives approximate values of resistance for some conductors.

| Electrical conductor | Resistance |
|---|---|
| 1 m of overhead power supply cable | 0.00005–0.0005 Ω |
| 1 m loudspeaker cable | ~0.05 Ω |
| 1 m of AV (audio-visual) cable | ≈5 Ω |
| A loudspeaker | 7 Ω |
| Light bulb filament (cold) | 10 Ω |
| Light bulb filament (hot) | 150 Ω |
| Human body (hand to feet) | 1 kΩ |
| Voltmeter | 100 MΩ |

*Table 2 Typical values of resistance*

**Resistors** are components made to provide a fixed value of resistance in a circuit. They come in all shapes and sizes (Figure 22). The cheapest type is a film of carbon coated onto an insulator. These can range in resistance from milliohms to megohms, although they cannot carry large currents. Other resistors are coiled wire, which can carry larger currents without melting. Variable resistors are also commonly used.

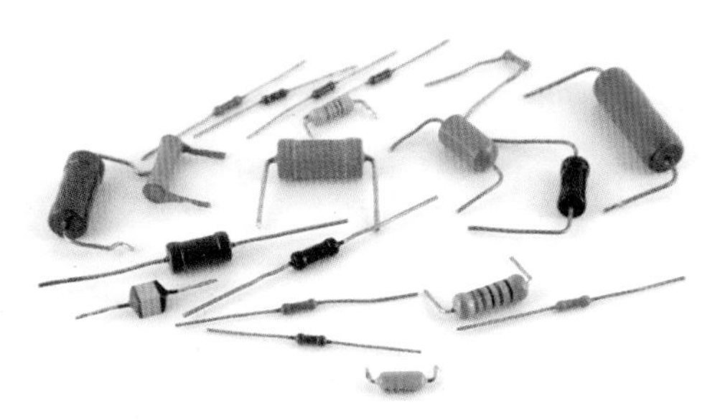

***Figure 22*** *Different types of resistor*

### Your resistance

The electrical resistance of the human body is given as approximately 1 kΩ in Table 2, but this is just a typical value. The actual resistance depends on the person, the path taken by the electric current, and whether the skin is wet or dry. Suppose that a person gets an electric shock from the 240 V mains. The current passing through them could be

$$I = \frac{V}{R} = \frac{240}{1000} = 0.24\,\text{A}$$

This is enough to cause a painful shock, and possibly enough to stop them breathing.

### QUESTIONS

**19.** In a stereo music system it is important that the resistance of the speaker cables is low. Why?

**20.** Use the information in Table 2 on the previous page to calculate the potential difference across 1 km of power supply cable when the current is 10 A.

**21.** The potential difference across a heating element is 4.0 V when the current is 6.0 A. What will the potential difference be when the current is 9.0 A? What have you assumed?

### KEY IDEAS

- Resistance, $R$, is a measure of an electrical component's opposition to current, $R = \frac{V}{I}$.
- As a current passes through a resistor, energy is transferred as heat.
- Ohm's law holds for metal conductors provided the temperature and other physical conditions are constant: the current through a component is proportional to the potential difference across it.

## 13.5 CURRENT–VOLTAGE CHARACTERISTICS

Non-ohmic conductors have $I$–$V$ graphs that are not straight lines. We will consider the behaviour of

- an incandescent (filament) lamp
- a semiconductor diode.

The shape of the graph for an incandescent filament lamp is shown in Figure 23a. The filament lamp is clearly non-ohmic since the resistance varies with current. This is because of the temperature change as the current increases.

A semiconductor **diode** has a very high resistance below about 0.7 V, but a low resistance above 0.7 V. The shape of the graph is shown in Figure 23b.

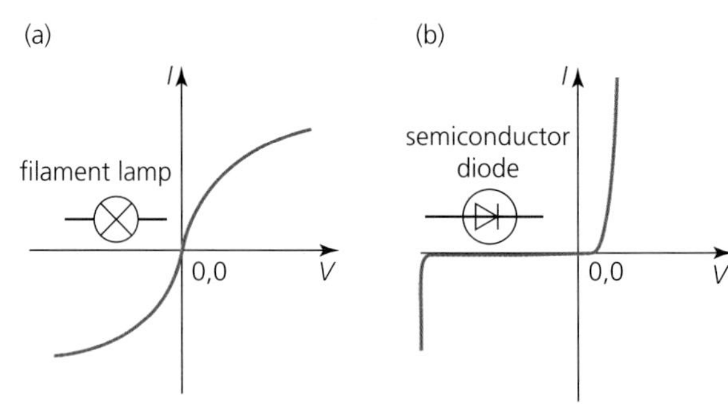


***Figure 23*** *The I–V graphs for (a) a filament lamp and (b) a semiconductor diode. Note that I–V characteristics are usually, but not always, plotted with current on the y-axis. Be aware of this especially if you are asked to calculate resistance.*

Most electrical components work in the same way no matter which way round they are connected in a circuit. However, a diode has the unusual property that its **polarity** makes a difference – in other words, it matters which way round it is connected in the circuit. A diode is normally connected with the cathode on the negative side of the circuit (Figures 24 and 25). This is called 'forward-biased' and

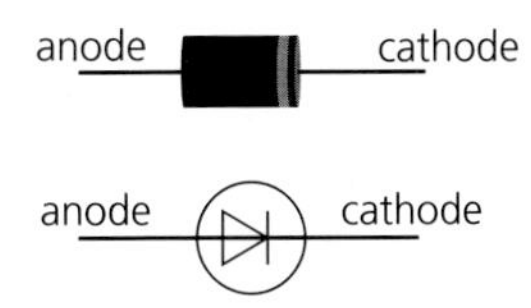


*Figure 24* *A diode has a band marked on it to mark the cathode. The circuit symbol for a diode clearly shows the direction of conventional current conducted by the diode.*

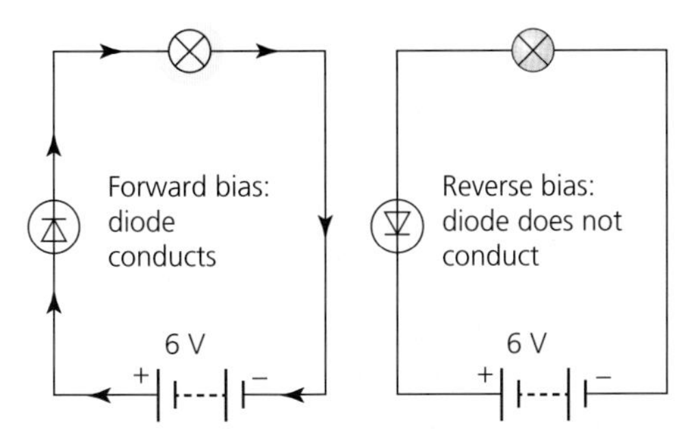


*Figure 25* *Forward bias and reverse bias of a diode*

corresponds to the upper-right section of the *I*–*V* graph in Figure 23b. A diode connected the opposite way round is said to be 'reverse-biased' and it will not conduct until the potential difference across it reaches its breakdown voltage of 50 to 100 V – see the lower-left section of Figure 23b. The diode will then be permanently damaged.

Figure 26 shows different types of **light-emitting diode** (**LED**), and the circuit symbol for an LED. Like other types of diode, LEDs need to be forward-biased to conduct. When they conduct, they emit light of a particular colour. Figure 27 shows the *I*–*V* characteristics for different colour LEDs.

The diodes are constructed from a semiconductor with other materials added in a process called doping. The doping affects the wavelength of light emitted, and the *I*–*V* characteristic.

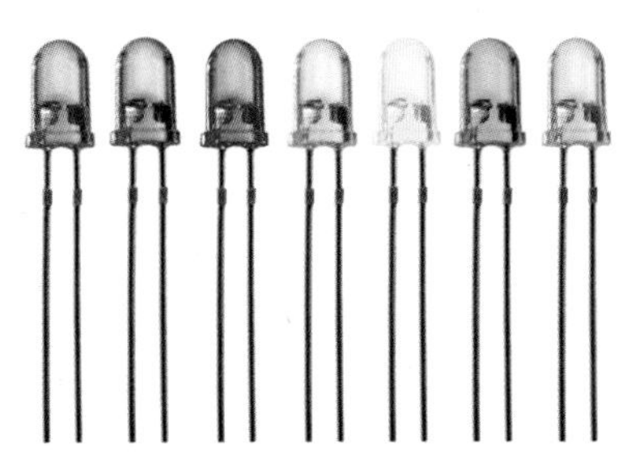

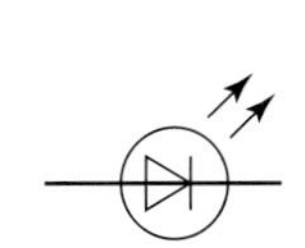

*Figure 26* *Light-emitting diodes (LEDs) and the circuit symbol*

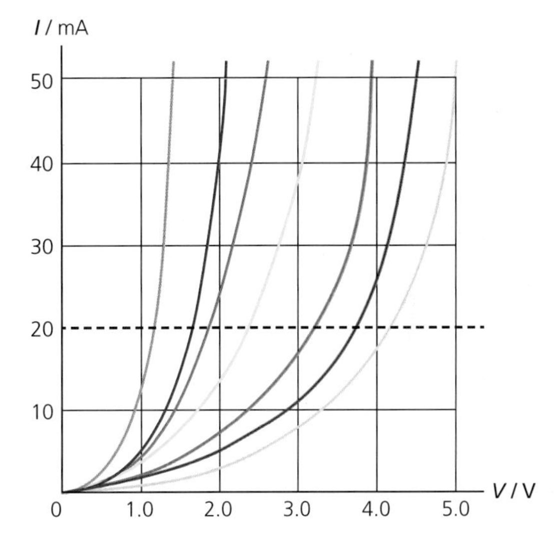


*Figure 27* *The I–V characteristics for forward-biased LEDs of different colours*

## QUESTIONS

**22.** This question refers to the LED characteristics in Figure 27.

**a.** What is the resistance of a green LED at 2.0 V?

**b.** What electrical power would the green LED have at 2.0 V?

**c.** What would you expect to happen to a blue LED if it was placed in a circuit with a potential difference of 12 V across it?

**d.** A 12 V supply is connected in parallel with red, yellow and green diodes (see Figure 28). The resistors in series with each LED are needed to limit the current through the LED and so prevent it being damaged.

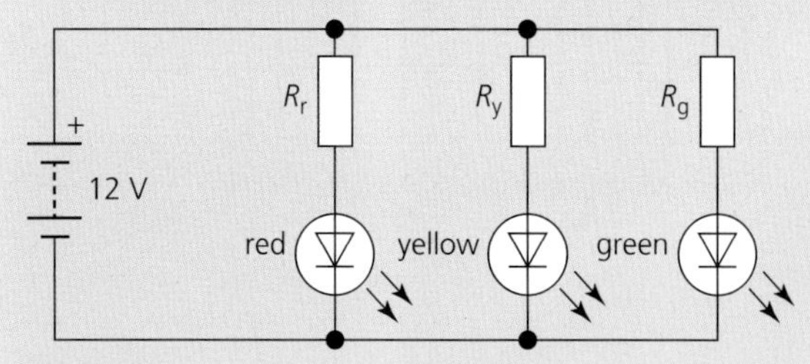


*Figure 28*

**i.** The current through each LED is to be limited to 20 mA. At this value of current, what will be the pd across each LED? (Refer to Figure 27.)

**ii.** Calculate the required value of each resistor.

**iii.** The same LEDs are connected with a pd of 2.0 V across each one. What would you expect to happen?

## ASSIGNMENT 2: FINDING THE *I–V* CHARACTERISTICS OF ELECTRICAL COMPONENTS

**(MS 1.1, MS 3.1, MS 3.2, MS 3.4, PS 2.2, PS 2.3, PS 2.4, PS 3.1, PS 3.2, PS 3.3, PS 4.1)**

The aim of this experiment is to investigate how the potential difference across a component affects the current that flows through it, and therefore to see whether a component is an ohmic or non-ohmic conductor. We need to be able to vary, and measure, the potential difference across the component and measure the resulting current that flows through it. A suitable circuit is shown in Figure A1.

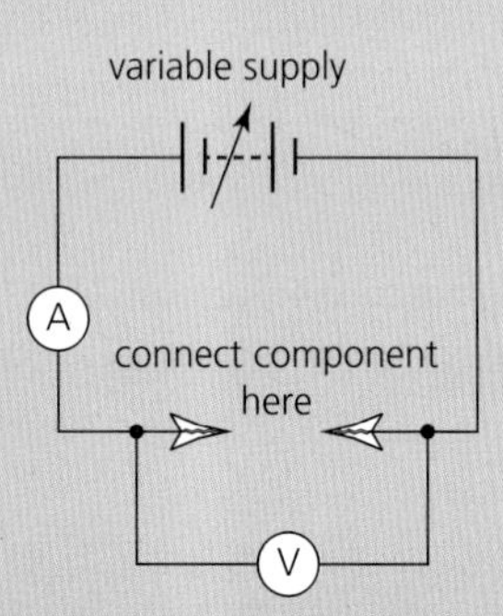


*Figure A1*

The component we will consider is a 1 m length of uninsulated nichrome wire. This must be prevented from coiling up on itself and short-circuiting. The wire can be taped to a meter ruler or made into a coil by wrapping it carefully around a pencil. Since it is important to keep the temperature constant, spreading it out into a straight line on a ruler will be a better solution, as it will help to dissipate the heat caused by the current through the wire.

It is important to keep the current low, so as to keep the temperature constant. The resistance of 1.00 m of 0.21 mm diameter nichrome wire is 31.5 Ω. So, to keep a maximum current of less than 0.2 A, the potential difference will have to be $V = IR - 0.2 \times 31.5 = 6.3$ V. We should aim to take around 10 readings between 0 and 6 V.

Sample results are shown in Table A1.

| Potential difference / V | Current / A |
|---|---|
| 0.0 | 0.02 |
| 0.5 | 0.04 |
| 1.0 | 0.05 |
| 1.5 | 0.07 |
| 2.0 | 0.09 |
| 2.5 | 0.11 |
| 3.0 | 0.12 |
| 3.5 | 0.14 |
| 4.0 | 0.16 |
| 4.5 | 0.18 |
| 5.0 | 0.19 |
| 5.5 | 0.21 |
| 6.0 | 0.23 |

*Table A1*

Other components can be tested in a similar circuit. An incandescent filament lamp will be marked with its working potential difference and the electrical power at that voltage. Typically, this will be (12 V, 36 W). Readings should be taken at values of potential difference between 0 V and 12 V.

The polarity of the potential difference is important for some components, in particular diodes, because a diode only conducts in one direction. With a semiconductor diode, readings should be taken from 0 V to above 0.7 V, and with reverse polarity from 0 V to −10 V.

**Questions**

**A1** Plot a graph of potential difference on the $y$-axis versus current on the $x$-axis, for this 1 m length of nichrome wire.

**A2** Use the graph to find the resistance of the wire. Up to what value of current does this hold?

**A3** Explain whether this is an ohmic or non-ohmic conductor.

**A4** Suggest one source of error in this experiment that is apparent from the graph.

# 13.6 COMBINING RESISTORS

In many circuits there are combinations of resistors in parallel and in series. In order to calculate the current through the circuit, we need a way of calculating the total resistance of a combination of resistors.

## Resistors in series

The pd across the chain of resistors is the sum of the pds across each of the resistors.

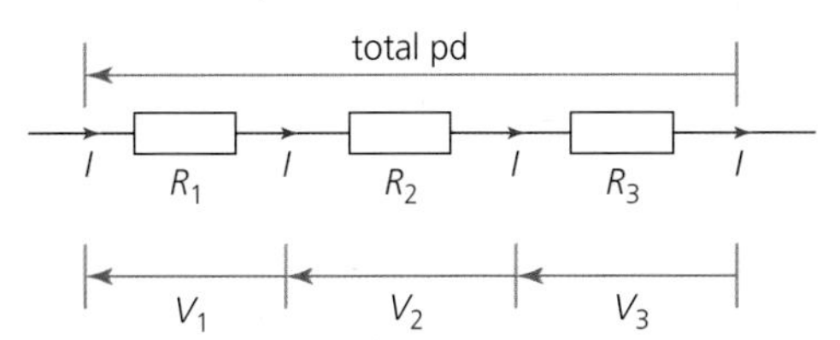


***Figure 29*** *Potential difference across resistors in series*

The same current passes through all resistors in a series circuit. The potential difference across the whole series is the sum of the potential differences across each resistor (Figure 29). This arrangement could be replaced by a single resistor of resistance $R_T$ that would pass the same current for the same total potential difference:

$$\text{total resistance} = \frac{\text{total pd}}{\text{current}}$$

$$R_T = \frac{V_1 + V_2 + V_3 + \cdots}{I}$$

$$= \frac{V_1}{I} + \frac{V_2}{I} + \frac{V_3}{I} + \cdots$$

$$R_T = R_1 + R_2 + R_3 + \cdots$$

The total resistance of a number of resistors in series is the arithmetic sum of the individual resistances.

## Resistors in parallel

We can assume that the connecting wires in circuit diagrams have no resistance. Therefore, the potential difference across each resistor in parallel is the same (Figure 30).

Parallel resistors have the same pd across them. The total current $I$ is the sum of currents through each of the resistors.

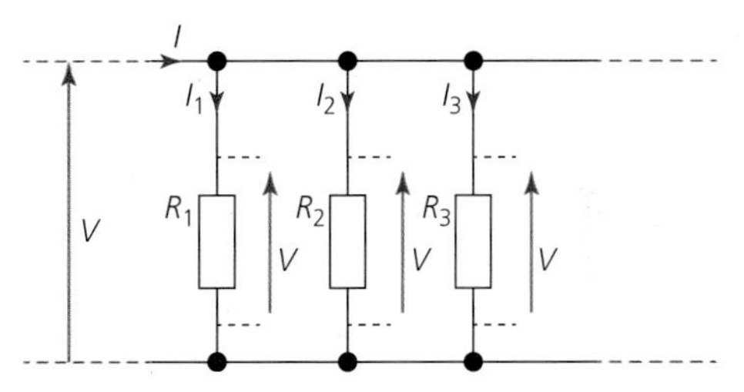


***Figure 30*** *Potential difference across resistors in parallel*

$$\text{total resistance} = \frac{\text{pd}}{\text{total current}}$$

$$R_T = \frac{V}{I}$$

Rearranging,

$$I = \frac{V}{R_T}$$

But the total current $I$ is the sum of the currents through each resistor. Therefore, we find that

$$I = I_1 + I_2 + I_3 + \cdots$$

$$= \frac{V}{R_1} + \frac{V}{R_2} + \frac{V}{R_3} + \cdots$$

so

$$\frac{V}{R_T} = \frac{V}{R_1} + \frac{V}{R_2} + \frac{V}{R_3} + \cdots$$

Dividing throughout by $V$ gives

$$\frac{1}{R_T} = \frac{1}{R_1} + \frac{1}{R_2} + \frac{1}{R_3} + \cdots$$

For just two resistors, the formula

$$R_T = \frac{R_1 \times R_2}{R_1 + R_2}$$

is often easier to use, but it does not extend simply for more resistors.

## Worked example 1

A heater is to be made from 3 Ω and 6 Ω heating elements (A and B, respectively) operated from 12 V battery. Different powers could be obtained by using just one element or by combining the two in series or in parallel. What would be the power in each case?

There are four possible values, as shown in Figure 31 and summarised in the table.

| | Total resistance | Power = $V^2/R$ |
|---|---|---|
| 1 single A | 3 Ω | 144/3 = 48 W |
| 2 single B | 6 Ω | 144/6 = 24 W |
| 3 in series | 6 + 3 = 9 Ω | 144/9 = 16 W |
| 4 in parallel | $\frac{1}{R_T} = \frac{1}{6} + \frac{1}{3} = \frac{1}{2}$ | |
| | $R_T = 2\ \Omega$ | 144/2 = 72 W |

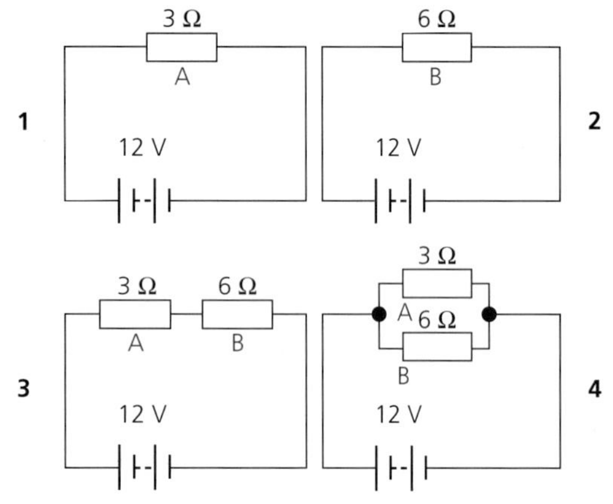


*Figure 31*

## Worked example 2

Real circuits can consist of a mixture of series and parallel elements. Suppose that a series arrangement of resistors, say heating elements, is connected in parallel with another heating element (Figure 32) Calculate the current $I$.

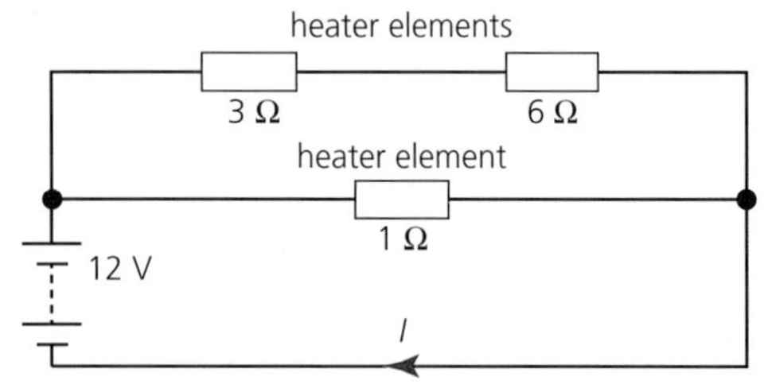


***Figure 32*** *A series–parallel combination*

First, deal with the series elements. These simply add together, so the resistance of the top branch is 9 Ω.

Then add the parallel parts together. The total resistance is given by

$$\frac{1}{R_T} = \frac{1}{R_1} + \frac{1}{R_2}$$

$$= \frac{1}{9\,\Omega} + \frac{1}{1\,\Omega} = \frac{1+9}{9\,\Omega} = \frac{10}{9\,\Omega}$$

$$R_T = \frac{9\,\Omega}{10} = 0.90\,\Omega$$

so

$$I = \frac{V}{R_T}$$

$$= \frac{12\,\text{V}}{0.90\,\Omega} = 13.3\,\text{A}$$

### QUESTIONS

**23.** Find the current in each of the circuits in Figure 33.

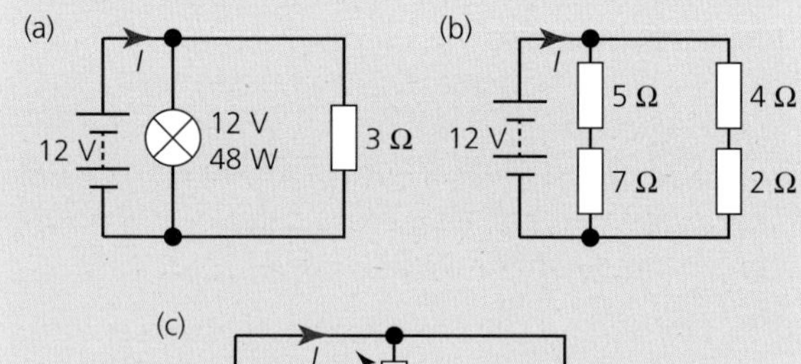


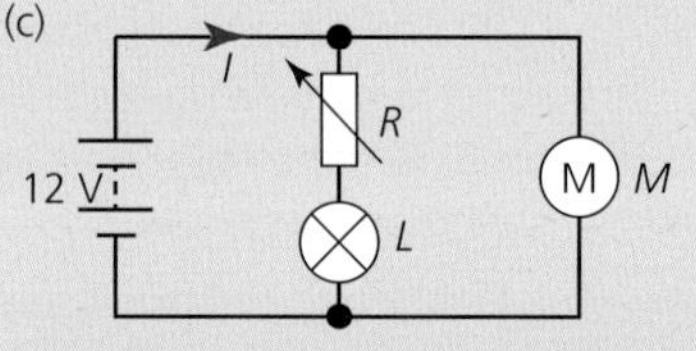


$R$ = dimmer resistor, resistance set at 10 Ω
$L$ = dashboard light, resistance 5 Ω at this setting
$M$ = heater fan motor, resistance 30 Ω

*Figure 33*

**24.** Find the potential difference marked $V$ in each of the circuits in Figure 34.

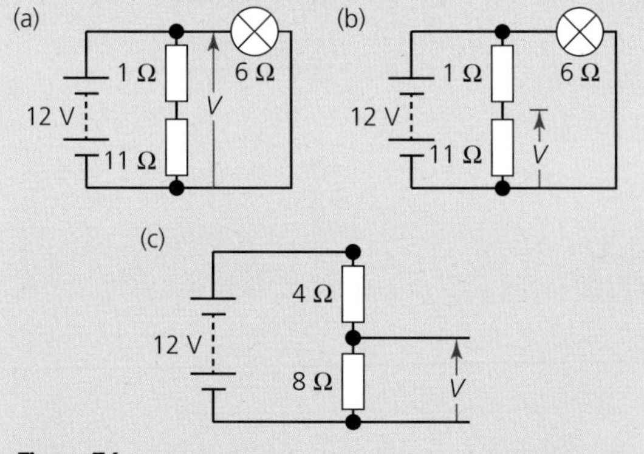


*Figure 34*

**KEY IDEAS**

- For resistors in series: $R_T = R_1 + R_2 + R_3 + \cdots$
- For resistors in parallel: $\frac{1}{R_T} = \frac{1}{R_1} + \frac{1}{R_2} + \frac{1}{R_3} + \cdots$
- For just two resistors in parallel: $R_T = \frac{R_1 \times R_2}{R_1 + R_2}$

## 13.7 POWER AND RESISTANCE

The batteries in an electric vehicle could all be connected in series or in parallel. The total energy storage in either case would be the same, but the potential difference across the batteries would be different. The series arrangement would lead to a higher potential difference and would therefore be more dangerous for service engineers (Figure 35), because high voltages can cause severe electric shocks. The same power could be delivered to the motor by connecting the batteries in parallel. Since power is $P = IV$, a lower potential difference would necessitate a much bigger current. This current would cause heating of the connecting wires. Would this be a problem?

To resolve this, we need to calculate the power dissipated in the non-negligible resistance, $R$, of the connecting wires. It is often convenient to use equations that relate power and resistance directly.

We know that

$$P = IV \quad \text{and} \quad V = IR$$

***Figure 35*** *"I need to connect lots of batteries together for use in a prototype electric car. Should I connect them all in series and risk a dangerously high voltage? Or would it be better to put them in parallel, and have very high currents?"*

Eliminating $V$ gives

$$P = I \times IR$$
$$P = I^2R$$

Eliminating $I$ instead gives

$$P = \frac{V}{R} \times V$$
$$P = \frac{V^2}{R}$$

The alternative arrangements are shown in Figure 36. Suppose that the wires connecting the battery to the motor have a resistance of $R = 0.005\ \Omega$ and that the power required by the motor is 6 kW.

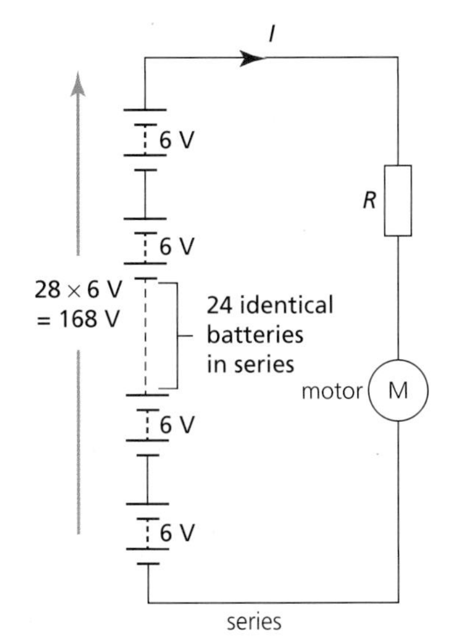


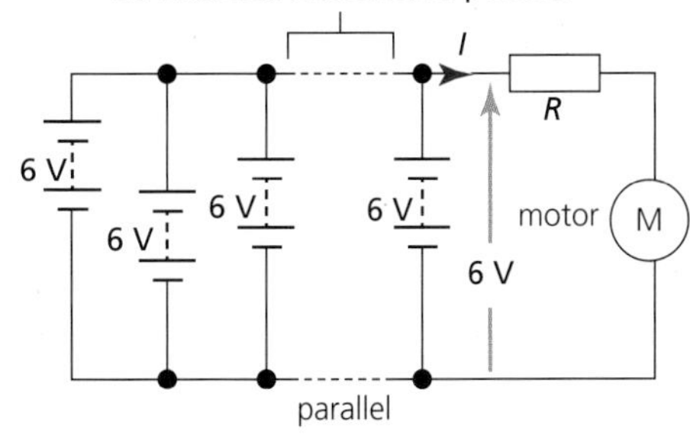


***Figure 36*** *Alternative battery arrangements*

In the *series* case, the potential difference of the battery is 28 × 6 V = 168 V. To deliver 6 kW requires a current of

$$I = \frac{P}{V}$$
$$= \frac{6000\,\text{W}}{168\,\text{V}} = 35.7\,\text{A}$$

Then the power lost in the wires is

$$P = I^2R$$
$$= (35.7\,\text{A})^2 \times 0.005\,\Omega = 6.4\,\text{W}$$

The efficiency of power transfer is

$$\text{efficiency} = \frac{6000\,\text{W} - 6.4\,\text{W}}{6000\,\text{W}}$$
$$= 0.999 \text{ or } 99.9\%$$

In the *parallel* case, the potential difference of the batteries is 6 V. To deliver 6 kW requires a current of

$$I = \frac{P}{V}$$
$$= \frac{6000\,\text{W}}{6\,\text{V}} = 1000\,\text{A}$$

Then the power lost in the wires is

$$P = I^2R$$
$$= (1000\,\text{A})^2 \times 0.005\,\Omega = 5000\,\text{W}$$

The efficiency of power transfer is:

$$\text{efficiency} = \frac{6000\,\text{W} - 5000\,\text{W}}{6000\,\text{W}}$$
$$= 0.167 \text{ or } 16.7\%$$

It is clear that the series arrangement is preferable, despite the risk of electric shock. The dramatic difference in efficiency (16.7% compared with 99.9%) could be overcome by making the connecting wires much thicker, but this would add weight and cost more.

**QUESTIONS**

**25.** The engineer suggests a compromise arrangement of batteries in series and parallel, as shown in Figure 37.

**a.** Calculate the efficiency of energy transfer for the same system using this arrangement of batteries.

**b.** Comment on the safety and suitability of the proposal.

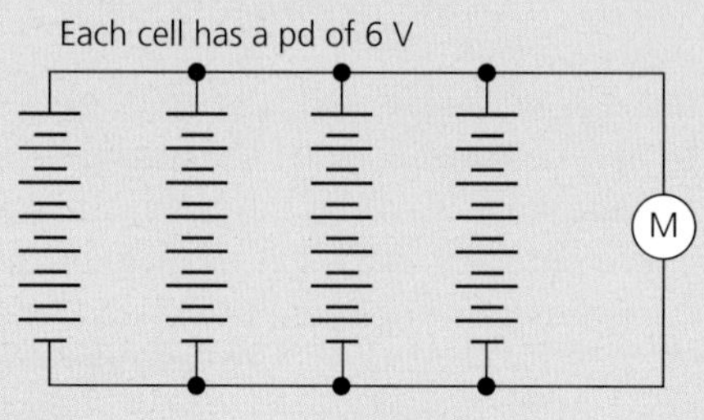


*Figure 37*

**KEY IDEAS**

› Electrical power, $P$, can be calculated using the following equations:

$$P = IV = I^2R = \frac{V^2}{R}$$

## 13.8 RESISTANCE AND RESISTIVITY

The resistance of a length of wire depends on its length, $l$, and its cross-sectional area, $A$. Using the model of resistance arising from collisions between electrons and lattice ions (section 13.4), we can deduce the following (Figure 38):

› Resistance is proportional to length, because doubling the length would double the chance of collision for any given electron (think of two identical resistors in series):

$$R \propto l$$

› Resistance is inversely proportional to cross-sectional area, because doubling the area would double the number of moving electrons, thereby doubling the current and halving the resistance (as with two identical resistors in parallel):

$$R \propto \frac{1}{A}$$

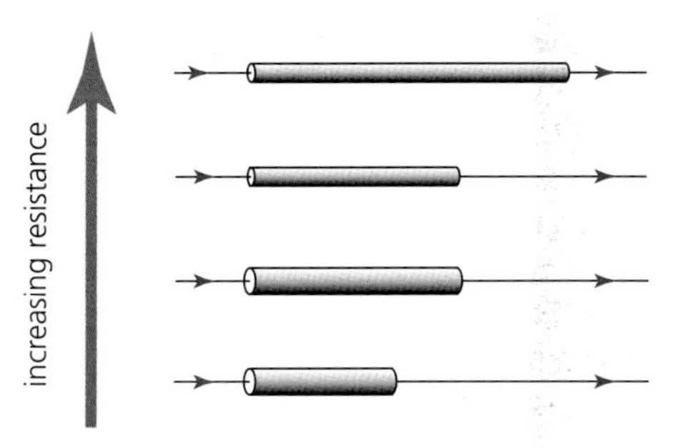


***Figure 38*** *Resistance and dimensions*

Combining these relationships gives

$$R \propto \frac{l}{A}$$

We can write this as

$$R = \rho \times \frac{l}{A}$$

where $\rho$ is a constant called **resistivity**, which depends on the material used and on the temperature.

If you set $l$ = 1 m and $A$ = 1 m$^2$, then the resistivity, $\rho$, can be thought of as the resistance of a standard size sample of material of unit length and unit area of cross-section. Rearranging the equation above gives an equation for resistivity $\rho$:

$$\rho = \frac{RA}{l}$$

Resistivity has the unit Ω m.

Good conductors are bad resistors and vice versa. We sometimes talk about 'conductance', which is the inverse of resistance, and when comparing materials use the concept of **conductivity**:

$$\text{conductivity} = \frac{1}{\text{resistivity}}$$

Conductivity has the same unit as $\frac{1}{\rho}$, that is, $\Omega^{-1}\ m^{-1}$, but instead of this the unit siemens per metre, S $m^{-1}$, is used. (The siemens is the unit of conductance, $\Omega^{-1}$, originally given the name 'mho' ('ohm' reversed) but renamed after the German scientist Werner von Siemens.)

Table 3 gives typical resistivity values of different types of material.

| Type of material | Order of magnitude of resistivity / Ω m |
|---|---|
| Metal | $10^{-7}$ |
| Semiconductor | ~$10^{2}$ |
| Insulator | ~$10^{10}$ |

***Table 3*** *Typical resistivity values*

## Worked example 1

Nichrome wire can be used in a heating element for a toaster. A 100 cm length of wire of diameter 1.22 mm has a resistance of 0.50 Ω. What is the resistivity of nichrome?

$$\text{Area} = \pi r^2$$
$$= \pi - (0.61 - 10^{-3}\ \text{m})^2$$
$$= 1.17 - 10^{-6}\ \text{m}^2$$

so

$$\rho = \frac{RA}{l}$$
$$= \frac{0.50\,\Omega \times 1.17 \times 10^{-6}\ \text{m}^2}{1.00\,\text{m}}$$
$$= 5.8 \times 10^{-7}\ \Omega\text{m}$$

## Worked example 2

We can use values of resistivity to evaluate alternative materials for use in connecting cables. Suppose the connecting cable from a storage battery in a solar cell array to a house is 3.5 m long. The wire must have a resistance of less than 0.001 Ω. How would the diameters and masses of copper and aluminium wires compare? (See Table 4 for data.)

| Material | Resistivity / Ω m | Density / $kgm^{-3}$ |
|---|---|---|
| Copper | $1.7 \times 10^{-8}$ | 8930 |
| Aluminium | $2.7 \times 10^{-8}$ | 2710 |

***Table 4*** *Resistivity and density data*

For copper:

$$\rho = \frac{RA}{l}$$
$$A = \frac{\rho l}{R}$$
$$= \frac{1.7 \times 10^{-8}\ \Omega\text{m} \times 3.5\,\text{m}}{0.001\,\Omega}$$
$$= 5.95 \times 10^{-5}\ \text{m}^2$$

and

$$A = \pi r^2$$
$$r^2 = \frac{5.95 \times 10^{-5}\ \text{m}^2}{\pi}$$
$$r = 0.00435\,\text{m}$$

so $\quad \text{diameter} = 0.0087\,\text{m} = 8.7\,\text{mm}$

Also

$$\text{mass} = \text{density} \times \text{volume}$$
$$= 8930\,\text{kg}\,\text{m}^{-3} \times 5.95 \times 10^{-5}\,\text{m}^2 \times 3.5\,\text{m}$$
$$= 1.9\,\text{kg}$$

Similarly, for aluminium:

$$A = \frac{\rho l}{R}$$
$$= \frac{2.7 \times 10^{-8}\,\Omega\text{m} \times 3.5\,\text{m}}{0.001\,\Omega}$$
$$= 9.5 \times 10^{-5}\,\text{m}^2$$

giving radius = 0.0055 m, diameter = 11 mm and mass = 0.90 kg.

The poorer conduction of aluminium can be compensated for by making the aluminium cable about 25% thicker than copper cable. The aluminium cable would still be lighter. Although this results in a mass saving of more than 50%, the manufacturer would need to take into account the cost of manufacture and other properties (ease of connecting, flexibility, corrosion resistance, etc.) of the materials before making a decision.

## QUESTIONS

**26. a.** Two connecting wires, X and Y, are made of copper. X is twice the length of Y and it has twice the diameter. Which wire has greater resistance?

**b.** Calculate the resistance of Y if it is 2.00 m long and its diameter is 0.90 mm. (Refer to resistivity data in Table 4.)

**c.** What is the resistance of X?

**27.** A sample of resistance wire has a resistivity of $5.00 \times 10^{-7}\,\Omega$ m and a diameter of 0.50 mm. Calculate the length of wire needed to make a resistance of 10.0 Ω.

## KEY IDEAS

- The resistance of a specimen of material of length $l$ and cross-sectional area $A$ is

$$R = \frac{\rho l}{A}$$

where $\rho$ is the resistivity of the material.

- Resistivity is measured in Ω m.

## REQUIRED PRACTICAL: APPARATUS AND TECHNIQUES

### Determination of the resistivity of a wire

The aim of this practical is to measure the resistivity of a metal, in the form of a wire. Since

$$R = \rho \times \frac{l}{A}$$

a series of readings of resistance, $R$, for different lengths of wire, $l$, will allow a graph to be plotted of resistance ($y$-axis) versus length (on the $x$-axis). This should yield a straight line, with a gradient that is numerically equal to $\rho/A$. A measurement of the cross-sectional area, $A$, of the wire will allow calculation of the resistivity, $\rho$.

The practical gives you the opportunity to show that you can:

- use appropriate analogue apparatus to record a range of measurements (to include length/distance) and to interpolate between scale markings
- use appropriate digital instruments, including electrical multimeters, to obtain a range of measurements (to include current, voltage and resistance)
- use callipers and micrometers for small distances, using digital or vernier scales
- correctly construct circuits from circuit diagrams using DC power supplies, cells and a range of circuit components, including those where polarity is important.

**Apparatus**

The circuit in Figure P1 is used to measure the resistance. A micrometer screw guage such as that in Figure P2 is used to measure the diameter of the wire.

Refer back to Figure 11 in Chapter 1 for a reminder of how to read the scale.

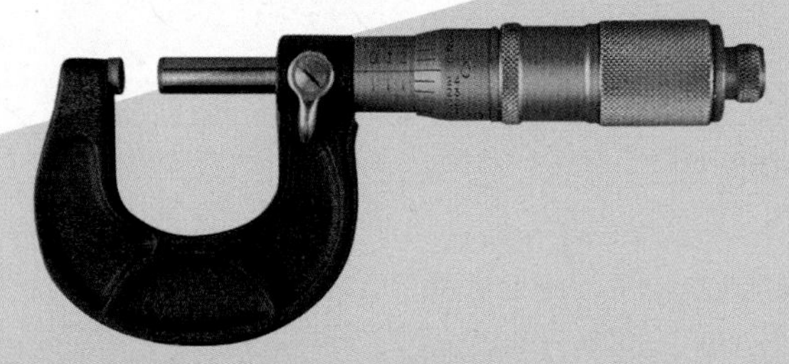

*Figure P2* A micrometer screw gauge

**Technique: Measuring resistance**

In order to measure the resistance of the metal wire, you need to measure the current through the wire and the potential difference across it. A variable resistor is used to vary the potential difference. If you choose a value that is too low, it may lead to significant uncertainties in potential difference and current. If you choose a value that is too high, the current will cause heating in the wire, which will affect the resistance.

The procedure is to set the crocodile clips 1.000 m apart (Figure P1) and take readings of the current and potential difference. The ratio of these readings $V/I$ gives the resistance. The clips are then moved to 0.900 m apart and the readings are repeated. This process is repeated down to 10 cm or so, taking care to adjust the potential difference to limit the current.

When there is only a short length of wire between the clips, the current could be large, which could cause the wire to get hot. This is a safety hazard with a risk of burning your fingers. In extreme cases, the wire could melt rapidly and spit, so goggles should be worn.

It is sensible to repeat the whole process several times, since there is a random error in reading the

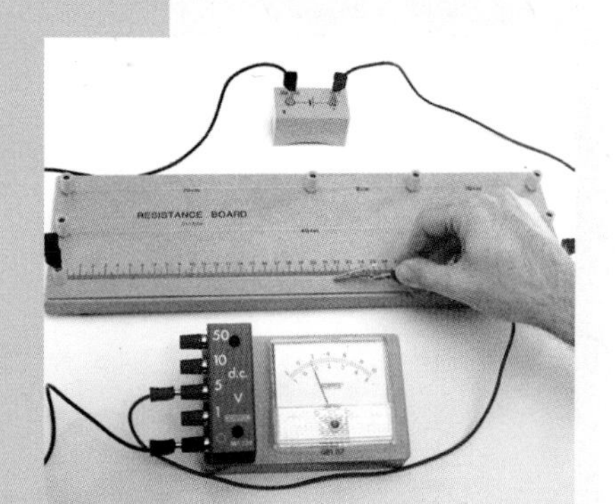


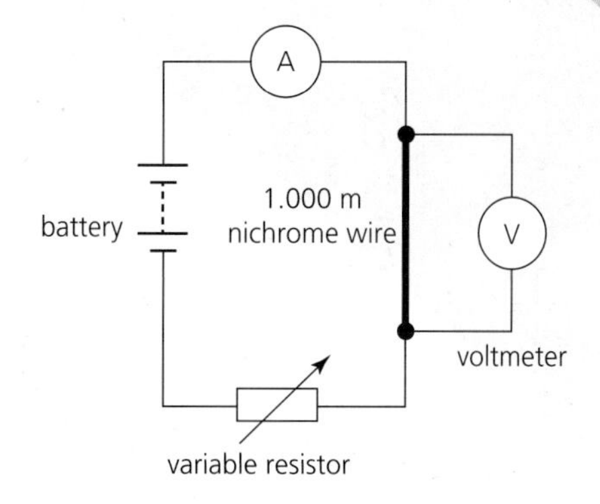


*Figure P1* The arrangement and the circuit. The nichrome wire is attached to a metre rule, for example with clear sticky tape, and with a crocodile clip at each end. The length of wire between the crocodile clips should be exactly 1.000 m.

| Length / m ± 0.001 m | Current / A ± 0.01 A | | | Potential difference / V ± 0.1 V | | | Resistance / Ω (mean value) |
|---|---|---|---|---|---|---|---|
| | 1 | 2 | 3 | 1 | 2 | 3 | |
| 1.000 | 0.30 | 0.31 | 0.34 | 10.1 | 9.9 | 10.0 | 31.6 |

*Table P1*

length between the crocodile clips, and the resistance of the connections that are made may vary each time.

A possible structure for the results table and a sample first set of results is shown in Table P1.

A graph of resistance against length is drawn. A sample one is shown in Figure P3.

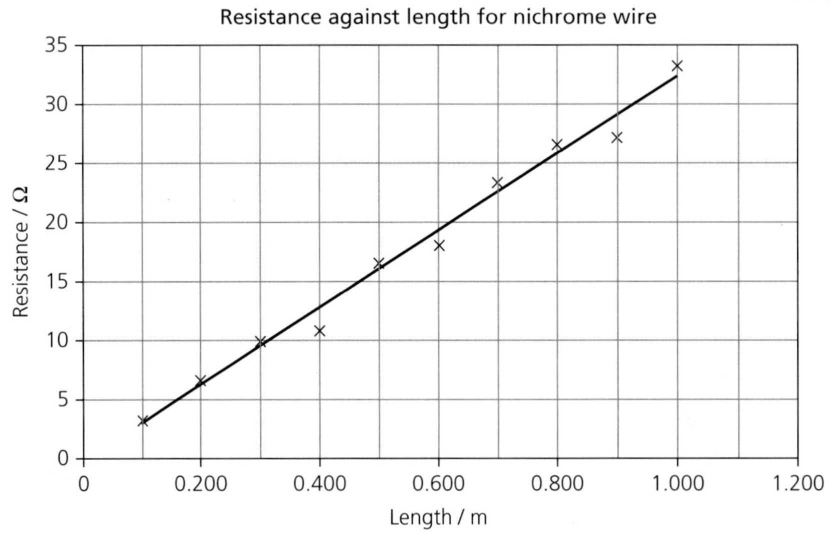


*Figure P3*

**Technique: Measuring cross-sectional area**

Several readings of the diameter of the wire should be taken using a micrometer (Figure P2). Some sample readings are shown in Table P2. The readings were taken at three different points along the wire and in two perpendicular directions.

| Diameter / mm | | |
|---|---|---|
| 0.20 | 0.18 | 0.22 |
| 0.19 | 0.20 | 0.24 |

*Table P2*

**QUESTIONS**

**P1** Calculate the gradient of the graph in Figure P3.

**P2** Find the mean cross-sectional area of the wire from the data in Table P2.

**P3** Use these values to calculate the resistivity of nichrome.

**P4** Estimate the uncertainty in your value for the cross-sectional area.

**P5** Estimate the uncertainty in your value for the gradient of the graph.

**P6** Re-state the resistivity of nichrome, complete with your estimate of its uncertainty.

## PRACTICE QUESTIONS

1. Figure Q1 shows two resistors, $R_1$ and $R_2$, connected in series with a battery of emf 12 V and negligible internal resistance.

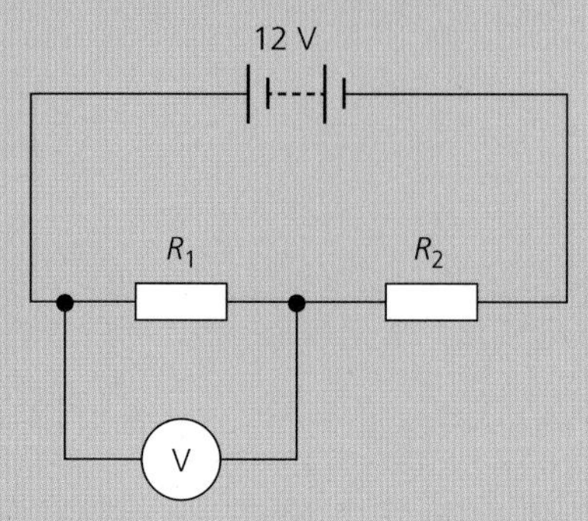


*Figure Q1*

The reading on the voltmeter is 8.0 V and the resistance of $R_2$ is 60 Ω.

**a.** Calculate the current in the circuit.

**b.** Calculate the resistance of $R_1$.

**c.** Calculate the charge passing through the battery in 2.0 minutes. Give an appropriate unit for your answer.

*AQA June 2012 Unit 1 Q7 (part)*

2. X and Y are two lamps. X is rated at 12 V, 36 W and Y at 4.5 V, 2.0 W.

**a.** Calculate the current in each lamp when it is operated at its correct working voltage.

**b.** The two lamps are connected in the circuit shown in Figure Q2. The battery has an emf of 24 V and negligible internal resistance. The resistors, $R_1$ and $R_2$, are chosen so that the lamps are operating at their correct working voltage.

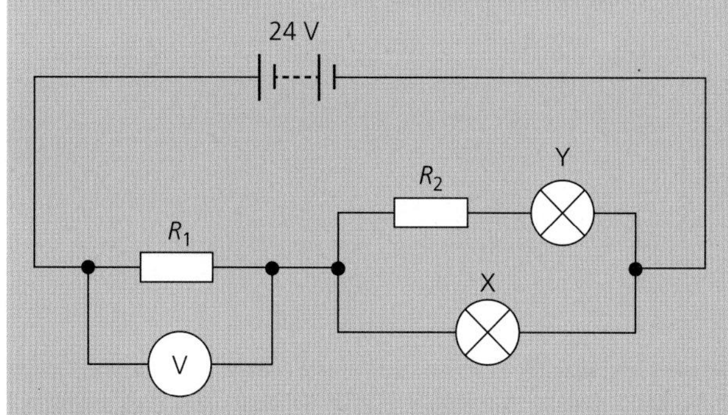


*Figure Q2*

**i.** Calculate the pd across $R_1$.

**ii.** Calculate the current in $R_1$.

**iii.** Calculate the resistance of $R_1$.

**iv.** Calculate the pd across $R_2$.

**v.** Calculate the resistance of $R_2$.

**c.** The filament of the lamp in X breaks and the lamp no longer conducts. It is observed that the voltmeter reading decreases and lamp Y glows more brightly.

**i.** Explain without calculation why the voltmeter reading decreases.

**ii.** Explain without calculation why the lamp Y glows more brightly.

*AQA Jan 2012 Unit 1 Q6*

3. The circuit shown in Figure Q3 shows an arrangement of resistors, W, X, Y and Z, connected to a battery of negligible internal resistance.

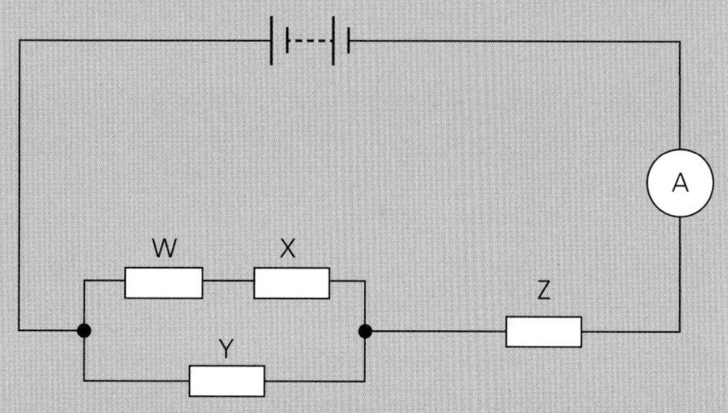


*Figure Q3*

The emf of the battery is 10 V and the reading on the ammeter is 2.0 A.

**a. i.** Calculate the total resistance of the circuit.

**ii.** The resistors W, X, Y and Z all have the same resistance. Show that your answer to part **a i** is consistent with the resistance of each resistor being 3.0 Ω.

**b. i.** Calculate the current through resistor Y.

**ii.** Calculate the pd across resistor W.

*AQA June 2011 Unit 1 Q7*

4. A voltmeter and an ammeter are to be used to measure the resistance of a component X in a circuit. Which row in Table Q1 accurately describes the experimental set-up?

| | Voltmeter | Ammeter |
|---|---|---|
| A | Connected in parallel with X, so needs a low resistance | Connected in series with X, so needs a low resistance |
| B | Connected in parallel with X, so needs a high resistance | Connected in series with X, so needs a low resistance |
| C | Connected in series with X, so needs a low resistance | Connected in parallel with X, so needs a low resistance |
| D | Connected in parallel with X, so needs a high resistance | Connected in series with X, so needs a high resistance |

*Table Q1*

5. Three identical resistors of resistance 20 Ω are used in different combinations. Which of the following resistances cannot be made with these three resistors?

A 60 Ω

B 6.67 Ω

C 30 Ω

D 45 Ω

6. Electrical power is measured in watts. Which of these units is **not** equivalent to the watt?

A $N\ m\ s^{-1}$

B $A^2\ \Omega$

C $V\ \Omega^{-1}$

D A V

7. A heater has an electrical power of *P* watts. A second, identical, heater is then wired in parallel with the first one. If the same power supply is used, which of A to D will be the power of the combination?

A $P$

B $P^2$

C $2P$

D $\frac{P}{2}$

8. If the heaters in question 7 are wired in series rather than in parallel, the power of the combination would be:

A $P$

B $P^2$

C $2P$

D $\frac{P}{2}$

9. The manufacturer's data for a digital ammeter says, "Accuracy ±4%. Maximum resolution 10 mA". This means that a reading of 1.55 A will have an uncertainty of:

A ±0.07 A

B ±0.04 A

C ±0.06 A

D ±0.01 A

## Stretch and challenge

10. A friend of yours, Justine Time, is planning to manufacture a new high-powered kettle, which she hopes will boil water twice as fast as the 2 kW kettle she currently uses. She has a coil of metal, which she thinks is nichrome, and she wants to use this as the heating element. She wants your advice on several issues.

a. How can she tell if the metal is nichrome? She knows that the resistivity of nichrome is about $10^{-6}\ \Omega$ m and she has basic laboratory apparatus, including an ammeter and a voltmeter, but not a resistance meter. Describe the measurements she must make to check whether or not this metal coil is made from nichrome.

b. The diameter of the wire making up the coil is 0.5 mm. What length of wire will she need to achieve her desired kettle power?

c. Are there any practical problems or safety issues with her plan? If so, what can she do about them?

# 14 ELECTRICITY 2

## PRIOR KNOWLEDGE

You will be familiar with the concepts of current, potential difference and resistance. You will know how to use meters to measure these in a circuit, and will understand the difference between series and parallel circuits.

## LEARNING OBJECTIVES

In this chapter you will learn what is meant by the electromotive force of a power supply and the effect of its internal resistance. You will learn about the effect of temperature on the resistance of circuit components, and about the design and applications of potential divider circuits. You will also meet the concept of superconductivity.

**(Specification 3.5.1.3 part, 3.5.1.5, 3.5.1.6)**

A satellite photograph of the Earth at night (Figure 1) is an illuminated map showing the distribution of electrical power in the world. Europe and the USA are ablaze with artificial light. Africa is largely dark. Throughout much of that continent electricity is a rare and valuable commodity. In east Africa, less than a quarter of the people in Kenya, Rwanda and Tanzania have an electrical supply, while in sub-Saharan Africa as a whole 600 million people have no access to electrical power.

***Figure 1*** *A composite satellite image of the Earth at night*

Solar power could change all that. Small-scale installations of photovoltaic cells (Figure 2) coupled with rechargeable batteries have a huge impact on village life in rural Africa. As well as providing light at night, solar panels also provide power for irrigation pumps that deliver clean water. This improves the village crops, reduces the risk of disease and relieves women of the daily burden of carrying water.

***Figure 2*** *Solar panels enable laptops and mobile phones to be recharged. Solar power is becoming increasingly important in African village life.*

Some experts predict that solar power will be the world's number one method of generating electricity by the year 2050. African sunshine may be contributing to UK electrical power long before that. A scheme to generate 2 GW of electrical power from a huge Tunisian solar power plant has been proposed. This could be supplying 2.5 million British homes by 2017.

# 14.1 ELECTROMOTIVE FORCE AND INTERNAL RESISTANCE

All electrical power supplies produce a potential difference between a pair of terminals, when connected into a circuit. A typical AA cell claims a potential difference of 1.5 V at its terminals, and, if you measure the potential difference supplied by a new cell with a digital voltmeter, it will read 1.5 V. But this is true only when the cell is not being used to supply a current. As soon as the cell is connected into a circuit, perhaps to light a torch bulb, the potential difference shown on the voltmeter will decrease (Figure 3). It seems that some potential difference has gone missing. This is often referred to as the 'lost volts'.

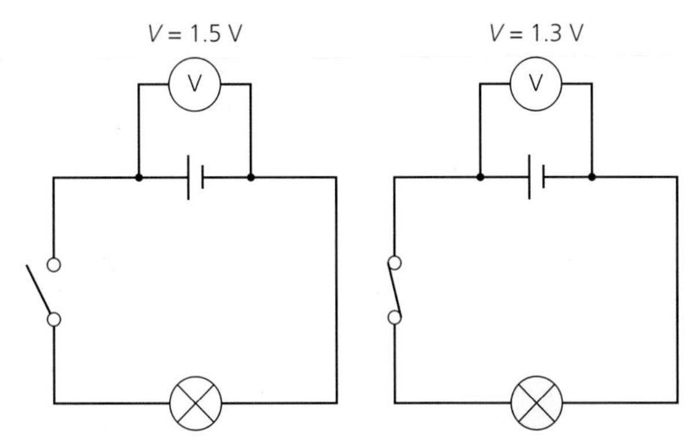


***Figure 3*** *When the switch is open, no current flows and the voltmeter reads 1.5 V. When the switch is closed, current flows and the voltmeter reading decreases to 1.3 V.*

The 'missing' potential difference (pd) is across the resistance of the cell itself. It is not possible to make a power supply (a cell or generator, for example) that does not have its own electrical resistance. This is known as the **internal resistance** ($r$) of the power supply.

An ideal power supply produces a potential difference that we call the **electromotive force** or **emf** (see section 13.3 of Chapter 13). It is this emf, symbol $\mathcal{E}$, that gives each coulomb of charge the energy required to keep the current flowing:

$$\mathcal{E} = \frac{E}{Q}$$

where $\mathcal{E}$ = emf, $E$ = energy transferred to charge and $Q$ = charge. However, when current passes through any real power supply, it transfers some energy by heating the internal resistance of the cell. So, although we cannot measure it directly, there is a potential difference $V_{\text{int res}}$ across the internal resistance. The potential difference that we can measure at the terminals, $V_{\text{term}}$, is less than the emf of the cell:

$$V_{\text{term}} = \mathcal{E} - V_{\text{int res}}$$

When the current through the cell is $I$, the potential difference across the internal resistance is equal to $Ir$, and therefore

$$V_{\text{term}} = \mathcal{E} - Ir$$

In circuit diagrams and calculations, we can treat a real power supply as if it were an ideal power supply, of emf $\mathcal{E}$, in series with an internal resistor, $r$, as in Figure 4.

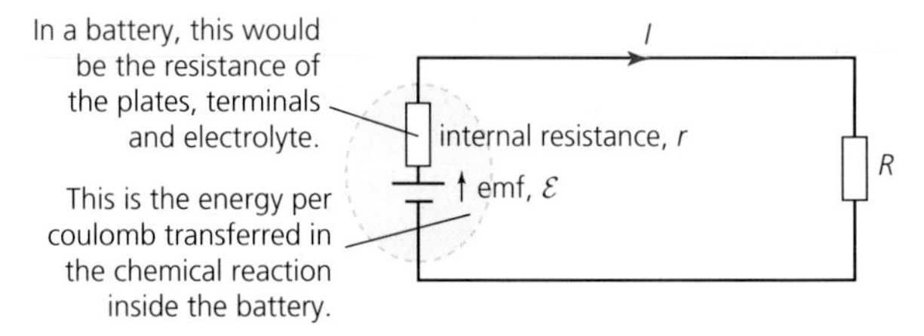


***Figure 4*** *Internal resistance and emf*

If a supply produces a current $I$ in an external circuit of total resistance $R$ (as in Figure 4), then $V_{\text{term}} = IR$ and the equation $V_{\text{term}} = \mathcal{E} - Ir$ becomes

$$IR = \mathcal{E} - Ir$$

or

$$\mathcal{E} = I(R + r)$$

which is

$$\mathcal{E} = V_{\text{term}} + V_{\text{int res}}$$

This is a statement of the conservation of energy in the circuit, or Kirchhoff's second law (see section 13.3 of Chapter 13).

The terminal potential difference of the power supply, $V_{\text{term}}$, depends on how much current, $I$, is flowing. The larger the current, the larger the potential difference across the internal resistance, $Ir$, and therefore the lower the potential measured at the terminals of the power supply. The potential difference measured at the terminals is only exactly equal to the emf when there is no current flowing. Table 1 summarises the differences between emf and terminal pd.

| | Emf | Terminal pd |
|---|---|---|
| **Meaning, in terms of energy** | energy per coulomb produced by cell (power supply) | energy per coulomb delivered to circuit components |
| **How it is measured** | voltmeter reading across cell terminals when no current is flowing | voltmeter reading across cell terminals when current is flowing |

***Table 1*** *The differences between emf and terminal pd*

## Worked example 1

The service battery in an electric vehicle is used to operate the lights, fans, radio, etc. (see Figure 5).

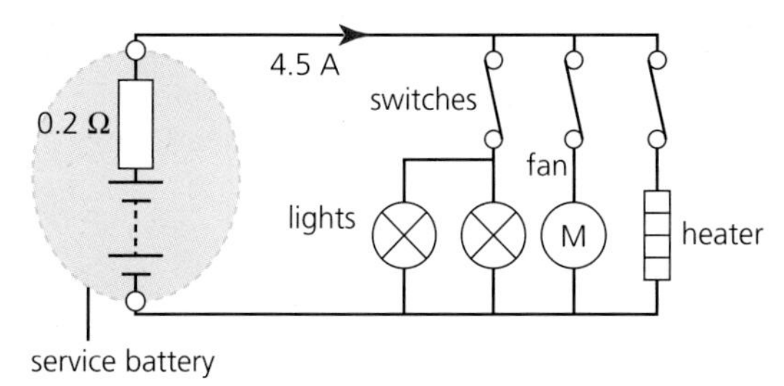


***Figure 5***

The current through the service battery varies, depending on which devices are switched on. Suppose 4.5 A flows through a battery with an internal resistance of 0.2 Ω and an emf of 12 V. How efficient is the battery at transferring energy to this current?

We have

$$V_{\text{intres}} = Ir$$
$$= 4.5\,\text{A} \times 0.2\,\Omega$$
$$= 0.9\,\text{V}$$

and

$$V_{\text{term}} = \mathcal{E} - V_{\text{intres}}$$
$$= 12\,\text{V} - 0.9\,\text{V}$$
$$= 11.1\,\text{V}$$

For every coulomb of charge that flows through it, the battery produces 12 J, but only delivers 11.1 J. The efficiency is therefore

$$\text{efficiency} = \frac{11.1\,\text{J}}{12\,\text{J}} = 0.925 = 92.5\%$$

## Worked example 2

A photovoltaic cell (solar cell) acts as a source of emf. The potential difference measured across the terminals of the cell is 5.0 V when 1.2 mA flows but drops to 1.8 V when 3.0 mA flows. What is the internal resistance of the cell?

Using $\mathcal{E} = V_{\text{term}} + Ir$ for the two situations, we have two equations

$$\mathcal{E} = 5.0 + 0.0012 \times r$$
$$\mathcal{E} = 1.8 + 0.0030 \times r$$

Eliminating $\mathcal{E}$ from these gives

$$5.0 + 0.0012 \times r = 1.8 + 0.0030 \times r$$
$$5.0 - 1.8 = 0.0030 \times r - 0.0012 \times r$$
$$3.2 = 0.0018 \times r$$

and so the internal resistance of the cell is

$$r = \frac{3.2}{0.0018} = 1777\,\Omega = 1.8\,\text{k}\Omega$$

## QUESTIONS

1. a. A cell has an emf of 1.2 V. It is used to provide a current of 1.0 A. How much energy will the cell transfer in one minute?
   b. If the cell has an internal resistance of 0.1 Ω, what will the potential difference at the terminals be?
   c. What is the efficiency of energy transfer to the circuit?
2. You may have noticed that the headlights go dim and the stereo fades when the starter motor of a petrol-engined car is switched on. Explain why this happens.
3. The current through an electric vehicle's main storage battery is 120 A, the terminal pd is 168 V and the internal resistance 0.1 Ω. What is the emf?
4. A car battery delivers 112 W of power to a 7.0 Ω resistor. The emf of the battery is 30 V. What is the internal resistance?

## ASSIGNMENT 1: INVESTIGATING THE EFFICIENCY OF ELECTRICAL ENERGY TRANSFER

**(MS 3.1, MS 3.2, MS 3.9, PS 2.2, PS 3.1, PS 3.2)**

When a power supply delivers current, some energy is transferred as heat in its internal resistance. This energy is dissipated within the power supply, and is not available to be useful in the external circuit. This is a source of inefficiency. Whether we are using a battery to power a torch, an amplifier to run loudspeakers, or a generator to drive a pump, we want to make the energy transfer as efficient as possible, and to minimise the energy wasted in the internal resistance.

We can model the real situation by the circuit in Figure A1. The load resistor $R$ represents the intended application.

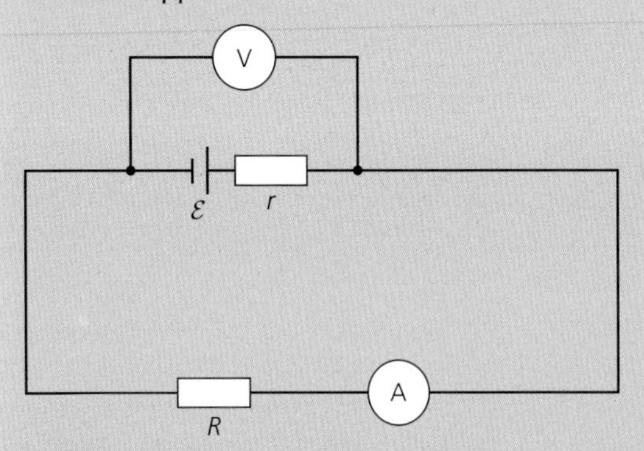


*Figure A1*

Rearranging the previous equation for the emf, the current in the circuit in Figure A1 is

$$I = \frac{\mathcal{E}}{(R + r)}$$

The power dissipated in the internal resistance is

$$I^2 r = r \times \frac{\mathcal{E}^2}{(R + r)^2}$$

The power in the external resistor is

$$P = R \times \frac{\mathcal{E}^2}{(R + r)^2}$$

We want to find out how efficient our circuit will be in delivering power to the external resistance. Is there an optimum value for $R$? There is more than one way to tackle this problem. We will consider two methods.

Method 1 asks you to make a mathematical model using a spreadsheet.
Method 2 involves calculus to find the maximum value of a function. If you are studying A-level Mathematics and you have used the quotient rule for differentiation, read through this method.

### Method 1: Using a spreadsheet and a graph

The power in the external resistor is given by

$$P = R\mathcal{E}^2/(R + r)^2$$

The variation of $P$ with $R$ can be modelled by using a spreadsheet to carry out many calculations with different values of $R$. This is your task in this assignment.

### Questions

**A1** Set up a table in the spreadsheet, as shown in Figure A2, with one column for external resistance $R$ and one for power $P$. Put values for $R$ into column A and put in column B the formula for the power in $R$. You also need to put in values for emf $\mathcal{E}$ and internal resistance $r$. Here these are in cells E3 and E2. Experiment with the values of $R$, $\mathcal{E}$ and $r$.

| | A | B | C | D | E |
|---|---|---|---|---|---|
| 1 | External resistor R | Power in R | Gradient | | |
| 2 | 10 | = (\$E\$3²)*A2/((A2 + \$E\$2)^2) | = (B3-B2)/(A3-A2) | IntRes | 50 |
| 3 | 20 | = (\$E\$3²)*A3/((A3 + \$E\$2)^2) | = (B4-B3)/(A4-A3) | EMF | 12 |
| 4 | 30 | = (\$E\$3²)*A4/((A4 + \$E\$2)^2) | = (B5-B4)/(A5-A4) | | |
| 5 | 40 | = (\$E\$3²)*A5/((A5 + \$E\$2)^2) | = (B6-B5)/(A6-A5) | | |

***Figure A2*** *Cell references with \$ signs in them are absolute; they always refer to the same cell. Cell references without \$ signs in them are relative.*

**A2** Find which value of $R$ gives the maximum value of power by plotting a graph of power $P$ versus resistance $R$. Use the spreadsheet to plot a scatter graph of column A against column B. If there is a 'best' value for $R$, where the power in $R$ is a maximum, the graph will change direction at that point, and the gradient of the graph will be zero. (See Figure A3 under Method 2 below.) A third column of the spreadsheet, column C, can be used to calculate the gradient. This is done by finding the difference in the $y$-value (in this case $B$) divided by the difference in the $x$-value (in this case $R$). Look in this column for the sign of the gradient changing from positive to negative.

**A3** What is the condition for maximum power?

**A4** What are the practical implications of this result? For example, what would it mean for an amplifier driving a loudspeaker?

## Method 2: Using calculus

### Stretch and challenge

We need an equation that links the useful power, $P$, to the value of the external resistor, $R$. We can then differentiate this to find out how $P$ changes as $R$ varies.

In calculus, $\mathrm{d}P/\mathrm{d}R$ is the rate of change of $P$ with $R$. This is the gradient of a graph of power ($y$-axis) against external resistance ($x$-axis). We are looking for the value of $R$ that makes $P$ a maximum. At the maximum, the gradient will be equal to zero (Figure A3).

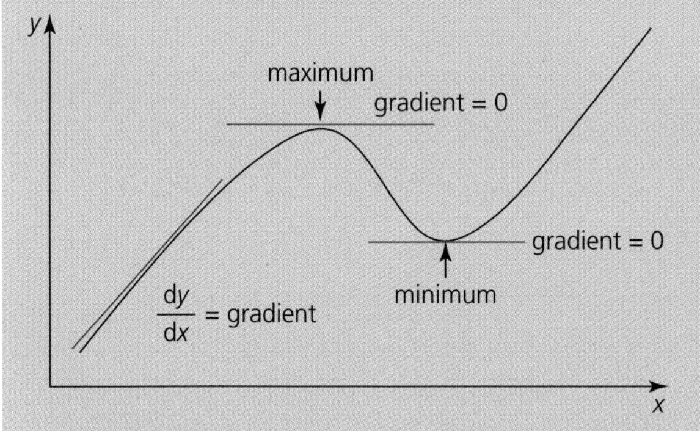


***Figure A3*** *The rate of change of a function is zero at a maximum and at a minimum.*

We differentiate the expression for power in the external resistor using the quotient rule. This rule is: if $y = \frac{u}{v}$, where $u$ and $v$ are functions of $x$, then

$$\frac{\mathrm{d}y}{\mathrm{d}x} = \frac{v\,\mathrm{d}u/\mathrm{d}x - u\,\mathrm{d}v/\mathrm{d}x}{v^2}$$

So in our case

$$\frac{\mathrm{d}P}{\mathrm{d}R} = \frac{v\,\mathrm{d}u/\mathrm{d}R - u\,\mathrm{d}v/\mathrm{d}R}{v^2}$$

Start with the expression for power:

$$P = \frac{R\mathcal{E}^2}{(R+r)^2}$$

We put $u = R\mathcal{E}^2$ and $v = (R + r)^2$. Then

$$\frac{\mathrm{d}P}{\mathrm{d}R} = \frac{(R+r)^2\mathcal{E}^2 - 2R\mathcal{E}^2(R+r)}{(R+r)^4}$$

The expression for $\mathrm{d}P/\mathrm{d}R$ will equal zero at a maximum (Figure A3). Equating $\mathrm{d}P/\mathrm{d}R$ to zero gives

$$(R + r)^2\mathcal{E}^2 = 2R\mathcal{E}^2(R + r)$$

Cancelling the common factors gives

$$(R + r) = 2R$$

$$R = r$$

This is the condition for the power to be a maximum.

### A further challenge

Actually, from the maths above, when $R = r$, leading to zero gradient, the power could be a maximum *or a minimum* (see Figure A3). To confirm that it is a maximum, we need to consider the gradient of the curve on either side of the point where $R = r$. If the gradient is positive for $R < r$ and negative for $R > r$, the graph will have a maximum at $R = r$. We examine the sign of the gradient when the resistance $R$ is a small amount, $\Delta r$, more than or less than $r$. Thus

$$\text{gradient} = \frac{\mathrm{d}P}{\mathrm{d}R} = \frac{(R+r)^2\mathcal{E}^2 - 2R\mathcal{E}^2(R+r)}{(R+r)^4}$$

For $R < r$, $R = r - \Delta r$, the gradient is

$$\frac{\mathrm{d}P}{\mathrm{d}R} = \frac{(r-\Delta r+r)^2\mathcal{E}^2 - 2(r-\Delta r)\mathcal{E}^2(r-\Delta r+r)}{(r-\Delta r+r)^4}$$

This is positive if

$$(2r - \Delta r)^2\mathcal{E}^2 > 2(r - \Delta r)\mathcal{E}^2(2r - \Delta r)$$

Cancelling common factors gives

$$(2r - \Delta r) > 2(r - \Delta r)$$

$$2r - \Delta r > 2r - 2\Delta r$$

which must always be true, so the gradient is positive here.

For $R > r$, $R = r + \Delta r$, the gradient is

$$\frac{dP}{dR} = \frac{(r + \Delta r + r)^2 \mathcal{E}^2 - 2(r + \Delta r)\mathcal{E}^2(r + \Delta r + r)}{(r + \Delta r + r)^4}$$

This is positive if

$$(2r + \Delta r)^2 \mathcal{E}^2 > 2(r + \Delta r)\mathcal{E}^2(2r + \Delta r)$$

Cancelling common factors gives

$$(2r + \Delta r) > 2(r + \Delta r)$$

$$2r + \Delta r > 2r + 2\Delta r$$

which cannot ever be true, so the gradient is negative here.

Since the gradient goes from positive to negative, as $R$ increases there must be a *maximum* at $R = r$.

## Measuring the emf of a cell

The emf is equal to the terminal potential difference when there is no current. When a high-resistance digital voltmeter is connected directly across the terminals, the terminal pd measured will be almost equal to the emf. A digital voltmeter might have a 'lead-to-lead' resistance of 10 MΩ. Suppose this was used to measure the potential difference at the terminals of a rechargeable cell, emf 1.2 V and internal resistance 30 mΩ.

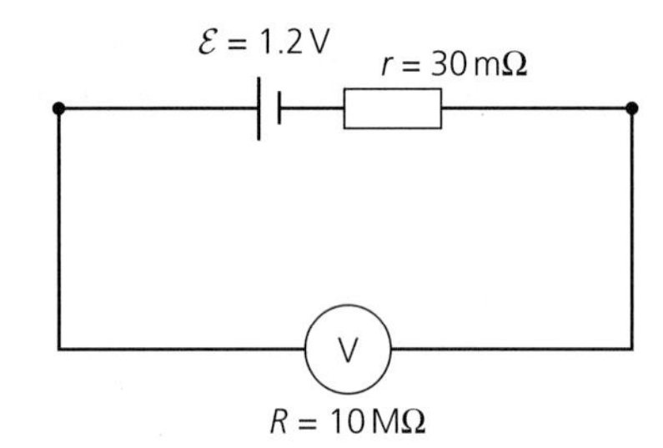


***Figure 6*** *Measuring the pd across a cell on open circuit*

The total resistance of the circuit (Figure 6) equals 10 MΩ + 30 mΩ, which is effectively 10 MΩ, so the current through the circuit is

$$I = \frac{V_{term}}{R} = \frac{1.2}{10 \times 10^6} = 1.2 \times 10^{-7}\text{ A}$$

The potential difference across the internal resistor is therefore

$$V_{int\ res} = Ir = 1.2 \times 10^{-7} \times 30 \times 10^{-3}$$
$$= 3.6 \times 10^{-9}\text{ V or } 3.6\text{ nV}$$

This is a difference from the true reading of only $3 \times 10^{-7}$ %.

### QUESTIONS

5. **a.** What would the percentage error be if the digital voltmeter used in Figure 6 were used to measure the emf of a solar panel (a series of connected solar cells) whose emf is actually 21.3 V with an internal resistance of 6.9 Ω?

   **b.** An analogue voltmeter is used instead, so as to observe a changing potential difference. It has a resistance of 200 Ω. How would this affect your answer to part **a**?

### KEY IDEAS

- The electromotive force (emf) $\mathcal{E}$ is the energy per unit charge transferred by a source:

$$\mathcal{E} = \frac{E}{Q}$$

- The terminal potential difference (pd) $V_{term}$ is less than the emf $\mathcal{E}$ because of the potential difference across the internal resistance, $r$, of the source:

$$\mathcal{E} = V_{term} + Ir$$
$$\mathcal{E} = I(R + r)$$

- The electromotive force is equal to the terminal potential difference only when there is no current.

## REQUIRED PRACTICAL: APPARATUS AND TECHNIQUES

### Investigation of the emf and internal resistance of a cell

The aim of this practical is to find the value of the emf and the internal resistance of a cell, by measuring the variation of terminal pd with current.

The practical gives you the opportunity to show that you can:

- use appropriate digital instruments, including electrical multimeters, to obtain a range of measurements (current, voltage)
- correctly construct circuits from circuit diagrams using DC power supplies, cells and a range of circuit components, including those where polarity is important
- design, construct and check circuits using DC power supplies, cells, and a range of circuit components.

### Apparatus

The circuit shown in Figure P1 is used.

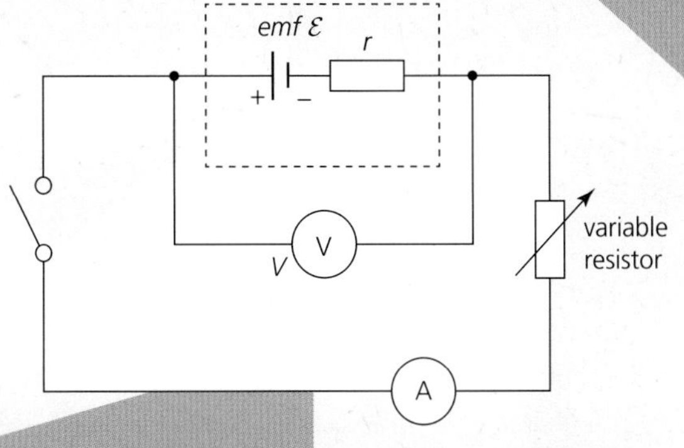


*Figure P1* The circuit

It is not essential to use digital meters, although these are likely to give higher precision. The variable resistor $R$ is a rheostat (Figure P2).

*Figure P2* A rheostat has a sliding connection to vary the amount of resistance wire in the circuit.

### Technique

The current through the circuit is varied by altering the variable resistor. A series of readings is taken of current, $I$, and terminal potential difference, $V$.

The circuit should be switched off whenever readings are not being taken. This is particularly important for low values of the variable resistor. This is because a relatively large current is flowing in the circuit, which will have two effects:

- The current will tend to heat up the internal resistor and change its resistance.
- The battery may begin to run down, which would also increase its internal resistance.

Note that the internal resistance of a rechargeable cell can be very small. If such a cell's terminals were to be short-circuited, perhaps with a piece of copper wire, this could lead to a very high current. It is possible that the cell might become so hot that the casing might burst quite violently. In this case, or in any case where the electric current is large, it is a good idea to connect a fixed series resistor into the circuit, thereby limiting the maximum current.

Comparing $V = \mathcal{E} - Ir$ with the equation of a straight line $y = mx + c$, we see that if $V$ is plotted on the $y$-axis and $I$ on the $x$-axis, the graph should be a straight line with gradient $-r$ and intercept $\mathcal{E}$.

The readings shown in Table P1 were taken using an AA cell.

| Current / A | Terminal pd / V |
|---|---|
| 0.75 | 0.75 |
| 0.50 | 1.00 |
| 0.38 | 1.13 |
| 0.30 | 1.20 |
| 0.25 | 1.25 |
| 0.21 | 1.29 |
| 0.19 | 1.31 |
| 0.17 | 1.33 |
| 0.15 | 1.35 |
| 0.14 | 1.36 |

*Table P1*

### QUESTIONS

**P1** Plot a graph using the data in Table P1 and use it to calculate the emf, $\mathcal{E}$, and the internal resistance, $r$.

**P2** Estimate the uncertainty in the readings of current and potential difference.

**P3** This experiment could be carried out for any power supply. If you were to do this experiment to find the emf and internal resistance of a solar (photovoltaic) cell, explain what further steps you would need to take to get repeatable results.

## 14.2 RESISTANCE AND TEMPERATURE

A solar cell, or photovoltaic cell, transfers light energy to electrical energy (Figure 7). The electrical power output depends on the intensity of light energy falling on the cell each second. The efficiency of the energy transfer depends on the type of cell and the conditions. Solar cells become less efficient at higher temperatures. One reason for this is that the electrical resistance, particularly of the metal wires and connectors, tends to increase with temperature.

***Figure 7*** *The latest solar panels (arrays of solar cells) can achieve efficiencies of up to 46% under ideal conditions.*

Resistivity in metals is caused by the interaction of moving electrons (the current) with the fixed positive ions of the metal. As the temperature of the metal increases, the electrons lose more energy in these interactions, so the resistance of a metal component will increase as it gets hotter. The relationship is approximately linear, that is

change in resistivity $\propto$ change in temperature

This holds for most metals, at least for small temperature changes, although the change in resistivity per unit temperature rise depends on the particular metal (Figure 8).

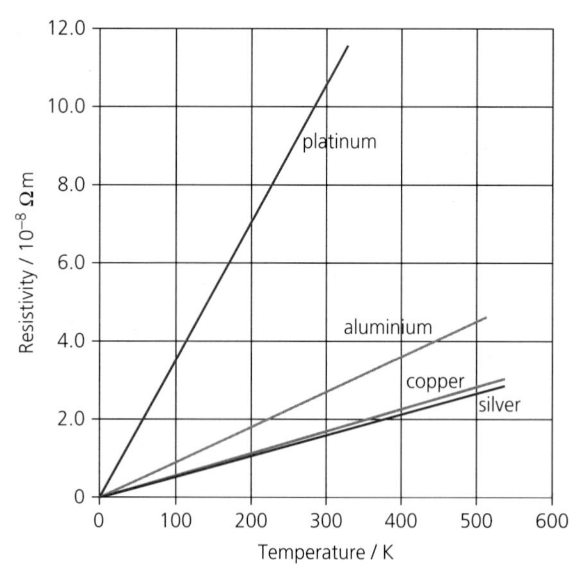


***Figure 8*** *Change in resistivity with temperature for different metals. Note that the scale on the x-axis shows absolute temperature (see section 14.4), that is, in kelvin (K) rather than degrees Celsius (°C). Remember that to convert from Celsius to kelvin, you must add 273. A change in temperature of 1 °C is exactly equivalent to a change of 1 K.*

### QUESTIONS

6. Use your knowledge of metals and the graph in Figure 8 to support your answers to the following questions.
   a. Which metal would you choose as a connector in an electronic circuit? Explain.
   b. Which metal would you use as a temperature sensor in an electrical resistance thermometer? Explain.

### KEY IDEAS

- The resistivity of metals increases as the temperature rises, because the conduction electrons lose more energy in collisions with positive ions.
- For a particular metal,

  change in resistivity $\propto$ change in temperature

## ASSIGNMENT 2: INVESTIGATING THE CURRENT SURGE THROUGH A FILAMENT LIGHT BULB

**(MS 0.1, MS 0.2, MS 0.4, MS 0.5, MS 1.1, MS 1.4, MS 2.3, MS 3.1, MS 3.5, PS 1.2, PS 2.1, PS 3.1, PS 3.2)**

***Figure A1*** *Filament bulbs sometimes 'hum'. The coils generate magnetic fields, which can set the filament vibrating.*

The current that flows through the tungsten filament of an incandescent light bulb (Figure A1), when it is working normally, at its stated potential and power output, can easily be calculated. For a 100 W incandescent light bulb that is designed to operate in a UK home,

$$\text{current} = \frac{\text{power}}{\text{potential difference}} = \frac{P}{V} = \frac{100}{240} = 0.42\text{ A}$$

The resistance is then

$$R = \frac{V}{I} = \frac{240}{0.42} = 580\,\Omega$$

This high resistance is due to several factors:

- The surprisingly long length of the filament. There is around 1.5 m of tungsten wire in a standard 100 W light bulb. It is formed into a coil of more than 1000 turns, so that it will fit in!
- The small cross-sectional area (remember that $R = \rho l/A$). The wire used has a diameter of less than 50 μm.
- The very high operating temperature. The filament of an old-fashioned light bulb is probably the hottest thing in your house. The working temperature is around 2600 °C (around 2900 K).

### Questions

**A1** The resistance of the light bulb described above is measured with a multimeter on the resistance setting, when the bulb is off, and found to be 45 Ω. Estimate the rate of change of resistance with temperature for the tungsten filament, that is, (change in resistance)/(change in temperature), assuming this is linear. This will be in units of ohm per degree Celsius ($\Omega\,°C^{-1}$) or ohm per kelvin ($\Omega\,K^{-1}$), which are exactly equivalent.

**A2** Someone suggests using the filament, with a resistance meter, as a thermometer. What do you think of the idea?

**A3** Using the figures given above, calculate the resistivity of tungsten at

**a.** room temperature (bulb off)

**b.** its operating temperature (bulb on).

When the filament light bulb is first turned on, its resistance is low, and a large current will flow, up to 10 times the current that flows when the bulb is at working temperature. This transient current surge can be investigated using a data logger in conjunction with a current sensor (see Figure A2).

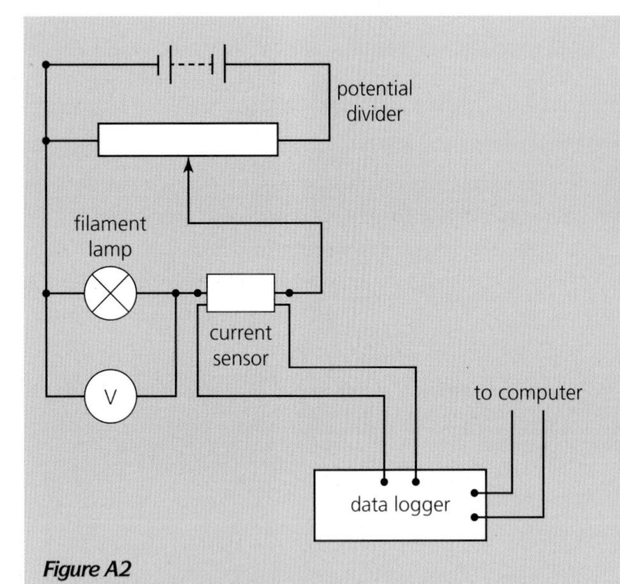


*Figure A2*

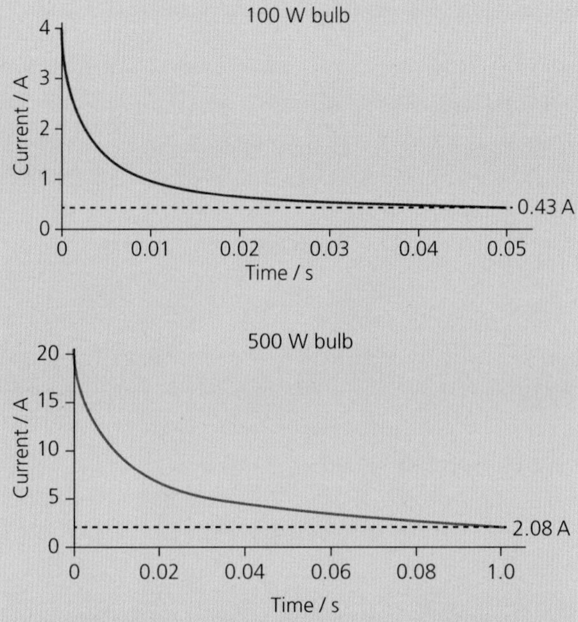


***Figure A3*** *The current surge through two different filament bulbs*

The range of the current sensor has to be chosen carefully. As always, the most sensitive range available is preferable, though the current through the bulb must not exceed the maximum on the chosen range. Remember that the initial surge current might be 10 times the steady current. The sampling rate needs to be chosen. The interesting part of the experiment is over in less than two minutes, and the samples should be around 10 ms apart. So the memory in the data logger needs to hold about $2 \times 600 \times 100 = 12\,000$ data points.

## Questions

**A4** Figure A3 shows the results of two experiments to monitor the current through a bulb as it is switched on in a mains (240 V) circuit. Describe and explain the common shape of the graphs.

**A5** The two graphs in Figure A3 show the transient behaviour of two filament bulbs of similar construction, though they have different electrical power ratings. What are the differences between the two graphs? Can you suggest any reason for the difference in behaviour? You should make your answer quantitative.

**A6** When incandescent bulbs fail, that is, the filament breaks, it is often in the first second after turning them on. Suggest why this might be.

**A7** Owen suggests that the initial current surge is wasteful and expensive in terms of the greater amount of electrical energy used. He says that it would be cheaper to leave the lights switched on all the time. Do some quick calculations to persuade him that he is wrong!

# 14.3 SEMICONDUCTORS

Materials that are good electrical conductors, such as metals, have charges that can move relatively freely. Insulators, such as glass or ceramics, have no mobile charges. There is an intermediate class of materials in which there are a limited number of mobile charges. Materials such as these are known as **semiconductors**. Examples are silicon and germanium.

Table 2 shows the resistivity of various materials at room temperature.

| Material | Resistivity / Ω m at 20 °C | |
|---|---|---|
| Silver | $1.59 \times 10^{-8}$ | conductors |
| Copper | $1.68 \times 10^{-8}$ | conductors |
| Gold | $2.44 \times 10^{-8}$ | conductors |
| Aluminium | $2.82 \times 10^{-8}$ | conductors |
| Tungsten | $5.60 \times 10^{-8}$ | conductors |
| Platinum | $1.06 \times 10^{-7}$ | conductors |
| Constantan | $4.90 \times 10^{-7}$ | conductors |
| Nichrome | $1.10 \times 10^{-6}$ | conductors |
| Gallium arsenide | $1.00 \times 10^{-3}$ to $1.00 \times 10^{8}$ | semiconductors |
| Germanium | $4.60 \times 10^{-1}$ | semiconductors |
| Silicon | $6.40 \times 10^{2}$ | semiconductors |
| Drinking water | $2.00 \times 10^{1}$ to $2.00 \times 10^{3}$ | insulators |
| Deionised water | $1.80 \times 10^{5}$ | insulators |
| Glass | $10.0 \times 10^{10}$ to $10.0 \times 10^{14}$ | insulators |
| Hard rubber | $1.00 \times 10^{13}$ | insulators |
| Wood (oven dry) | $1.00 \times 10^{14}$ to $1.00 \times 10^{16}$ | insulators |
| Air | $1.30 \times 10^{16}$ to $3.30 \times 10^{16}$ | insulators |

*Table 2* *The resistivity of some materials at a temperature of 20 °C*

As a semiconductor is heated, some of the vibrational energy is transferred to atomic electrons, some of which become free to move through the material. As the temperature increases, the number of mobile charge carriers increases, so the resistivity of a semiconductor tends to *decrease* as the temperature increases. The rate of change in resistivity with temperature is much greater than that for a metal (that is, it is a large negative number).

## Thermistors

Semiconductors are used to make electrical components called **thermistors** (Figures 9 and 10). These are resistors designed to change their resistance with temperature. Thermistors whose resistance drops as temperature increases (Figure 11) are known as NTC (negative temperature coefficient) thermistors.

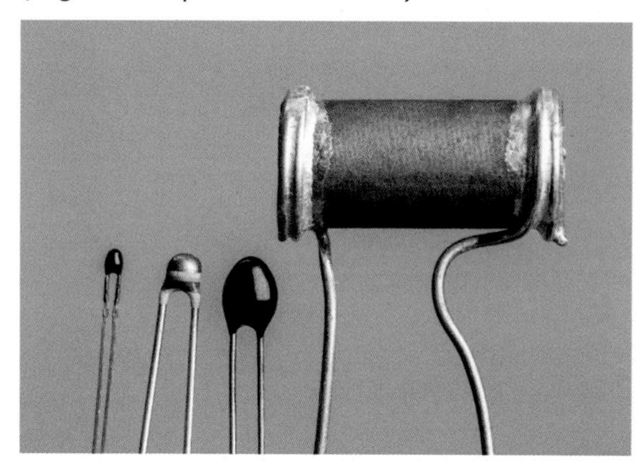

***Figure 9*** *Some different types of thermistors*

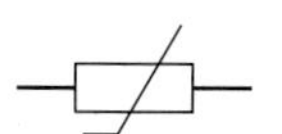

***Figure 10*** *The circuit symbol for a thermistor*

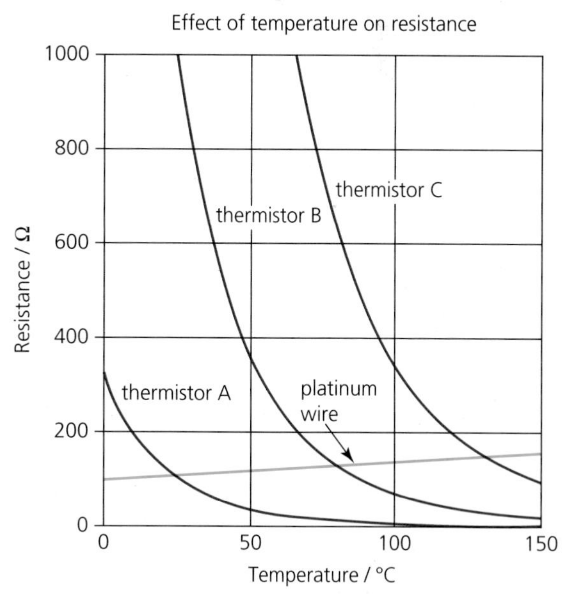


***Figure 11*** *The fall in resistance of different NTC thermistors with temperature, and the rise in resistance of platinum metal for comparison*

There are several important uses for NTC thermistors:

- as thermometers, in cars for example, to check that the engine coolant is not overheating
- as thermostats in domestic appliances, such as fridges, toasters, hairdryers and coffee makers, to sense and control the temperature
- in circuits to prevent a current surge when the circuit is first turned on, because NTC thermistors have a high resistance when they are cold, but as current flows through the thermistor it will heat up, its resistance will drop and more current will flow.

## Light-dependent resistors (LDR)

There are some semiconductors, including silicon and germanium, whose resistivity depends on the intensity of the light to which they are exposed. Electrons in the semiconductor absorb energy from light, and, as the light intensity increases, more electrons become mobile and the resistivity decreases. A circuit component designed so that its resistance changes with incident light intensity is called a **light-dependent resistor** (LDR). The circuit symbol for an LDR is shown in Figure 12. Light-dependent components are often made from semiconductors mixed with small amounts of another substance, such as gallium arsenide or cadmium sulfide.

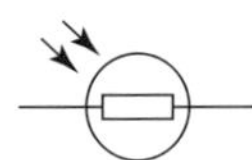

***Figure 12*** *The circuit symbol for an LDR. Sometimes the symbol is drawn without the circle.*

LDRs are used to automatically turn on lights, for example, street lighting, night lights and in offices, when it becomes dark. They are also used in motion-activated security sensors (Figure 13) and in alarm devices to detect when a light beam has been interrupted. A type of intruder alarm system, for example, uses laser beams that are directed at LDRs. If the beam is interrupted, the light falling on the LDR is reduced and its resistance increases, which triggers an alarm.

***Figure 13*** *The device beneath the security lamp contains a movement sensor and an LDR. When someone passes the sensor, a switch is triggered to turn on the lamp, but only if it is dark enough for the resistance of the LDR to be above a certain value.*

## QUESTIONS

7. An NTC thermistor of type A shown in Figure 11 is connected in series with a cell of emf 1.5 V and negligible internal resistance. What current will flow in the circuit when the temperature of the thermistor is
   a. 0 °C
   b. 25 °C?
8. Which of the thermistors shown on the graph in Figure 11 would make the best thermometer to replace a mercury-in-glass thermometer (range 0 to 100 °C)? What else would you want to know about the thermistor before you used it as a thermometer?
9. The resistance of an LDR in the dark is 1 MΩ. In bright light, the resistance drops to 5 kΩ. The LDR is connected in a circuit with a cell of emf 1.5 V and negligible internal resistance, and a sensitive ammeter. What current would you expect to measure on the ammeter:
   a. when the LDR is in the dark
   b. when a strong light is incident upon the LDR?

## KEY IDEAS

- The resistivity of semiconductors tends to decrease as the temperature rises, owing to the increase in number of mobile charge carriers.
- An NTC thermistor is a semiconductor device that has a high resistance at low temperatures. Its resistance decreases as its temperature increases.
- A light-dependent resistor (LDR) is a semiconductor device that has a high resistance in the dark. Its resistance increases as the light intensity falling on it increases.

## ASSIGNMENT 3: INVESTIGATING AN NTC THERMISTOR

**(MS 2.1, MS 3.1, MS 3.2, MS 3.12, PS 1.2, PS 2.1, PS 3.1, PS 3.2, PS 4.1)**

In order to find how the resistance of an NTC thermistor is affected by temperature, the thermistor is heated in a beaker of distilled water (Figure A1). An electric immersion heater (or Bunsen burner) may be used to heat the water. Stirring the water is important to make sure that the temperature is uniform throughout the beaker.

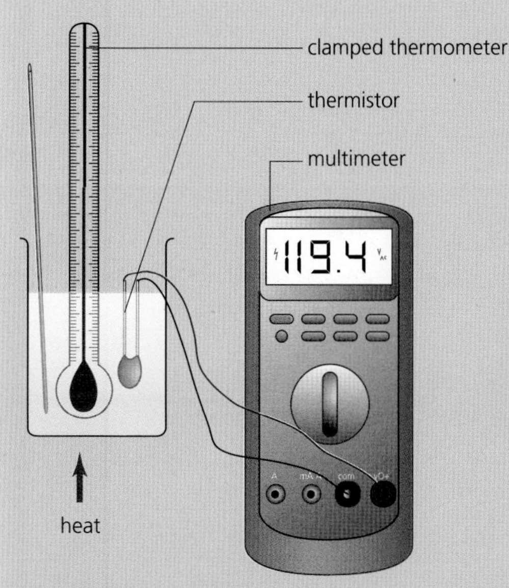


*Figure A1*

Readings of temperature and resistance may be taken as the water heats up, but it is easier to keep the temperature uniform as the water cools – since cooling is (usually) a slower process than heating.

A sample set of results is given in Table A1.

| Temperature / °C | Resistance / Ω |
|---|---|
| 100 | 12.3 |
| 90 | 15.5 |
| 80 | 19.8 |
| 70 | 25.5 |
| 60 | 33.5 |
| 50 | 44.7 |
| 40 | 60.7 |
| 30 | 84.2 |
| 20 | 119.4 |
| 10 | 173.6 |

*Table A1*

### Questions

**A1** Plot a graph of temperature ($x$-axis) against resistance ($y$-axis).

**A2** The resistance of the thermistor when it is in the air in the laboratory is 100 Ω. What is the temperature in the laboratory?

**A3** Discuss the errors that might arise in the experiment and describe the steps that could be taken to make sure that temperature and resistance measurements were as accurate as possible.

### Stretch and challenge

**A4** Theory suggests that the relationship between resistance, $R$, and absolute (kelvin) temperature, $T$, is an exponential formula:

$$R \propto e^{1/T}$$

How could you test whether or not this is true?

# 14.4 SUPERCONDUCTORS

When a metal is cooled, the vibrations of the positive ions are reduced, conduction electrons lose less energy in collisions with the ions and so the metal's electrical resistivity decreases (section 14.2). In the early 20th century, physicists were speculating about what would happen to electrical resistance at very low temperatures, specifically as the temperature approached **absolute zero**, that is, zero on the 'absolute temperature' scale or **kelvin scale**, 0 K, equivalent to −273.15 °C. Some scientists, including Lord Kelvin, argued that the electrons would become less mobile and perhaps stop moving altogether. This would mean that all materials would be perfect insulators at 0 K. Others suggested that electrical resistance might continue to decrease as the temperature dropped, and perhaps the resistance might disappear altogether. Kamerlingh Onnes, professor of a cryogenics laboratory in Leiden, held the latter view. In 1908 Onnes had won the race to liquefy helium gas (see Figure 14). In 1911 he was using liquid helium to investigate the electrical conductivity of mercury at low temperatures, down to as low as 4 K (−269 °C).

**Figure 14** *At the turn of the 20th century, low-temperature research and the race to liquefy gases must have seemed highly theoretical and rather remote from everyday concerns. But the practical 'spin-offs' have been impressive, ranging from superconductivity to the vacuum (thermos) flask. This picture shows James Dewar, who invented the vacuum flask to help him liquefy hydrogen. Dewar is celebrated in this humorous form of verse written by Edmund Clerihew Bentley (and now named a 'clerihew' after him):* "Sir James Dewar / Is a better man than you are / None of you asses / Can liquefy gases."

Onnes made a remarkable discovery – when the mercury was cooled to just 4.2 degrees above absolute zero, its electrical resistance abruptly and totally disappeared. Onnes realised that this was not just a continuation of the gradual reduction of resistance on cooling. It was a *change of phase*, like freezing or boiling, that occurred at a specific temperature, known as the **critical temperature**, $T_c$. Below its critical temperature, mercury becomes a **superconductor**, a material that has absolutely no electrical resistance whatsoever (Figure 15). An electric current, once started, in a circuit made of superconducting material would persist indefinitely, without the need of a power supply.

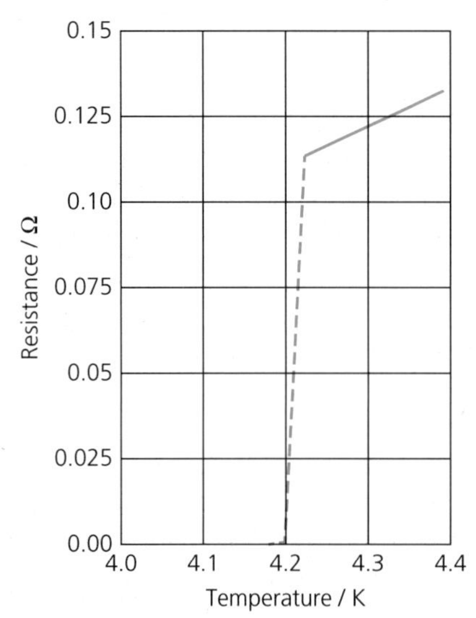


**Figure 15** *A replotting of Onnes' results showing the superconductive transition of mercury*

After 1911, other metals, for example lead and tin, were shown to have superconducting properties, though the transitions happened at different critical temperatures. It later became apparent that it was not only metals that could show superconducting properties. In 1986, superconductivity was discovered in the ceramic material yttrium barium copper oxide, known as YBCO (Figure 16). This excited tremendous interest, as its critical temperature was around 100 k. Although still quite cold by everyday standards, this is well above the temperature at which nitrogen liquefies. Liquid nitrogen is much cheaper and easier to work with than liquid helium. This has raised the possibility of finding a material with a critical temperature above 'room' temperature, one that would be superconducting without cooling. This would bring enormous practical benefits, for example cables with no power losses and smaller, faster computers that did not overheat.

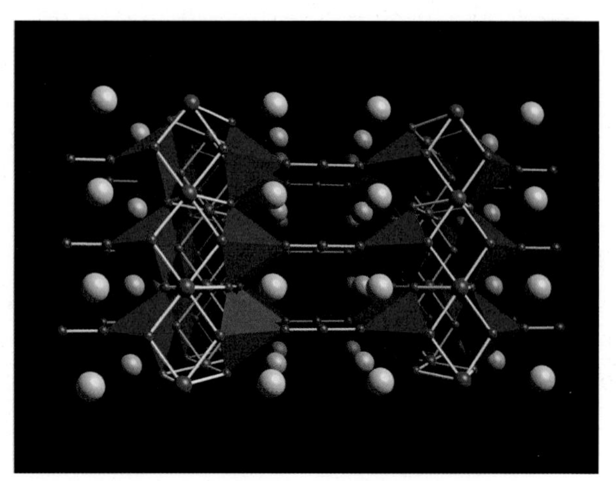

**Figure 16** *The structure of a YBCO crystal that is superconducting below 125 K. Room-temperature superconductivity was achieved in 2014 in a modified YBCO crystal. Laser pulses were used to alter the structure. Unfortunately, the superconductivity lasted only for a few nanoseconds.*

The mechanism of conduction in a superconductor is now understood to be completely different from that in an ordinary conductor. Electrons move around in pairs, with energy being passed from one to the other and back again, but the electrons do not interact at all with the positive ions in the crystal lattice. Electrons are not scattered by the lattice and do not transfer energy to it, so there is no resistance to their motion.

## Applications of superconductors

Superconductors have the property of excluding external magnetic fields. This behaviour, known as the Meissner effect, was discovered in 1933. If a superconducting material is placed in a magnetic field, and then cooled below its critical temperature, it will eject the magnetic field from itself (except in the outer few nanometres). The superconductor in effect 'repels' the magnetic field and so can lead to magnetic levitation (Figure 17).

***Figure 17*** *The Meissner effect*

In another form of magnetic levitation, powerful electromagnets with superconducting windings create a magnetic field strong enough to suspend a train above the track, reducing friction enormously (Figure 18).

***Figure 18*** *Maglev (magnetic levitation) trains have proved controversial. The Shanghai Transrapid shown here is the fastest commercial train in operation and has a top speed of 430 km $h^{-1}$. But only two such trains have been built anywhere in the world.*

Powerful superconducting electromagnets are also used to steer protons around the huge circular tunnel at the LHC (Large Hadron Collider) at CERN (see Figure 4 in Chapter 3). But they are also used much closer to home – they produce the strong magnetic fields inside MRI (magnetic resonance imaging) scanners (Figure 19). Large concentric coils of superconducting material, niobium titanite, are kept cool using liquid helium. A tank containing around 1000 litres (1 $m^3$) is needed, and it has to be kept below 4 K. The coils and helium, cooled below 4 K, are thermally sealed at the time of construction, and should stay at that temperature for their operating lifetime of about 10 years. It would take very little energy (a few microjoules) to raise the temperature of the helium above the critical temperature of the coils. The coils would then lose their superconductivity, heat up rapidly and very quickly boil the helium. There are safety measures in place to prevent this happening.

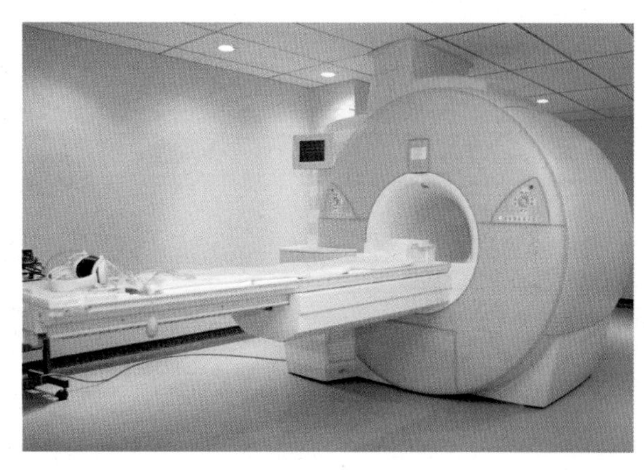

***Figure 19*** *MRI scanners require strong magnetic fields that can only be generated efficiently by superconducting electromagnets.*

Electric motors and generators also rely on coils of wire to carry current and generate magnetic fields. Energy is transferred as heat in these coils due to their electrical resistance. Coils made from superconducting material, with no resistance and so no heating effect, would directly save energy. Superconducting materials are capable of carrying high currents through a small area, so the coils would be thinner and much lighter, which would also improve the efficiency of a motor or generator.

The world's first commercial superconducting power transmission cable came into use in 2014 in Essen, Germany. The cable is cooled by liquid nitrogen and carries a current five times larger than that in the traditional copper cables it replaced. The energy losses are minimal. Around 7.5% of the UK's electrical energy is currently wasted as heat in the wires and transformers of the distribution system. If the transmission cables could one day be replaced by superconducting cables, the energy savings would be huge. There would be less need to transmit power at very high voltages and so fewer transformers would be required.

## QUESTIONS

10. Explain why there is enormous interest in finding a superconductor with a critical temperature above about 290 K.
11. Why is it an advantage to use superconducting coils in the electromagnets of an MRI scanner?

## KEY IDEAS

- Superconducting materials have no electrical resistivity when cooled below their critical temperature.
- The critical temperature is different for different superconducting materials.
- Superconducting cables can carry large currents with no energy losses. They are used in the windings of electromagnets, motors and generators.

# 14.5 THE POTENTIAL DIVIDER

### Getting the right potential difference

Power supplies, such as solar cells or batteries, tend to generate a fixed potential difference. This will not necessarily be the right size for our application, so we need a way of providing a variable potential difference. We can use a **potential divider** to select an appropriate fraction of the power supply's terminal potential difference. The potential divider is, at its simplest, just two resistors connected in series across the power supply (Figure 20).

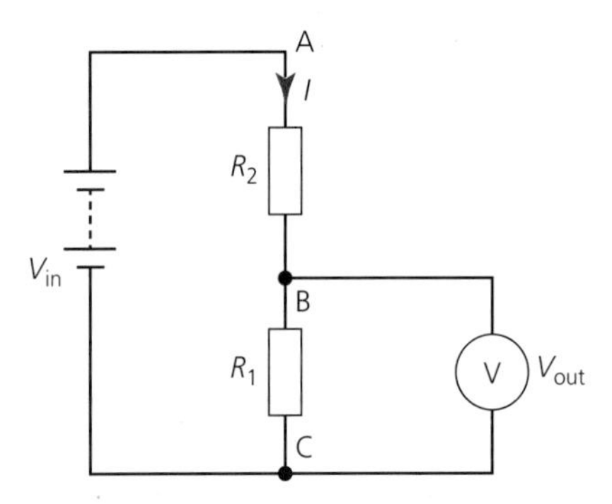


***Figure 20*** *A potential divider circuit*

If the voltmeter in Figure 20 has a very high resistance, we can assume that no current flows through it. So then the same current, $I$, flows through both resistors:

$$I = \frac{V}{R} = \frac{V_{in}}{R_1 + R_2}$$

In Figure 20, $V_{out}$ is the pd across resistor $R_1$:

$$V_{out} = IR_1$$

Substituting for $I$:

$$V_{out} = \frac{V_{in}}{R_1 + R_2} \times R_1$$
$$= V_{in} \times \frac{R_1}{R_1 + R_2}$$

The voltage is divided in the same ratio as the resistors:

$$\frac{V_{out}}{V_{in}} = \frac{\text{output resistance}}{\text{total resistance}}$$

### Worked example 1

Suppose we want to run a 6 V device from a 12 V supply (Figure 21).

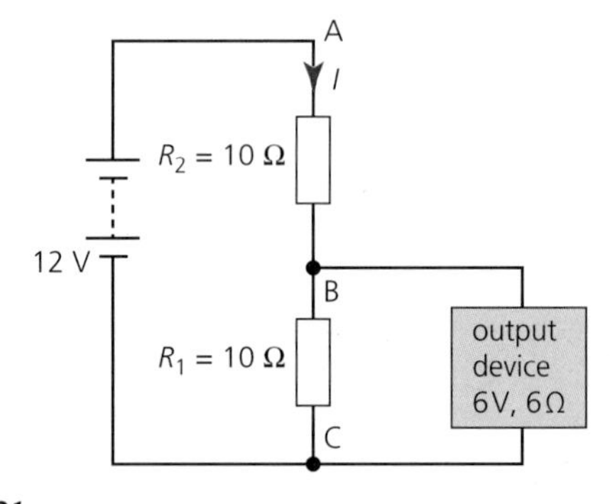


***Figure 21***

In this case, the resistance $R_1$ is no longer just the 10 Ω resistor of the potential divider. It is connected in parallel with the output device, which has a resistance of 6 Ω. The resistance of this combination is

$$\frac{1}{R_1} = \frac{1}{10} + \frac{1}{6} = \frac{(6 + 10)}{60} = \frac{16}{60} = \frac{4}{15}$$

so $$R_1 = \frac{15}{4} = 3.75\ \Omega$$

The potential divider is now effectively made up of two resistors, $R_2 = 10\ \Omega$ and $R_1 = 3.75\ \Omega$. The output potential difference (across the device) will be

$$V_{out} = V_{in} \times \frac{R_1}{(R_1 + R_2)}$$

$$= 12 \times \frac{3.75}{(3.75 + 10)} = 3.3\ \text{V}$$

This is not sufficient to operate the device properly.

We need to adjust $R_1$ so that the combination of it and the output device has the same resistance as $R_2$. Or we could make $R_2$ a variable resistor, and turn the resistance down to $3.75\ \Omega$, when it will equal the resistance of the combination of $R_1$ and the output device. The use of a variable resistor as $R_2$ in a potential divider is a common arrangement in sensor circuits, which will be considered in the next section.

## QUESTIONS

**12.** Write down the output voltage of the potential dividers in Figure 22.

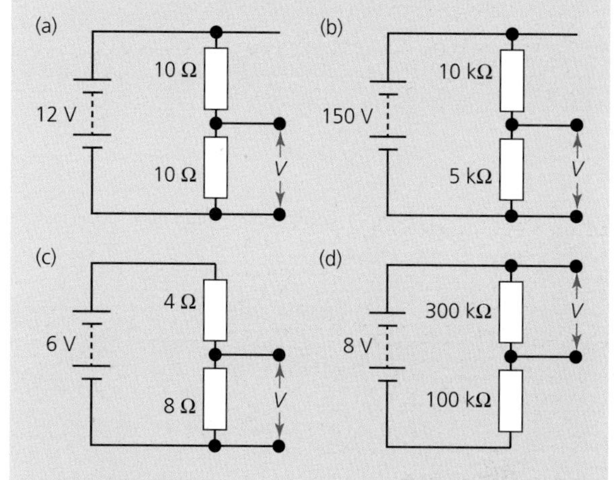


*Figure 22*

**13.** Suppose that you plan to apply the potential from the circuits in Figure 22 to a device with a resistance of 10 Ω. How would that affect your answers to question 12?

## ASSIGNMENT 4: ANALYSING THE USE OF A POTENTIAL DIVIDER

**(MS 3.1, MS 3.2, PS 1.1, PS 1.2, PS 2.1, PS 2.2, PS 3.1, PS 3.2, PS 4.1)**

In school physics laboratories, students often use large rheostats (see the Required Practical) to vary the potential difference across a component. They can be connected in two different ways.

- As a variable resistor (Figures A1a and A2a). As the sliding contact is moved, the effective resistance of the rheostat is changed. As the resistance changes, the current through the circuit is changed, which in turn alters the potential difference across the component (in this case a buzzer).
- As a potential divider (Figures A1b and A2b). As the sliding contact is moved, it changes the ratio of the resistances $R_1$ and $R_2$, which varies the potential difference across the buzzer.

We want to find out which method gives the biggest variation of the potential difference across the buzzer.

(a)

(b)

***Figure A1*** *Two ways of connecting a rheostat*

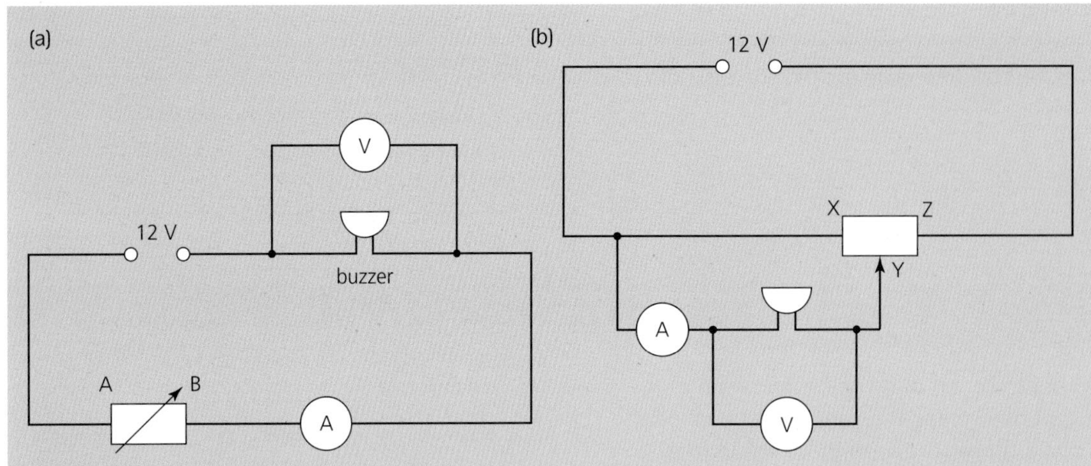


***Figure A2*** *The two circuit arrangements*

Suppose that the cell has an emf of 12 V and an internal resistance of 1 Ω and that the resistance of the buzzer is 100 Ω. The rheostat has a resistance that is variable between 0.5 Ω and 16 Ω.

### Questions

**A1** For each circuit arrangement (a) and (b) in Figure A2, calculate the maximum and minimum potential difference across the buzzer. Which is the better arrangement?

### Stretch and challenge

**A2** Use a spreadsheet to plot graphs showing how the potential difference across the buzzer varies as the resistance of the rheostat is changed from 0.5 to 16 Ω, in arrangements (a) and (b).

**A3** Use your spreadsheet to investigate how the two arrangements would work if the resistance of the buzzer was:

**1.** 1 Ω

**2.** 1000 Ω

It would be helpful to plot graphs of resistance against the potential difference across the buzzer.

**A4** Write a brief paragraph to summarise your findings.

## Sensors in potential divider circuits

Electrical methods of measuring are increasingly important. One of the main advantages is that the measuring instrument can be 'remote', so, for example, you do not need to be able to see the central heating boiler to know if the water is too hot or see the fuel tank to know if the fuel in your car has fallen to a low level. Such measurements make use of **sensors**. Electrical readings can be logged automatically, or used to trigger a light, a heater or an alarm.

Many sensors rely on a change of resistance to make their measurement. These sensors are often used in a potential divider circuit. In order to measure temperature, an NTC thermistor is used as one of the resistors in the potential divider. As the temperature increases, the resistance of the thermistor falls (Figure 23), so that there is a smaller fraction of the potential difference across it. The temperature at which $V_{out}$ switches a device connected to it can be set by adjusting the value of the variable resistor.

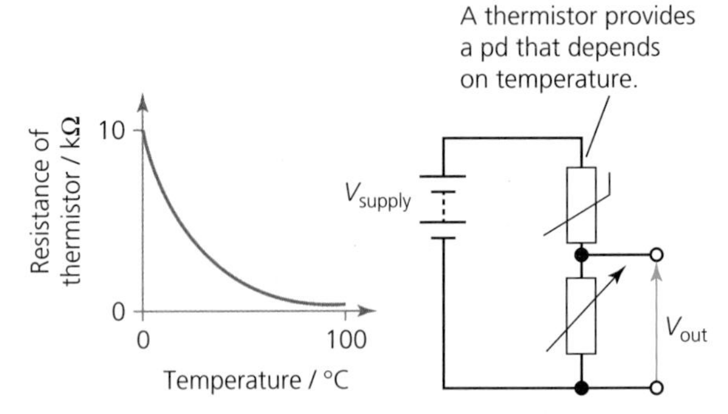


***Figure 23*** *The potential difference of the supply $V_{supply}$ is shared across the thermistor and the variable resistor. As the temperature increases, $V_{out}$ increases.*

Despite the curved resistance–temperature graph, the thermistor can be quite a useful 'linear' sensor. Over a limited temperature range – when the resistance of the thermistor is around the same size as the resistor – the potential divider's output rises roughly linearly with increasing temperature. A potential divider circuit such as that in Figure 23 could be used to turn on heaters if the temperature drops too low (Figure 24). This 'turn-on' temperature is adjustable using the variable resistor. A similar circuit, using an LDR rather than a thermistor, can control lighting levels.

*Figure 24* A thermistor in a potential divider circuit can be used to control the temperature of a greenhouse.

### Worked example 2

An electrically operated garage door is convenient but potentially dangerous. It needs to stop descending if someone is in the way. Design a circuit using an LDR in a potential divider circuit that gives a high voltage when a light beam is blocked. The resistance of the LDR is about 50 kΩ in the dark.

We need to use the LDR in series with a resistor of comparable resistance to that of the LDR in the dark (see Figure 25); or this could be a variable resistor of suitable range. A light beam is shone onto the LDR, which keeps its resistance low. Most of the potential difference is then across the series resistor, and the output voltage is low, that is, close to 0 V. When the beam is interrupted, the resistance of the LDR goes high. Now most of the potential is across the LDR and the output voltage is high – closer to the supply voltage, 9 V in Figure 25. This 'high' voltage can be used to operate a switch that turns the motor off.

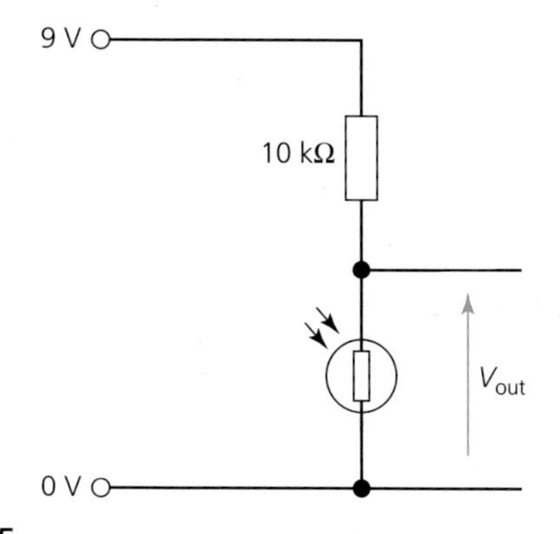


*Figure 25*

## QUESTIONS

**14.** A student wants to design a biscuit tin with an alarm that sounds when the tin is opened. Explain how she could achieve this using an appropriate sensor and a potential divider.

**15.** Design a circuit that gives an output voltage of 3 V when the temperature of the room is 25 °C. Refer to Figure 11 for relevant data.

### Stretch and challenge

**16.** The thermistor in the circuit in Figure 26 has a resistance–temperature relation as shown in Figure 27. The circuit is to be used as a sensor to measure the temperature of hot water in a domestic central heating system.

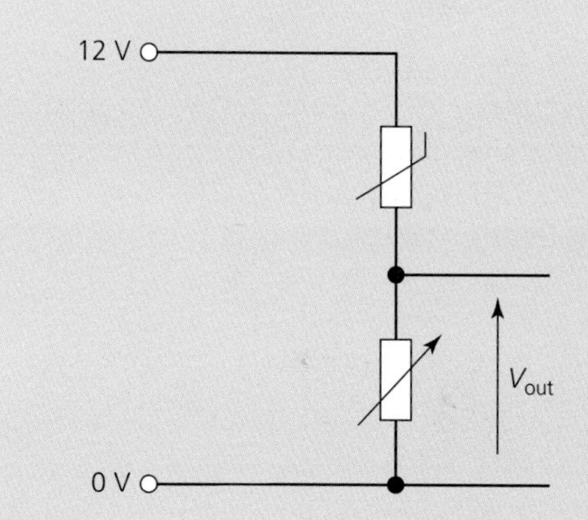


*Figure 26*

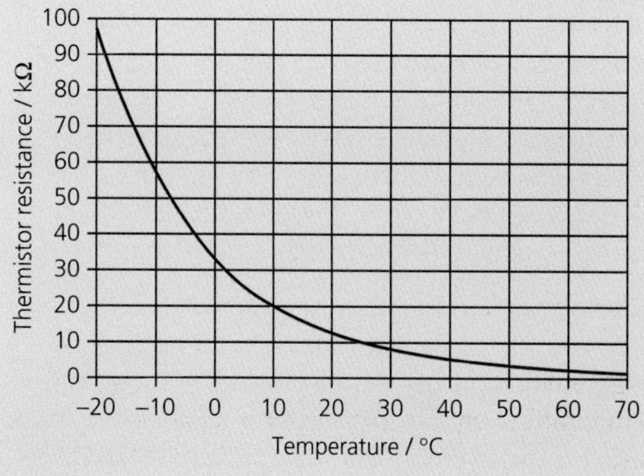


*Figure 27*

a. The thermistor is placed in a container with melting ice and water. The variable resistor is adjusted so that the output voltage is 1 V. What will be the resistance of the variable resistor?

b. What is the current flowing through the potential divider at that point?

c. The thermistor is then placed in a beaker of hot water. After a few minutes the voltage $V_{out}$ is read as 5 V. What is the temperature of the water?

d. What difference would it make to the readings if the variable resistor was set at 30 kΩ? Would this be an improvement? (Consider the sensitivity of the sensor: on average, how many volts per °C in each case?)

## KEY IDEAS

- A potential divider is an arrangement of resistors designed to share the potential difference of a power supply in a given ratio.
- It can be used with a power source of fixed emf to provide a variable potential output.
- The output voltage of a potential divider is

$$V_{out} = V_{in} \times \frac{R_1}{(R_1 + R_2)}$$

  where $V_{out}$ is the potential difference across $R_1$ (see Figure 20).
- A sensor, such as an LDR or thermistor, can be used in a potential divider circuit (as $R_1$ or $R_2$) to give an output potential difference that varies with light or temperature.
- A variable resistor as the series resistor in such a sensor circuit allows choice of the output pd obtained at a certain light intensity or certain temperature.

## PRACTICE QUESTIONS

1. The circuit in Figure Q1 shows a battery of electromotive force (emf) 12 V and internal resistance 1.5 Ω connected to a 2.0 Ω resistor in parallel with an unknown resistor, *R*. The battery supplies a current of 4.2 A.

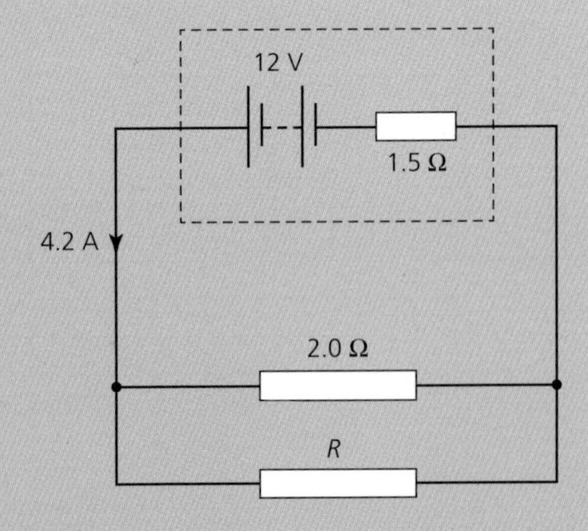


*Figure Q1*

a. i. Show that the potential difference (pd) across the internal resistance is 6.3 V.

ii. Calculate the pd across the 2.0 Ω resistor.

iii. Calculate the current in the 2.0 Ω resistor.

iv. Determine the current in *R*.

v. Calculate the resistance of *R*.

vi. Calculate the total resistance of the circuit.

b. The battery converts chemical energy into electrical energy that is then dissipated in the internal resistance and the two external resistors.

i. Using appropriate data values that you have calculated, copy and complete Table Q1 by calculating the rate of energy dissipation in each resistor.

| Resistor | Rate of energy dissipation / W |
| --- | --- |
| Internal resistance | |
| 2.0 Ω | |
| $R$ | |

*Table Q1*

ii. Hence show that energy is conserved in the circuit.

*AQA Unit 1 June 2013 Q6*

2. Figure Q2 shows a 6.0 V battery of negligible internal resistance connected in series to a light-dependent resistor (LDR), a variable resistor and a fixed resistor, $R$.

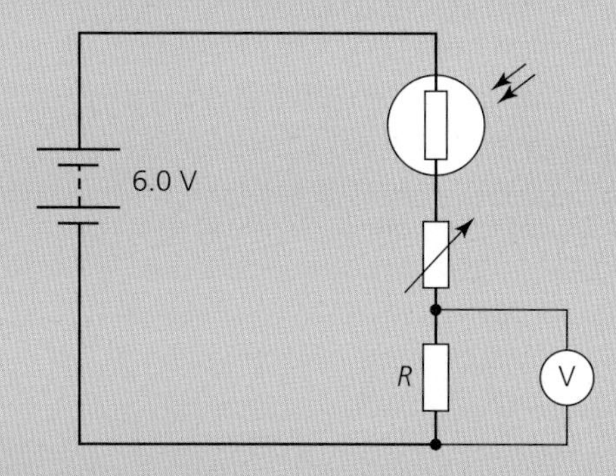


*Figure Q2*

a. For a particular light intensity, the resistance of the LDR is 50 kΩ. The resistance of $R$ is 5.0 kΩ and the variable resistor is set to a value of 35 kΩ.

i. Calculate the current in the circuit.

ii. Calculate the reading on the voltmeter.

b. State and explain what happens to the reading on the voltmeter if the intensity of the light incident on the LDR increases.

c. For a certain application, at a particular light intensity, the pd across $R$ needs to be 0.75 V. The resistance of the LDR at this intensity is 5.0 kΩ. Calculate the required resistance of the variable resistor in this situation.

*AQA Unit 1 June 2013 Q7*

3. Figure Q3 shows a 12 V battery of negligible internal resistance connected to a combination of three resistors and a thermistor.

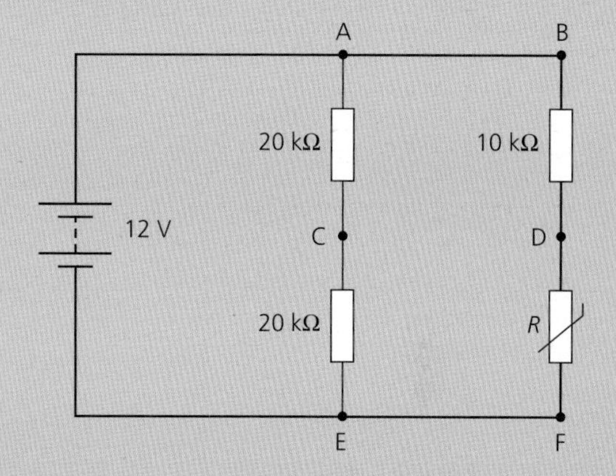


*Figure Q3*

a. When the resistance of the thermistor is 5.0 kΩ

i. calculate the total resistance of the circuit

ii. calculate the current in the battery.

b. A high-resistance voltmeter is used to measure the potential difference (pd) between points A–C, D–F and C–D in turn. State the reading of the voltmeter at each of the three positions.

c. The thermistor is heated so that its resistance decreases. State and explain the effect this has on the voltmeter reading in the following positions.

i. A–C

ii. D–F

*AQA Unit 1 January 2013 Q6*

4. Figure Q4 shows three graphs of variation of resistance with temperature.

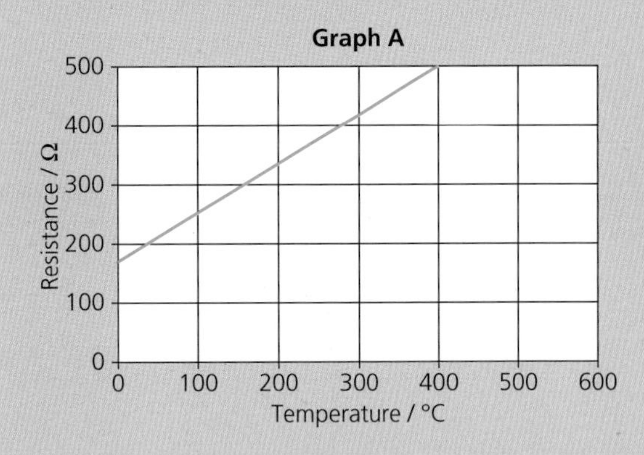


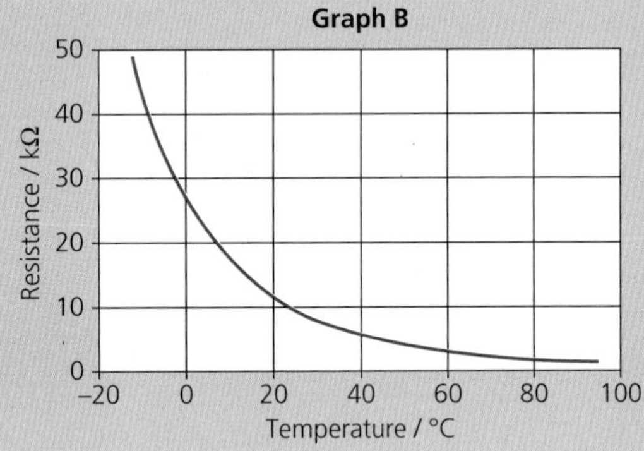


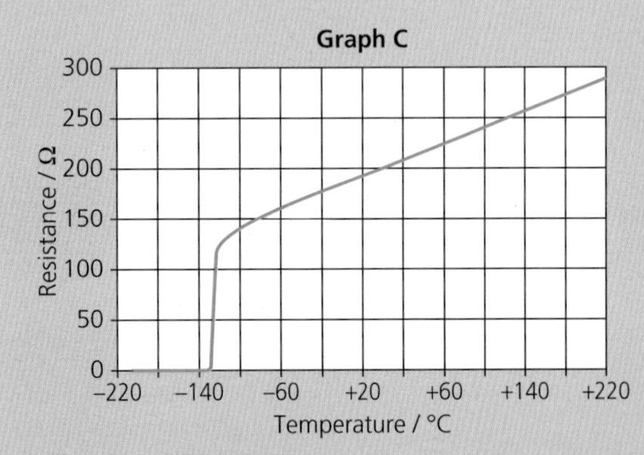


*Figure Q4*

Which graph shows the behaviour of

a. an NTC semiconductor thermistor

b. a superconductor

c. a length of platinum wire?

5. Which of the following statements A to D about the emf of a power supply is **not** true?

**A** The emf is less than the terminal potential difference because of the voltage dropped across the internal resistor.

**B** The emf is the energy transferred to each charge by the power supply.

**C** A larger emf will always lead to a larger potential difference at the terminals.

**D** The emf is equal to the terminal potential when no current flows through the power supply.

6. A potential divider is constructed by connecting two 1 kΩ resistors across a 12 V cell of negligible internal resistance. A 2 kΩ resistor is connected in parallel with one of the 1 kΩ resistors. The potential difference across the 2 kΩ resistor is:

**A** 6 V

**B** 12 V

**C** 4.8 V

**D** 7.4 V

7. Superconducting magnets are used in magnetic resonance imaging (MRI) to produce a very strong magnetic field. Some possible advantages and disadvantages of using superconducting coils are listed in Table Q2. Select the row that correctly shows the main advantage and main disadvantage.

| | Advantage | Disadvantage |
|---|---|---|
| A | Produces a strong magnetic field with very little current | Has to be cooled to absolute zero |
| B | No energy is lost in overcoming electrical resistance | Has to be cooled using liquid helium |
| C | The coils exclude any external electromagnetic interference | Has to be cooled to absolute zero |
| D | Superconducting coils produce a very uniform field | Has to be cooled using liquid helium |

*Table Q2*

8. Which of the circuits in Figure Q5 would give a high voltage output when the temperature was low?

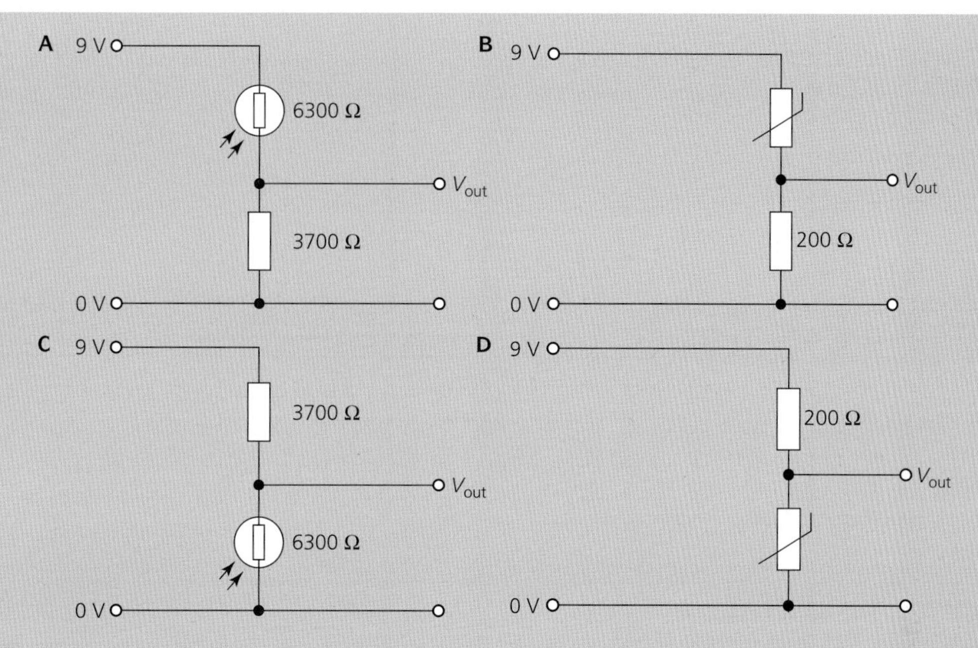


*Figure Q5*

9. Read this extract, taken from a presentation by the Superconductivity Council, and answer the questions that follow.

**Superconductors offer the promise of important major advances in efficiency and performance in electric power generation, transmission and storage; medical instrumentation; wireless communications; computing; and transportation, that will result in societal advances that are cost effective and environmentally friendly.**

**Superconductors differ fundamentally in quantum physics behaviour from conventional materials in the manner by which electrons, or electric currents, move through the material. It is these differences that give rise to the unique properties and performance benefits that differentiate superconductors from all other known conductors.**

1. **Zero resistance to direct current**
2. **Extremely high current carrying density**
3. **Extremely low resistance at high frequencies**
4. **Extremely low signal dispersion**
5. **High sensitivity to magnetic field**
6. **Exclusion of externally applied magnetic field**
7. **Close to speed of light signal transmission**

From *Present and Future Applications: Superconductivity*,
CSC Council on Superconductivity © 2009 CCAS

a. In 1911 the Dutch physicist Kamerlingh Onnes was investigating the conductivity of mercury at low temperatures when he discovered superconductivity. Most metals conduct better at lower temperatures. Explain why this is, and explain what was remarkable about Onnes's results for mercury.

b. Teams of scientists around the world are searching for a material that has a critical temperature above 273 K (0 °C). Explain why this would be such a great prize.

c. Explain why superconducting coils are used in MRI scanners.

d. Electric motors rated at 750 kW and above consume 25% of all electricity generated in the USA. The use of high-temperature superconductors could substantially reduce energy losses. Explain how the use of superconducting windings can make the motor more efficient.

### Stretch and challenge

e. The USA generated $4.2 \times 10^{12}$ kWh of electrical energy in 2012. If the use of superconductors can improve the efficiency of a motor by 2%, how much energy might be saved in a year in the USA?

# ANSWERS TO IN-TEXT QUESTIONS

## 1 MEASURING THE UNIVERSE

1. $27 \times 10^6$ Hz (hertz) or 27 MHz
2. 149 600 000 000 m = $1.496 \times 10^{11}$ m
3. Time equals distance divided by speed. So for a classroom about 30 m long:
   time = $30/(3 \times 10^8) = 10^{-7}$ s = 100 ns or 0.1 μs
4. To avoid any changes in mass due to adsorption of gas molecules, finger marks, and so on.
5. This shape has the smallest surface area for its volume, except for a sphere ... and we would not want it to roll away! The metal alloy is very non-reactive, and does not oxidise.
6. Volume = $\pi r^2 h = 4.72 \times 10^{-5}$ m$^3$
   Mass = 1 kg
   So density = $1/(4.72 \times 10^{-5}) = 2.119 \times 10^4$ kg m$^{-3}$
7. Advantage: convenient, easy to use, familiar to all.

   Disadvantage: arm length is variable, and you might have to buy a length of carpet from a man with short arms!
8. The speed of light is now defined as this value, so this is one value that has no uncertainty.
9. 1 Pa = 1 kg m$^{-1}$ s$^{-2}$
10. Work = force × distance, so 1 J = 1 kg m$^2$ s$^{-2}$
11. **a.** No. There is no reason why the readings should fluctuate.

    **b.** Now it is worth taking lots of readings, or a 'big' reading of lots of ball bearings, to find an average.
12. The wire could vary in diameter along its length, and it might not be perfectly circular.
13. **a.** You could be weighing an amount of sugar to the nearest 0.01 g, so the readings are precise, but you forget to subtract the mass of the container. (Or the balance did not read 0.0 g before the start of your readings – a zero error.)

    **b.** You could measure the textbook, for example, with a ruler marked in cm. Your answer would be accurate, but not precise.
14. Pulse about 80 per minute, so 0.75 s apart.
    Resolution of time keeper = ±0.75 s

    The time of one oscillation = 2/0.75 = 2.67 pulses; so 3 pulses ± 1 pulse (2.25 ± 0.75 s), which is about 33% percentage uncertainty.

    The time of 10 oscillations = 20/0.75 = 27 pulses ± 1 pulse, giving a time of one oscillation as 2.7 pulses ± 0.1 pulse, which is a percentage uncertainty of 0.1/2.7 = 3.7%.
15. Find the mass of 100 blackcurrants and divide by 100.
16. It makes no sense to give a result more precisely than the original measurement. You should write 121.3 ± 0.1 g, unless the scatter of results was larger.
17. **a.** Width = 19.2 ± 0.1 cm

    **b.** The book is much wider than it is thick. There is a much greater percentage uncertainty.

    **c.** Measure the thickness (not including the covers) and divide by the number of leaves (half the number of pages). It would be the same percentage uncertainty as the thickness of the book measurement.

**18.** The measurements of length, breadth, and so on contribute 15% to the uncertainty; the measurement of mass contributes only 1%. It would be better to improve the resolution of your length measurement.

**19.** Measure the length of the cube three times at different parts of the cube, and find the mean. Repeat for the other dimensions.

**20.** Suppose you use 100 ml of ethanol in a measuring cylinder. Resolution is at best 1 ml, probably 2 ml when you take into account the curved surface of the water (meniscus). Weigh on a top-pan balance giving 140 ± 1.0 g, and with the empty container 60 ± 1.0 g, so the mass of the ethanol is 140 – 60 = 80 ± 2 g.

Density = $m/V$ = 80/100 = 0.8 g ml$^{-1}$ or 0.8 g cm$^{-3}$ or 800 kg m$^{-3}$. Uncertainty in volume = 2%. Uncertainty in mass = 2/80 = 2.5%. So uncertainty in density = 4.5%

Ethanol density = 800 ± 36 kg m$^{-3}$ (or round to 800 ± 40 kg m$^{-3}$)

**21.** There is a systematic error that affects all the readings from 60 drops onwards, perhaps due to a zero error on the balance, or to an additional unnoticed mass.

# 2 INSIDE THE ATOM

**1.** **a.** This showed that the particles in cathode rays were present in all metals.

**b.** This showed that the 'electrons' were the same in all metals.

**2.** The charges in the atoms in his body were repelling the charges in the wall. In one sense, the General does not even touch the wall.

**3.** A football has a radius of about 11 cm, call it 0.1 m. Electron orbits have a radius that is 100 000 × the nuclear radius, which would give a radius of 10 km: the orbit would take them, for example, through Hyde Park in Central London.

**4.** $\dfrac{+1.602 \times 10^{-19}\ \text{C}}{1.6726 \times 10^{-27}\ \text{kg}} = 9.578 \times 10^{7}\ \text{C kg}^{-1}$

**5.** 92 protons, 92 electrons and 235 – 92 = 143 neutrons

**6.** 146 neutrons in the nucleus. It is slightly more massive.

**7.** hydrogen +9.54 × 10$^{7}$ C kg$^{-1}$; helium +4.80 × 10$^{7}$ C kg$^{-1}$; lithium +4.11 × 10$^{7}$ C kg$^{-1}$; beryllium +4.27 × 10$^{7}$ C kg$^{-1}$; boron +4.36 × 10$^{7}$ C kg$^{-1}$

**8.** A larger nucleus means that some protons will be outside the range of the attractive nuclear force, but still repel each other due to electric force. More neutrons are needed to hold the nucleus together.

**9.** Short range (< 3 fm) attraction between all nucleons; very short range (< 0.5 fm) repulsion between all nucleons.

**10.** The electric and magnetic fields were not strong enough to deflect the alpha rays by very much due to its relatively high momentum.

**11.** Charge = +2$e$; mass = 16u assuming the mass of the electrons are negligible; so specific charge = 2 × 1.602 × 10$^{-19}$ / 16 × 1.66 × 10$^{-27}$ = 1.2 × 10$^{7}$ C kg$^{-1}$

**12.** 5 000 000 ÷ 10 = 500 000 ion pairs

**13.** Outside the body an alpha particle would not penetrate the outer layers of skin (although very close contact would cause burns).

**14.** $^{241}_{95}\text{Am} \rightarrow {}^{237}_{93}\text{Np} + {}^{4}_{2}\alpha$ (or $^{4}_{2}\text{He}$)

**15.** $^{14}_{6}\text{C} \rightarrow {}^{14}_{7}\text{N} + {}^{0}_{-1}\beta$ (or $^{0}_{-1}\text{e}$)

**16.** Alpha particles are heavily ionising, have a relatively large momentum, and are not significantly deflected by collisions with atomic electrons; therefore they make straight, thick tracks. Beta particles have far fewer collisions in the same distance and are less massive, and are easily deflected by collisions with atomic electrons, especially when they have slowed down; therefore they make thinner tracks, often with a change of direction, especially at the end of the track.

**17.** $^{23}_{12}\text{Mg} \rightarrow {}^{23}_{11}\text{Na} + {}^{0}_{+1}\beta$ (or $^{0}_{+1}\text{e}$)

A proton in the parent nucleus has changed into a neutron in the daughter, so that the proton number has decreased by one, but the nucleon number has remained the same.

**18.** Proton: strong nuclear, weak nuclear, electromagnetic, gravity

Neutron: strong nuclear, weak nuclear, gravity

**19.** Protons repel each other by the electromagnetic force. They attract each other by the strong nuclear force. Electromagnetic repulsion has a much longer range than the strong nuclear attraction, which drops to zero when nucleons are more than 3 fm apart. Two protons, one on either side of a large nucleus, will still repel each other through the electromagnetic force, but are too far apart for the strong interaction to hold them together. So, the nucleus becomes unstable.

**20.** Probably, yes. The weak interaction is very short range ($10^{-18}$ m) but its strength is $10^{-5}$ of the strong interaction, while gravity has a strength of only $10^{-39}$ that of the strong interaction. Also, the weak interaction is the only interaction that can change the nature of a particle (for example, $n \rightarrow p$ in beta decay).

**21. a.** Gravity plays very little part in atoms and nuclei. The gravitational force between the electron and the proton is around $10^{-37}$ times the electromagnetic force. Gravity would have to become hugely stronger to have any major impact.

Stronger gravity on the scale of the Universe could mean smaller stars. A denser star would mean more fusion reactions happening more quickly. Stars would use up their supply of hydrogen more quickly and have a shorter lifetime, perhaps too short for life to evolve on any suitable nearby planet. At the end of their lives, stars might form neutron stars and black holes more readily.

Stars would attract one another more strongly, so galaxies could be smaller and more dense. Galaxies in their turn would attract one another more strongly, so that we might be living in a contracting Universe, or it may have contracted too fast to allow us to be here at all.

**b.** Nuclei would be held together less strongly; larger nuclei would not be stable against fission. The periodic table would have fewer elements. There may be just hydrogen if the strong force is not large enough to support fusion. There would be no stars in that case.

**c.** If the weak interaction was much weaker, radioactive decay would be less likely, the fusion of two protons (vital in stars) might not occur and there would be no stars. A longer range and stronger weak interaction might lead to more decay, with the Universe made of hydrogen (protons).

Note: It has been argued that the relative 'strength' of the fundamental forces is fine-tuned; any small variations would result in a very different Universe where life could not arise. However, it has been possible to create a theoretical model of a Universe that has no weak force at all, and still has stars, galaxies, etc. (Naturally, other constants have to be altered to make this work.)

# 3 ANTIMATTER AND NEUTRINOS

**1.** $100 \times 10^3 \times 1.6 \times 10^{-19} = 1.6 \times 10^{-14}$ J

**2. a.** Energy released = $200 \times 10^6 \times 1.6 \times 10^{-19}$ J = $3.2 \times 10^{-11}$ J

11 600 kWh = $11\,600 \times 1000 \times 60 \times 60 =$ $4.176 \times 10^{10}$ J needed

$(4.176 \times 10^{10}$ J$)/(3.2 \times 10^{-11}$ J$) =$ $1.3 \times 10^{21}$ reactions

**b.** The mass of uranium used for each reaction is approximately $235 \times 1.66 \times 10^{-27}$ kg but we need $1.3 \times 10^{21}$ of these, so that is 0.0005 kg, or half a gram.

**c.** This is an underestimate because we have assumed that the energy transfers are all 100% efficient. But nuclear power stations are at best 35% efficient and there are further losses in generation and transmission before it gets to the household.

**3. a.** 48 J

**b.** Enough to power an LED light bulb for 10 s or to fire a cherry tomato 200 to 300 m in the air!

**c.** Physicists were amazed that this energy could be obtained just from one proton.

**4. a.** $1.88 \times 10^{-28}$ kg

**b.** 0.114 u or approximately 0.1 u, about one-tenth the mass of a proton

**5. a.** 200 MeV = $3.2 \times 10^{-11}$ J, which is 64 million times as much as the chemical reaction.

**b.** 26 MeV = $4.16 \times 10^{-12}$ J, about 8 million times the energy from the chemical reaction.

**c.** Chemical energy/kg =

$(0.5 \times 10^{-18})/(2.99 \times 10^{-26}) =$ $1.7 \times 10^{7}$ J kg$^{-1}$

Fission energy/kg =

$(3.2 \times 10^{-11})/(235 \times 1.67 \times 10^{-27}) =$ $8 \times 10^{13}$ J kg$^{-1}$

Fusion energy/kg =

$(4.16 \times 10^{-12})/(4 \times 1.67 \times 10^{-27}) =$ $62 \times 10^{13}$ J kg$^{-1}$

**6.**

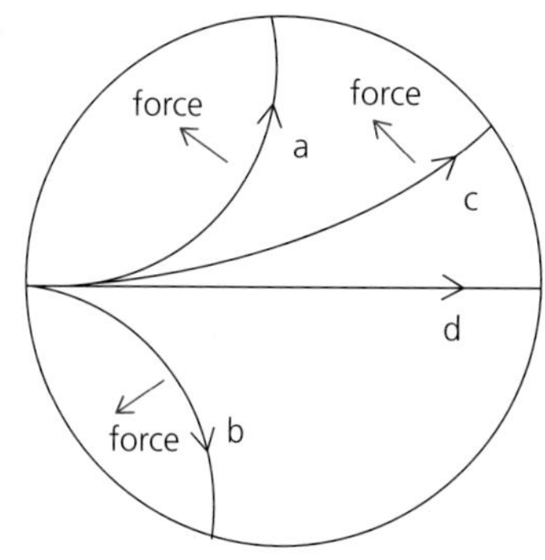


**7.** It is difficult to store antimatter in a container made of matter since it would be annihilated if it came into contact with the container.

**8.** $E = hc/\lambda$
$= (6.63 \times 10^{-34} \times 3 \times 10^{8})/(480 \times 10^{-9})$
$= 4.14 \times 10^{-19}$ J $= 2.59$ eV

**9.** **a.** 511 keV = 511 000 × $1.6 \times 10^{-19}$
$= 8.176 \times 10^{-14}$ J

$f = E/h = 1.23 \times 10^{20}$ Hz

**b.** $\lambda = c/f = (3 \times 10^{8})/(1.23 \times 10^{20})$
$= 2.43 \times 10^{-12}$ m

**10.** **a.** $f = c/\lambda = (3 \times 10^{8})/(532 \times 10^{-9})$
$= 5.64 \times 10^{14}$ Hz

**b.** $E = hf = 3.74 \times 10^{-19}$ J $= 2.34$ eV

**c.** Each photon carries $3.74 \times 10^{-19}$ J. The energy through 1 $cm^2$ is $50 \times 10^{-6}$ J in 1 s. Therefore, there must be $1.34 \times 10^{14}$ photons per second.

**11.** $2.47 \times 10^{20}$ Hz

**12.** Charge has to be conserved. Making a single electron would create charge, whereas in pair production there is no change in net charge.

**13.** He was correct in saying that a neutrino is emitted in beta decay and that it is electrically neutral. However, its mass is much smaller than that of the electron, and may be zero (although the latest measurements suggest not).

**14.** The neutrino is not charged, whereas the electron is.

**15.** High hydrogen content means lots of protons for the reaction with antineutrinos.

**16.** A nuclear reactor releases lots of β-particles and (hence) antineutrinos from fission reactions and from the decay of radioisotopes created by fission.

**17.** A positron, an electron neutrino and muon antineutrino.

# 4 THE STANDARD MODEL OF PARTICLE PHYSICS

**1.** **a.** $1 \neq 0 + (-1)$; forbidden; charge is not conserved

**b.** $(-1) + (1) = (-1) + (1)$; charge is conserved

**c.** $(-1) + (1) \neq (-1) + 1 + (-1)$; forbidden; charge is not conserved

**2.** None are forbidden:

**a.** $1 = 1 + 0$; baryon number is conserved

**b.** $1 + 1 = 1 + 1 + 1 + (-1)$; baryon number is conserved

**c.** $1 + 1 = 1 + 1 + 0 + 0$; baryon number is conserved

**3.** **a.** $0 + 0 \neq -1 + (-1)$; forbidden; strangeness is not conserved

**b.** $0 + 0 = 1 + (-1)$; strangeness is conserved

**c.** $1 + 0 \neq 0 + (-1)$; forbidden; strangeness is not conserved

**d.** $0 = 1 + (-1)$; strangeness is conserved

**e.** $0 \neq 0 + (-1)$; forbidden; strangeness is not conserved

**4.** The kaon decay contravenes the conservation of strangeness, and therefore has to take place by the weak interaction. Such decays have much longer lifetimes.

**5.** **a.** $Q$ ✓ $B$ ✓ $S$ ✗; forbidden; strangeness not conserved

**b.** $Q$ ✓ $B$ ✗ $S$ ✓; forbidden; baryon number not conserved

**c.** $Q$ ✓ $B$ ✓ $S$ ✓; allowed

**d.** $Q$ ✓ $B$ ✓ $S$ ✓; allowed

**e.** $Q$ ✓ $B$ ✗ $S$ ✓; forbidden; baryon number not conserved

**f.** $Q$ ✓ $B$ ✗ $S$ ✗; forbidden; baryon number and strangeness not conserved

**6.** **a.** not allowed; neither $L_e$ nor $L_\mu$ is conserved

**b.** allowed

**c.** allowed

**7.** Students' own answers

**8.**

| | u | d | d | neutron |
|---|---|---|---|---|
| $Q$ | 2/3 | −1/3 | −1/3 | 0 |
| $B$ | 1/3 | 1/3 | 1/3 | 1 |
| $S$ | 0 | 0 | 0 | 0 |

| | $\bar{u}$ | $\bar{d}$ | $\bar{d}$ | antineutron |
|---|---|---|---|---|
| $Q$ | −2/3 | 1/3 | 1/3 | 0 |
| $B$ | −1/3 | −1/3 | −1/3 | −1 |
| $S$ | 0 | 0 | 0 | 0 |

**9.** A meson is a quark (strangeness 0 or −1) and an antiquark (strangeness 0 or + 1) combined. A baryon is made up of three quarks, so can have strangeness of any sum of 0, + 1 and −1, e.g. 0 + 1 + 1, or (−1) + (−1) + (−1).

**10.**

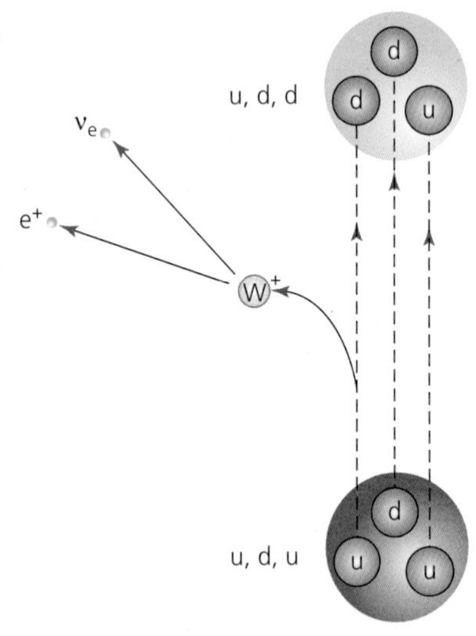


**11.** Gravity and electromagnetic forces both have infinite range. Their bosons are massless. Gravity is only attractive, whereas electromagnetism is attractive and repulsive because two different charges, plus and minus, exist.

**12.** They only interact with matter via the weak interaction (very short range) and gravity (a weak force).

**13.** **a.** electromagnetism

**b.** strong interaction

**c.** weak interaction

# 5 WAVES

**1.** A wave does store energy. For example, a wave on a string stores energy in the form of potential energy (gravitational potential energy and/or elastic potential in the string) and kinetic energy. But it is not a complete answer. A brick thrown through the air is also a mobile energy store. The student could have said: "A periodic/regular disturbance which is a mobile energy store." Or at least something about transporting energy ... but not matter.

**2.**

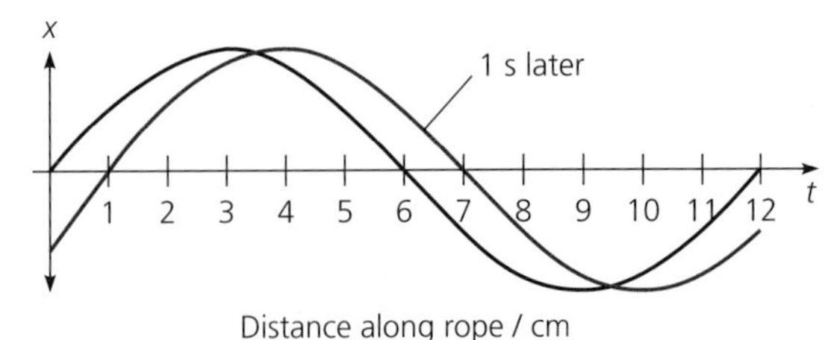


**3.** There is no force between adjacent people. No energy is being transferred along the wave.

**4.** **a.** The particles in a solid are closer together, and the forces between them are stronger.

**b.** Sound is a mechanical wave – there has to be something to vibrate! Light can travel through a vacuum but sound cannot.

**5.** It is difficult to shake a liquid from side to side; the layers of water shear, and slide past one another. In fact, there are 'internal' waves deep below the ocean surface. These long-wavelength waves appear at the boundaries of the layers of water of different density.

**6.** You need to show that the air moves backwards and forwards when the sound waves pass. A candle flame in front of a loudspeaker works well at low frequency. Perhaps you could use a video camera, and slow playback to slow the action down at the higher frequencies.

**7.** Place a source of sound in a bell jar and use a vacuum pump to remove the air. For best results the sound source should make as little contact with the jar as possible, so the sound is not transported out of the jar. It is usually suspended from a thread.

**8.** All EM waves move at the same speed in a vacuum.

**9.** **a.**

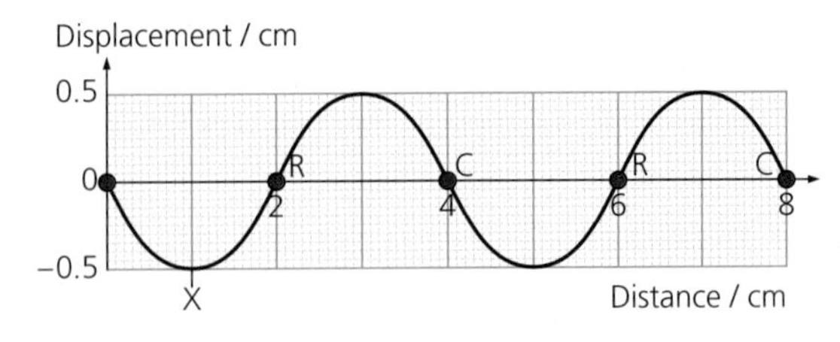


**b.** See diagram.

**c.** See diagram.

**d.** Instantaneously at rest at its maximum leftward displacement. Will then move right, towards its equilibrium position.

**10.** $\frac{3}{2}\lambda = 150$ m so $\lambda = 100$ m

**11. a.**

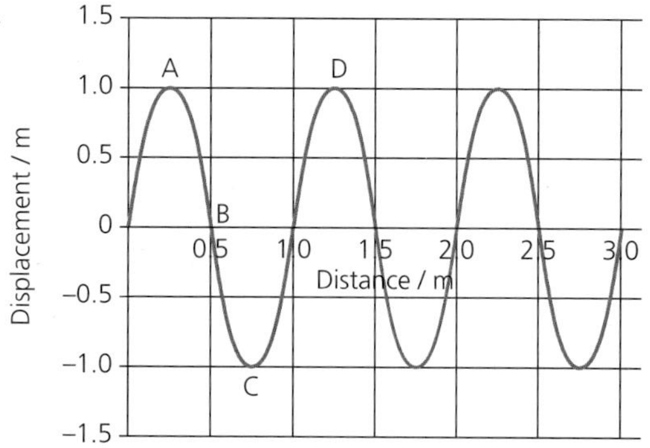


**b.** See diagram.

**12. a.** $2\pi$ radians = 360°, so $\pi/6$ = 30°

**b.** This is 1/12 of a cycle, so 1/12 = 0.25 m

**13.** Period = 1/frequency = 1/256 = 3.91 ms

**14.** In one hour there are 60 × 60 = 3600 seconds, so frequency = 1/3600 = $2.78 \times 10^{-4}$ Hz

**15. a.** ¼ **b.** 90° **c.** $\pi/2$

**16.** 330/256 = 1.29 m

**17.** $3 \times 10^8/1500$ = 200 kHz

**18. a.** $f = 1/T$ = 500 Hz

**b.** The gradient of the dispacement–time graph at any point gives the velocity.

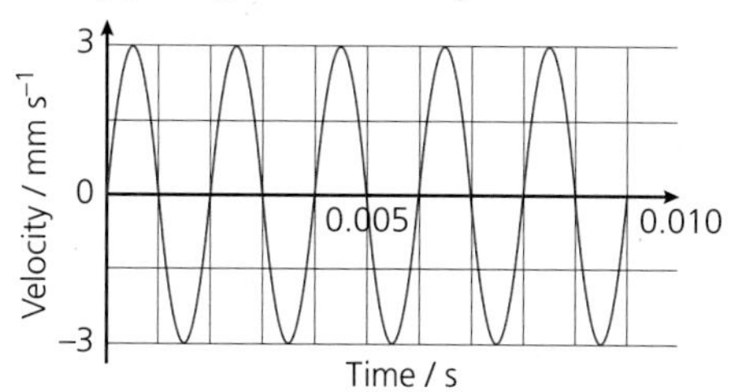


**19.** Sound cannot be polarised as it is a longitudinal wave; the vibrations are only in one direction.

**20.** Liquid-crystal displays emit polarised light. If my Polaroid sunglasses only admit light that is polarised in a perpendicular plane, no light will get through.

Light reflected from the road and from the sky will be partially polarised. Stress patterns in the windscreen can rotate the plane of polarisation; wearing Polaroid sunglasses reveals these areas of stress.

**21.** Some of the light in the reflection is polarised parallel to the surface of the sea. Wearing polarising glasses that only admit vertically polarised light will cut out the glare. Tilting your head (and the glasses) changes the plane of polarisation that the glasses admit and so allows more or less of the reflected light into your eyes.

**22.** If the car windscreen and headlights had opposite polarisations, you would not see the headlights of approaching cars. This might not be a good idea!

A polarising windscreen alone might cause problems. If you wore Polaroid sunglasses and tilted your head, no light would reach you through the windscreen.

**23.** There are two forms of the same sugar molecule. These have exactly the same atoms but arranged in a slightly different way; these are called stereoisomers. Natural sugar contains only one form of this molecule, which rotates the plane of polarisation of light to the right. Artificial sugars have both types of molecule, so there is no net rotation. Bacteria have evolved to digest only the natural 'right-handed' sugar, so they leave behind the other 'left-handed' type.

**24.**

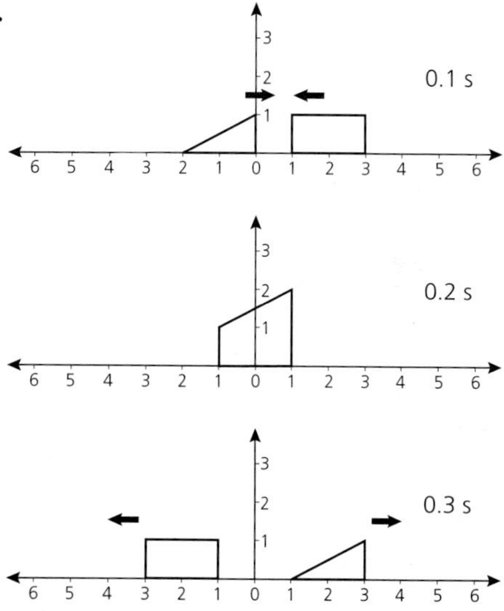


**25.** Estimates: length of string is about 75 cm; mass is about 5 g; tension is about 100 N.

$$f = \frac{1}{2 \times 0.75}\sqrt{\frac{100}{0.005 / 0.75}}$$
$$= 0.66 \times 10\sqrt{150}$$
$$\approx 80 \text{ Hz}$$

**26.** The frequency of the first harmonic will be the same, but the note is not pure – there are additional harmonics present. The strengths of the harmonics present are different for a violin than for a guitar.

**27.** Frequency is proportional to $1/l$, and $l$ probably shortens to about ¼ of its length; therefore $f$ increases × 4 to 800 Hz.

**28.** Twice the frequency. Touching it creates a node in the middle, not an antinode, so the second harmonic will be heard.

**29.** $f \propto \frac{1}{\sqrt{\mu}}$, so $f$ decreases by a factor of $\frac{1}{\sqrt{2}}$. But $f \propto \sqrt{T}$, so tension will need to double to keep $f$ the same.

**30.** Frequency is inversely proportional to length. So quarter-length strings will play notes four times the frequency. The tension in the strings could be reduced by a factor of × 1/16; or the mass per unit length could be increased by × 16; or she could increase the mass per unit length by a factor of × 4, and decrease the tension to × 1/4 of its usual value.

**31.** **a.** For the first harmonic the length of the pipe is equal to $\lambda/2$. So $\lambda = 60$ cm = 0.60 m. $f = c/\lambda =$ 330/0.60 = 550 Hz

**b.** If the pipe is closed at one end, for the first harmonic the pipe length is equal to $\lambda/4$ so $\lambda = 120$ cm = 1.20 m.

$f = c/\lambda = 330/1.20 = 275$ Hz

**32.** **a.**

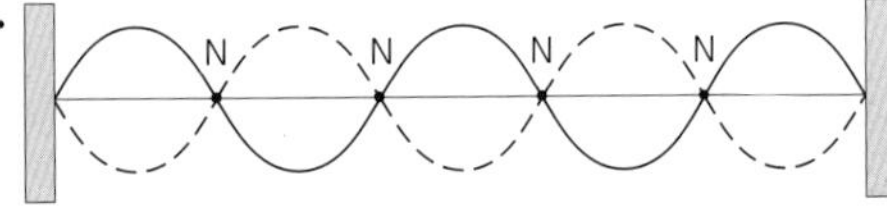


**b.**

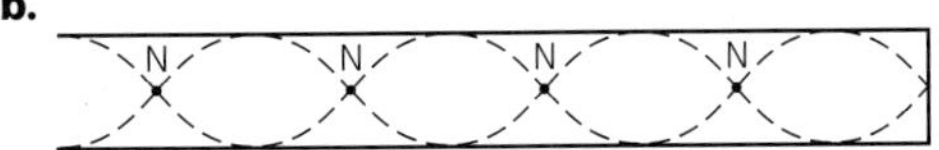


## 6 DIFFRACTION AND INTERFERENCE

**1.** The microphone would detect a series of quiet and loud points. A quiet point would occur when there was a difference of an odd number of half-wavelengths between the distance travelled by the two sound waves. A loud point would occur when there was a difference of a whole number of wavelengths between the two waves. For example, in the middle of the speakers each wave has travelled 5.0 m and there would be a loud point. At 4.75 m from one speaker, and 5.25 m from the other, the path difference is half a wavelength and the microphone would detect a quiet point.

**2.** **a.** $\lambda = ws/D = 0.5 \times 3.0/5.0 = 1.5/5.0 = 0.3$ m
$f = c/k = 330/0.3 = 1100$ Hz or 1.1 kHz

**b.** Since $w = \lambda D/s$, the three possible changes are: increase the wavelength (lower frequency); increase the distance between the observer and the loudspeakers; or move the speakers closer together.

**3.** The speakers are not playing the same note. In stereo music reproduction the speakers play different versions of the same recording, thereby giving the impression of depth. Reflections from the walls and ceilings would also make it difficult to find a place where the sound exactly cancelled out.

**4.** **a.** Assuming you are about 1 m away, then $w = D\lambda/s = 1 \times \lambda/1 = \lambda$, so $w = \lambda = c/f =$ 330/1000 = 0.33 cm.

**b.** Not as many reflections from walls and so on to complicate the interference pattern.

**c.** You hear a low-frequency note as well as the 1 kHz signal. The two sound waves will sometimes be in phase, giving constructive interference, and sometimes completely out of phase, giving destructive interference. This is a phenomenon known as 'beats'.

**5.** Because fringe separation depends on wavelength, and the wavelength range in the electromagnetic spectrum goes from hundreds of kilometres to billionths of a metre.

**6.** Put the slits closer together; move the screen further away.

**7.** **a.** The car passes through positions of cancellation and positions of strong signal.

**b.** Using $c = f\lambda$ gives $\lambda = 100$ m. Using $s/\lambda = D/w$ gives $w = 100 \times 10\,000/1000 = 1000$, so the car passes through a maximum every 1.0 km.

**8.** There are areas where the signal cannot penetrate sufficiently, and areas where two or more signals (because of reflections) interfere destructively.

**9.** The reflector increases the signal strength by reflecting back a wave that constructively interferes. The reflecting sheet also stops the aerial picking up stray waves that have bounced off buildings behind the aerial. This prevents 'ghosting'.

**10.** The aerial receives two radio waves, one direct from the transmitter and the other reflected from the aircraft. The two waves interfere. As the aircraft flies over, the path difference between the two radio waves changes, so that they are sometimes in phase and sometimes out of phase with each other. This leads to a fluctuating signal.

**11.** Nodes would be $\frac{1}{2}\lambda$ apart.
$\lambda = (3 \times 10^8)/(2.4 \times 10^9) = 0.125$ m $= 12.5$ cm
Nodes would be 6.25 cm apart.

**12.** Low-frequency sounds have longer wavelengths. They can diffract more around the head and the shadowing effect is reduced.

**13.** The wavelength of radio waves used for TV is $c/f = (3 \times 10^8 \text{ m s}^{-1})/(600 \times 10^6) = 0.5$ m. The wavelength of medium-wave radio waves is $(3 \times 10^8 \text{ m s}^{-1})/(1000 \times 10^3) = 300$ m. There is much more diffraction with the longer 'medium-wave' radio and so reception is possible in the 'shadow' of a hill.

**14.** Blue light has a shorter wavelength, so there will be less diffraction. The central maximum will be narrower and the secondary fringes will be narrow and close to the central maximum.

**15.** The wavelength of these microwaves is $\lambda = c/f = (3 \times 10^8 \text{ m s}^{-1})/(10 \times 10^9 \text{ Hz}) = 0.03$ m or 3 cm.

**a.** The gap is much larger than the wavelength, so there would be little diffraction, and microwaves would be detected mainly in the region opposite the gap.

**b.** The gap is now a similar size to the wavelength. The microwaves would be significantly diffracted and would be detected well inside the region of geometric shadow.

**16.** The interference pattern for two slits has the diffraction pattern for one slit superimposed on it (see diagram). In some places the light from slit A and the light from slit B have a path difference of a whole number of wavelengths, and are in phase. If the two-slit interference was acting alone, there should be a maximum, but light is also being diffracted by each single slit. When the single-slit diffraction pattern has a minimum, there will be no bright interference fringe.

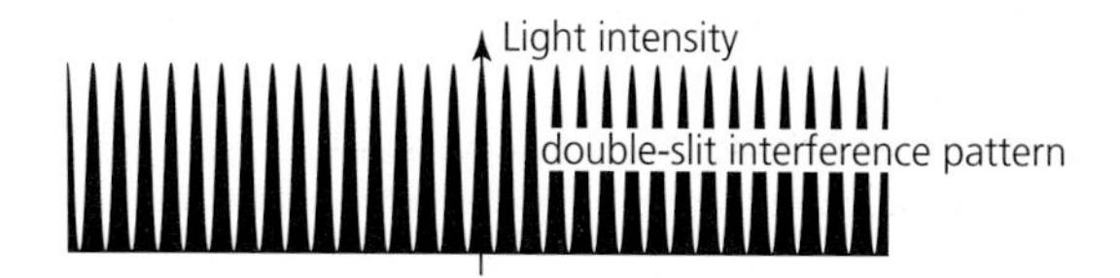


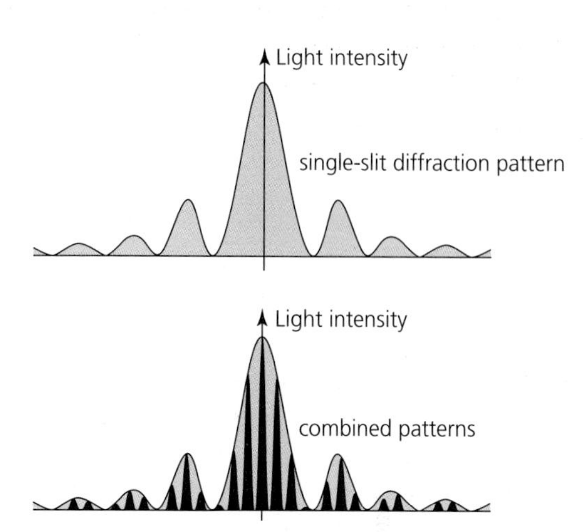


**17.** Radio waves have a much longer wavelength than light waves, and so a radio telescope needs to have a much larger aperture to reduce diffraction problems and be able to 'see' more detail.

**18.** In the central maximum, there is no path difference between the light paths from each slit. All wavelengths are in phase and so all constructively interfere.

**19.** The distance between each line is $d = 1/200\,000 = 5 \times 10^{-6}$ m. The maximum number of diffracted beams can be found by putting $\theta = 90°$, that is, $\sin\theta = 1$, into the equation $n = (d \sin\theta)/\lambda$. This gives $n = (5 \times 10^{-6} \times 1)/(600 \times 10^{-9}) = 8.33$. So there are eight beams on each side of the central maximum, so $8 \times 2 = 16$, plus the zero-order central maximum, making 17 diffracted beams in total.

**20. a.** A CD stores data in lines (actually a spiral track) of pits. These act as lines in a reflection diffraction grating.

**b.** Shine a white light on the CD (or use a laser). Choose a colour and estimate the wavelength (between 400 nm for violet and 700 nm for deep red) and estimate the angle, $\theta$, between it and the incident light beam.

Use $d = n\lambda/\sin\theta$. For example, for red, $d = 650 \text{ nm}/\sin 25° \approx 1.5$ μm, that is, about 670 000 lines per metre, or 660 lines per mm (actual separation = 1.6 μm).

**c.** DVD result ≈ 0.74 μm. Tracks are closer together, as more data needs to be stored.

**21. a.** Wavelength of sound = 330 m s$^{-1}$/330 Hz ≈ 1 m, so for the first order, $d = \lambda/\sin\theta$. To get $\theta \approx 30°$, need $d = 2$ m. A fence with gaps every 2 m would work.

**b.** A column loudspeaker, used for speech in a public space, is similar to a grating as there are many coherent sources, one above the other. The sound spreads out horizontally but not much vertically so the sound does not bounce off the roof too much.

# 7 REFRACTION AND OPTICAL FIBRES

**1.** Light obeys the law of reflection, but because the surface is not smooth, the normal lines will not be parallel to each other. The rays of light from your face are scattered in many different directions (see diagram). This is referred to as diffuse reflection.

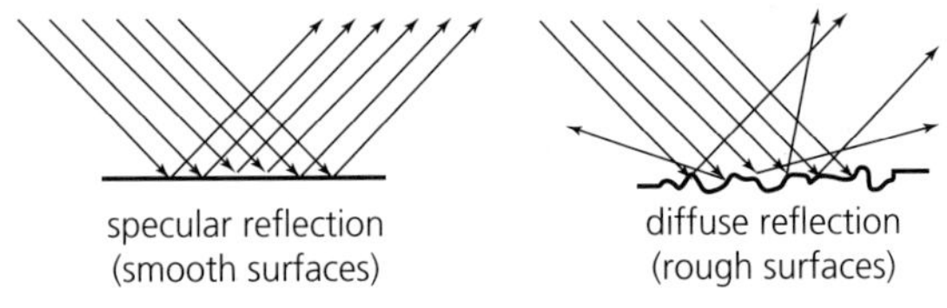


**2.** With lateral inversion, no true inversion has taken place, because each point on the image is actually opposite the corresponding point on the object. It is just that you are facing the mirror. If you stood with your back to the mirror and raised your right hand, so would your image.

**3.** The amount of light reflected, or transmitted, depends on the refractive indices of the two materials involved, which depend on

- the wavelength of light
- the angle of incidence and
- the polarisation of the incident light beam.

(For ordinary glass at normal incidence, about 4% of the incident light is reflected from the first air-to-glass boundary. There will be another reflection at the glass-to-air boundary. A small fraction of the incident light will undergo several reflections, which is what causes a repeated edge to the image that is noticeable in a thick mirror.)

**4.**

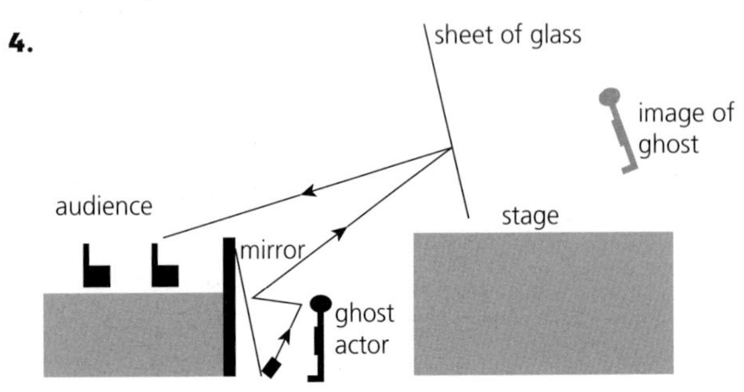


**5.**

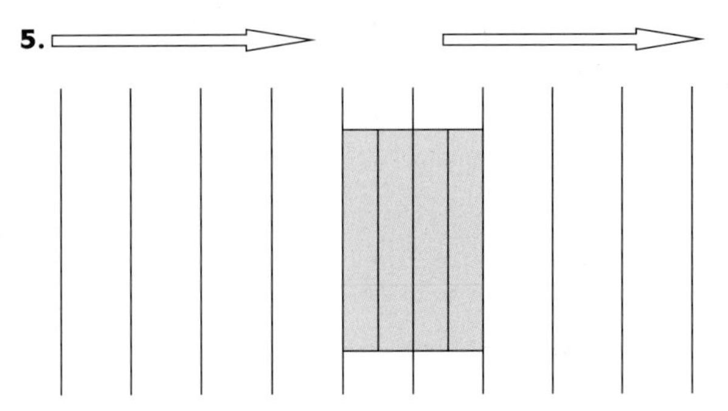

**6.**

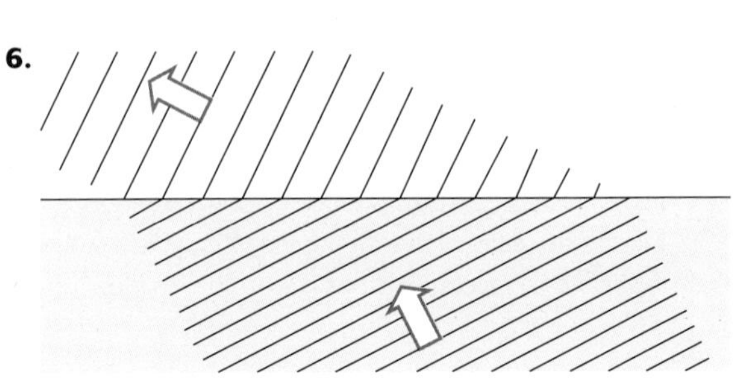

**7.**

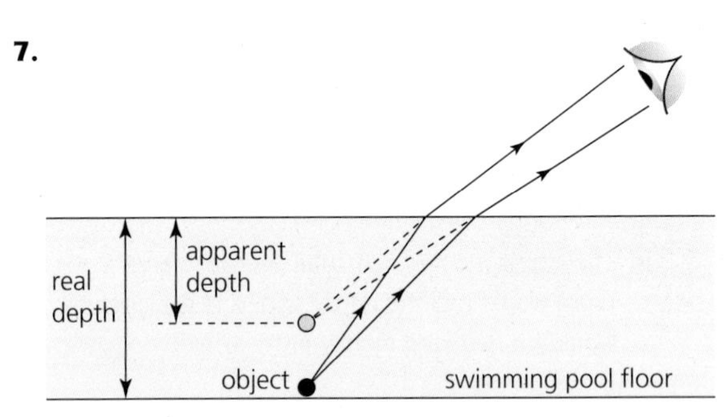


**8.** As the waves reach the beach, the water gets shallower and the wave speeds drops. The wavelength decreases. Refraction gradually changes the wave's direction. See diagram.

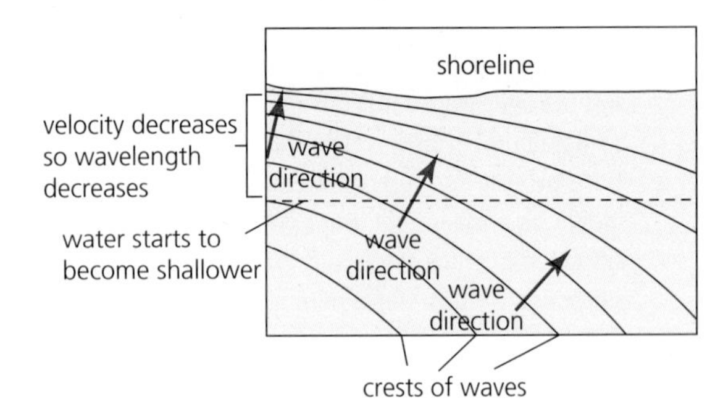

**9.** Frequency is determined by the source and cannot change. Imagine what would happen if it did. Think about light waves entering glass – sunlight through a window, for example. Suppose the frequency in glass was less than in air. More waves per second would enter the glass than would leave it, contravening the principle of conservation of energy.

**10. a.**

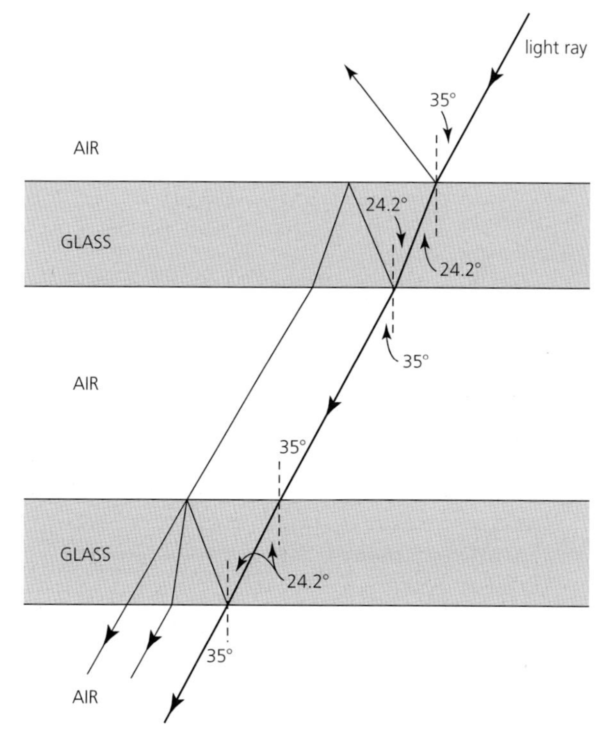


**b.** $n_1 \sin \theta_1 = n_2 \sin \theta_2$, so $\sin \theta_2 = (1/1.4) \sin 35° = 0.41$, so $\theta_2 = 24.2°$

**c.** See diagram.

**d.** See diagram. Objects seen through the window may seem to have blurred or multiple edges. The partial reflection is weak, so that it only becomes obvious for a brightly lit object with a dark background, say a streetlight at night. Thicker window glass would increase this effect.

**11. a.**

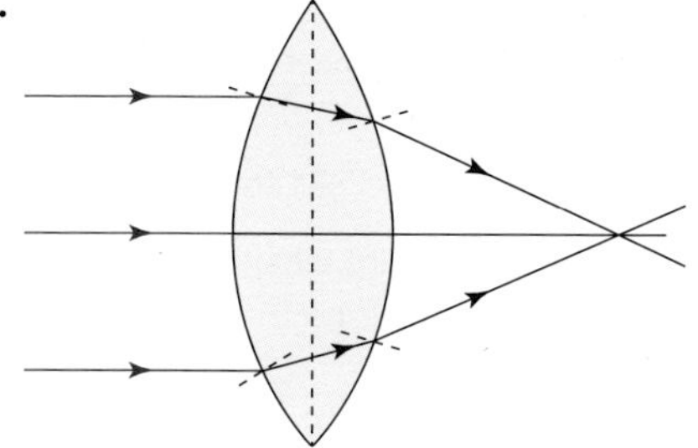

**b.** Light rays are refracted towards the normal as they enter the lens and away from the normal as they leave. The curvature (shape) of the lens means that parallel light rays will converge to a point.

**c.** A higher refractive index would slow the light more and refract it through a larger angle, at both surfaces of the lens, bringing the light rays to a focus in a shorter distance.

**d.** Blue light would be refracted more as it entered the lens and as it left. The blue rays would come to focus nearer to the lens than the red rays.

**e.** The lens would disperse the colours in white light. Different colours would come to a focus at different points, meaning that the image will have coloured edges. (This is known as chromatic aberration.)

**f.** Reflected light is not dispersed, and all colours (wavelengths) are reflected in the same way, whereas refraction depends on wavelength.

**12. a.**

| Medium for light path through the eye | Refractive index | Ratio $n_2/n_1$ |
|---|---|---|
| Air | 1.0003 | |
| | | 1.38 |
| Cornea | 1.376 | |
| | | 0.97 |
| Aqueous humour | 1.336 | |
| | | 1.05 |
| Lens | 1.4 | |
| | | 0.96 |
| Vitreous humour | 1.337 | |

**b.** At the cornea.

**c.** The ratio of (refractive index of cornea)/(refractive index of water) = 1.376/1.33 = 1.034, much less than the ratio between the cornea and air, so less refraction takes place and light rays will not come to a focus. Goggles trap air and restore the air–cornea boundary.

**13. a.** The ratio of the refractive indices is 1.40/1.60 = 0.875.

**b.** Using Snell's law, $(n_1/n_2) \sin \theta_1 = \sin \theta_2$, a ray incident at 50° would refract at 42.1°, so it would strike the wall at (90° – 42.1°) = 47.9°.

**14. a.** The critical angle for diamond is given by $\sin \theta_c = 1/2.42 = 0.413$, so $\theta_c = 24.4°$.

**b.** This low critical angle means that light will reflect many times inside a diamond. Even when light strikes a face at an angle less than the critical angle, much of the light will be internally reflected. The net result will be a small amount of light escaping from each face of the diamond, making it sparkle.

**15. a.**

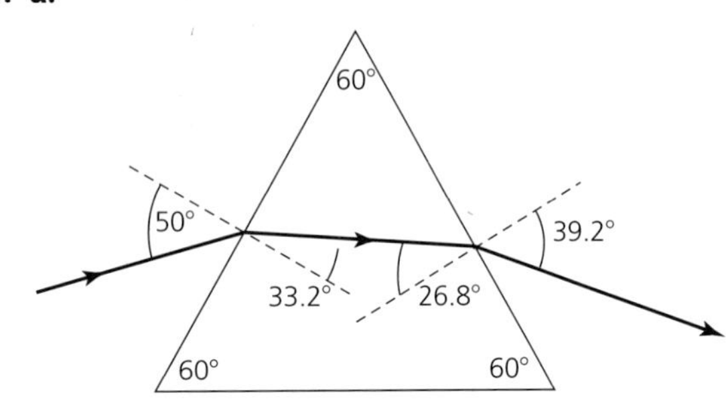


$\theta_2 = \sin^{-1}[(1/1.4) \sin 50°] = 33.2°$

$\theta_4 = \sin^{-1}[1.4 \sin 26.8°] = 39.2°$

**b.**

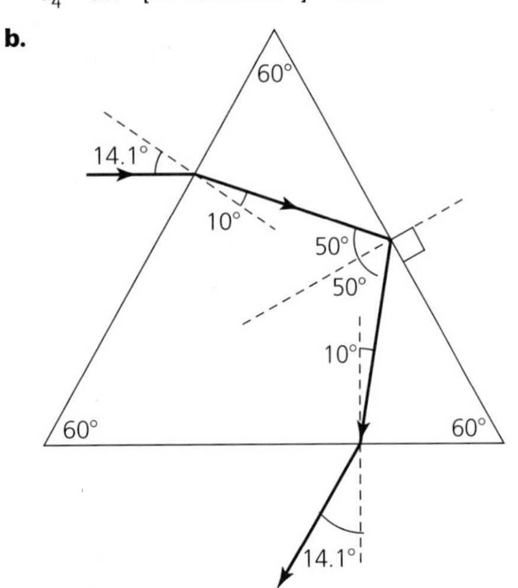


If the angle of incidence on the glass–air interface is 50°, then by the geometry of the prism the angle of refraction at the air–glass interface must have been 10°. The original angle of incidence (air to glass) is given by $\sin \theta_1 = 1.4 \sin 10°$, so $\theta_1 = 14.1°$.

The angle of incidence at the glass–air boundary (50°) is greater than $\theta_c$ for glass, which is given by $\theta_c = \sin^{-1}(1/1.4) = 45.6°$. So the ray is totally internally reflected at the glass–air boundary. This reflected ray strikes the bottom (horizontal) face of the prism at an angle of incidence of 10°. This is less than the critical angle, so the ray is refracted out of the prism at an angle of $\sin^{-1}(1.4 \sin 10°) = 14.1°$.

**16.** The critical angle at the boundary would be almost 90°, because $\sin c = n_1/n_2$ would be very nearly 1, so the difference in path length between the longest path (when light is incident at the core – cladding boundary at the critical angle; $i = c$) and the straight-through path is as small as possible, and hence the time delay between different modes is also as small as possible.

**17.** The speed of light in the fibre is $v = c/n = 1.875 \times 10^8$ m s$^{-1}$, and Figure 28 shows that, for every unit along the fibre, the critical angle path is 1.0667 units. So times are: straight-through path of 1 km, time $= s/v = 5.333 \times 10^{-6}$ s; longest path, time $= 1.0667$ km$/v = 5.689 \times 10^{-6}$ s. The time delay per kilometre is therefore 0.356 μs.

**18.** From the answer to question 17, over a 10 km cable there would be a time delay of 3.56 μs. You could not transmit pulses with a period longer than this. The maximum frequency is therefore $f = 1/T =$ 281 kHz. (This would be unacceptably low – real fibre-optic cables have very close refractive indices for core and cladding to minimise the time delays.)

**19.** Using $n = c/v$ and time = distance/speed, the times for the two colours are 164.04 and 164.00 μs. The difference of arrival times is therefore 0.04 μs, or 40 ns.

**20.** Between 800 and 900 nm, around 1300 nm and between 1500 and 1700 nm.

**21.** As a result of dispersion, pulses broaden, that is, they spread out over time. The further they travel, the more spread out they become, and eventually they will merge with the preceding or following pulse.

# 8 SPECTRA, PHOTONS AND WAVE-PARTICLE DUALITY

**1. a.** Rigel has the higher surface temperature since it appears blue, rather than red as for Betelgeuse. Blue light has a shorter wavelength, which indicates a higher temperature.

**b.** The size of the star and its distance from us.

**2.** Objects with a high surface temperature emit strongly across the whole spectrum and appear white to our eyes.

**3.** Flash of lightning (blue-white, ≈30 000 K); surface of the Sun (≈5800 K); filament of a car headlight bulb (white, ≈3000 K); heating element of a toaster (red, ≈900 K); central heating radiator (infrared, ≈340 K). The objects with the highest surface temperature emit more of their light at short wavelengths.

**4.** No, the Moon's surface is not hot; it is simply reflecting the white light from the Sun. The link between temperature and wavelength applies to light that is *emitted* from a surface, not reflected light.

**5.** In an old-fashioned (cathode ray tube) television set, electrons accelerate across the tube and collide with atoms in the phosphor coating on the back of the screen. Electrons in the atoms of the phosphor coating gain energy from these collisions and are excited to higher energy levels. As these electrons return to lower energy levels, they emit light. In some cases the atoms in the phosphor coating are ionised. Electrons from higher energy levels then fall into the gaps left by the emitted electron, and light is emitted.

**6.** During the day, atoms in the stars absorb energy from light. This lifts some of the electrons into higher energy levels. As the electrons return to lower energy levels, they emit light. When all the atoms are in the ground state, that is, all their electrons are in their lowest energy levels, no more light is emitted. Torch-light excites some of the electrons back into higher energy levels. The gradual nature of the emission suggests that electrons can spend some time in the excited states, and do not instantly return to the ground state.

**7. a.** 3 eV is not enough to excite the atomic electron to a higher energy level. No energy is transferred to the atomic electron, and the incident electron scatters off the atom without losing any energy. This is called elastic scattering.

**b.** 11 eV is more than enough to excite the atomic electron to the next energy level. If that happened, the incident electron would have $11 - [-3.4 - (-13.6)] = 0.8$ eV left in the form of kinetic energy.

**c.** 15 eV is enough energy to ionise the atom, totally removing the electron. There would be $15 - 13.6 = 1.4$ eV left, which would be shared between the two electrons as kinetic energy.

**8. a.** $E = hc/\lambda = (6.63 \times 10^{-34}\text{ J s} \times 3 \times 10^{8}\text{ m s}^{-1})/(400 \times 10^{-9}\text{ m}) = 4.97 \times 10^{-19}\text{ J}$

Divide by $1.6 \times 10^{-19}$ to convert the answer to electronvolt, giving 3.11 eV. This is the smallest amount of energy that can be transferred at this wavelength: it is the energy of one photon of this violet light.

**b.** Electromagnetic radiation energy can only be absorbed in multiples of this value, so the next largest energy transfer = $2 \times 3.11 = 6.22$ eV.

**c.** The one after that equals 9.33 eV and so on.

**9. a.** The Planck constant is a very small number, so that typical photon energies are a few electronvolts, which is a very small energy on a human scale.

**b.** Suppose that the Planck constant were so large that photon energies were of the order of millijoules (mJ) or joules (J), rather than electronvolts (eV). A joule of energy is sufficient to lift a 1 N weight through a height of 1 m. You would be able to feel the photons arriving. Having a torch shone in your face might be a more painful experience than it is at the moment.

(*Aside*: If the Planck constant only increased by a few powers of 10, then it is unlikely that any atoms would be left in the Universe. The ionisation energy depends on $1/h^2$, so atoms would lose their electrons easily. The atomic radius also depends on $h^2$, so any surviving atoms would be huge!)

**10. a.** Three (from level 3 to 2, from level 3 to 1, and from level 2 to 1).

**b.** From level 3 to 1: $\Delta E = -1.5 - (-13.6) = 12.1\text{ eV} = 1.94 \times 10^{-18}\text{ J}$, so

$$f = \frac{\Delta E}{h} = \frac{1.94 \times 10^{-18}}{6.63 \times 10^{-34}} = 2.93 \times 10^{15}\text{ Hz}$$

Similar calculations give frequencies of $4.59 \times 10^{14}$ Hz and $2.47 \times 10^{15}$ Hz.

**11.** The largest possible energy change is 13.6 eV. This gives $\lambda = 91$ nm.

**12. a.** Most of the UV radiation that leads to a suntan is absorbed by the phosphor and the glass tube.

**b.** Use more mercury atoms in the tube, and possibly increase the tube current, so that there are many collisions between high-energy electrons and mercury atoms. This will lead to more UV emission. Do not use a phosphor that absorbs UV, and use quartz, rather than glass, for the tube. Though, since UV leads to premature aging and cancer of the skin, perhaps none of these is a good plan!

**13.** The phosphors are carefully chosen to give a balance of red, green and blue light, to give the impression of white. So-called 'daylight' lamps have phosphors whose emission spectrum is broadly similar to the Sun's.

**14.** The gaps between electron energy levels (in a gas), or between energy bands (in a solid or liquid) determine which photons can pass through and which can be absorbed. If the photon's energy matches an energy gap, it will be absorbed, and the material will be opaque to that wavelength. Ordinary glass has no energy gaps that match photon energies in the visible range, so it is transparent to visible light.

**15.** The lines of colour that form the emission spectrum reveal the energy levels corresponding to the allowed orbits in the atom.

**16.** The electrons in atoms that absorb light must be in a lower energy state before they absorb any photons. Cooler atoms will have fewer excited electrons.

**17.** Without strong sunlight, the heat of the flame will excite some of the electrons in the sodium atoms to a higher energy level. These emit yellow light as they return to the lower energy level. With strong sunlight shining through the flame, there is more absorption in the sodium vapour. This makes the lines look darker.

**18.** The photoelectric effect is the emission of electrons from the surface of a metal. Shining light on a metal may knock electrons out of the surface and so reduce the excess negative charge. If the cap was positively charged, that is, if there was an electron deficiency, any photoelectric effect would simply increase the positive charge (though this would be difficult since the electrons would be attracted more strongly to the surface of the metal in this case).

**19.** Photons of ultraviolet light have sufficient energy to remove electrons from the zinc, thus removing the excess negative charge, which allows the gold leaf to fall.

**20.** A photon of red light does not have enough energy to eject an electron from the surface of the zinc. Bright red light simply means more low-energy 'red' photons, none of which can cause photoelectric emission.

**21.** The photoelectric effect depends crucially on the frequency of incident light. Below the threshold frequency, no photoemission occurs, no matter how intense the light. If the light has a frequency higher than the threshold frequency, photoemission will occur (and the brighter the light, the more electrons will be ejected, but Table 1 does not show this).

**22.** The coconuts in their holders represent electrons held into the surface of a metal. The ping-pong balls represent low-energy photons of light that are below the threshold frequency. These do not have enough energy to remove the coconut (electron) from the holder (surface). The bullets represent high-energy photons of light that are above the threshold frequency. A single bullet (photon) has sufficient energy to knock a coconut (electron) from its holder (surface).

**23.** **a.** There is no effect.

**b.** There is no effect.

**c.** Faint blue light would cause some photoemission.

**d.** Bright blue light would cause more photoemission – more electrons would be ejected from the surface.

**e.** There are no electrons in **a** and **b**. In **c** and **d**, the maximum kinetic energy of the emitted electrons is not affected by the brightness of the light.

**24.** Walking speed, $v \approx 2 \text{ m s}^{-1}$

Mass, $m \approx 60$ kg

Momentum, $p = mv = 120 \text{ kg m s}^{-1}$

$\lambda = h/p = 6.6 \times 10^{-34}/120 = 5.5 \times 10^{-36}$ m

Diffraction is only noticeable if the gap or obstacle is approximately the same as the wavelength $\lambda$, so that would need an extremely small doorway.

**25.** The wavelengths are too large to be diffracted by the crystal structure.

**26.** It gets smaller (because the wavelength gets shorter).

## 9 THE EQUATIONS OF MOTION

**1.** Momentum and electric current are vectors. Density, temperature, and power are scalars.

**2.** Average speed = $(45 \times 1.6 \times 1000)/(70 \times 60) = 17 \text{ m s}^{-1}$

Average velocity = $(14.4 \times 1.6 \times 1000)/(70 \times 60) = 5.5 \text{ m s}^{-1}$ on a bearing (estimated) of 280°

**3.** No, this is not possible. (You would need to find a route shorter than a straight line!)

**4.** Velocity is a vector quantity. The direction is important. Since a runner's direction is constantly changing as she runs around the track, velocity is also constantly changing.

**5.** **a.** average speed = (total distance)/time = 10 000/1813.74 = 5.513 m $s^{-1}$

**b.** average velocity = displacement/time = 0 m $s^{-1}$

**6.** She is running at a constant velocity of 11 m $s^{-1}$.

**7.** The runner with the highest average speed wins. They will cover the distance in the shortest time.

**8.** It is the instantaneous speed at the moment of take-off that determines the length of the jump. The average speed of the run-up is not important.

**9.** **a.** $-2 + (-2) = -4$ m $s^{-1}$

**b.** The answer is the same as in part **a** because B is at rest relative to O.

**c.** $(-1) + (-2) + (-2) + (-2) = -7$ m $s^{-1}$

**10.** No, not in any absolute sense. An object can only be stationary when compared to something else.

**11.** **a.** At 10 m $s^{-1}$. Initially, the ball, you and the guard are all at rest in the same frame of reference.

**b.** At 20 m $s^{-1}$

**c.** At 10 + 40 = 50 m $s^{-1}$

**12.** Exactly where you started. You were at rest with respect to the Earth when you jumped up. Nothing has happened to change your horizontal velocity.

**13.** Galileo wrote "A Dialogue Concerning the two chief world systems" in which he staged the debate between the heliocentric (Sun-centred), Copernican solar system and the geocentric (Earth-centred), Aristotelian model favoured by the Church. In it he imagined a ship sailing smoothly along on calm seas at steady speed. He imagined someone on the ship watching water drip from a bottle, butterflies being released on deck and fish swimming in a tank and claimed that these would behave in exactly the same way as if the ship were stationary. This argument served to show that the objections to the movement of the Earth were unfounded. (The geocentric argument in the book was made by a character named Simplicio, which may not have endeared Galileo to the Church.)

**14.**

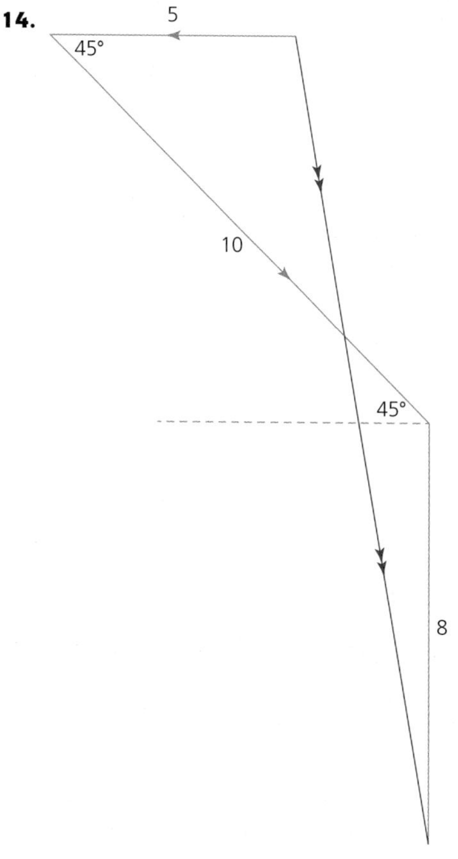


Take 15(.2) paces on a bearing of 172°.

**15.** **a.** 4.0 m $s^{-1}$ at 49° to the river bank

**b.** 27.8 minutes

**c.** 30° to the river bank

**d.** 55.7 minutes

**16.**

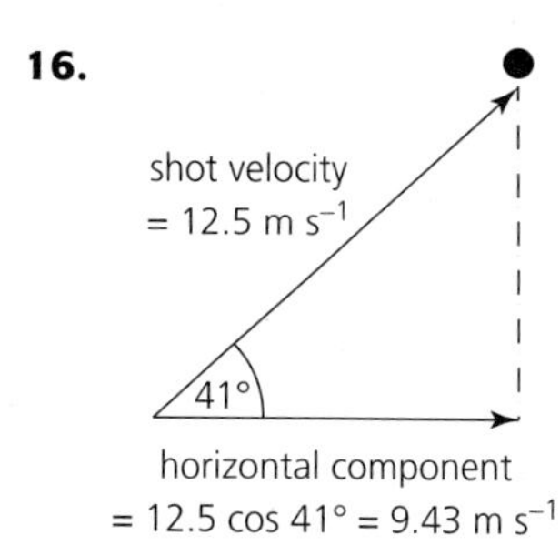

**17. a.**

| | Angle to the horizontal / degree | Horizontal velocity / m s$^{-1}$ | Vertical velocity / m s$^{-1}$ |
|---|---|---|---|
| i | 20 | 18.8 | 6.8 |
| ii | 30 | 17.3 | 10.0 |
| iii | 40 | 15.3 | 12.9 |
| iv | 50 | 12.9 | 15.3 |
| v | 60 | 10.0 | 17.3 |
| vi | 70 | 6.8 | 18.8 |
| vii | 80 | 3.5 | 19.7 |

**b.** The optimum angle to achieve the greatest range can be deduced to be about midway between 0 and 90 (in fact it is 45°). The reasoning is that, at 20°, for example, although the horizontal velocity is greatest, the small vertical velocity means that the time of flight will be short. And at 80° the high vertical velocity means that the ball stays in the air for a longer time but the horizontal velocity is low so that the ball does not travel very far.

**18.** $a = \frac{\Delta v}{\Delta t} = \frac{32 - 0}{0.020} = 1.6 \times 10^3 \, \text{m s}^{-2}$

**19.** Assume that a car starts from rest and reaches 30 miles per hour in around five seconds. 30 mph = 13.3 m s$^{-1}$, so acceleration = 13.3/5 ≈ 2.7 m s$^{-2}$. Any value in the region of 1 m s$^{-2}$ to 5 m s$^{-2}$ is reasonable.

**20.** Just after the ball is dropped, its velocity and acceleration are downwards. After the ball has bounced, the velocity is upwards but the acceleration is still downwards (due to gravity).

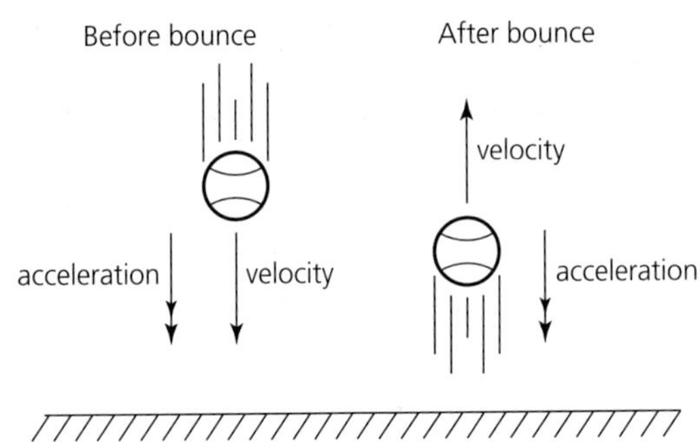


**21. a.** Distance travelled cannot be negative, because direction is not relevant. The graph cannot have a negative gradient, as this would mean the distance travelled is getting less as the journey progresses.

**b.** The graph would be the same as Figure 23 as far as $t$ = 5 s. Then, rather than decrease, it would increase in an upward curved line to reach a maximum value of 16 m at $t$ = 10 s.

**22.** Average velocity = (−4−6)/5 = −2 m s$^{-1}$. The actual (instantaneous) velocity at $t$ = 5 s is less than the average. At $t$ = 10 s the actual velocity is greater than the average.

**23. a.**

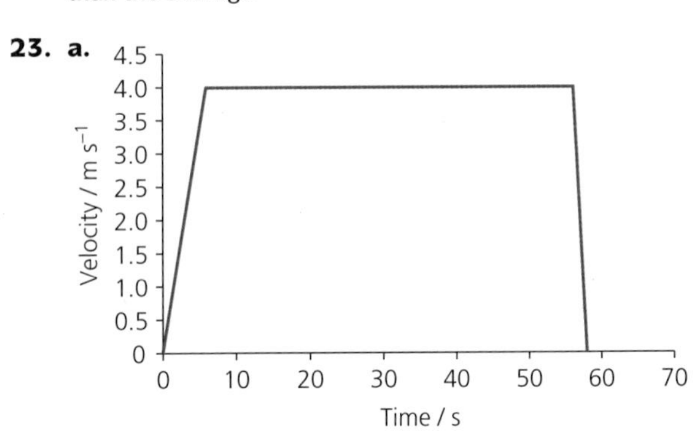


Uniform acceleration has been assumed at the beginning and end.

**b.**

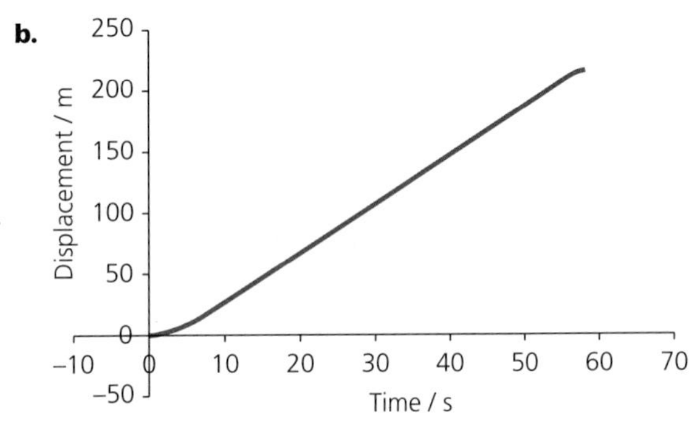


**24. a.** Acceleration is the slope of the graph.
$a = \frac{8}{2} = 4 \, \text{m s}^{-2}$

**b.** Distance covered is the area under the graph. Do this in two stages, from 0–2 s and from 2–8 s.
The total distance = $\frac{1}{2}$ (2 × 8) + (8 × 6) = 56 m

**25.**

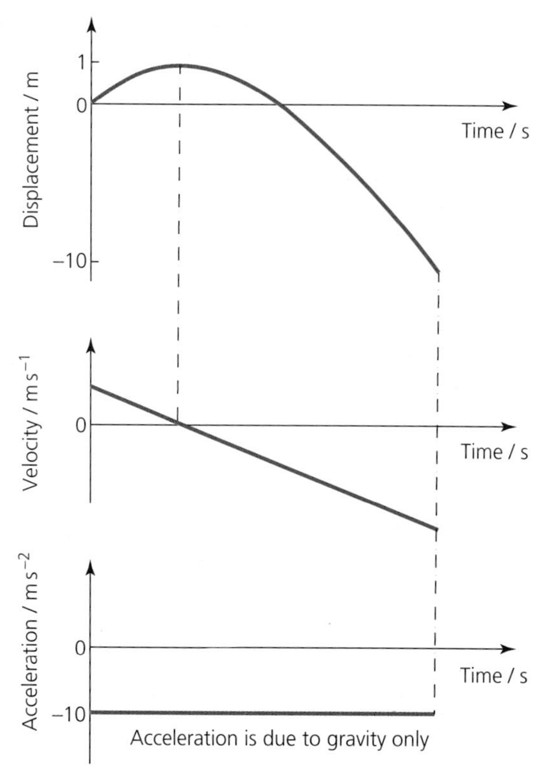


**26.** Use $a = \frac{v - u}{t} = \frac{75 - 0}{0.0005} = 150\,000\ \text{m s}^{-2}$. (This is 15 000 times the acceleration due to gravity!)

**27. a.** $s = 120$ m, $u = 0$, $v = 15$ m s–1, $t = ?$, $a = ?$

Use $v^2 = u^2 + 2as$

$225 = 0 + 2 \times a \times 120$

$\frac{225}{240} = a$

$a = 0.94\ \text{m s}^{-2}$

**b.** $s = ?$, $u = 15\ \text{m s}^{-1}$, $v = 0$, $t = 30$ s, $a = ?$

To find $a$, use $v = u + at$

$0 = 15 + a \times 30$

$-15 = 30a$

$a = -0.5\ \text{m s}^{-2}$

To find s, use $s = \frac{1}{2}(u + v)t$

$s = \frac{1}{2} \times (15 + 0) \times 30 = 225$ m

So the total distance travelled = 225 + 120 = 345 m

**28.** For the whole jump (up and down), $s = 0$, $u = ?$, $v = ?$, $t = ?$, $a = -9.81\ \text{m s}^{-2}$

There are too many unknowns here. We need at least three values to solve the equations. So divide the jump into two parts, up and down.

Up: At the top of the jump, the instantaneous velocity is zero. Then

$s = 3$, $u = ?$, $v = 0$, $t = ?$, $a = -9.81\ \text{m s}^{-2}$

Use $v^2 = u^2 + 2as$, giving
$0 = u^2 + 2 \times (-9.81) \times 3$
$u^2 = 58.9$
$u = 7.67\ \text{m s}^{-1}$

This is the take-off speed. We can use this to find $t$, using $t = \frac{v - u}{a}$, so

$$t = \frac{0 - 7.67}{-9.81} = \frac{-7.67}{-9.81} = 0.78\ \text{s}$$

Down: We can treat this in the same way, except that now $u = 0$ rather than v, but as the second part takes exactly the same time as the first, $t = 0.78$ s.

Total time in the air = 1.56 s

**29. a.** Use $s = ut + \frac{1}{2}at^2$, but because $u = 0\ \text{m s}^{-1}$, $s = \frac{1}{2}at^2$, so $s = \frac{1}{2} \times 9.81 \times 3.5^2 = 60$ m

**b.** If the potholer was using a stopwatch, he might manage $\pm 0.1$ s, which is approximately 3% uncertainty, so $60 \pm 2$ m.

**c.** At a depth of 60 m, the sound would take about 60/340 = 0.18 s to reach the potholer. This is comparable to the uncertainty, so the speed of sound should be taken into account.

**30. a.** The bag falls with you at the same rate, $g$, as this is independent of mass. So to anyone in the lift, the bag appears to 'float'.

**b.** Zero. They are falling too, at the same rate as you. So the contact force between you and the scales drops to zero.

**31.** To reduce drag and thereby achieve a higher top speed.

**32.** A mouse has a relatively large surface area to weight ratio. Air resistance is equal to the weight at a low speed. The mouse's terminal velocity is small enough not to be fatal. This is not the case for the rat, the man nor indeed the horse, for which air resistance would equal weight at progressively higher speeds. Terminal velocity would be precisely that!

**33.**

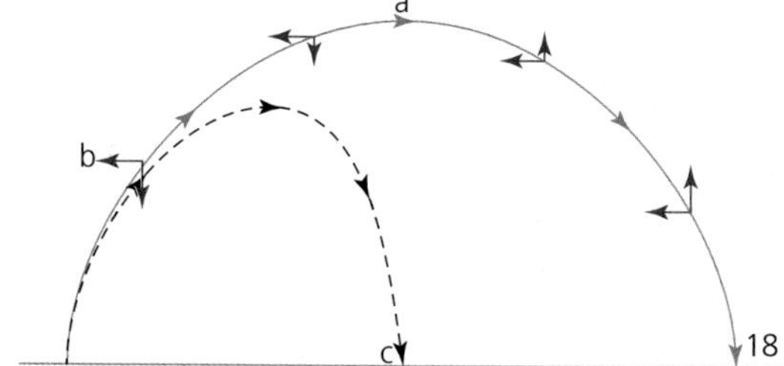

**34.** At the top of the jump: $v = 0\ \text{m s}^{-1}$, $u = 4.8\ \text{m s}^{-1}$, acceleration is due to gravity and acts down, $a = -9.81\ \text{m s}^{-2}$.
Rearrange $v^2 = u^2 + 2as$ to give

$$s = \frac{v^2 - u^2}{2a} = \frac{0 - (4.8)^2}{2 \times (-9.81)} = \frac{-23.0}{-19.6} = 1.17\text{m}$$

This is the height gained by the centre of mass, so add 0.95 m to give a height of 2.12 m for the centre of mass. (A top-class athlete will jump with his or her centre of mass as close to the bar as possible.)

**35. a.** Consider vertical motion first. Use $s = ut + \frac{1}{2}at^2$ to find the time of flight. Since the jumpers move horizontally at first, $u = 0\ \text{m s}^{-1}$, so

$$t = \sqrt{\frac{2s}{a}} = \sqrt{\frac{2 \times 26.7}{9.81}} = 2.33\ \text{s}$$

Horizontal distance must be at least 8.22 m, so horizontal velocity must be at least

$$\frac{8.22}{2.33} = 3.52\ \text{m s}^{-1}$$

**b.** Horizontal velocity is approximately constant = $3.52\ \text{m s}^{-1}$
Vertical velocity is

$$v = u + at = 0 + 9.81 \times 2.33$$
$$= 22.9\text{ms}^{-1}$$

Use scale drawing (or Pythagoras's theorem) to combine these.

Final velocity = $23.2\ \text{m s}^{-1}$ at 81.2° to the horizontal.

**36. a.** Horizontal component = 100 cos 45° = $70.7\ \text{m s}^{-1}$
Vertical component = 100 sin 45° = $70.7\ \text{m s}^{-1}$

**b.** The vertical component sets the time of flight, so the larger the vertical component, the longer the flight lasts. The range is equal to the horizontal component × time. Too steep a trajectory leads to a flight that lasts a longer time, but travels nowhere. Too shallow a trajectory leads to a flight that does not last long enough to get anywhere.

**c.** Firing from a height means that the time of flight is extended; the maximum range is achieved by using a smaller angle.

# 10 FORCES IN BALANCE

**1.** The cables are in tension – pulled down by the weight of the road, and held up by the towers.

The towers are in compression – pulled down by the cables, and pushed up by the ground.

**2. a.** $W = mg = 80.0 \times 9.81 = 785\ \text{N}$

**b.** $g_{\text{Moon}} = \frac{128}{80.0} = 1.60\ \text{N kg}^{-1}$

**3.**

contact force from river bank

weight of bridge

contact force from person (= weight of person)

**4.** Prof. H is right, since without friction between your shoes and the floor, you could not gain any horizontal speed. Of course, if you did manage to get moving, you would just keep moving, so maybe Prof. G has a point, though school corridors would be a nightmare.

**5.** On a straight, level road, increasing the weight would increase the normal contact force and therefore friction would increase, making skidding less likely. If the car was turning a corner, more mass would need a larger frictional force to push it round the corner, so more likely to skid.

**6.** Actual contact area could be 0.01% of 4 cm$^2$, say $4 \times 10^{-8}\ \text{m}^2$.

$$\text{pressure} = \frac{\text{weight}}{\text{area}}$$

weight = density × $g$ × $V$
$= 8000 \times 9.81 \times 8 \times 10^{-6} = 0.628\ \text{N}$

so pressure at points of contact = $0.628/(4 \times 10^{-8})$
$= 16 \times 10^6\ \text{Pa} = 16\ \text{MPa}$

**7. a.** $F = 0.02 \times 50 \times 9.81 = 9.8\ \text{N}$

**b.** Because the normal contact force is now less than the weight. Surfaces are not pressed together as hard.

**8.**

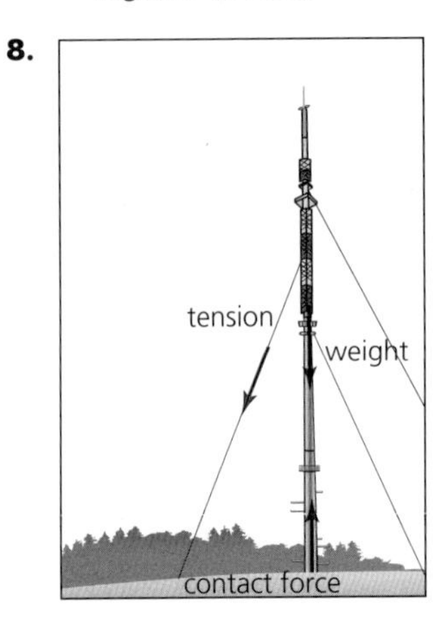

**9.** Boats must displace their own weight of water in order to float. In this case, the volume of water that the boat must displace is eight times the volume of steel used in its construction.

**10.** Liquids do not compress very much, so that the density of sea water is similar at all depths. But gases do compress. The density of the air is much greater at sea level, than, say, at 15 km up in the atmosphere.

**11.** Probably the vertical forces of weight (acting downwards) and contact force from the floor and/or chair (acting upwards) ... that could be it, unless you are outside in the wind.

**12.** Buoyancy (upthrust) (upwards), weight (downwards) and drag (downwards).

**13.** Weight (downwards), contact force from the road (upwards), force from tow-bar (forwards) and drag (air resistance and rolling resistance) (backwards).

**14.** These forces should be marked: lift (upwards), weight (downwards), thrust (forwards) and drag (backwards).

**15. a.** At maximum thrust, the aircraft will accelerate until the drag equals the thrust. The drag, and hence the value of the top speed, will depend on the shape of the aircraft, the air density (so altitude) and the weather conditions.

**b.** The lift depends on the density of air (and so the altitude) and the speed of the aircraft. As the aircraft climbs, the air gets less dense, reducing lift. Eventually, even at top speed, the lift is equal to the weight and the aircraft cannot climb further (other factors are the angle of attack of the wings, the area of the wings, the shape of the wings and aircraft, and the weight of the aircraft).

**16.**

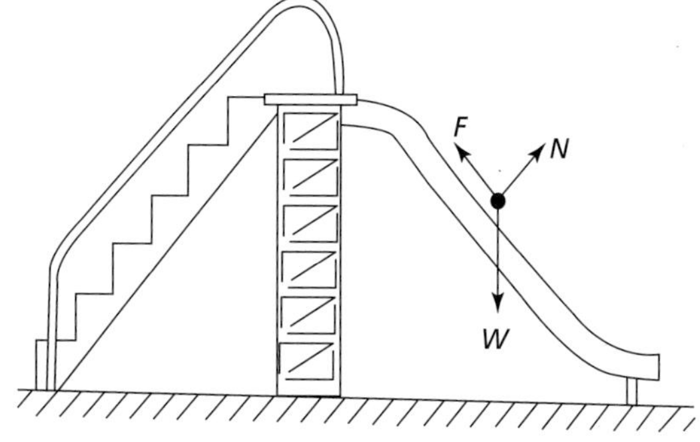


$W$ = child weight
$N$ = normal contact force
$F$ = friction acting on child

**17.**

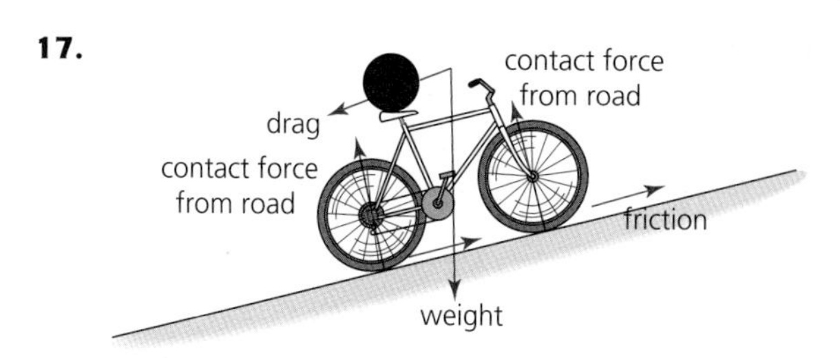


**18.**



**19. a.** Tension in cords, contact force of floor on baby's right foot, and weight of baby.

**b.**

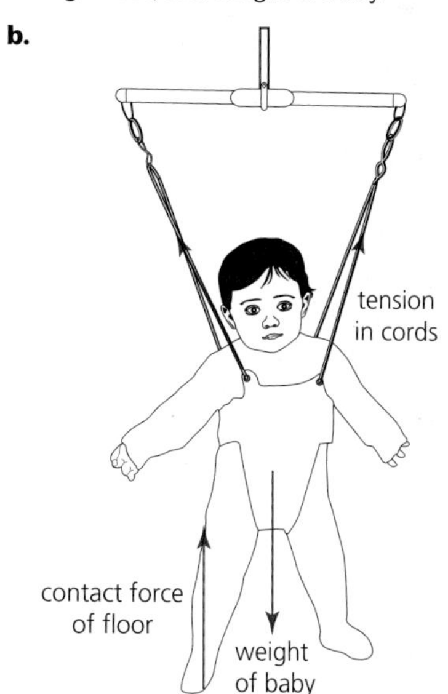


**c.** Tension in cords, and tension in elastic/rubber suspending it.

**d.**

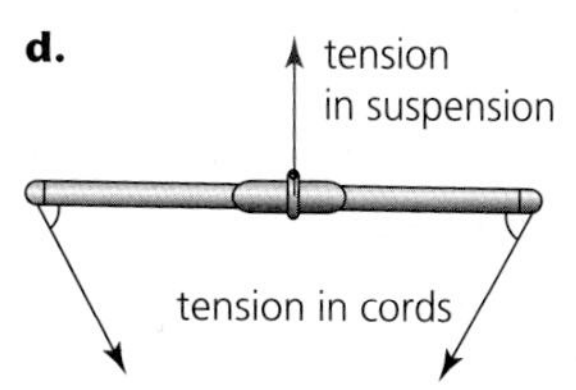


**e.** First, treating the whole baby plus harness as one object, the tension force $F$ in the elastic/rubber suspending the bar is:

$F = 12g + 2g - 10\text{ N} = 14 \times 9.81 - 10 = 127\text{ N}$

The component of the tension $T$ in the two cords, acting vertically downwards, is $2T\cos 30°$. This must equal 127 N (upwards). So

$$2T\cos 30° = 127$$

$$T = \frac{127}{2\cos 30°} = 73.3\text{ N}$$

**20. a.**

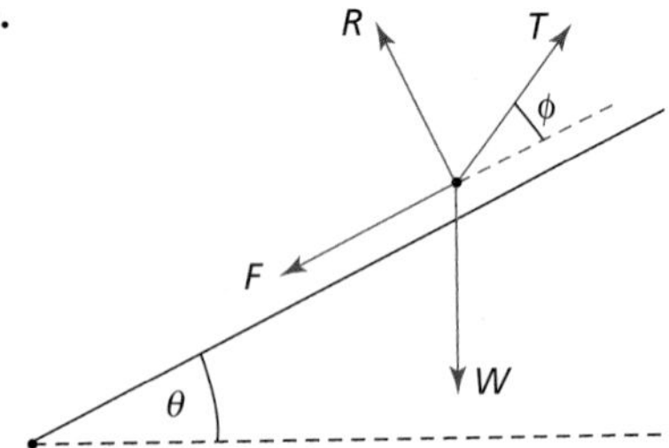


$T$ = tension in tow rope
$F$ = friction
$R$ = normal contact force
$W$ = skier's weight

**b.** Parallel to the slope (upwards positive) the forces are $T\cos\phi$, $-W\sin\theta$, $-F$.

Perpendicular to the slope (upwards positive) the forces are $R$, $T\sin\phi$, $-W\cos\theta$.

**c.** For equilibrium along the slope:

$$F + W\sin\theta = T\cos\phi$$

For equilibrium perpendicular to the slope:

$$R + T\sin\phi = W\cos\theta$$

**21.** $1000g \times \cos 15° = N = 9480\text{ N}$

$1000g \times \sin 15° = F = 2540\text{ N}$

**22. a.** Moments can be equal even if the forces (weights) are not: the lighter person needs to sit further from the pivot.

**b.** A larger-diameter handle provides a greater turning effect (see diagram) and hence a greater force on the screw.

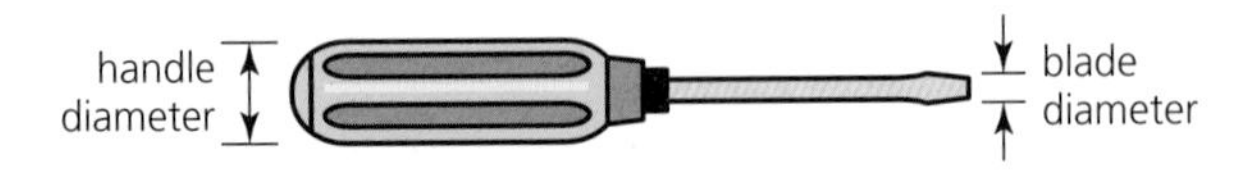


Moment of couple = applied force × handle diameter

= force on screw × screwdriver blade diameter

If the diameter of the handle is 5 times that of the blade, your applied force will be magnified 5 times.

**c.** There was no power steering. As in **b**, the large radius of the steering wheel amplified the turning effect.

**23.** Say Aristotle has a mass of 80 kg, and assume that he can exert a force equal to his weight and that for this very rough estimate $g = 10\text{ N kg}^{-1}$. Then

$$\frac{\text{force required}}{\text{force exerted}} = \frac{6 \times 10^{24}}{80} = 7 \times 10^{22}$$

so that ratio is needed on either side of the lever.

If the Earth sits at one end of the lever, and the fulcrum is one Earth radius away, the other side of the lever must be $7 \times 10^{22}$ Earth radii = $5 \times 10^{29}$ m long (about 10 million times the Galactic radius).

**24.** If the strong person pushes, say, 1 cm from hinge, and the weak person pushes, say, 75 cm away from the hinge, then the strong person needs a force of over 75 times that of the weak person to 'win'. This is very unlikely!

**25. a.**

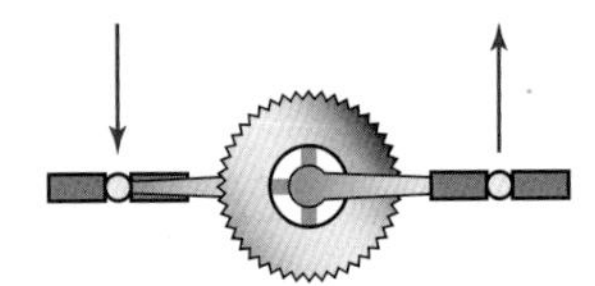

**b.**

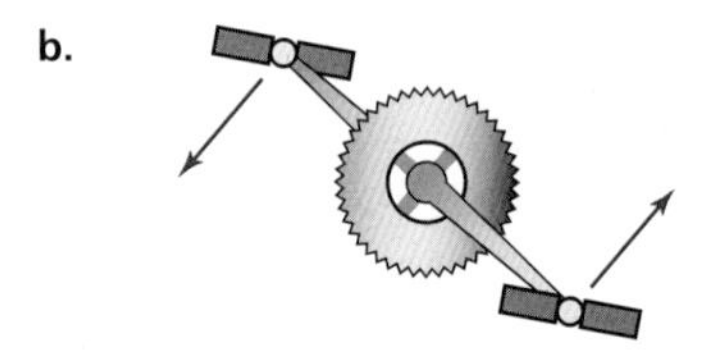

(You would need to have cleats, that is, pedals fastened to your shoes, to be able to exert a force upwards on the pedal.)

**26.** $W \times 1.5 = 20\,000 \times 3.2 = 42\,700\text{ N}$

**27.** The transverse force of the wind tends to tip the board over. The resistive force of the water on the daggerboard helps to balance this.

**28. a.** Treating this as a couple:

moment = force × perpendicular distance

$= 10\text{ N} \times 0.30\text{ m} = 3.0\text{ N m}$

**b.** moment = force × perpendicular distance

$= 50\text{ N} \times 0.20\text{ m} = 10\text{ N m}$

**c.** moment = force × perpendicular distance

$= 5.0\text{ N} \times 0.70\text{ m} = 3.5\text{ N m}$

**29.** **a.** anticlockwise moment = clockwise moment

$X \times 2.5 = 600 \times 2.0$

$X = 1200/2.5 = 480$ N

**b.** anticlockwise moment = clockwise moment

$1000 \times 10 = X \times 2.0$

$X = 5000$ N

**c.** anticlockwise moment = clockwise moment

$0.1 \times 7.5 + 0.1 \times 2.5 = 0.1 \times x$

$x = \frac{(0.75 + 0.25)}{0.1} = 10$ cm

**30.** As the mast is lifted, the angle increases from 25°. The component of $T$ perpendicular to the mast also increases. The moment of the weight decreases, because the perpendicular distance to X decreases.

# 11 FORCES AND MOTION

**1.** In a head-on collision, the front of the car is rapidly brought to a halt. Any rear-seat passengers will continue at their original velocity until acted on by a force, in accordance with Newton's first law. It may take a force equal to the weight of an elephant, say 35 000 N, to stop them. Without rear-seat safety belts, this force could be provided by the front-seat passengers.

**2.** When a car is hit from behind, it is accelerated forwards. The car seats will accelerate with it, pushing a passenger's body forwards. Without an external force to accelerate it, a passenger's head will retain its original velocity, lagging behind the rest of the body, until the muscles of the neck pull it forwards. This results in 'whiplash' injuries. A head restraint provides an external force and accelerates the head at the same rate as the body.

**3.** **a.** $a = \frac{(25.9 - 0)}{\Delta t} = 13 \text{ m s}^{-2}$,
so $\Delta t = 2$ s approximately

**b.** We need to estimate the mass of a cheetah – perhaps a little less than an adult human, say 50 kg. So $F = 50 \times 13 = 650$ N.
An acceleration of 13 m $s^{-2}$ will mean more force on the cheetah than would arise from its own weight, so cheetahs have bones with a larger cross-sectional area to make them stronger.

**4.** A bicycle goes from 0 to 15 km $h^{-1}$ in less than 10 m. Using $v^2 = u^2 + 2as$ gives acceleration ≈ 1 m $s^{-2}$ (professional cyclists achieve almost double this). So force = $ma$ = 100 kg × 1 m $s^{-2}$ = 100 N.

**5.** Using $F = ma$, $F = 10 \text{ kg} \times 200 \text{ m s}^{-2} = 2000$ N. This is close to the force required to beat the world weightlifting record. It would be impossible for an adult to provide this force to restrain a baby in a crash.

**6.** $s = 23 - 9 = 14$ m, $v = 0$ m $s^{-1}$ and $u = \frac{(48.3 \times 1000)}{3600} = 13.4 \text{ m s}^{-1}$
Using $v^2 = u^2 + 2as$,
$a = (v^2 - u^2)/2s = \frac{(-13.4)^2}{28} = -6.4 \text{ m s}^{-2}$
Since $F = ma$, $F = 1500 \times 6.4 = 9600$ N

**7.** The crash barrier takes time to deform. This extends the duration of the collision. Force = rate of change of momentum (Newton's second law), so if the momentum change takes place over a longer time, the force is reduced.

**8.** **a. i.**

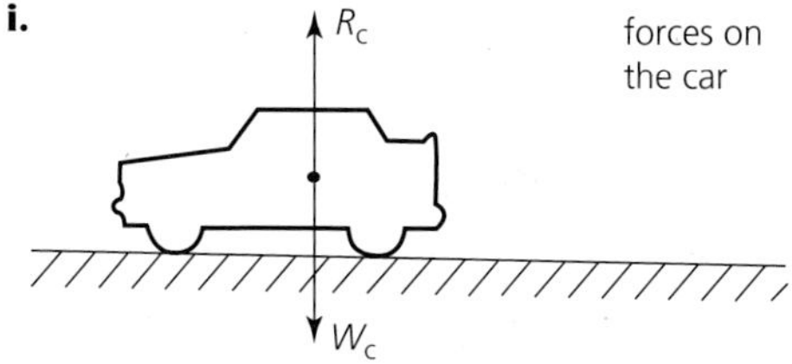


$W_C$ = gravitational pull of the Earth on the car (the car's weight)

$R_C$ = contact force of the Earth's surface on the car

**ii.**

$W_E$
$R_E$
forces on the Earth

$W_E$ = gravitational pull of the car on the Earth

$R_E$ = contact force of the car on the Earth's surface

**b. i.** Because the car and the Earth are in equilibrium, Newton's first law says that there must be no net force, so $R_C = W_C$ and $R_E = W_E$.

**ii.** Newton's third law says $R_C = R_E$ and $W_E = W_C$. The forces in these equations act on different objects.

**9.** A car's wheels are turned by the force provided by the engine. The wheels push the road backwards via the friction between the tyres and the road surface. The road exerts an equal but opposite force on the car (Newton's third law). It is this force that pushes the car forwards.

**10.** **a.** The momentum of the bullet is 0.04 kg × 300 m s$^{-1}$ = 12 kg m s$^{-1}$. If the victim has a mass of 60 kg, then, in the absence of any external forces, such as friction, the victim should move backwards at a speed of 12/60 = 1/5 = 0.2 m s$^{-1}$. This is hardly flying backwards, and in reality the external forces would prevent much movement.

**b.** The shooter should recoil with the same momentum as the victim, in the opposite direction.

**11.** **a.** She should throw the spanner away from the spacecraft. She would gain an equal but opposite momentum towards the spacecraft.

**b.** Momentum of the spanner equals 1 × 4 = 4 kg m s$^{-1}$. The astronaut has an equal but opposite momentum, 4 = 60 × $v$, so $v$ = 4/60 = 1/15 = 0.07 m s$^{-1}$. (This is quite slow but she should get back to the spacecraft before her oxygen is used up!)

**12.** **a.** Newton's third law. As skater A pushes skater B away, skater A will experience an equal but opposite force from skater B.

**b.** Newton's second law. If you knew the force on the skater and their mass, you could calculate the acceleration using $F = ma$.

**c.** Newton's first law. An object will travel at constant velocity, unless acted upon by an external, resultant force.

**13.** Consider the vertical motion to begin with. At first, the car has no vertical momentum. But as it falls, it accelerates, gaining momentum. This does not contravene the conservation of momentum, as gravity is an external force. If we included the Earth in the 'system', then gravity becomes an internal force, and we have to consider the momentum of the Earth, which is equal and opposite to the car's momentum at all times. The Earth actually comes up to meet the car, but since the Earth's mass is so much greater than the car's mass, its velocity will be very much smaller than that of the car.

A similar argument can be made for the horizontal momentum. As the car accelerates forwards, it pushes the Earth backwards. Again, the total momentum is always conserved.

**14.** As the rockets fire, the spacecraft accelerates in the opposite direction. The total momentum of the gas ejected from the back of the spacecraft during the burn is 0.1 kg × 2000 m s$^{-1}$ = 200 kg m s$^{-1}$. The spacecraft will gain this extra momentum in the other direction: 200 = 1000 × $v$, so it will travel 0.2 m s$^{-1}$ faster after the burn. To sum up: the spacecraft moves at a constant velocity of 10.0 m s$^{-1}$ until the rockets fire, when it accelerates uniformly to a speed of 10.2 m s$^{-1}$, and then it maintains this new velocity.

**15.** The eggs need to be accelerated and decelerated gradually. 'Soft' hands are needed, that is, the eggs have to be caught with your arms at full stretch and then slowed down gradually. The change in momentum is unavoidable, but the rate of change of momentum can be reduced. The impulse (force × time) is fixed by the speed and the mass of the egg, but a smaller force applied over a longer time will provide the same impulse, and there will be less chance of the yolk being on you!

**16.** The basketball has a large mass compared to the Powerball. When the basketball bounces on the floor, it changes direction and so collides with the Powerball. In this collision, the basketball will transfer some of its momentum to the Powerball, which has a relatively low mass and therefore gains a high velocity.

**17.** The speed of the ball before hitting the racket can be estimated as 200 km h$^{-1}$.
This is (200 × 10$^3$)/(60 × 60) m s$^{-1}$. Converting its mass to kilograms (60 × 10$^{-3}$ kg), its momentum before it hits the racket is
(60 × 10$^{-3}$ × 200 × 10$^3$)/(60 × 60) kg m s$^{-1}$ = 3.33 kg m s$^{-1}$.
Assume momentum afterwards is the same magnitude, but in the opposite direction.
The change in momentum
= 2 × 3.33 = 6.66 kg m s$^{-1}$. The impact time is about 5 ms, so the force = 6.66/0.005 = 1300 N.

**18.** **a.** Elastic. If the collisions were inelastic, the molecules would lose kinetic energy, move more slowly and so the gas would cool spontaneously.

**b.** Inelastic. A tennis ball loses height with each bounce and is therefore losing kinetic energy.

**c.** Inelastic. Some of the kinetic energy of the electron is transferred to the energy of the X-ray.

**19. a.** Momentum before collision = mass of pellet × velocity = $0.5v$ kg m s$^{-1}$

Momentum after collision = $150.5 \times 0.2 =$ 30.1 kg m s$^{-1}$

These must be equal, so $v = 30.1/0.5 = 60$ m s$^{-1}$

**b.** The uncertainty in the mass of the pellet and glider could be around ±0.1 g, but this is less than 0.1%. The uncertainty in measuring the velocity of the glider depends on the precision of the measurement of the flag. Estimate this as ±1 mm, for a flag of length 5 cm, so approximately 2%.

**c.** We have assumed that there were no external forces acting so that conservation of momentum applies.

**d.** Ensure that the rifle is held firmly in a clamp and aimed at the Plasticine. (A laser beam could help you to align the rifle.) You should ensure that no other students can enter the firing area. You could wear goggles if there is a risk of ricochet.

Repeating the readings would help to reduce the random error.

**20.** Kinetic energy is proportional to velocity squared, so doubling the velocity means that the car has four times the kinetic energy, which would be transferred, partly to the child, in a collision.

**21. a.** In city traffic, you are accelerating and braking frequently. Energy is being transferred from the fuel to increase the kinetic energy of the car, which you then transfer by heating the car's brakes. Since kinetic energy is proportional to mass, a lighter car reduces the energy required and hence the cost. During motorway driving, there are (it is hoped) long periods of driving at constant speed. (Energy is being transferred as work against resistive forces, mainly drag. A car with a smaller frontal area and more streamlined shape will reduce energy demands and hence cost.)

**b.** Reducing your speed helps in both cases. In the city, since kinetic energy is proportional to $v^2$, cutting your speed reduces the kinetic energy that you lose each time you brake. On the motorway, resistive forces also depend on $v^2$, so cutting your speed will reduce the work done against these forces.

**22. a.** Every minute 200 kg of potatoes are accelerated from 0 to 3 m s$^{-1}$, which is
$E_k = \frac{1}{2}mv^2 = \frac{1}{2} \times 200 \times 3^2 = 900$ J

**b.** $900/60 = 15$ J s$^{-1}$

**c.** The belt will be working against friction, so energy will be transferred as heat, and not all the energy will go into moving the potatoes.

**23.** The minimum energy is the gravitational potential energy at the height to be gained:

$E_p = mg\Delta h = 150 \times 9.81 \times 20 = 29\,400$ J

**24.** Suppose that the height between floors is 5 m and the escalator has steps 0.3 m high, so that there are 16 steps between floors. Suppose the escalator has one person on every step. Taking their average mass to be 60 kg, the escalator has to lift 1000 kg (neglecting its own mass). Say it takes 20 s to move between floors. The gain in gravitational potential energy is
$E_p = mg\,\Delta h = 1000 \times 9.81 \times 5 = 49\,000$ J
But this is in 20 s, so in 1 s energy needed ≈ 2500 J.

**25.** Impulse is equal to change in momentum. First estimate the mass of the child, say 30 kg, and the height fallen as 2 m.

Loss of potential energy = $mg\,\Delta h$, which has been transferred to kinetic energy = $\frac{1}{2}mv^2$. Equating these gives $v^2 = 2 \times 9.81 \times 2 \approx 40$, so $v \approx$ 6.3 m s$^{-1}$.

Just before hitting the trampoline, the child's momentum is $30 \times 6.3$ kg m s$^{-1}$; and just after it, neglecting the effect of any external force, the momentum will be $-30 \times 6.3$ kg m s$^{-1}$, equal but in the other direction (assuming an elastic collision).

Impulse = change in momentum = $2 \times 30 \times 6.3 \approx$ 380 kg m s$^{-1}$ = 380 N s

**26. a.** Power = force × velocity, so

$F = P/v = 112\,000/8.9 = 12.6$ kN at 20 mph

$F = P/v = 112\,000/17.8 = 6.3$ kN at 40 mph

**b.** To find acceleration, we need to use $a = F/m$, but $F$ is the resultant force. We need to know the resistive force acting to find the resultant force.

**27. a.** Drag increases with speed (proportional to $v^2$ at higher speeds). Eventually, the resistive forces are as large as the maximum force that the vehicle can produce. The vehicle can no longer accelerate.

**b.** Maximum force (torque) provided by the engine.

By improving the shape (aerodynamics) of the vehicle.

**28.** **a.** Yes. Lower gear means greater torque and so greater force on the road and lower velocity (power = $Fv$; at constant power if $F$ increases, $v$ decreases).

**b.** No. As in **a**, the cyclist will pedal faster with reduced force whereas the bike will exert greater force at lower velocity.

**c.** No. Gears cannot add energy; in fact, due to friction, the output power is likely to decrease slightly.

**29.** The typical weight of an adult is, say, 650 N; height of a flight of stairs is, say, 3 m. So potential energy gained = work done in climbing stairs = $mg\,\Delta h$ = $650 \times 3 = 1950$ J.

$$\frac{1100000}{1950} = 564 \text{ times}$$

**30.** **a.** Kinetic energy of the air (wind)

**b.** $\frac{20\text{kW}}{0.6} = 33$ kW

**c.** To get all the kinetic energy would mean actually stopping the wind that blew through the rotor blades. This would build up as a high-pressure air 'wall' and the wind turbine would stop working.

**31.** **a.** Input energy = food; output energy = kinetic energy and any gain in gravitational potential energy

**b.** Heat in the human 'engine', friction at the pedals, gears, axles and so on

**c.** You would probably need a closed laboratory with a rolling road or a fixed frame in a wind tunnel, for each bike. All energy inputs would be monitored closely (incidental heating, food, and so on). Different designs could be examined under identical circumstances.

**d.** The low road position means that traffic cannot see you and you are at the same level as most exhaust pipes. It is difficult to climb hills, and carrying panniers can be difficult.

**32.** 3.0 gallons = 13.64 litres, so the power input is $13.64 \times 38.6 = 526.5$ MJ per hour

Output power = 30 kW = $(30 \times 60 \times 60)$ kJ per hour = 108 MJ per hour

Efficiency = $\frac{108}{526.5} = 20.5\%$

**33.** Simply comparing the energy used is not enough. You need to take into account the number of passengers and the distance travelled. You would need to compare the energy used (joules) per passenger per kilometre, so joules per passenger kilometre.

# 12 THE STRENGTH OF MATERIALS

**1.** 1 kg = 1000 g = $10^3$ g,
so 1000 kg = $1000 \times 10^3 = 10^6$ g

1 m = 100 cm = $10^2$ cm
1 $m^3$ = $100 \times 100 \times 100$ $cm^3$ = $10^6$ $cm^3$

So 1000 kg $m^{-3}$ = $10^6$ g/$10^6$ $cm^3$ = 1 g $cm^{-3}$

**2.** Volume = mass/density
= 1/21 450 = $47 \times 10^{-6}$ $m^3$

**3.** Cylinder volume = $\pi r^2 h = \pi \times (0.25)^2 \times 1.0 = 0.196$ $m^3$

Since mass = volume × density, then mass of water = $0.196 \times 1000 = 196$ kg

**4.** Sugar comes in 1 kg bags that measure about 20 cm × 10 cm × 5 cm = 1000 $cm^3$. This gives an average density of 1 kg $\text{litre}^{-1}$ or 1 g $cm^{-3}$ or 1000 kg $m^{-3}$. (The published density of granulated sugar is 1.59 g $cm^{-3}$, so there is a significant amount of space between the granules.)

**5.** A pack of butter has a mass of 250 g and measures about 10 cm × 6 cm × 4.5 cm = 270 $cm^3$, which gives a density of 250/270 = 0.93 g $cm^{-3}$. (The actual density of butter is 0.94 g $cm^{-3}$ ± 15% depending on water content.)

**6.** **a.** $F = A\sigma = \pi \times (1 \times 10^{-3})^2 \times 85 \times 10^6 = 267$ N

(so a fish of mass approximately 27 kg!)

**b.** $F = 267 \times 5 = 1335$ N

$A = F/\sigma = 1335/(85 \times 10^6)$
$= 15.7 \times 10^{-6}$ $m^2$, so diameter = 4.5 mm

(or by ratios)

**7.** Strain = extension/length = $(7.5 \times 10^{-2})/50 = 1.5 \times 10^{-3}$

**8.** **a.** Copper

**b.** Blu-tack, Plasticine, clay

**c.** Rubber band

**d.** Glass, ceramics

**e.** Rubbers, alloys, composite materials

**9.** The limit of proportionality is when Hooke's law no longer applies (where the force–extension graph ceases to be a straight line through the origin). The elastic limit is the highest stress before permanent deformation takes place.

**10.** **a.** $k = F/\Delta l = (0.2 \times 9.81)/(15 \times 10^{-3}) = 131\ \text{N m}^{-1}$

**b.** $E = \frac{1}{2}k(\Delta l)^2 = 0.5 \times 131 \times (0.035)^2 = 0.080\ \text{J}$

**c.** $mg\,\Delta h = 0.080\ \text{J}$, so $\Delta h = 0.08/(0.2 \times 9.81) = 4.1\ \text{cm}$

**11.** **a.** Maximum load $\approx$ 100 N, an extension of 5 cm would be manageable and this would give a scale of 1 mm = 2 N or 200 g. Need $k = 100/0.05 = 2000\ \text{N m}^{-1}$.

**b.** Uncertainty of about 200 g, so $5 \pm 0.2$ kg, which is a percentage uncertainty of 4%.

**c.** A high spring constant would mean that a large mass could be weighed without stretching the spring too far, but this rather 'stiff' spring would not be very sensitive. So the student is right.

**12.** **a.** The material absorbs the energy by deforming, doing work against internal forces. Work = force × distance, so a long extension at a relatively high force will do a lot of work.

The area below the stress–strain curve is a measure of toughness. So high strength (high stress before breaking) combined with long ductile region (large strain before breaking) lead to high toughness.

**b.** Sketch should show straight-line region only near the origin, and breaking point at high value of stress and high value of strain. The actual shape for synthetic rubber is shown here.

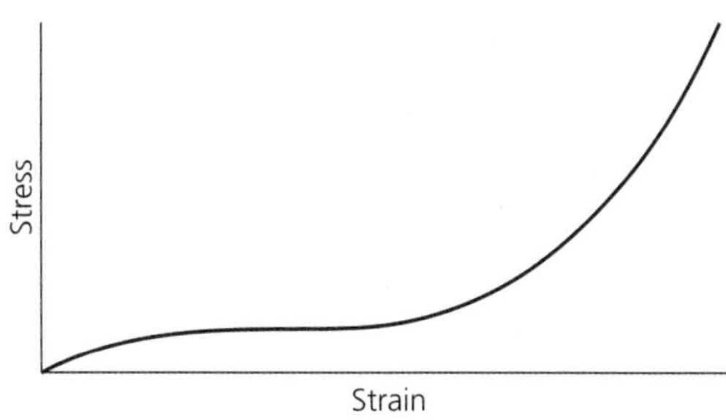


**13.** **a.** B **b.** A **c.** C **d.** C **e.** A

**f.** B (and C to some extent)

**14.** Specific strength of Kevlar is 2.15 MPa $\text{kg}^{-1}\ \text{m}^3$.

Specific stiffness of Kevlar is 86.1 MPa $\text{kg}^{-1}\ \text{m}^3$.

Specific strength of carbon fibre is 1.09 MPa $\text{kg}^{-1}\ \text{m}^3$.

Specific stiffness of carbon fibre is 154 MPa $\text{kg}^{-1}\ \text{m}^3$.

Kevlar is the better choice.

Although carbon fibre is stiffer, it is not as strong.

**15.** **a.** Lower resilience means a larger hysteresis loop, that is, more energy is transferred as heat in the material.

**b.** The lower resilience of silk cords would mean that more energy is dissipated, so there would be fewer bounces than using rubber cords. Because silk is stiffer, it would mean that the cords would stretch less and would decelerate the jumper more quickly.

**16.** The graph for the cold ball is a hysteresis loop with a large area; the graph for the warm ball is a hysteresis loop with a smaller area. Initially the ball is cold and not very bouncy, so each collision is inelastic and energy is transferred to the ball as heat.

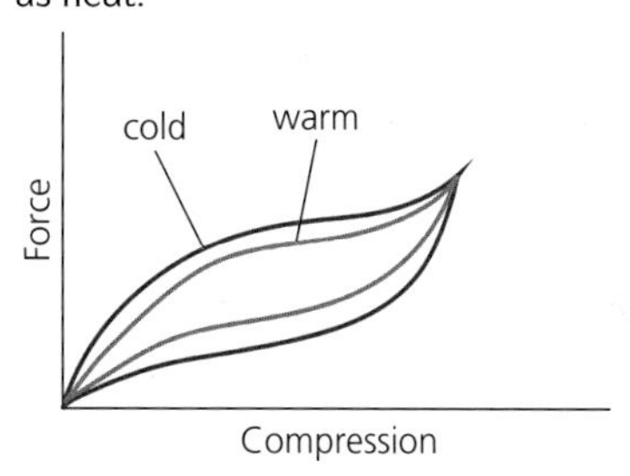


**17.** Increasing the mass of the jumper means that the cords stretch further, and more energy is stored as elastic strain energy. When the bag is dropped at the bottom of the jump, the mass is decreased, and so the jumper rebounds much higher, possibly crashing into the platform they left only a brief moment earlier.

## 13 ELECTRICITY 1

**1.** $1000\,000 \times 60 \times 60 = 3.6 \times 10^9\ \text{J} = 3.6\ \text{GJ}$

**2.** **a.** $100 \times 24 \times 60 \times 60 = 8.6 \times 10^6\ \text{J}$
$0.1 \times 24 = 2.4\ \text{kWh}$

**b.** $(18/100) \times 8.6 \times 10^6 = 1.5 \times 10^6\ \text{J}$
$(18/100) \times 2.4 = 0.43\ \text{kWh}$

**3.** **a.** $26.4 \times 1000 \times 60 \times 60 = 95\ \text{MJ}$

**b.** $(95 \times 10^6)/(3 \times 60 \times 60)$ or $26.4/3 = 8.8\ \text{kW}$

**4.** The fluorescent tube uses 80 W power. You would need only half of that with the new LED tube, so 40 W. Estimate that the light is on for 3 hours per day. Energy per day = 0.040 kW × 3 h = 0.12 kWh per day. Multiplying by 365 gives 44 kWh per year, which is about 10p × 44 = £4.40 per year.

**5.** $1/(1.6 \times 10^{-19}) = 6.25 \times 10^{18}$ electrons

**6.** The energy to make the electrons in an insulator mobile is large on an atomic scale, around 3 eV per electron. A huge number of electrons are needed to form a current (1 μA means that $6 \times 10^{12}$ electrons pass a point every second). Our rubbing of two surfaces together may transfer energy of the order of a few joules – more than enough to free electrons ($1 \text{ eV} = 1.6 \times 10^{-19}$ J).

**7.** $0.25 \times 60 \times 60 = 900$ C

**8.** $(0.1 \times 10^{-3})/(1.6 \times 10^{-19}) = 6.25 \times 10^{14}$ electrons

**9.** **a.** $I_1 = I_2 + I_3 = I_4$

**b. i.** Not true. Current does not get 'used' as it flows through a resistor.

**ii.** Not always true. (Only true if the two resistors have the same value.)

**iii.** Not true. Current does not get 'used' as it flows through a resistor.

**iv.** True. The current divided to pass through the parallel section but has joined back together.

**v.** True. The current divided to pass through the parallel section but has joined back together.

**10.** $E = QV = 30 \times 2 = 60$ J

**11.** **a.** $12 - 4.0 = 8.0$ V

**b.** 8.0 joules per coulomb; charge $Q = It = 3.0 \times 20 \times 60 = 3600$ C in 20 minutes; so $8.0 \times 3600 = 28\,800$ J in 20 minutes

**c.** Power = energy per second = $28\,800/(20 \times 60) = 24$ W

**12.** **a.** 1.5 V across each

**b.** $0.30 - 0.05 - 0.15 = 0.10$ A

**c. i.–iii.** 1.5 J in each

**d.** Energy $E = Q \times V$. Current $I$ is charge $Q$ per second. So energy per second is $I \times V$.

**i.** $0.05 \times 1.5 = 0.075$ J

**ii.** $0.15 \times 1.5 = 0.225$ J

**iii.** $0.10 \times 1.5 = 0.15$ J

**iv.** 0.45 J

**e.** $1.5 \times 0.30 = 0.45$ J

**13.** $I = P/V = 1400/240 = 5.8$ A, so a 13 A fuse

**14.** $I = P/V = 7000/240 = 29$ A. Standard three-pin plugs are limited to 13 A. There are also issues of the water and electrical connections.

**15.** If the toaster takes 2 min = 120 seconds, then

energy transferred $= P \times t = I \times V \times t$
$= 5 \times 240 \times 120 = 144\,000$ J $= 144$ kJ

**16.** Although an individual electron moves slowly round the circuit, all the electrons in the circuit begin to move (and so transfer energy) almost instantaneously when you press the switch. (The change in the electric field travels at the speed of light; there will be more on this in Book 2.)

**17.** The velocities are random and there is no net movement of charge in any particular direction.

**18.** More current means more electrons passing a point every second, more collisions and so more vibrational energy in the lattice, leading to a higher temperature.

**19.** So that the power delivered (by the amplifier) is not dissipated by heating the wires.

**20.** $R$ is in the range 0.00005 Ω to 0.0005 Ω. For $R = 0.00005$ Ω:

$V = IR = 10 \times 1000 \times 0.00005 = 0.5$ V

So the pd will be in the range 0.5 V to 5 V.

**21.** $R = V/I = 4.0/6.0 = 0.67\,\Omega$

$V = IR = 9.0 \times 0.67 = 6.0$ V

Assumed that $R$ will stay the same. That is probably not true, as $R$ may increase with temperature.

**22.** **a.** $R = V/I = 2.0/0.008 = 250\,\Omega$

**b.** $P = IV = 8 \text{ mA} \times 2.0 \text{ V} = 16$ mW

**c.** A large current would flow, permanently damaging the LED.

**d. i.** pd = 1.5, 2.0 and 3.0 V for the red, yellow and green LEDs, respectively

**ii.** Let the potential difference across the resistor, $R_r$, in the red LED branch be $V_r$. Then $V_r + 1.5 = 12$. So $V_r = 10.5$ V.

For a current of 20 mA

$R_r = V_r/0.020 = 10.5/0.020 = 525\,\Omega$

Similarly for the other branches.

The resistors need to be 525 Ω (red), 500 Ω (yellow) and 450 Ω (green).

**iii.** At 2.0 V the red LED would have a large current characteristic, probably damaging it. The yellow LED would operate correctly, but the green LED would only have a low power and be hardly lit.

**23.** **a.** $R_{lamp} = V^2/P = 12^2/48 = 3\ \Omega$
$1/R_{total} = 1/3 + 1/3 = 2/3$
$R_{total} = 3/2 = 1.5\ \Omega$
$I = V/R_{total} = 12/1.5 = 8.0$ A

**b.** 12 Ω in parallel with 6 Ω
$1/R_{total} = 1/12 + 1/6 = 3/12$
$R_{total} = 12/3 = 4\ \Omega$
$I = V/R_{total} = 12/4 = 3.0$ A

**c.** 30 Ω in parallel with 15 Ω
$1/R_{total} = 1/30 + 1/15 = 3/30$
$R_{total} = 30/3 = 10\ \Omega$
$I = V/R_{total} = 12/10 = 1.2$ A

**24.** **a.** 12 V

**b.** 11 V

**c.** 8 V

**25.** **a.** Battery pd is 42 V, giving a current of 143 A. Power loss in the wires is 102 W. Efficiency is 98%.

**b.** This pd is safer and loss of efficiency is small.

**26.** **a.** Y

**b.** $R = \rho l/A = 1.68 + 10^{-8} \times 2.00 / (\pi \times (0.45 \times 10^{-3})^2) = 0.053\ \Omega$

**c.** $R = \rho(2l)/\pi(2r)^2$. So the resistance of X is $\frac{1}{2}$ that of Y, 0.026 Ω.

**27.** $A = \pi(0.25 \times 10^{-3})^2 = 1.96 \times 10^{-7}\ \text{m}^2$

$R = \rho l/A$

so $l = AR/\rho$

$$= \frac{1.96 \times 10^{-7} \times 10.0}{5.00 \times 10^{-7}}$$

$= 3.93$ m

# 14 ELECTRICITY 2

**1.** **a.** 1.2 V is 1.2 joules per coulomb. In 1 minute, $60 \times 1.0\ \text{A} = 60$ C will flow, so $60 \times 1.2 = 72$ J.

**b.** 'Lost volts' $= Ir = 1 \times 0.1 = 0.1$ V, so terminal pd $= 1.2 - 0.1 = 1.1$ V

**c.** Energy 'lost' as heat in the internal resistance $= I^2r = 1 \times 1 \times 0.1 = 0.1$ J every second, so 6 J in 1 minute. Efficiency $= (72 - 6)/72 = 92\%$

**2.** The headlights and stereo are designed to work at 12 V. When the starter motor operates, a large current flows from the battery, around 30 A. This means that there is a significant voltage drop across the internal resistance, so the terminal voltage drops, from close to 12 V, to a much lower value.

**3.** The emf $= 168 + 120 \times 0.1 = 180$ V

**4.** The equation for power is $P = I^2R$. So $I^2 = P/R = 112/7.0 = 16$, giving $I = 4.0$ A.
But $V_{term} = IR = 4.0 \times 7.0 = 28$ V, so 2.0 V dropped across $r$, giving $r = V_{int\ res}/I = 2/4 = 0.5\ \Omega$.

**5.** **a.** $V_{term}/R = I = 2.13\ \mu\text{A}$. So the potential difference across the internal resistance is $V_{int\ res} = Ir = 2.13 \times 10^{-6} \times 6.9 = 14.7\ \mu\text{V}$, a difference from the true value of $7 \times 10^{-5}$ %.

**b.** Lower resistance means more current drawn through the voltmeter. So $V/R = 21.3/207 = 0.103$ A, so pd across $r$ is 0.71 V. Voltmeter would read $21.3 - 0.71 = 20.6$ V, a percentage error of 3%.

**6.** **a.** Silver, copper and aluminium are all good conductors, whose resistance changes slowly with temperature. Copper is cheaper than silver, and a better conductor than aluminium. So copper is used in circuitry (except where its high density gives a weight problem, for example, in the transmission lines for the National Grid).

**b.** We need a metal whose resistivity changes quickly with temperature, such as platinum. The linear response of platinum makes it ideal.

**7.** **a.** At 0 °C, the resistance of the thermistor is 300 Ω. Then $I = V/R = 1.5/300 = 5$ mA.

**b.** At 25 °C, the resistance of the thermistor is 100 Ω. Then $I = 1.5/100 = 15$ mA.

**8.** Thermistor B is the best choice. It covers the range between 0 and 100 °C, and has a wider variation of resistance in that range than A, so it should give a more precise reading. The physical size of the thermistor is important. A small thermometer can be used to measure temperatures with better spatial resolution, and also better resolution in time, since a smaller temperature probe will take less time to respond to fluctuating temperatures.

**9.** **a.** $I = V/R = 1.5/(1 \times 10^6) = 1.5\ \mu\text{A}$

**b.** $I = V/R = 1.5/5000 = 0.3$ mA

**10.** Refrigeration is expensive, especially if liquid helium is needed. A room-temperature superconductor would mean that connecting cables would have zero resistance and therefore lose no energy and generate no heat.

**11.** Generating a strong magnetic field with an electromagnet needs large currents. This would normally lead to high energy losses and high temperatures due to ohmic heating.

**12.** **a.** 6 V

**b.** 75 V

**c.** 4 V

**d.** 6 V

**13.** Adding a 10 Ω resistor in parallel will reduce the resistance of that part of the potential divider, and its share of the potential difference would therefore drop. The potential difference with a device connected would be less than that measured with a voltmeter.

**14.** She could use the sensor circuit shown in Figure 25, but with the LDR and the fixed resistor swapped over. The LDR has to be inside the biscuit tin. In the dark, the resistance of the LDR is very high and so the potential $V_{out}$ is close to 0 V (low). If the lid is lifted, light will fall on the LDR, its resistance will fall, most of the potential difference will be across the fixed resistor and $V_{out}$ would go high.

**15.** The circuit should be a potential divider with $R_1$ as a thermistor, as shown. Thermistor A from Figure 11 would be suitable: its resistance is about 100 Ω at 25 °C. $R_2$ should be 100 Ω, and the power supply 6 V.

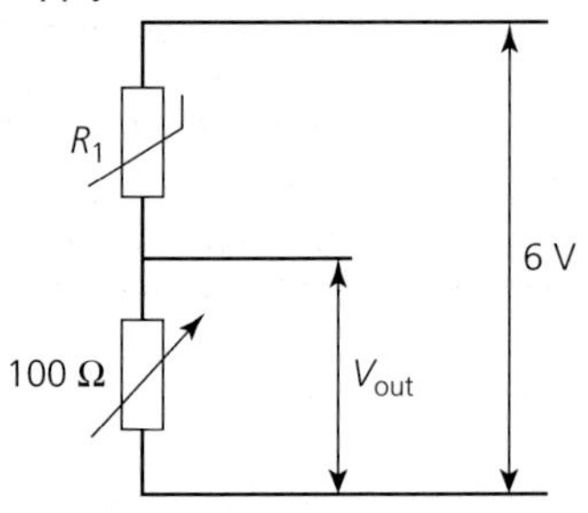


**16.** **a.** Use $V_{out} = V_{in} \times R_V/(R_V + R_T)$, so
$1 = 12 \times R_V/(R_V + R_T)$
giving $R_V = (1/11) \times R_T$.

At 0 °C, from the graph $R_T = 33$ kΩ,
so $R_V = 3$ kΩ.

**b.** $I = V/(R_V + R_T) = 12$ V/36 kΩ = 0.3 mA

**c.** Use $V_{out} = V_{in} \times R_V/(R_V + R_T)$, with $R_V = 3$ kΩ and $V_{out} = 5$ V. So $5 = 12 \times R_V/(R_V + R_T)$ giving $R_T = (7/5)R_V = 4$ kΩ. From the graph, the temperature is approximately 45 °C.

**d.** At 0 °C, $V_{out} = 12 \times 30/63 = 5.7$ V.

At 45 °C, $V_{out} = 12 \times 30/34 = 10.6$ V.

So the range of output voltage would be approximately 5 V compared with 4 V.

There would be a slight improvement in the sensitivity, on average 0.11 V per °C compared with 0.09 V per °C (and a lower current, so less energy used).

# GLOSSARY

**Absolute zero** Zero on the *kelvin scale* of temperature: the lowest possible temperature, at which there is zero kinetic energy of the particles (except for quantum-mechanical motion).

**Absorption** A process in which a photon of light may interact with an electron in an atom, in glass for example, and be absorbed.

**Absorption spectrum** Dark lines at specific frequencies on an otherwise continuous bright spectrum.

**Acceleration** The rate of change of velocity: the change in velocity in unit time; unit m $s^{-2}$.

**Accurate** Describes a reading that is very close to the true value.

**Activity** The number of emissions per second from a radioactive source, measured in becquerels (Bq); 1 Bq = 1 emission per second.

**Aerofoil** The curved shape of a wing as seen in cross-section.

**Air resistance** A force that acts to oppose motion through the air.

**Ampere (amp)** The SI unit used to measure electrical current; symbol A.

**Amplitude** The maximum height of a wave, or the largest displacement from equilibrium.

**Angle of incidence** The angle with the normal (perpendicular to the surface) at which a ray strikes a surface.

**Antimatter** The term used to describe matter composed of antiparticles; it is created by high-energy collisions or particle decays but exists only in small amounts in the observable Universe.

**Antineutrino** The antiparticle of a neutrino; emitted during the process of beta decay.

**Antinode** A point on a *stationary wave* where the displacement is maximum.

**Antiparticle** A particle that is identical in mass to another, more common particle, but opposite in other properties, e.g. charge, baryon number, strangeness. A bar over the particle's symbol indicates that it is an antiparticle.

**Antiphase** Two points on a wave, or points on two waves, are in antiphase if their vibrations are 180° out of phase with each other.

**Archimedes' principle** A law stating that the upthrust on an object is equal to the weight of fluid displaced by the object.

**Atomic mass unit** A small unit of mass used in nuclear physics defined as 1/12 of the mass of a carbon-12 atom; symbol u.

**Attenuation** The reduction of intensity as a wave travels through a material.

**Attenuation coefficient** A measure of the attenuation of waves caused by the absorption of energy as they travel through a medium; measured in *decibels* per kilometre.

**Baryon** A particle (a type of *hadron*) made up of three quarks, e.g. a proton.

**Base units** The seven basic SI units of measurement: metre, kilogram, second, ampere, kelvin, mole, candela.

**Bending losses** Attenuation in an optical fibre when light strikes the cladding at an angle smaller than the critical angle, causing it to not be reflected back into the core.

**Bohr model** A model of atomic structure in which electrons travel round the nucleus in a number of orbits at fixed distances from the nucleus.

**Breaking stress** The maximum stress (force per unit area) that a material can withstand before it breaks.

**Brittle** Describes a material that does not deform plastically before it fractures.

**Cathode rays** The rays emitted from the cathode in a low-pressure gas tube; shown to be fast-moving electrons by J.J. Thompson.

**Centre of gravity** The point at which the weight of an object can be taken to act; an object will balance if it is supported at its centre of gravity.

**Centre of mass** The point at which the mass of an object can be taken to be concentrated; in a uniform gravitational field, this is the same point as the *centre of gravity*.

**Charge** A fundamental property of matter; charged objects experience electrostatic forces. Charge may be positive, zero or negative; unit *coulomb*, C.

**Chromatic dispersion** The splitting of waves into different component wavelengths during refraction, e.g. producing a coloured spectrum by a prism; also called material dispersion.

**Cladding** The outer covering of an optical fibre that confines the light to the core.

**Closed polygon of forces** The result of the vector addition of all the individual forces on an object in *equilibrium* (zero resultant force).

**Coefficient of friction** The ratio of the frictional force (the force required to start or maintain relative motion) between two surfaces to the *normal contact force* between them.

**Coherent** Describes waves that are in a constant phase relationship (and, in the case of transverse waves, have the same polarisation); the waves necessarily have the same frequency.

**Component** A vector can be considered made up of two perpendicular components; the vertical component, for example, is the part of the vector that acts in a vertical direction. Also, an electrical device in a circuit in which energy is transferred.

**Compression** An object in compression is under the influence of forces that tend to squash it.

**Conduction (electrical)** The flow of electric charges through a material.

**Conductor (electrical)** A material that allows an electric current to pass through it.

**Conductivity** A measure of how well a material conducts electricity; the inverse of resistivity; unit $\Omega^{-1}$ $m^{-1}$ or S $m^{-1}$.

**Conservation of charge** The total charge before and after a reaction between particles is always the same. Also, the rate of flow of charge into a circuit junction must equal the rate of flow of charge out (Kirchhoff's first law).

**Constructive interference** Interference between waves that results in an increased amplitude.

**Contact force** An electromagnetic force exerted when two solid surfaces touch each other; also known as the 'reaction'.

**Core** The inner layer of an optical fibre, made from very pure glass, along which light is transmitted.

**Cosmic rays** High-energy particles from space that travel close to the speed of light.

**Coulomb** The unit of measurement of electric charge, symbol C, equal to the quantity of charge transferred in one second by a current of one ampere.

**Couple** Two equal, opposite and parallel forces, not in the same line; causes a turning effect.

**Critical angle** For a ray of light moving from an optically dense medium to a less dense medium, the maximum angle of incidence at which refraction takes place; above the critical angle all the light is internally reflected.

**Critical temperature** The temperature below which a material becomes superconducting, i.e. its electrical resistivity drops to zero. (The term 'critical temperature' can also refer to a gas, when it is the highest temperature at which a gas can be liquefied.)

**Current** Electric current, $I$, is the charge that passes a point in unit time: $I = \Delta Q/\Delta t$; unit A.

**Daughter nucleus** The nucleus that remains after radioactive decay has taken place.

**De Broglie wavelength** Particles exhibit wave-like behaviour with a wavelength equal to $h/p$, where $h$ is the Planck constant and $p$ is the particle's momentum. This is known as the De Broglie wavelength.

**Decay** The change that occurs in an unstable nucleus when it emits radioactivity.

**Decibel** A unit of measurement for the ratio of intensities of sound; abbreviated to dB.

**Density** The mass per unit volume; symbol $\lambda$; unit kg $m^{-3}$.

**Derived units** Units of measurement that can be defined in terms of the seven *base units* of the SI system, e.g. 1 newton (N) = 1 kg m $s^{-2}$ .

**Destructive interference** Interference of waves leading to decreased wave amplitude, or complete cancellation.

**Diffraction** The spreading out of waves after passing through an opening or around an obstacle, to occupy areas that would otherwise be in shadow.

**Diffraction grating** A grating of many narrow slits, each causing diffraction. These diffracted waves each then interfere.

**Diffraction grating formula** The formula used to relate the path difference of diffracted beams to the distance between each slit, $d$, in a diffraction grating: $d \sin \theta = n\lambda$.

**Diode** An electric circuit device made of semiconducting material that has low resistance for current in one direction (when forward biased) but very high resistance for current in the opposite direction (when reverse biased).

**Displacement** A vector quantity describing the difference in position of two points; SI unit m.

**Dissipation** A process in which energy is transferred without accomplishing useful work.

**Drag** A resistive force, such as air resistance, which acts to oppose motion in a fluid.

**Ductile** Describes materials that show extended plastic deformation and become elongated under tension; a ductile metal can be drawn out into wires.

**Dynamic friction** The frictional force between two surfaces in relative motion; it acts to oppose the sliding.

**Efficiency** The ratio of useful energy transferred (or work done) to total input energy; or the ratio of useful output power to input power. Efficiency is always less than 1.

**Elastic** Describes a material that returns to its original dimensions after a deforming force is removed. Also, describes a collision between objects in which no kinetic energy is transferred to other forms.

**Elastic limit** The maximum tensile force at which a material behaves elastically, i.e. returns to its original dimensions after the force is removed.

**Elastic strain energy** The potential energy stored in an elastic material that has been stretched; it is equal to the work done in stretching.

**Elasticity** The property of a material that returns to its original dimensions after a deforming force is removed.

**Electromagnetic (EM) spectrum** Electromagnetic waves ordered according to wavelength and frequency, ranging from low-frequency radio waves to high-frequency gamma rays.

**Electromagnetic force** The force that acts between all charged particles, for example holding atoms together and binding atoms into molecules.

**Electromotive force (emf)** The energy transferred by a power source to each coulomb of charge; equal to the potential difference at the terminals of the source when no current is flowing; unit volt, V.

**Electronvolt** A unit of energy, equal to the energy transferred when an electron moves through a potential difference of 1 V; abbreviated to eV.

**Electrostatic** Describes electric phenomena such as static charges, in which there is no continuous flow of electric charge.

**Emission spectrum** The range of wavelengths emitted by a luminous source; it may be a line spectrum, a band spectrum or a continuous spectrum.

**Energy level diagram** A graphic representation of the possible values of orbital electron energy, in the Bohr model of the atom.

**Energy quantum (plural quanta)** Discrete packet of radiation energy; now known as a photon.

**Equilibrium** A state describing an object that is stationary or moving at a constant velocity.

**Exchange particle** A *virtual particle* that transfers force between two subatomic particles; for example the virtual photon is the exchange particle that transfers electromagnetic force; also called a gauge boson.

**Excitation** The process of raising an atom to a higher energy level, for example by a collision with a free electron or the absorption of a photon raising an orbital electron to a higher orbit.

**Extension** The increase in length of a material caused by a tensile force.

**Flavour** A word for the type of quark; may be up, down, top, bottom, strange or charm.

**Fluorescence** The emission of ultraviolet light by a material as a result of its atoms being excited or ionised.

**Fracture** The breaking of brittle materials.

**Frame of reference** A set of imaginary axes that define the position of a body in space and time.

**Free fall** A state in which the only force acting on an object is the gravitational force.

**Free-body diagram** A simplified picture of a physical situation that shows all the relevant forces acting on a body.

**Frequency** The number of oscillations, or waves, in one second; unit hertz, Hz.

**Frictional force** A resistive force between two solid surfaces in contact with each other, which opposes relative motion.

**Fringe separation** The distance between adjacent light and dark fringes in a light interference pattern.

**Fringes** Bands of maxima and minima intensity produced by light interference.

**Fundamental particle** A particle without any known internal structure, e.g. the electron.

**Fuse** Electrical device for limiting the size of a current in a circuit.

**Gauge boson** A *virtual particle* that transfers force between two subatomic particles; also called an *exchange particle*.

**Gluon** The gauge boson that transfers the strong nuclear force between quarks.

**Gravitational field strength** The gravitational force that acts on an object per unit mass; unit N $kg^{-1}$.

**Gravitational potential energy** The energy stored by a mass due to its position in a gravitational field; in a uniform field of field strength $g$, the gravitational potential energy of a mass $m$ that is raised by a distance D$h$ is given by $E_p = mg\,\Delta h$.

**Graviton** The gauge boson that transfers the force of gravity between particles with mass.

**Gravity** The force of attraction between all objects due to their mass; the strength of gravitational attraction depends on the mass of the bodies and the distance between them.

**Ground state** An atom in its ground state has all its electrons in their lowest possible energy level; it cannot emit any photons.

**Hadron** Any particle composed of quarks; hadrons are divided into two subgroups – mesons (with a quark and an antiquark) and baryons (with three quarks, or three antiquarks).

**Harmonics** The different frequencies at which stationary waves are set up in a system, e.g. on a stretched string.

**Hooke's law** A law stating that, for an object under tension such as a wire or a spring, the extension is proportional to the applied force.

**Huygens' construction** A geometrical construction showing that every point on a wave front may itself be regarded as a source of secondary waves.

**Hysteresis** When a physical quantity, such as the extension of a sample of material under tension, depends on what has previously happened to the sample.

**Impulse** The impulse of a force, $F$, is equal to $Ft$, where $t$ is the time for which the force acts. Impulse has units of Newton seconds, Ns.

**In phase** Two points on a wave, or points on two waves of the same frequency, are in phase if they are at the same point in their cycle at the same time.

**Inelastic** Describes a collision between objects in which some or all of the kinetic energy is transferred to other forms.

**Inertia** The resistance of any object to a change in its state of motion.

**Instantaneous velocity** The rate of change of displacement, as measured over a very small time interval.

**Insulator (electrical)** A material with very few mobile charges, which therefore has a very high electrical resistivity.

**Interference** The superposition of waves that leads to a change in amplitude; may be constructive or destructive.

**Internal resistance** The intrinsic electrical resistance between the terminals of a power supply, which reduces the terminal potential difference when a current is delivered.

**Ion** An atom or molecule that has gained or lost one or more electrons.

**Ion pair** The positive and negative ions produced when enough energy is applied to a neutral atom.

**Ionisation** The removal of electrons from an atom, or the addition of electrons to an atom.

**Isotope** Atoms of an element can exist in different forms, called isotopes, which have the same number of protons and electrons but different numbers of neutrons.

**Kelvin scale** The 'absolute temperature' scale that has its zero at *absolute zero*; a temperature difference of 1 kelvin (1 K) is equal to a temperature difference of 1 °C. Temperature $T$ in K = temperature $\theta$ in Celsius + 273.15.

**Kilowatt-hour** The energy transferred by an electrical appliance of power 1 kW used for 1 h.

**Kinetic energy** The energy of a mass $m$ moving with velocity $v$, given by $E_k = \frac{1}{2}mv^2$.

**Kirchhoff's first law** A law that states that the sum of currents at a junction is zero (charge is conserved).

**Kirchhoff's second law** A law that states that the sum of the emfs is equal to the sum of the pds around a closed circuit loop (energy is conserved).

**Laws of reflection** The principles explaining what happens to light when it hits a reflective surface: (1) when measured from the normal, the angle of incidence equals the angle of reflection; (2) the incident ray, reflected ray and the normal all lie in the same plane.

**Lepton** One of a family of fundamental particles; the electron, muon, tau particle and neutrino are all leptons.

**Lepton number** Leptons have lepton number 1 and antileptons have lepton number −1. Electron lepton number, muon lepton number and tau lepton number are all conserved in any particle reaction.

**Lift** The upward force that keeps aircraft in the air, resulting from the aircraft wings travelling at speed through the air.

**Light-dependent resistor (LDR)** A device in an electric circuit whose resistance falls as the light intensity falling on it increases.

**Light-emitting diode (LED)** A semiconductor diode that emits light when it conducts.

**Limit of proportionality** The maximum force on a spring or elastic material for which Hooke's law applies; for greater forces the extension is no longer proportional to the force.

**Limiting static friction** The maximum frictional force between two surfaces; an applied force greater than this will produce motion.

**Line spectrum** An emission spectrum that is a series of sharp bright lines of specific frequency, against a dark background.

**Linear momentum** The product of the mass and the velocity of an object moving in a straight line; $p = mv$; a vector quantity; unit kg m $s^{-1}$.

**Linear motion** Motion along a straight line; the 'equations of linear motion' are applicable to linear motion with a constant acceleration.

**Longitudinal wave** A type of wave for which the direction of oscillation is parallel to the direction of propagation (travel) of the wave, e.g. a sound wave.

**Magnitude** A measurement of the size of an object; a scalar quantity.

**Mass–energy** The combined mass and energy of a subatomic particle; the total mass–energy in an interaction is conserved.

**Material dispersion** The splitting of waves into different component wavelengths when passing through a material, such as an optical fibre, because of their differing refractive indices; also called chromatic dispersion.

**Mechanics** The study of forces and motion.

**Medium** A material through which a wave passes.

**Meson** A particle (a type of *hadron*) made up of a quark and an antiquark.

**Modal dispersion** The broadening of light pulses transmitted by an optical fibre, because of rays travelling along different paths in the fibre having different journey times.

**Moment of a couple** The turning effect of a couple: the product of either of the forces of a couple by the perpendicular distance between them; it is either clockwise or anticlockwise; unit N m; also known as a torque.

**Moment of a force** The turning effect of a force; the moment of a force about a point is equal to the product of the force and the perpendicular distance from the line of action of the force to the point; it is either clockwise or anticlockwise; unit N m.

**Momentum** See *linear momentum*.

**Monomode** Describes an optical fibre with a very narrow core, designed to carry a single ray; usually used for transmission over longer distances.

**Multimode** Describes an optical fibre designed to carry multiple light rays or 'modes' at the same time; usually used for transmission over shorter distances.

**Muon** An unstable heavy lepton, with charge $-1e$ and mass 200 times greater than that of an electron.

**Neutron number** The number of neutrons, $N$, in a nucleus; $N = A - Z$.

**Newton** The SI unit of force; 1 newton (1 N) is the force that will accelerate a mass of 1 kg at 1 m $s^{-2}$.

**Newton's first law** A law stating that an object will remain at rest, or continue to move with uniform velocity, unless acted on by an external, resultant force.

**Newton's second law** A law stating that the rate of change of momentum of a body is directly proportional to the external, resultant force acting on it. The change of momentum takes place in the direction of that force.

**Newton's third law** A law stating that when body A exerts a force on body B, then body B exerts an equal but opposite force on body A.

**No-parallax** Describes a method for locating an image in a plane mirror.

**Node** A point on a *stationary wave* where the displacement is zero.

**Non-uniform acceleration** Varying acceleration: the velocity changes by different amounts over equal time intervals.

**Normal** A line drawn or imagined at right angles to a surface, for example, in optics, at the point where the incident ray meets it.

**Normal contact force** An electromagnetic force exerted when two solid surfaces touch each other; when no friction is involved, this is 'normal': at right angles to the surfaces.

**Nuclear atom** A model of the structure of an atom proposed by Rutherford, in which electrons orbit a positively charged massive nucleus.

**Nucleon** Any proton or neutron in an atomic nucleus.

**Nucleon number** The total number of protons and neutrons in a nucleus; also referred to as the mass number; symbol $A$.

**Nucleus** The positively charged, dense matter at the centre of every atom; consists of protons and neutrons.

**Ohm** Unit of electrical resistance; symbol Ω.

**Ohm's law** A law that states that the current through a conductor is proportional to the potential difference across it, provided the temperature and other physical conditions stay constant.

**Ohmic conductors** Materials that follow Ohm's law, e.g. a metal at constant temperature.

**Optically dense** A material with a high *refractive index*, in which light travels at a significantly lower speed than in a vacuum.

**Outlier** An experimental observation that lies outside the main group of results.

**Pair production** A process in which a photon's energy is manifested as the masses of an electron and a positron.

**Parallel** Describes an electric circuit arrangement in which components are connected across one another.

**Parent nucleus** The nucleus of a radioisotope that emits radiation and decays to the daughter nucleus.

**Path difference** The difference in the distance travelled, for example by two waves, in which case it is often expressed as a fraction of the wavelength, $\lambda$.

**Period** The time taken to complete one complete cycle of an oscillation or rotation; symbol $T$.

**Phase difference** A measure of the difference in motion of two points on a wave, or of points on two waves of the same frequency, in terms of a fraction of a cycle; usually expressed as an angle where $360°$ is one full cycle; e.g. two points in antiphase (with equal but opposite displacements) have a phase difference of $180°$.

**Phosphorescence** The delayed or prolonged emission of visible light by a material, as a result of its atoms being left in an excited or ionised state.

**Photoelectric current** The number of electrons emitted per second from a surface by photoemission.

**Photoelectric effect** The liberation of electrons from a metal surface exposed to electromagnetic radiation of frequency above a minimum.

**Photon** A quantum of electromagnetic radiation; it carries an amount of energy, $E$, that depends on the frequency, $f$, of the radiation. $E = hf$, where $h$ is the Planck constant.

**Pivot** A point, or hinge, about which an object turns; also called a fulcrum.

**Planck constant** Electromagnetic radiation, of frequency $f$, is quantised in photons of energy, $E = hf$, where $h$ is the Planck constant. $h = 6.626 \times 10^{-34}$ Js.

**Plane polarised** Describes a transverse wave whose oscillations are confined to just one plane.

**Plastic** Describes a material that is permanently deformed after an applied force is removed.

**Polarised** See *plane polarised*.

**Polarising filter** A filter that polarises light by allowing transmission of only waves with oscillations in a particular plane.

**Polarity** An electrical component with polarity has a positive and a negative terminal, or pole, so the way it is connected in a circuit is important.

**Polymer** A material made up of long-chain molecules, e.g. rubber or Kevlar.

**Potential difference (pd)** The (electric) potential difference between two points in a conducting circuit is the energy transferred per unit charge moving between the points; symbol $V$; unit volt, V.

**Potential divider** A pair of resistors that divide input pd in the ratio of the resistors.

**Power** The rate at which energy is transferred or at which work is done, measured in joules per second, or watts, W.

**Precise** Describes a reading that can reliably be given to several significant figures.

**Principle of conservation of energy** A fundamental law stating that the total energy of a closed system is constant.

**Principle of moments** A law stating that if an object is in equilibrium, the sum of the clockwise moments about any point must equal the sum of the anticlockwise moments about that point.

**Principle of superposition** The principle that when similar waves meet the total displacement is the vector sum of the individual displacements.

**Principle of conservation of linear momentum** A fundamental law stating that the total linear momentum of a system does not change, provided that no net external force is acting on it.

**Progressive wave** A travelling wave, which transfers energy from one place to another.

**Projectile motion** Motion in two dimensions where the vertical motion is subject to gravity.

**Proton number** The number of protons in a nucleus, also called the atomic number; symbol $Z$.

**Pulse broadening** A feature of modal dispersion, in which short pulses sent from a transmitter spread out by the time that they reach the receiver, reducing the quality of the signal.

**Quantised** A quantity is quantised if it can take only discrete values; e.g. charge cannot take any value but has to be a multiple of the charge carried by an electron.

**Quantum (plural quanta)** A discrete amount of a physical quantity, such as energy or charge.

**Quantum number** A number that describes a quantity that can only take certain discrete values, e.g. lepton number in particle physics.

**Quark** A fundamental particle that is not observed in isolation but always in a combination of three (to make a baryon) or two (a quark–antiquark pair to make a meson); quarks have a baryon number of 1/3 and they carry fractional charge ($\pm 2/3$ or $\pm 1/3$ of the charge of the electron).

**Radian** The angle subtended when the arc length is equal to the radius; 1 rad = $57.3°$.

**Radioisotope** A form of a nucleus (see *isotope*) that is radioactive.

**Random error** A type of experimental error that fluctuates about the mean, causing readings to be randomly too high or too low.

**Real image** An image formed by the convergence of rays of light; it can be formed on a screen.

**Refraction** The change of direction of waves that results from a change of speed.

**Refractive index** A measure of how much light slows down in a medium, compared with its speed in a vacuum. The absolute refractive index of a material is the speed of light in a vacuum divided by the speed in the material; it is always greater than 1; it has no unit.

**Relative atomic mass** The average atomic mass of an element, taking into account the relative abundance of the isotopes of that element.

**Repeatable** Describes a measurement that is consistently achieved when an experimenter repeats it.

**Reproducible** Describes a measurement that can be reproduced in different laboratories using different techniques.

**Resilient** Describes a material that can undergo repeated deformations without transferring a significant amount of energy as internal energy.

**Resistance (electrical)** The resistance of an electrical component is defined as the ratio $V/I$, where $V$ is the potential difference across the component and $I$ is the current through the component.

**Resistivity** Resistivity, $\rho$, is a material property which quantifies the material's ability to oppose an electrical current. For a sample of material of resistance, $R$, cross-sectional area, $A$, and length, $l$, the resistivity is given by $\rho = AR/l$ ; unit $\Omega$ m.

**Resistor** A component in an electrical circuit that provides a fixed value of resistance.

**Resolution** The resolution of a measuring device is the smallest increment in the measured quantity that can be shown on the device. The resolution of an optical instrument is its ability to distinguish very close objects as separate.

**Resolving** The splitting up of a vector into components, usually two perpendicular components.

**Resultant** The vector sum of two or more vector quantities. A resultant force is the single force that has the same effect as all the forces acting on an object.

**Scalar** A physical quantity that has magnitude but no direction. Mass, speed, temperature and potential difference are examples of scalar quantities.

**Scattering** The dispersal of a stream of particles or light rays in different directions.

**Scientific notation** A system of writing large or small numbers as numbers from 1 to 10 multiplied by a power of 10; e.g. in scientific notation $5000 = 5 \times 10^3$ and $0.005 = 5 \times 10^{-3}$; also called standard form.

**Semiconductor** A type of material with electrical conductivity between that of metals and insulators; it has a limited number of mobile charges.

**Sensor** A device used to measure a quantity through electrical readings, often relying on a change of resistance; e.g. a *light-dependent resistor* or a *thermistor*.

**Series** Describes an electric circuit arrangement in which components are connected one after the other.

**SI** The international system of units of measurement, all defined in terms of seven *base units*.

**Snell's law** A law that states that the *refractive index* of a material is equal to the ratio of sin (angle of incidence) to sin (angle of refraction). The angles are measured from the normal.

**Spark chamber** A device for detecting the paths of high-energy particles.

**Specific charge** The charge–mass ratio of a particle; unit C $kg^{-1}$.

**Speed** A measure of how fast an object travels: distance travelled per unit time; unit m $s^{-1}$.

**Spring constant** The force needed to stretch a spring by unit extension: $k$ = force/extension; unit N $m^{-1}$; it is a measure of the *stiffness* of the spring.

**Standard form** See *scientific notation*.

**Statics** The branch of *mechanics* that deals with forces in equilibrium.

**Stationary wave** A non-progressive wave formed when two progressive waves of the same frequency travel in opposite directions and superpose, e.g. on a fixed stretched string; characterised by *nodes* and *antinodes*.

**Step-index fibres** Optical fibres made up of two layers of glass (core and cladding) of slightly different refractive indices.

**Stiffness** The resistance to extension of a material under tension.

**Strain** See *tensile strain*.

**Strangeness** A conserved, quantised quantity carried by the strange quark, which has strangeness $S = -1$.

**Stress** See *tensile stress*.

**Strong nuclear force** One of the fundamental forces; it acts between nucleons over a very short range and holds the nucleus together; also called the strong interaction.

**Superconductor** A conductor that has zero resistance at a temperature below its critical value.

**Systematic error** A type of experimental error that leads to results that are consistently too high or too low, due to factors such as instrument error or poor experimental design.

**Tangent** A straight line on a graph or plane that touches a curve without intersecting it.

**Tensile strain** The extension per unit length; has no unit.

**Tensile stress** The force per unit cross-sectional area that is tending to stretch an object.

**Tension** A force that acts to stretch an object, e.g. a tow rope would be 'in tension'; also called a tensile force.

**Terminal speed** The top speed of a falling object, reached when the opposing forces of weight and *drag* are equal.

**Thermistor** An electrical sensor that detects temperature change by a change in its resistance.

**Threshold frequency** The minimum frequency of incident electromagnetic radiation for which photoelectrons are liberated from the surface of a metal.

**Torque** A turning effect: the *moment of a couple* about a point; unit N m.

**Total internal reflection** When light is incident on a boundary between an *optically dense* medium and one that is less dense, for angles of incidence greater than the *critical angle* all the light is reflected back into the denser medium.

**Tough** Describes materials that can absorb a lot of energy before they break.

**Toughness** A measure of how much energy a material can absorb in deformation before breaking.

**Transverse waves** A type of wave for which the oscillations are at right angles to the direction of wave travel.

**Two-source interference formula** The formula for the fringe separation *w* in an interference pattern due to two coherent sources: $w = \lambda D/s$; where $\lambda$ is the wavelength, *D* is the distance from the sources to the detector, and *s* is the distance between the sources.

**Ultimate tensile stress** See *breaking stress*.

**Uncertainty** A measure of the precision of experimental results; may be expressed as a percentage of the value.

**Uniform acceleration** Constant acceleration: the velocity changes by the same amount over equal time intervals.

**Upthrust** The buoyancy force on an object in a fluid, due to the difference in fluid pressure on the top and bottom surfaces of the object.

**Vector** A physical quantity that has a direction as well as a magnitude. Force, velocity and electric field strength are examples of vector quantities.

**Velocity** The rate of change of *displacement*: the change of displacement in unit time; unit $m\ s^{-1}$.

**Vernier scale** A moveable scale that allows a fractional part on the main scale to be determined.

**Virtual image** An image caused by rays that do not converge; the image can be seen by the eye but not formed on a screen.

**Virtual particle** A very short-lived particle, only observable by its effects, such as an exchange particle (gauge boson) that mediates interactions between particles.

**Watt** A unit of power, equal to a rate of energy transfer of 1 joule per second; symbol W.

**Wave–particle duality** The concept that electromagnetic energy and matter particles both exhibit wave-like and particle-like properties.

**Wavelength** The distance travelled by a wave in one period of oscillation; the length in space of one cycle.

**Weak interaction** A fundamental force that acts between all particles over a very short range; it is responsible for radioactive beta decay.

**Work** The work done is the energy transferred when a force moves through a distance in the direction of the force; $W = Fs$.

**Work function** The minimum energy required to remove a photoelectron from the surface of a metal.

**Yield stress** The minimum *tensile stress* at which plastic deformation occurs.

**Young modulus** The Young Modulus, E, is a measure of the stiffness of a material that is under tensile stress. It is the ratio of tensile stress / tensile strain. It has units of Pascals, Pa.

# DATA SECTION

## FUNDAMENTAL CONSTANTS

| Quantity | Symbol | Value | Unit |
|---|---|---|---|
| speed of light in vacuo | $c$ | $3.00 \times 10^{8}$ | $\mathrm{m\,s^{-1}}$ |
| charge of electron (magnitude) | $e$ | $1.60 \times 10^{-19}$ | C |
| the Planck constant | $h$ | $6.63 \times 10^{-34}$ | J s |
| electron rest mass | $m_e$ | $9.11 \times 10^{-31}$ | kg |
| electron specific charge | $e/m_e$ | $1.76 \times 10^{11}$ | $\mathrm{C\,kg^{-1}}$ |
| proton rest mass | $m_p$ | $1.67(3) \times 10^{-27}$ | kg |
| proton specific charge | $e/m_p$ | $9.58 \times 10^{7}$ | $\mathrm{C\,kg^{-1}}$ |
| neutron rest mass | $m_n$ | $1.67(5) \times 10^{-27}$ | kg |
| gravitational field strength | $g$ | 9.81 | $\mathrm{N\,kg^{-1}}$ |
| acceleration due to gravity | $g$ | 9.81 | $\mathrm{m\,s^{-2}}$ |

## SI UNIT PREFIXES AND SYMBOLS

| Multiplication factor | | Prefix | Symbol |
|---|---|---|---|
| 1000 000 000 000 | $10^{12}$ | tera | T |
| 1000 000 000 | $10^{9}$ | giga | G |
| 1000 000 | $10^{6}$ | mega | M |
| 1000 | $10^{3}$ | kilo | k |
| 0.01 | $10^{-2}$ | centi | c |
| 0.001 | $10^{-3}$ | milli | m |
| 0.000 001 | $10^{-6}$ | micro | μ |
| 0.000 000 001 | $10^{-9}$ | nano | n |
| 0.000 000 000 001 | $10^{-12}$ | pico | p |
| 0.000 000 000 000 001 | $10^{-15}$ | femto | f |

## GEOMETRICAL EQUATIONS

circumference of circle = $2\pi r$

area of circle = $\pi r^2$

surface area of cylinder = $2\pi rh$

volume of cylinder = $\pi r^2 h$

surface area of sphere = $4\pi r^2$

volume of sphere = $\frac{4}{3}\pi r^3$

arc length = $r\theta$ ($\theta$ in radians)

$\sin^2\theta + \cos^2\theta = 1$

for small angles, $\sin\theta \approx \tan\theta \approx \theta$ in radians

equation of a straight line $y = mx + c$
($m$ = gradient; $c = y$ intercept)

## FUNDAMENTAL PARTICLES

| Class | Name | Symbol | Rest energy / MeV |
|---|---|---|---|
| photon | photon | $\gamma$ | 0 |
| lepton | (electron) neutrino | $\nu_e$ | 0 |
| | muon neutrino | $\nu_\mu$ | 0 |
| | electron | $e^\pm$ | 0.510999 |
| | muon | $\mu^\pm$ | 105.659 |
| meson | $\pi$ meson (pion) | $\pi^\pm$ | 139.576 |
| | | $\pi^0$ | 134.972 |
| | K meson (kaon) | $K^\pm$ | 493.821 |
| | | $K^0$ | 497.762 |
| baryon | proton | p | 938.257 |
| | neutron | n | 939.551 |

## PROPERTIES OF QUARKS

| Type | Charge | Baryon number | Strangeness |
|---|---|---|---|
| u | $+\frac{2}{3}e$ | $+\frac{1}{3}$ | 0 |
| d | $-\frac{1}{3}e$ | $+\frac{1}{3}$ | 0 |
| s | $-\frac{1}{3}e$ | $+\frac{1}{3}$ | −1 |

Antiquarks $\bar{u}$, $\bar{d}$, $\bar{s}$ have opposite signs.

## PROPERTIES OF LEPTONS

| | | | | | Lepton number |
|---|---|---|---|---|---|
| particles | $e^-$ | $\nu_e$ | $\mu^-$ | $\nu_\mu$ | +1 |
| antiparticles | $e^+$ | $\bar{\nu}_\mu$ | $\mu^+$ | $\bar{\nu}_\mu$ | −1 |

## PHOTONS AND ENERGY LEVELS

photon energy $E = hf = hc/\lambda$

in photoelectric effect $hf = \phi + E_{k\,(max)}$

energy levels $hf = E_1 - E_2$

de Broglie wavelength $\lambda = h/p = h/mv$

## WAVES

wave speed $c = f\lambda$

$f = 1/T$ (where $T$ = period)

for first harmonic on a string

$f = (1/2l)\sqrt{(T/\mu)}$

interference fringe spacing $w = \lambda D/s$

diffraction grating formula $d\sin\theta = n\lambda$

refractive index of substance $n = c/c_s$

law of refraction $n_1 \sin\theta_1 = n_2 \sin\theta_2$

for critical angle $\sin\theta_c = n_2/n_1$ for $n_1 > n_2$

## MECHANICS

$v = \Delta s/\Delta t$

$a = \Delta v/\Delta t$

$v = u + at$

$s = \frac{1}{2}(u + v)t$

$v^2 = u^2 + 2as$

$s = ut + \frac{1}{2}at^2$

momentum $p = mv$

$F = \Delta(mv)/\Delta t$

for constant $m$, $F = ma$

impulse $F\,\Delta t = \Delta(mv)$

work done $W = Fs\cos\theta$

$E_k = \frac{1}{2}mv^2$

$\Delta E_p = mg\,\Delta h$

power $P = \Delta W/\Delta t = Fv$

moment $= Fd$

efficiency = useful output power/input power

## MATERIALS

density $\rho = m/V$

Hooke's law $F = k\,\Delta l$

tensile stress $= F/A$

tensile strain $= \Delta l/l$

Young modulus = tensile stress/tensile strain

energy stored $E = \frac{1}{2} F\,\Delta l$

## ELECTRICITY

$I = \Delta Q/\Delta t$

$V = W/Q$

$R = V/I$

resistivity $\rho = RA/l$ (where $l$ = length)

resistors in series $R_{tot} = R_1 + R_2 + R_3 + \ldots$

resistors in parallel $R_{tot} = 1/R_1 + 1/R_2 + 1/R_3 + \ldots$

$P = VI = I^2R = V^2/R$

emf $\varepsilon = E/Q$

$\varepsilon = I(R + r)$

# INDEX

## D



## E

## S



## T

# ACKNOWLEDGEMENTS


**Practical work in Physics**

p2, top: Peter Ginter/Getty Images; p2, bottom: indukas/Thinkstock; p3: Phil Boorman/Science Photo Library; p4, top: www.philipharris.co.uk; p5, bottom, Martyn F Chillmaid/Science Photo Library;

**Chapter 1**

p5: NASA; p7, Fig 2: Gts/Shutterstock; p8, Fig 3: NASA; p10: Professor Peter Fowler/Science Photo Library; p13: Richard Griffin/Shutterstock; p17: GIPhotostock/Science Photo Library

**Chapter 2**

p24, Fig 1: David Parker/Science Photo Library; p24, Fig 2: Ethan Daniels/Shutterstock; p25, Fig 3: World History Archive/Alamy; p25, Fig 4: Science Museum/Science & Society Picture Library; p25, Fig 5: Andrew Lambert Photography/Science Photo Library; p28: www.philipharris.co.uk; p38: NASA; p39, Fig 27: teekaygee/Shutterstock; p39, Fig 28: Vasily Smirnov/Shutterstock; p39, Fig 29: argus/Shutterstock; p40: Pictorial Press/Alamy

**Chapter 3**

p44: Sven Lindstrom Ice Cube NSF; p45: Lawrence Berkeley Laboratory/Science Photo Library; p46: David Parker/Science Photo Library; p47, top: American Institute of Physics/Science Photo Library; p47, bottom: Westminster Abbey; p48: Carl Anderson/Science Photo Library; p49: Excellent Backgrounds/Shutterstock; p50: Cern/Science Photo Library; p51: Dr John Mazziotta Et Al/Neurology/Science Photo Library; p55, Fig 14: Kamiokande Observatory, ICRR (Institute for Cosmic Ray Research), The University of Tokyo; p55, Fig 15: Dept of Physics, University of Bristol/Science Photo Library

**Chapter 4**

p61, Fig 1: Denis Balibouse/AFP/Getty Images; p61, Fig 2: James King-Holmes/Science Photo Library; p64: Cern/Science Photo Library; p71: Brookhaven National Laboratory/Science Photo Library; p74: C Powell, P Fowler & D Perkins/Science Photo Library; p76: Lawrence Berkeley Laboratory/Science Photo Library; p77, Fig A2: Science & Society Picture Library; p77, Fig A3: drserg/Shutterstock

**Chapter 5**

p82, Fig 1: AFP/Getty Images; p83, Fig 2: Kasza/Shutterstock; p84, Fig A1: Universal Education, Universal Images Group via Getty Images; p84, Fig 5: BBC; p86, Fig 9: NASA; p91, Fig A1: Minerva Studio/Shutterstock; p91, Fig A2: Otto Gruele Jr/Getty Images; p94, Fig 23 left: Science Photos, Right: Science Photos; p95, Fig 26: Alfred Pasieka/Science Photo Library; p97, Fig 27: Global Warming Images/Alamy; p98, Fig 29: Dmitry Naumov/Shutterstock; p.98, Fig 32: Andrew Lambert Photography/Science Photo Library; p98, Fig 31: Furtseff/Shutterstock; p100, Fig 37: Bernard Richardson; p101: Scientifica/Maynard & Boucher/Visuals Unlimited/Science Photo Library; p102, Fig P3: Andrew Lambert Photography; p103, Fig 39: Igor Goloviov/Shutterstock; p107: Dave King/Getty Images

**Chapter 6**

p112, Fig 1: Michaelpuche/Shutterstock; p.112, Fig 2: Cinemafestival/Shutterstock; p113, Fig 3: Marceclemens/Shutterstock; p115, Fig 7: Berenice Abbott/Science Photo Library; p121: Uber Images/Shutterstock; p122, Fig 15: Christian Delbert/Shutterstock; p123, Fig 18: GIPhotostock/Science Photo Library; p123, Fig 19: Edward Kinsman/Science Photo Library; p125, Fig 25: American Spirit/Shutterstock; p125, Fig 26: Dr Juerg Alean/Science Photo Library; p127, Fig 30: GIPhotostock/Science Photo Library; p127, Fig 31: NASA JPL/University of Colorado; p128, Fig 33: Konstantin Shaklein/Alamy; p128, Fig 34: Corbin O'Grady Studio/Science Photo Library; p128, Fig 35: Science Source/Science Photo Library; p130, Fig P2: GIPhotostock/Science Photo Library; p134, Fig Q5: Doug Shnurr; p134, Fig Q6: SolvinZanki/Nature Picture Library;

**Chapter 7**

p135, Fig 1: Caida/Science Photo Library; p136, Fig 3: GIPhotostock/Science Photo Library; p137, Fig 6: Dusica Paripovic/Getty Images; p137, Fig 8: World History Archive/Alamy; p138, Fig 10: George Silk/The LIFE Picture Collection/Getty; p138, Fig 11: Berenice Abbott/Science Photo Library; p139, Fig 13: GIPhotostock/Science Photo Library; p139, Fig 14: MilanB/Shutterstock; p140, Fig 17: Richard Wainscot/Alamy; p146, Fig 27: Baloncici/Shutterstock; p147, Fig A2: Sarah2/Shutterstock; p149, Fig 31: Laschon Maximilian/Shutterstock

**Chapter 8**

p160, Fig 2: Alma Observatory; p161, Fig 4: BSIP SA/Alamy; p.161, Fig 5: Shaiith/Shutterstock; p162, Fig 6: NASA; p162, Fig 7: © Science Photo Library; p162, Fig 8: Roger Ressmeyer/Corbis; p163, Fig 9: zhu difeng/Shutterstock; p163, Fig 10: Library of Congress/Science Photo Library; 165, Fig 14: Ekkamai Chaikanta/Shutterstock; p166, Fig A2 left: Dept. of Physics, Imperial College/Science Photo Library; center/right: Ted Kinsman/Science Photo Library; p169, Fig 19: Lucy Roth; p170, Fig A1 Left: SSPL/Getty Images, Right: Chones/Shutterstock; p171, Fig A3: Lucy Roth; p172, Fig 22: Physics Dept., Imperial College/Science Photo Library; p173, Fig 24: Interfoto/Alamy; p180, Fig 27: Steve Gschmeissner/Science Photo Library; p180, Fig 28: Andrew Lambert Photography/Science Photo Library; p182: Philip Harris; p184, Fig Q2: Physics Dept., Imperial College/Science Photo Library

**Chapter 9**

p187, Fig 1: FRANCK FIFE/AFP/Getty Images; p187, background: David Warren/Alamy; p191, Fig 8: SoleilC/Shutterstock; p193, Fig 14: Roberto Tetsuo Okamura/

Shutterstock; p193, Fig 16: Lilyana Vynogradova/Shutterstock; p194, Fig 19: Herbert Kratky/ Shutterstock; p197, Fig A1: Will Rodrigues/Shutterstock; p204, Fig A1: Pasco Scientific; p205, Fig 29: NASA; p205, Fig 30: NASA; p205, Fig 31: Neil Mitchell/ Shutterstock; p209, Fig 33: Paolo Bona/Shutterstock; p209, Fig 34: CSER, Sheffield Hallam University; p209, Fig 35: joggiebotma/istock; p211, Fig A1: Georgios Kollidas Shutterstock; p212, Fig 36: Keystone Pictures USA/Alamy; p213, Fig 38: Lauren Winters/Visuals Unlimited Inc/ Science Photo Library; p213, Fig 39: Henglein and Steets/Getty Images

**Chapter 10**

p222: PHB.cz (Richard Semik)/ Shutterstock.com; p222, Fig 1: Stocker1970/Shutterstock; p224, Fig 5 Top: Ivaschenkco Roman/ Shutterstock, Bottom: Kletr/ Shutterstock; p225, Fig 9: Manfred Kage/Science Photo Library; p226, Fig 10: Gorillaimages/Shutterstock; p226, Fig 11: Nik Taylor Sport/ Alamy; p227, Fig 12: Tim Boyles/ Getty Images; p227, Fig 13: Bertold Werkmann/Shutterstock; p229, Fig 16: Luca Nicolotti/ Stocktrek Images/Getty Images; p230, Fig 18: Dejan Gileski/ Shutterstock; p232, Fig 4: Dan Kenyon/Getty Images; p 233, Fig 25: one-image Photography/Alamy; p234, Fig 29: Kuttig – People – 2/ Alamy; p237, fig A1: Max Earey; p238, Fig 32: Alexander Nikiforov/ Thinkstock; p240, Fig 37: Dima Fadeev/Shutterstock; p241, Fig 41: Scott Biscuiti/Science Photo Library; p242, Fig 44: CCI Archives/Science Photo Library, p248, Fig Q7: Sam Kovak/Alamy

**Chapter 11**

p250, Fig 1: Iain Masterton/Alamy; p251, Fig 2: NASA JPL/Caltech; p253, Fig 4: Shutterstock/Hetman Bohdan; p254, Fig 6: Shutterstock/ cobrado; p254, Fig 7: Sean Gallup/ Getty; p257, Fig 11: Shutterstock/J Chaikom; p259, Fig 13: Shutterstock/Neale Cousland; p260, Fig A1: Shutterstock/TENGLAO; p261, Fig 16: Misso.Anu/Science Photo Library; p262, Fig 18: Romilly Lockyer/Getty; p264, Fig 21: Henning Dalhoff/Science Photo Library; p266, Fig 22: Shutterstock/ Pal2iyawit; p267, Fig 23: Creative Commons License; p268, Fig 24: Shutterstock/Christian Lagerek; p270, Fig 25: Shuttersetock/ VanderWolf Images; p270, Fig 26: Shutterstock/ChameleonsEye

**Chapter 12**

p278, Fig 1: Shutterstock/Cliff Collings; P278, Fig 2: AF Archive/ Alamy; p279, Fig 3: Shutterstock/ Greg Epperson; p280, Fig 5: Radharc Images/Alamy; p282, Fig 9: Adam Hart-Davies/Science Photo Library; p283, Fig 12: Maxbmx/ Getty Images; p284, p285: Philip Harris; Fig 13: Shutterstock/ terekhov igor; p291, Fig A1: Dave Stamboulis/Alamy; p291, Fig A2: Shutterstock/Wavebreakmedia

**Chapter 13**

p298, Fig 2: iStock/Peter de Kievith; p300, Fig 5: Shutterstock/ Anna Omelchenko; p301, Fig 6: iStock/Maria Bobrova; p302, Fig 8: SPL; p302, Fig 9 Left: shutterstock/Oleksiy Mark, Right: shutterstock/Kvadrat; p307, Fig A1: Martyn Chillmaid; p307, Fig A2: Phillip Harris; P308, Fig 20: istock/iceninephoto; p310, Fig 22: Martyn Chillmaid; p311, Fig 24: shutterstock/Krasowit; p315, Fig 35: shutterstock/Danie Nel: p319, Fig P1: Philip Harris; p319, Fig P2: shutterstock/magicoven

**Chapter 14**

p323, Fig 1: shutterstock/ pio3; p323, Fig 2: shutterstock/ Daleen Loest; p329, Fig P2: shutterstock/sNike; p330, Fig 7: shutterstock/zstock; p331, Fig A1: istock/doug4537; p333, Fig 9: Martyn Chillmaid; p334, Fig 13: shutterstock/Mark William Richardson; p336, Fig 14: Science Photo Library/RIA Novosti; p336, Fig 16: SPL/Alfred Pasieka; p337, Fig 17: SPL/Patrick Gaillardin/ Look at Sciences; p337, Fig 18: shutterstock/cyo bo; p337, Fig 19: Istock c0 skynesher; p314, Fig 24: shutterstock/CreativeNature R.Zwerver

# NOTES